Praise for

Transatlantic

"A fluid, breezy style that reads as smoothly as a good novel. . . . Fox captures the experience of the sailing ships to perfection, buttressing his wonderful descriptions with excerpts from firsthand accounts."

—Titanic Book Site.com

"The great Atlantic steamships were worlds unto themselves. They came to occupy, as well, an important part of the history and folklore of travel, and still stand as a symbol of the coming-of-age of an empire (the British), an industry (tourism), and an ethos (the taming of technology for leisure and luxury). . . . [*Transatlantic*] is captivating reading." —*Wall Street Journal*

"A lively, well-researched history of the race, technological and commercial, to send steam-powered vessels across the pond."

—*Kirkus Reviews*

"One of the innumerable merits of Stephen Fox's absorbing history of transatlantic steamship travel is the way he balances the romance and reality. He vividly re-creates the texture of shipboard life. . . . It is a terrific celebration of the commercial spirit."

—*Sunday Telegraph* (London)

"In his splendid book *Transatlantic*, Stephen Fox offers a definitive history of commercial navigation across the North Atlantic Ocean, from the introduction of steam power in the 1820s to the early years of the twentieth century. . . . A book that will be welcomed by maritime historians as well as by armchair sailors." —*Pittsburgh Tribune-Review*

"Fox has fashioned a comprehensive and informative book."

—*Publishers Weekly*

ALSO BY STEPHEN FOX

John Muir and His Legacy: The American Conservation Movement

The Mirror Makers: A History of American Advertising and Its Creators

Blood and Power: Organized Crime in Twentieth-Century America

*Big Leagues: Professional Baseball, Football,
and Basketball in National Memory*

The Guardian of Boston: William Monroe Trotter

TRANSATLANTIC

TRANSATLANTIC

Samuel Cunard, Isambard Brunel,
and the Great Atlantic Steamships

STEPHEN FOX

Perennial
An Imprint of HarperCollinsPublishers

To the memory of

PHYLLIS RUTH BLAKELEY (1922–1986)

Provincial Archivist for Nova Scotia
Biographer of Samuel Cunard

A hardcover edition of this book was published in 2003 by HarperCollins Publishers.

TRANSATLANTIC. Copyright © 2003 by Stephen Fox. All rights reserved. Printed in the United States of America. No part of this book may be used or reproduced in any manner whatsoever without written permission except in the case of brief quotations embodied in critical articles and reviews. For information address HarperCollins Publishers Inc., 10 East 53rd Street, New York, NY 10022.

HarperCollins books may be purchased for educational, business, or sales promotional use. For information please write: Special Markets Department, HarperCollins Publishers Inc., 10 East 53rd Street, New York, NY 10022.

First Perennial edition published 2004.

Designed by Kris Tobiassen

The Library of Congress has catalogued the hardcover edition as follows:
Fox, Stephen.
    Transatlantic : Samuel Cunard, Isambard Brunel, and the great Atlantic steamships / Stephen Fox.—1st ed.
        p.   cm.
    Includes index.
    ISBN 0-06-019595-9
    1. Steamboat lines—North Atlantic Ocean—History. 2. Ocean liners—North Atlantic Ocean—History. 3. North Atlantic Ocean. I. Title.

HE945.A2F68   2003
387.5'09163'109034—dc21   2002191928

ISBN 0-06-095549-X (pbk.)

04 05 06 07 08 ❖/RRD 10 9 8 7 6 5 4 3 2 1

Contents

Prologue

THE NORTH ATLANTIC OCEAN
AND THE *BRITANNIA*

From Liverpool, on the River Mersey, a ship bound for a port in the northeastern United States heads west eighty miles across the Irish Sea, and then—when clear of Holyhead—turns sharply south into St. George's Channel. The ship navigates carefully through St. George's, which funnels currents and storms from larger contiguous seas into a narrowing, unpredictable passage squeezed between England and Ireland. She moves southwesterly along the Irish coast, skirting the Old Head of Kinsale and other jutting headlands, to reach (but avoid) Cape Clear and the Fastnet Rocks at the bottom of Ireland. To this point the ship has gone about 300 miles since departing Liverpool. From Cape Clear the ocean stretches out unimpeded to the western horizon and far beyond. Starting there, the great circle route to America arcs across nearly 3,000 miles of the North Atlantic: one of the most varied, troublesome ocean voyages in the world.

Over its entire course, the great circle route veers gradually southward from fifty-three to forty degrees north latitude. Giant spirals of wind and weather gust far above and perpendicular to the ocean's surface, rotating across twenty or more degrees of latitude in counterclockwise systems that generally hit the great circle to America in the southern half of their spins. Prevailing winds in that swatch of ocean therefore come from the west and southwest, fighting any westbound ship. The weather is typically unsettled, with odd, sudden shifts in temperature, pressure, wind speed, and direction. Systems collide and combine and bounce around. Long, high, stately deepwater waves march along over hundreds of miles of ocean. On occasion, several wave components may converge momentarily, producing a rogue wave much bigger than any of its parts—up to four times the height of an average North Atlantic wave, sometimes even 100 feet high or more.

The weather, seldom agreeable, turns worse in winter. From October through March, the days are cold, short, and dark, the sun low in the sky, the sea a turbulent dark gray. At midocean, abrupt winter gales may quickly reach speeds of sixty-five knots, with steep waves of 40 to 60 feet and smaller combers curling out in extended cycles up to 400 feet from crest to crest. White flecks—from spindrift and foam atop the breaking waves—stand out in sharp relief against the slates of sea and sky: the natural world pared down to stark monochromes of gray and white, pretty but indifferent in its muted danger.

Toward the end of the voyage, well over 2,000 miles from Liverpool—just as crews may be growing tired and irritable, with flagging attention and potential lapses in discipline, and passengers bored or restive, and food or fuel perhaps running low—the ship enters the notorious graveyard of the North Atlantic. This most hazardous sector of the great circle is encountered precisely when the ship's company may already be stretched tight and vulnerable.

Here the winter gale season yields, with no relief, to iceberg season, which generally starts in January and then peaks from April to July. Far to the north, in the west of Greenland, glaciers flow down the coastal mountains to the sea, annually calving thousands of icebergs into the Davis Strait. The bergs float slowly southward in the Labrador Current. At unpredictable times in the following year, the surviving remnants—about four hundred icebergs each season—reach the shipping lanes off Newfoundland. A typical splinter or castle berg weighs over 100,000 tons and stands about 150 feet high by 300 feet long, above water; in extraordinary cases both dimensions may be doubled and more. Smaller bergs, growlers, field ice, and floes can pose more hidden danger to ships: lower in profile, sometimes barely above the ocean surface, they are harder to see and avoid. When a ship enters the iceberg zone, temperatures dip and the air smells different, and lookouts get edgy.

The icebergs further roil the tumultuous Gulf Stream. The strongest of all ocean circulations, and one of the most startling, mysterious discoveries made by the early European explorers of the New World, the Gulf Stream was first charted by Benjamin Franklin in 1769. Trade winds pile up water along the continental edge of South America near the equator and send it "downhill" through the quickening channel between Cuba and the Florida Keys. The stream then runs up the eastern seaboard of the United States, thirty to fifty miles wide, at two to six miles an hour. Its warm water and rapid course inhibit the growth of phytoplankton, creating a vivid swath of deep, clear, pure blue against the greener, grayer surrounding ocean. Off the North Carolina coast, it divides into smaller substreams, which loop and meander. Even so, when the Gulf Stream reaches the area south of Nova Scotia, it is still hauling over 150 million cubic meters of water each second—some 10,000 times the volume of the Mississippi River.

East of the Grand Banks of Newfoundland, and squarely athwart the great circle route to America, the warm Gulf Stream collides with the cold Labrador Current, producing the most extreme temperature differences in any ocean. This volatile mix has vast consequences in the air and water. Some are even whimsical (icebergs spinning slowly on their vertical axes, like huge, silent, snowy carousels). Most are more serious: mists, gales, squalls, driving rain, and churning waves. At its worst, the atmospheric mingling of cold Canadian air and warm, moist Gulf Stream air may generate the sudden winter hurricanes called bombs, or rapidly intensifying cyclones. An abrupt drop in atmospheric pressure at the center of a comma-shaped cloud mass, usually in January or February, can unpredictably generate winds of hurricane force. With little warning, the bomb just explodes.

The most widespread result of this massive convergence of cold and warmth is dense, persistent fogs. Off Cape Race, at the southeastern tip of Newfoundland, from April through September at least twelve days of every month are shrouded in sea fog. At midsummer it is nearly constant. "These horrid fogs infest the air most part of the year," the English hydrographer John Purdy noted in 1817, "and will last eight or ten days successively, sometimes longer." The fogs can assume geometric, hedge-like forms, with vertical slabs at their eastern edges squared off by long, level top layers. An observer may watch a ship emerging slowly from one of these banks, revealing herself in sharply defined foot by foot, as though being dragged out of a gray cliff at water's edge.

In iceberg season, with visibility more crucial than ever, the Grand Banks fogs throw a dense, smothering blanket over the ice field. The wind dies down. The sea is lumpy and tumbling. Warning bells and foghorns are muffled, distorted, their direction and distance rendered unknowable. Both sight and hearing become untrustworthy. A constant condensing rain drips from the ship's superstructure. Shapes take on gigantic, unnatural proportions. A bird may resemble a sail. Ghosts and mirages float by. Peering hopelessly through the thick white smoke, a lookout can mistake an iceberg for a ship, or ship for iceberg. The circumstances are gloomy, anxious, strange, and very dangerous.

The great circle route runs along the Newfoundland coast, with its rocky headlands and variable currents driving now toward shore, then out to sea. "The uncertainty requires the greatest caution," John Purdy warned in 1817. Farther west, about a hundred miles off Nova Scotia, lurks Sable Island and its shifting shoals and sandbars. Moving steadily eastward, sometimes at a mile every four years (and therefore impossible to chart precisely), and often invisible in fog, Sable has sunk many ships. The fog can then persist all the way down the American coast to Boston and New York.

By the nineteenth century, through various accidents of history, this most dangerous sea passage had also become the most trafficked long ocean route in the world. The burgeoning imperatives of trade, empire, and human migration between the hemispheres would not give way, even to the North Atlantic Ocean in winter. Some reliable means of making this roughest transatlantic crossing—in all seasons, and in reasonable speed, safety, comfort, and economy—had to be devised. It posed a fundamental challenge to the newly inventive, progressive spirit of the age.

*T*he first enduring steamship service between England and America began when the wooden paddle wheeler *Britannia,* of the new Cunard Line, left Liverpool for Halifax and Boston on July 4, 1840. With Samuel Cunard, founder of the line, on board, the *Britannia* labored across the ocean against head winds and adverse currents. Ten days out, an iceberg was sighted in the near distance: a reminder of the North Atlantic's perils. The ship was scheduled to depart Liverpool a few days before the fourth, and she was therefore expected in Boston by the fourteenth. When that day passed without the *Britannia,* people in Boston started worrying. A steamship was supposed to make faster, more reliable ocean passages than a sailing ship; that was the whole point of adding the steam engine and paddles.

Four days went by amid swelling anxieties. Finally, at ten o'clock on the night of Saturday, the eighteenth, the *Britannia* glided into Boston Harbor. Despite the late hour she was greeted by fireworks and huzzahs—and a general sense of relief. She had made the passage in fourteen and a half days, just one-third of the time consumed by the most recent sailing packet from England to Boston. On Sunday, Sam Cunard received eighteen hundred invitations to dinner. "No event which has occurred since the commencement of the present century," the Reverend Ezra Gannett told his prominent congregation that morning, "seems to me to have involved more important consequences to this city."

The man and his ship dominated Boston for the next two weeks. Thousands of people came to inspect the *Britannia* at the new Cunard wharf; Bostonians had never seen such a ship. At the time, a typical coasting steamboat ran about 160 feet long and 400 tons, a typical ocean-sailing packet about 150 feet and 700 tons. (In nautical idiom, a "ton" measured not weight or displacement but interior space, calculated in different ways but, in 1840, supposed to equal forty cubic feet.) The *Britannia* was much bigger than contemporary ship norms at 207 feet and 1,156 tons, with steam power in due proportion. She had been built of African oak and yellow pine at Robert Duncan's shipyard on the River Clyde in Scotland; the celebrated Robert Napier of Glasgow had provided her 403-horsepower engine. From a distance she looked like a quite large sailing ship, dominated by three masts and sails in conventional riggings. But at midship two black-and-gold paddle boxes extended almost twelve feet out to each side, with a single smokestack just aft, painted burnt-red with a black ring at the top: a future signature of Cunard ships. The *Britannia* was, properly speaking, neither steamship nor sailing ship but a hybrid of the two.

Much of the main deck was left flat and clear to give the crew unobstructed room for handling sails and rigging. Deckhouses at midship provided quarters for the officers, sailors, and the ship's cows. A raised, exposed bridge between the paddle boxes and above the deckhouses allowed the captain and his adjutants the sight lines and free access they needed to run the ship. (Years later, after paddle wheels had yielded to screw propellers for propulsion, the term "bridge" remained to designate where a steamship's officers stood and gave orders.) Aft of the mainmast, another deckhouse held the passengers' dining saloon, the largest room on the ship at thirty-six feet by fourteen feet; it also functioned as a sitting room and assembly hall. At the stern, a raised platform gave the helmsman and his wheel a quite wet, windy place from which to steer the ship.

From the saloon, stairs descended to the gentlemen's and ladies' cabins and lounges. Men and women were consigned to separate sections, linked by a passage that allowed decorous contact without risking the weather up on deck. A typical "stateroom" measured about twelve by six feet, tightly packed with two bunk beds, jugs and basins for washing and emergencies, a small mirror on the wall, a water carafe and glasses, a day sofa, and pegs for hanging clothes. A porthole or oil lamp provided dim light. "All these rooms are highly finished," a Bostonian noted, "without any attempt to dazzle with tinsel." The undersides of the cabin floors were covered with a thick, coarse woolen cloth intended to seal off smells and heat from the holds and engine room. Passengers—the ship had room for up to one hundred men and twenty-four women, all in a single class—shared a few water closets and had no bathing facilities at all.

The provision deck below held quarters for the engineers and fire-men. They fed and tended the engine, the rhythmically beating heart of the ship. The machinery and coal bunkers at midship took up a third of the *Britannia's* length, leaving relatively little space for cargo on such an enormous vessel. The firemen shoveled coal into twelve furnaces firing four boilers feeding steam to the engine. Still brand-new, the engine and its moving parts shone like burnished silver. Two cylinders, six feet in diameter, drove nineteen-foot levers to turn the paddle wheel crankshaft. At full steam the paddles, nine feet wide and twenty-eight feet in diameter, could push the ship up to almost nine knots.

To a greater degree than anything previously seen in Boston, the *Britannia* was a ship and a building and a machine, all at once, on the grandest and most daring scale. One dazzled observer called her "the con-summation of human ingenuity," no less. She seemed to vault beyond the usual construction categories, gathering them into a novel kind of man-made artifact. Large, plush, and inventive, utterly modern but oddly familiar, beautiful from her soaring masts down to her gleaming engine room and yet promising such great practical significance, she left admir-ers in Boston mingling their superlatives. "She is truly a magnificent ves-sel," the *Evening Journal* declared, "—a floating palace."

Three days after Samuel Cunard's arrival, Boston threw a grand party for him attended by nearly 2,000 people. The toasts were so extended, the speeches so hyperbolic even by the rhetorical standards of the time, that it seems apparent that more than just a man and his ship was being celebrated. The Cunard Line was largely a British enterprise, based in Liverpool and launched by a mail contract from the British Admiralty. The first Cunard ship was pointedly named the *Britannia,* and she was commanded by Captain Henry Woodruff of the Royal Navy, not a civil-ian. In the previous sixty-five years, Britain and America had fought two bitter wars against each other and then had engaged in constant mutual insults and fierce squabbles over Canadian independence, boundaries, and fishing rights. Only a year earlier, the American state of Maine and the Canadian province of New Brunswick had nearly started a war. Now the Cunard Line inspired new hopes for friendlier ties between mother country and wayward child. In addition, Boston by 1840 was losing its former commercial and maritime eminence to New York; but both cities had competed for the glittering prize of becoming Cunard's American terminus, and Boston had apparently won. This coup perhaps augured a general resurgence for the city against its bumptious rival down on the Hudson. And, finally, that summer the United States was torn by an especially rancorous presidential contest, the log cabin and hard cider campaign of William Henry Harrison against the incumbent Martin Van Buren. With the *Britannia* on hand, at least, Boston's feuding Whigs and Democrats might briefly unite behind a promising new venture of general benefit.

The man himself remained a mute mystery. Nobody in Boston knew Sam Cunard well; he had a few business associates there, nothing more. He didn't talk much, and he had accepted few of those eighteen hundred dinner invitations. He was said to be an Englishman, or perhaps a Canadian, or maybe of American parentage. The city's keenly focused interest in him derived, in part, from simple unsatisfied curiosity: Who *was* this guy? A few skeptics remained doggedly unimpressed. "Mr. Cunard, a substantial, sensible Englishman, and not an Emperor, sits enthroned in state in the saloon of the *Britannia*," one doubter wrote to a local paper. "A proper self-respect will not warrant us in canonizing him." But this was only a cranky dissent, drowned out by a tidal wave of adulation and applause.

The "Cunard Festival" to honor him took place at the Maverick House hotel, near the Cunard wharf in East Boston. Planned while the *Britannia* was still at sea, the event was staged on a scale and opulence seldom previously seen in Boston. A temporary pavilion and awning stretched 200 feet along the front of the hotel. Pennants and flags of all nations snapped in the breeze. An elliptical arch spread across the Maverick's second story, resting on two abutments. One of these bore the British coat of arms and the name Watt, honoring the Scottish inventor who had improved the steam engine. The other showed the American arms and the name Fulton, the Hudson River steamboat pioneer. At the center of the arch, joining the two ancient national foes in symbol and reality, was the name Cunard in large gold letters—a premature tribute, mustered in brave hope and confidence, to the new but unproven steamship service.

On a raised platform sat the presiding officer, Josiah Quincy Jr., president of Harvard College and former mayor of Boston. To his right sat

Samuel Cunard, Senator Daniel Webster (the leader of Massachusetts Whigs), and other important men. To Quincy's left sat Captain Woodruff of the *Britannia,* George Bancroft (the Democratic boss of Massachusetts), and others of distinction. Overlooking them, outside the pavilion itself, on the hotel's porch and balconies and in upstairs windows, were hundreds of women dressed in their summer finery: the first time in memory that women had been invited to attend a public dinner in Boston (though not, actually, to be fed).

After a fancy meal, wines, and mounds of ice cream, the extended speeches and toasts sounded the framing issues of the day. Daniel Webster, a noted orator, spoke about the rippling impacts of steam power on civilization, commerce, war, and politics. George Bancroft welcomed the Cunard Line as "an omen of peace," sure to usher in a new era of friendly relations, and offered a hopeful toast: "Old England—She renounces the ambition of ruling the seas, and effects the nobler purpose of connecting continents." Cigars were passed around (but quickly put away in deference to the unfamiliar feminine presence). Someone offered a song in tribute to the new line:

> How timid and slow, but a few years ago,
> The world hobbled on in its motion.
> Old Europe seem'd far as the fixed Northern star,
> On the boundless expanse of the ocean;
> But tho' it was hard—at the word of Cunard
> *Britannia* herself is a rover.
> Old England a while, that fast anchor'd Isle,
> By steam is now here—half seas over.

Josiah Quincy introduced the man of the hour. "The enlightened foresight of Mr. Cunard, a citizen of Nova Scotia," he declared, "aided by the liberality of the British crown, has established a line of steam packets on a permanent basis." By advancing the interests of his own country, Cunard had incidentally conferred coveted gifts on America, and— Quincy hoped—Boston might now recover its old prosperity. The band impartially played both "Yankee Doodle" and "God Save the Queen." The climactic moment had arrived.

In this speechmaking age, when events happened live, in oral discourse, any distinguished man was expected to have an easy knack for facing and holding an audience. Grand oratory was a routine tool of persuasion and power. The crowd clapped and cheered hard for Cunard, waiting in curiosity and expectation for him to speak, watching him. He was a handsome man, apparently of middle age, a bit less than average height. He had a large, round head, balding at the crown, with a fringe of gray hair turning white, and closely trimmed muttonchop whiskers in the style of the day. He looked healthy and well knit, compact and tightly wound, and quite decisive around the mouth and eyes.

Cunard stood up and started speaking, inaudibly, before the last applause subsided. The newspaper reporters seated nearby could not hear him. He said only a few more words, and then—to the surprise of everyone—sat down. Josiah Quincy, rushing into the dead air, sprang up to say that Mr. Cunard had explained he was unaccustomed to public speaking and thus would make no speech, but he felt quite grateful to be so honored, given that the real credit for the new steamship line belonged to the British government. And that was that. His odd performance concluded, the hero remained triumphantly unknown in Boston.

A few days later, in much more modest circumstances, another dinner was held for men of the *Britannia*. Only Captain Woodruff and a few officers had been invited to the Cunard Festival; so a group of local machinists and mechanics, described as "respectable" by one newspaper, threw a small celebration for the ship's chief engineer, Peter Kenneth, and his mates at the Stackpole House, on Milk Street near the waterfront. The innkeeper, James Ryan, provided entertainment. (No Boston Irishmen like Ryan had taken visible part in the Cunard Festival.) All the guests on hand told a story or sang a song or ventured a sentiment. They raised their own toasts to the owners and hands of the *Britannia* ("Her successful voyage has proved that their capital and labor were most happily united") and, in engineer's vernacular, to the mother country ("She sometimes gets the steam up a little too high, but she finds an escape pipe when she visits her daughter"). The engine men, both ashore and from below decks, added their voices to the general spirit of determined reconciliation.

After two weeks, Cunard and his ship left Boston for Halifax and Liverpool on the afternoon of August 1, well loaded with eighty passengers. Spectators around the harbor were again frustrated; having arrived in nighttime darkness, the *Britannia* departed in a daytime thick fog that left her visible only from close at hand. She picked her way slowly along the Maine coast, taking three days to Halifax, shrouded in fog the whole way. The North Atlantic was again extending its typical welcome.

"The Atlantic to America is the worst navigation in the world," Sam Cunard pointed out years later, from a prudent and well-earned distance. "The westerly winds prevail very much, and you have ice and fog to contend with." Despite these daunting natural obstacles, however, he had launched his Atlantic steamship line and made it run through endless crises and troubles. "I originated this service at a great risk," he claimed in pardonable pride, dropping the humble pose of the Cunard Festival in Boston, "and at a time when no other party could be found to undertake it." And the result? "A beautiful line of communication between the eastern and western world."

*I*t was all there from the start. The major themes of transatlantic steamship history, to be echoed repeatedly over the next hundred years, first appeared in the summer of 1840 during the *Britannia's* maiden round trip, Liverpool to Boston and back.

The fearsome North Atlantic Ocean passage between Britain and America.

To meet and perhaps subdue this most difficult natural environment: the peerless shipbuilding and marine engineering along the River Clyde in Scotland, where most of the finest Atlantic liners of the nineteenth century would be designed and built.

The magical, transforming element of steam, the universal microchip of this era, and the utopian hopes of abolishing time and space it inspired.

The touchy ties among nations, especially between Britain and America, and the additional utopian hopes of international reconciliation forged by regular steamship service.

The stratifications of class and duty among the ship crews: officers and men, sailors and engineers, canvas and coal, above and below decks.

The recurring public wonder over every successive version of the newest, biggest, fastest steamship on the ocean, each likely to be described in turn as "a floating palace."

And at the center of it all, driving and organizing, the elusive figure of Samuel Cunard and the great transatlantic steamship line he founded.

PART ONE

The Packet Ship Era, 1820–1840

The Sailing Packets

*B*efore steamships started crossing the North Atlantic, the best way to travel between Europe and America was by the sailing ships called packets. Built and run mainly by Americans, the packet lines introduced new concepts and comfort levels for ocean voyages. They dominated the transatlantic traffic for decades, setting key precedents for the steamships that eventually replaced them. Along with their more famous contemporaries, the whaling and clipper ships, they comprised the golden age of American sail. Of these three types, the packets lasted the longest and made the most voyages and money for their owners and crews. Yet today whalers and clippers remain drenched in popular legend, while the packets are scarcely known beyond dedicated circles of ship buffs. No packet builder ever became as famous as Donald McKay with his clippers, and no novelist ever wrote a *Moby Dick* about the packets. They just did their jobs quietly and well, year after year, and then passed into the historical obscurity reserved for predictable competence.

A group of textile importers in New York started the first packet line. The main founder, Jeremiah Thompson, was an English immigrant from Yorkshire who had come to New York at age seventeen in 1801 to join his uncle in representing the family's woolen manufacturing business. From that base they engaged in shipping and shipowning with three local associates. These five men all lived near the waterfront at the southern tip of Manhattan. Four of them were Quakers. (Jeremiah Thompson, an active Friend, was an officer in the New York Manumission Society, dedicated to freeing slaves; but he also made a fortune by exporting raw cotton, grown in the American South by slave labor.)

Thompson had a breakthrough idea for improving ocean travel. At the time, a shipowner might advertise a ship's day of departure, but the captain would then wait until enough cargo and passengers had been loaded, and wind and weather seemed favorable, before weighing anchor. A passenger hoping to embark might have to hang around the docks, spending money on food and lodging and wasting time, for a week or more. Thompson, dealing in volatile markets for finished imports and raw exports, wanted faster, more reliable service. He conceived the notion of a transatlantic ship "line": several vessels under coordinated private management, sailing on known dates between established ports, and locked into an unchanging departure schedule for the foreseeable future.

In the fall of 1817, the Thompsons and their three associates placed a notice in New York's newspapers. "In order to furnish frequent and regular conveyances for GOODS and PASSENGERS," they announced, "the subscribers have undertaken to establish a line of vessels between NEW-YORK and LIVERPOOL, to sail from each place on a certain day in every month throughout the year." They listed the line's first four ships: three-masted and square-rigged, and larger than average size for their time at around 110 feet long and 400 tons. The *Pacific,* launched in 1807 and the oldest of the four, was especially fast; earlier that year she had made a run to Liverpool in only seventeen days. "These ships have all been built in New-York, of the best materials," the owners asserted. "They are known to be remarkably fast sailers, and their accommodations for passengers are uncommonly extensive and commodious." Thompson and his partners were promising a daring trifecta of speed, comfort, and predictability—qualities previously unknown on the North Atlantic.

The first two ships of the line sailed from New York and Liverpool in January 1818. For identification they showed a large black ball painted on their fore topsail, at the highest point of the first mast. The "Black Ball Line" at once earned a tight reputation for minding the calendar. Fighting winter gales, the *Pacific* made a slow return trip to New York of forty-eight days; she was then unloaded and reloaded in an impossibly short six days and left for Liverpool as scheduled on the fifth of April. Later that year, the Black Ball's *Courier* on leaving Liverpool met the *Pacific* coming in, and when approaching New York met the Black Baller *Amity* going out. The line added more ships, allowing two sailings a month each way. For any eastbound trip under twenty-two days or westbound run under thirty-five, Jeremiah Thompson gave the captain a new coat, with a dress for his wife. After two years, even *Niles' Weekly Register,* from the rival port of Baltimore, had to concede that the Black Ball ships were running with the speed and almost the regularity of a horse-drawn mail coach. "Such steadiness and despatch is truly astonishing," said the *Register*, "and, in a former age, would have been incredible."

Success brought competition. Atlantic packet lines started running from Philadelphia and Boston. Early in 1824, the Boston line's *Emerald* caught a rare easterly gale and rode it all the way home from Liverpool in an astonishing seventeen days, a westward record for years. In New York, the Red Star and Blue Swallowtail lines competed directly with Black Ball. Other new lines ran to London and to Le Havre on the northern coast of France. The sharp rivalry among all these lines added another new concept to transatlantic travel. Ship technologies in Europe and America had been essentially static for some two hundred years; conservative builders and owners resisted innovations and kept turning out the same old models. Packet competition kicked ship design into the progressive nineteenth century. Constructed mainly in shipyards along the East River in New York, ever bigger and fancier, the new packets became the largest and finest ships yet built in America, evolving more quickly than any other type of vessel.

Black Ball set the initial pace. The *Canada*, 132 feet and 525 tons, was launched in March 1823. "We have never examined a ship which was in all respects equal to her," said a local newspaper. Her dining cabin offered polished mahogany tables and pillars, sofas, and plush crimson draperies. The men's cabins, brightened by skylights of ground glass in the main deck overhead, had olive-colored damask silk curtains. In the ladies' cabins the curtains were fine blue silk. A year later, the Blue Swallowtail Line answered Black Ball with its own *York*. To the now-expected mahogany woodwork, the *York* added a library with a printed catalog, redwood pillars finished in imitation bronze, and venetian blinds in the cabin doors that allowed ventilation with privacy. Cabin washstands doubled as desks. A Turkish carpet covered the floor and muffled shipboard sounds. The ladies' lounge even featured a small piano flanked by large mirrors. "In the comfort and entertainment which the American ships afford," a Liverpool newspaper's account of the *York* acknowledged, ". . . their superiority over British vessels is most conspicuous."

While ship interiors became plusher and better equipped, the East River shipwrights puzzled over how to increase speed without losing cargo space. By slow degrees, the rounded bow and plump midship lines of the first packets gave way to faster ships with longer, thinner hulls and sharper bows and sterns. Shipwrights believed intuitively that speed also required a V-shaped hull, tapering down to a narrow keel at the bottom of the ship. These design tendencies all meant less payload and lower profits for a ship of a given length.

A solution to this tightening dilemma was discovered accidentally. In the early 1830s, Edward Knight Collins of New York started running coastal packets to New Orleans. Because the bar at the mouth of the Mississippi River required ships of shallow draft, the New Orleans packets were built with flat bottoms; speed was not deemed so important in the coastal traffic. But it turned out, to general surprise, that flat hulls did not make the ships any slower or harm their sailing qualities. A flat bottom also let a ship rest upright when grounded by a low tide.

In 1836 Collins launched his Dramatic Line of flat-bottomed packets to Liverpool, with ships named for famous theatrical figures. Flamboyant and excessive, indeed theatrical, and bold to the point of recklessness, Collins left the competition in his foaming wake. His *Shakespeare* ran 142 feet and 747 tons; his *Garrick, Sheridan,* and *Siddons,* 158 feet and 895 tons each; and the *Roscius,* at 168 feet, was the first New York packet to exceed 1,000 tons (and to cost $100,000). Collins also moved his passenger cabins from below deck, where they were subject to nauseating bilge odors, to a long deckhouse on the main deck stretching from the stern almost to the central mainmast. Up there, the cabins got more air and light—but without making the ship top-heavy or harming her behavior or safety. The cabins themselves were three times larger than those on the first Black Ball ships. The Dramatic Line's food, wines, and decor all set new standards of elegance. And the ships were fast. Over their first ten years, the New York packets had averaged twenty-four days out, thirty-eight days home—excellent times compared to those of previous ships. By 1839 the Collins packets, the swiftest in the trade, had cut those averages to twenty days, twelve hours and thirty days, twelve hours. Ocean travel had never before made such vaulting strides in only two decades.

*A*n ocean voyage, in this era or any other, had to work around three endemic aspects of the experience: seasickness, danger, and boredom. The worst bouts of mal de mer usually lifted after the first few days but could last longer, especially for women and in heavy seas. Any ship—in particular a sailing vessel—remained at the mercy of mighty natural forces, and on the heavily traveled North Atlantic might also collide with another ship or with an iceberg. Every day passengers had to find ways to kill time, a search that became more desperate and exhausted toward the end of the passage. "A sea voyage . . . is a sort of Purgatory under the best of circumstances," William Young of Halifax, Nova Scotia, wrote in his journal aboard a packet in 1839. "You can follow no regular employment and tho' not sick, I am never quite well enough for study. You can't write on account of the motion and one's reading is uninstructive and desultory." Young and his fellow passengers enjoyed clean berths, attentive service, abundant food and drink, and clear sailing. "And yet, from a sea voyage, Good Lord deliver me."

A packet voyage began with the captain. A safe passage depended absolutely on his skill, judgment, and tenacity. He had to make the daily computations with sextant and chronometer that established the ship's position and heading. At any time of day and night, he might order sailors aloft to set or take in sail. All steel and velvet, he was supposed to charm the passengers and, on occasion, bully the crew. His authority was final in all shipboard matters; yet he was ultimately just another mortal dealing with the unknowable mysteries of sailing the ocean. Every sailing ship displayed her own individual personality: what sailors called "ship sense," an ineffable quality of seeming alive, even of having consciousness. Under way, each seagoing amalgam of natural materials—of wood, hemp, and canvas—thrummed a unique vibration which the captain could feel tingling from the rudder to the wheel, and hear blowing overhead through the sails and rigging. The trick was to pay careful attention and work with the ship, not to dominate her. To the extent that a packet voyage responded to mere human will and intention, it came down to a captain and his ship getting along well.

Passengers would pay their one-way fares of $140 to secure a particular captain as much as for his particular ship or packet line. An especially popular commander—such as George Maxwell of the Black Ball, Nash DeCost of the Blue Swallowtail, or Nathaniel Palmer of the Dramatic—reliably attracted extra business. That meant more money for the captain himself, because he typically owned a one-eighth stake in his ship and received 5 percent of the freight and steerage charges and 25 percent of the cabin fares. The governments in Washington and London also paid him two cents for each American letter and two pence (four cents) for each British letter he carried. These extras brought his nominal annual salary of $360 up to as much as $5,000 a year, a plush income at the time. The packets therefore drew the services of the best captains on the Atlantic.

The crews were made up of sailors from many nations of Europe and North America. The cooks and stewards were usually black Americans, "clever mulattoes," according to James Fenimore Cooper, "who have caught the civilization of the kitchen." Passengers had most of their shipboard contact with the stewards, who served food, cleaned and fetched, and answered redundant questions about the weather and general course of the trip. The sailors kept to themselves, bunking in cramped quarters in the forecastle at the bow of the ship and conversing in the arcane, excluding patois of the sea. Passengers would marvel, from a distance, at the sailors' strength and agility as they danced around the rigging in all weathers and acted variously as tailor, carpenter, cooper, stevedore, clerk, and astronomer. It was easy to romanticize their often brief, dangerous lives. Captains drove the packets hard, always to the limit that sails and masts could bear, straining for speed. That meant constant action in the rigging (especially in bad weather), much bellowing and cursing, and sailors occasionally falling to their sudden deaths.

Everyone, ships and humans, remained at the indifferent mercy of the North Atlantic Ocean, in particular of the capricious wind. "We are pensioners of the wind," Ralph Waldo Emerson wrote in his journal at sea in 1833. "All our prosperity, enterprize, temper come and go with the fickle air. If the wind should forget to blow we must eat our masts." A ship could lie becalmed in midocean for a week or more, the wilted sails slapping irritatingly against the masts, the ship rising and falling helplessly on the endless swells. At the other extreme, too much wind brought its own delights. In April 1831 the *President* of the London Black X Line picked her way from New York through twelve straight days of cold, dense fogs and heavy, rolling seas. A fierce gale pushed waves almost up to her topmast. As the ship rolled back and forth on her bow-to-stern axis, water came over the five-foot bulwarks onto the deck, then into the cabins below. The captain, standing in water up to his knees, could not leave his post for twenty-four hours. The *President* limped into port after a hard passage of thirty-nine days. Other ships at journey's end might come within tantalizing sight of land and then have to spend days tacking back and forth along the coast, held at sea by contrary winds.

A recurring drama of initiation awaited those crossing for the first time. The first few days at sea might seem deceptively tranquil. No prior experience of the ocean from the vantage point of a beach or an offshore boat could adequately prepare a neophyte for the North Atlantic in full cry. In November 1835, Fanny Appleton of Boston (the future wife of Henry Wadsworth Longfellow) embarked on the packet *Francis Depau* for Le Havre. Eighteen years old, bright-eyed and curious, she dismissed the fourteen other passengers as "well meaning, uninteresting folks" but liked watching the sailors and the shifting sea. Commanders often managed to find time for pretty young women on board their ships; so Captain Henry Robinson gave her a puzzling lesson in navigation and "shooting the sun" with his sextant at noon to find the *Depau's* position. Five days out, running before a brisk following wind, the ship dashed along at ten knots, prancing like a sea horse. "What a glorious exhilaration in this fine sea-air, a reckless thought-defying sense of liberty and life," Appleton wrote in her journal, with the joy of a cosseted young woman now perceiving a wider world. "The exhilaration of our speed fills us with a mad glee . . . we run and shout."

The wind continued as the sea got rougher. Still bowling along at ten knots, the ship pitched and rolled and tossed. Appleton felt dizzy and exhausted, had trouble dressing, and—one week out—started longing for land. After a miserable night, she "wondered where the romance of the sea was found—certainly not below the deck." She wept in despair. The steward brought tea. The constant motion and cacophony of wind, sea, and shipboard sounds kept her from reading and writing. "Oh this eternity of noise and motion stupefying the brain, exhausting the body. Truly a ship-board life teaches one . . . humility: we are brought to our lowest ebb of self-respect." Her mood fluctuated wildly for the rest of the trip, depending on the weather. Like many sea diarists, she gradually ran short of fresh material and made briefer entries as the journey dragged on. Other passengers expressed surprise that she could still write so much about so little. "I am determined to prove one can write a Journal at sea," she vowed—and then left three straight days blank. ("Little worth recording," she noted.) The ship reached Le Havre after twenty-five days, none too soon for Fanny Appleton.

Most of the cabin passengers on packets were men: British textile merchants and army officers, and American businessmen. Thrown together at close quarters amid the Anglo-American political tensions of the time, they sometimes bristled at each other. More typically, as frequent transatlantic travelers they settled amicably into the shipboard routines they had come to know well. In this quite masculine atmosphere, isolated by time and circumstances, they could dress casually and indulge at will in boyish recreations. "We endeavored to amuse ourselves as best we could," one man noted, "and, for the want of work, turned boys again, and went to play." They shot rifles at random targets in the rigging, sang songs, made bets, played cards and games, held mock courts, told jokes and stories; they drank all day long. Even the tightly buttoned Ralph Waldo Emerson, on the Black Baller *New York* from Liverpool in 1833, succumbed to shipboard spirits. "These are the amusements of wise men in this sad place," he decided. "I tipple with all my heart here. May I not?" In midsummer 1832, the English actress Fanny Kemble, bemused and amused, watched the men on a packet in a festive Saturday-night mood of drinking, dancing, singing, and romping around on the quarter-deck at the stern. They toasted their absent wives and sweethearts, in the tradition of sailors. The captain proposed "The Ladies—God bless them." ("And the Lord deliver us!" someone added.)

The most promisingly pleasant aspect of packet life available to all cabin passengers, drunk or sober, was the food. A housed-over longboat on deck held a three-tiered menagerie: sheep and pigs on the bottom, then ducks and geese, and hens and chickens on the top. The ship's cow lived in another structure nearby, not happily. These animals provided fresh meat, eggs, milk, and cream for the laden plates in the dining saloon. On the Black Ball's *Europe* in 1833, breakfast consisted of ham, eggs, bacon, mutton cutlets, shadfish, rolls, and cognac. Dinner ran to three comparable courses, good pastries, seven kinds of alcohol, and dried fruits for dessert. The long, rolling swells of the North Atlantic left many passengers unable to eat or hold down their meals; but whenever circumstances allowed, especially during the early weeks of the voyage, the food and drink were generous.

Except for the Le Havre lines, which brought over thousands of German immigrants to America, the packets carried few steerage passengers until after 1840. During the first two decades, steerage business in general was just an afterthought; if a packet didn't fill the 'tween-deck's upper hold with fine freight, the ship's carpenter would fashion temporary bunks of rough, unplaned lumber, and the steerage would take on human cargo at twenty dollars a head. That fare paid only for the cramped bunk and a place on deck to cook. Steerage passengers had to bring their own food, pots and pans, and plates and utensils. The tightly packed steerage became a fetid horror at night and in bad weather. Two small ventilating hatches had to be closed in rain or heavy seas. The steerage air mingled a stifling bouquet of foul bilge water, rotting wood and ropes, and human sweat, vomit, and excrement. At times these closed-down conditions went on for days, getting worse until the storm lifted.

Everybody, cabin and steerage passengers alike, felt better up on deck. There one could breathe fresh air and take walks, play shuffleboard, and watch the sea and the other passengers. A dampening shipboard rhythm of ennui and lassitude gradually settled onto the company. Earnest intentions of reading, writing, and needlework were laid aside, and people lounged away the time. Even the most trivial daily events— meeting and (after due inquiry) identifying another ship, sighting a whale or porpoise, or discovering a stowaway—took on gripping, inordinate significance. Wagers were placed on the daily run and the date of arrival. After evening tea, the dining saloon might hold lectures, charades, rounds of whist, and singing. A bold man could even venture, trolling, into the ladies' cabin. "This snuggery affords tolerable convenience for a little flirtation," noted the Irish actor Tyrone Power, "if you are lucky enough to get one up."

On clear nights, far from the obscuring lamps of shore, and in the prevailing deprived mood of being easily diverted, passengers seized on natural light shows. The sun would fall into the western horizon, a blazing ball that slowly guttered out as though being submerged in the sea. The moon would rise, never before so distinctly. Stars filled the sky, brighter and denser than when seen from land. Whole new constellations revealed themselves to the naked eye. In every direction, the sky arched all the way down to the horizon. Shooting stars zoomed around this vast inverted bowl, and the aurora borealis looked deeper and more brilliant in its roses and purples. From the *Europe* in 1835, cruising along at nine knots, Anna Eliot Ticknor of Boston watched the phosphorescence tossed up by the bow. The foam at the stern glittered like diamonds. Distant waves broke and lit up. Porpoises darted around, leaving trails like seaborne comets. The sails and rigging glowed. It was all so wild and beautiful.

People stayed up late because sleeping was so difficult. After the sumptuous meals and lounges, the private sleeping quarters inevitably disappointed: small, unheated, dimly lit, and poorly ventilated. "About as big as that allowed to a pointer in a dog-kennel," the English novelist Frederick Marryat groused. "I thought that there was more finery than comfort." The berth, generously called a bed, offered thin sacking over boards and a hollow down the middle that was supposed to hold the occupant in place. Emerson's sides grew sore from rolling back and forth. "Oh for a bed!" Fanny Kemble keened. "A real bed! Any manner of bed, but a bed on shipboard!" Settled down for sleep, a passenger could not miss the remarkable, unexpected variety of noises on a wooden sailing ship. The waves kept up a steady background of thumps and pulses, generating constant small motions in the seams of the hull, flexing and twisting, which caused sharp creaking sounds in the jointed woodwork of bulkheads, cabin partitions, and steerage bunks. Passengers lay awake in their berths, trying not to hear. The animal barnyard overhead maintained a running, distressed commentary. The mates and sailors yelled back and forth, their footfalls thuddingly heavy at night, as they changed sails and scraped down the deck with screeching holystones. Wind whistled through the rigging. Morning could take a long time coming.

The English author and protosociologist Harriet Martineau wrote the fullest, most forgiving account of a packet voyage. In August 1834, thirty-two years old, she had just completed a popular series of short stories that improbably urged the beauties of classical economics. For over two years she had written so constantly that she could not spare time even to take a walk. With that work finally done, basking in her first great success, she booked passage for New York on the Red Star Line's *United States,* 140 feet and 650 tons. Martineau looked forward to a restful month on the ocean without mail, newspapers, or intruding strangers, and then would travel around America to report on that boisterous, unmatured experiment in democracy.

She cataloged the twenty-three cabin passengers on board: a Prussian physician, a New England preacher, a Boston merchant "with his sprightly and showy young wife," a high-spirited young South Carolinian returning from study in Germany, a newly married couple who kept to themselves, a Scottish army officer whose many crotchets amused the young people, an elderly widow, a Scottish lady of undisclosed age, and a young man from Yorkshire. The rest were English and American merchants, transatlantic veterans not deemed interesting enough by Martineau for detailed comment. With two or three exceptions, they all mingled congenially into a single traveling party.

The voyage began slowly, dawdling through calm days of little wind. The Americans, longing for home, became anxious. Martineau seized the welcome quiet time to think and be still. On the third day the wind freshened and the sea churned, leaving the dining saloon empty at dinner as most passengers remained seasick in their berths. The next morning, Martineau rose unsteadily but forced herself to dress and go up on deck to escape the bilious sights and activities below. Captain Nathan Holdredge took her to a seat by the rail. She looked out to sea, avoided noticing the invalids strewn around the deck, and felt better after half an hour. The wind was too strong for a large, flaring bonnet, but she tried a warm black silk cap, snugly fitted, which she recommended for any woman at sea.

An uncommon traveler, Martineau had two qualities that maintained her spirits on the ocean: a bottomless curiosity and delight in new experiences, and an absolute refusal to be discouraged by anything whatever. After six days of mostly unhelpful winds, the *United States* was still only three hundred miles from Liverpool; at that rate the voyage would take two months. No matter. "Our mode of life was very simple and quiet; to me, very delightful," she wrote. "A voyage is the most pleasant pastime I have ever known." After breakfast, the happiest meal of the day, she sat down to write a long article, the one major task she had set herself for the trip. The New England preacher would find her a place on deck, out of the wind and sun, and there she wrote through luncheon until two o'clock. Children from the steerage peered at the famous lady writer over her shoulder and from behind chests and casks. One particular man planted himself in front of her, arms akimbo, and stared at the point of her pen, transfixed by the mysterious act of female composition.

Finished writing for the day, she took her position at the rail and exulted in the passing scene. If she wanted to be left alone, she held a volume of Shakespeare. Otherwise someone joined her. "I strongly suspect," she later reflected, "that those who complain of the monotony of the ocean do not use their eyes as they do on land." She saw Portuguese men-of-war, flying fish, dolphins, and the web-footed birds called Mother Carey's chickens. A sail on the horizon brought everyone over to look and exclaim. Early one morning a distant ship made signals of distress. Great flutters of excitement; "the faces of the gentlemen began to wear, in anticipation, an expression of manly compassion." Captain Holdredge took in sail and hove to. The other ship, it turned out, had only lost her longitude bearing. Holdredge shouted it out, angry over losing valuable time for such a small matter, and ordered the sails up again.

The captain, kind and patient even with repeated, unanswerable questions, could never forget his mandate for maximum speed at all times. One day Martineau noticed another ship ahead on the same westerly course. She told the captain, who took a hard look through his telescope and then barked out sharp orders to the helmsman and crew. The other ship was the *Montreal* of the rival Black X Line. Smaller and slower, she had left England four days before the *United States*. An ocean race was on. "Our captain left the dinner-table three times this first day of the race, and was excessively anxious throughout. It was very exciting to us all." In three days the *United States* overtook the *Montreal* and left her far behind, slowly falling below the eastern horizon.

Most days were clear enough for visible sunsets. Everyone, from cabin and steerage alike, gathered on deck. A few climbed up into the rigging. People grew quiet except for pointing out particular features in the clouds or sea. As the sun went under, some of the party stood on tiptoes, reaching for one last glimpse. Then the normal talk and bustle resumed as walkers promenaded the deck, thirty paces up and back.

After evening tea, Martineau avoided the convivial cabin and found a place to herself at the stern. A true writer, an onlooker by nature, she craved a safe solitude from which to watch developments. She studied the wake behind, "a long train of pale fire," and the sails ahead, outlined against the sky and stars. A night fog might scud through, thick and moving fast, with occasional open spaces for the moon. Lost and engrossed, utterly content, sometimes she forgot that she was at sea. Snatches of old songs floated through her head from nowhere, and the first poems she had ever loved. "Such are the hours when all that one has ever known or thought that is beautiful comes back softly and mysteriously."

She did acknowledge a few discomforts at sea: rainy days that kept everyone below in stifling air, and prolonged calms that made tempers short, provoking rude behavior at dinner and accusations of cheating at shuffleboard. In midocean, a ferocious storm lasted all night, to the disquieting sounds of breaking glass and screaming women. Toward the end of the long voyage, the dried fruits got moldy, and the kitchen ran out of cider, ale, claret, and soda water. In general, though, Martineau denied the usual purported annoyances of ocean travel. She made a list, in her methodical way, of all such claimed aggravations, along with their (to her) satisfactory remedies.

1. Seasickness. ("An annoyance scarcely to be exaggerated while it lasts." No remedy.)

2. The damp, clammy feel of everything one touches. (Wear gloves, and clothes too worn to be spoiled. "In this latter device nearly the whole company were so accomplished that it was hard to say who excelled.")

3. Lack of room. (Put everything away in tight, orderly fashion.)

4. The candles flare, dribble wax, and look untidy. (Avoid looking at candles; go to the stern at night, which has its own, better lights.)

5. The seats and beds are too hard. (Have patience. Try air cushions.)

6. Freshwater use is limited. (Bathe in seawater, and drink cider at dinner.)

7. The cider may run out. (Switch to other beverages.)

8. The noise of sailors scraping the deck. (Again, patience; because the deck must be scraped.)

9. The clamor overhead when the sails are shifted at night. (Go back to sleep.)

10. Sour bread. (Eat biscuits.)

11. Getting sunburned. (Don't look in a mirror.)

Not even the North Atlantic Ocean could daunt such a temperament. (It should be noted that Martineau was partly deaf and therefore protected from the worst noises at sea.) After the restive final stretches of the passage, everyone's spirits rose as the ship approached America. People changed into their best clothes, not seen for weeks, in preparation for landing. The *United States* reached New York after forty-two days: "a long but agreeable voyage," she insisted.

*B*y the late 1830s, twenty packet ships were running from New York to Liverpool, twelve more to London, and sixteen to Le Havre. Every month, a dozen packets left New York for Europe and a dozen more arrived; an average of one ship every thirty hours, all year long, regardless of the wind and weather. The packets suffered occasional collisions and founderings at sea, but only two accidents caused any loss of life over the first two decades. The *Albion* of the Black Ball Line sank off Ireland in 1822, killing forty-six people, and four years later the *Crisis* of the Black X Line disappeared on a westbound run with her crew and about a dozen passengers. Those two disasters aside, the packets had compiled—for the time—a remarkable record of fast, safe, predictable transatlantic travel.

According to testimony from both sides of the ocean, Americans were building and running the finest sailing ships in the world. A London newspaper in 1834, after comparing the safety records of the New York packets and the British government's mail ships, urged the Admiralty to buy American vessels. In 1836, a committee of the British Parliament inquiring into the problem of shipwrecks presented evidence that American ships were better built than their British counterparts (and thus preferred by shippers and insurance agents), and that American commanders and officers were more educated and competent and American seamen more carefully selected, more efficient, and better paid—to the point that the best British sailors were defecting to American ships. American authorities could only happily agree. Matthew Maury, an American naval officer and one of the founders of oceanography, praised the New York packets in 1839 in language of patriotic but unchallenged hyperbole: "For strength, safety, fleetness and beauty; and for a combination of all the requisites of a good ship, in such admirable proportions, no nation can boast of vessels, public or private, comparable to them."

The packets became, in some measure, the victims of their own success. They had created the very notion of rapid technical improvement in transatlantic travel. Passengers came to expect bigger, faster ships every few years. The wind, however, could not be improved: it blew hard or not at all, from the east or west, but always beyond any human control. Sailing ships could only *depart* on a scheduled date. The time of arrival might then vary by weeks, depending on the ocean's vagaries. Steam power extended the possibility of keeping a ship on schedule, or nearly so, at both ends of the passage. But Americans became so proficient and applauded at turning out wooden sailing ships that, as time passed, they—in complacency and inertia—kept building those ships for too long, far past their technological prime. In Great Britain, especially in Scotland, other men were about to take over the leadership of transatlantic shipbuilding.

2 .

Steam on Water

Steam power drove both the Industrial Revolution and the progressive nineteenth century. Of all the thousands of inventions that have created the pervasive material modernity of the past two hundred years, the steam engine was the first cause, the prime mover, and sine qua non. Unlike muscle power, it never tired or slept or refused to obey. Unlike waterpower, its immediate predecessor, it ran in all seasons and weathers, always the same. Unlike the wind, it responded tractably to human will and imagination: turning on and off, modulating smoothly from the finest delicacy to greatest force, ever under responsive control. "It is impossible to contemplate, without a feeling of exultation, this wonder of modern art," the *Quarterly Review* of London declared in 1830. After first transforming mining, manufacturing, and transportation, from those bases the steam engine eventually reached into the smallest aspects of everyday life. Seen from the distant perspective of two centuries later, the great Steam Age looks like an unbroken, triumphal march.

Seen closer at hand, the application of steam power to any given field was a messy process overflowing with false starts and repeated, redundant discoveries. The most baffling aspect of inventing a steamboat, it turned out, did not involve the engine, fuel, boiler, or hull. Instead it came down to the propelling mechanism, the essential driving link between the steam engine and the water. The challenge of how to contrive a harnessing device that would let an engine power a boat forward, even against winds and tides, had no obvious, inevitable solution. Many lone tinkerers in

Europe and America took whacks at the puzzle and subsided in defeat. One such inventor figured out key practical breakthroughs and even built and ran an influential steamboat; but he was overwhelmed by unrelated forces beyond his ken, became discouraged, and died broke and unappreciated. Another pioneer took the work of this inventor and others without giving credit, later lied about it, and finally perjured and embarrassed himself; but he also thereby acquired great fame and fortune, and to this day retains a thumping historical reputation as the true father of steam navigation. The story has its ironies.

*T*he steam engine and steamboat both emerged from a visible chain of invention: a series of innovators, aware of earlier work in the field and consciously building on it, adding and subtracting and thus moving the whole process forward by small increments until the machine ran right. The final, laborious success when ultimately achieved was descended from many parents, leading to bitter quarrels and lawsuits over who should get the credit and rewards.

For thousands of years, unconnected individuals had puzzled over how to control and use the power of steam. Nothing important happened until Thomas Newcomen started a chain of invention in 1712. An ironmonger in southwestern England, Newcomen made tools for the tin miners of Cornwall. As mines were dug deeper, they were flooded with groundwater, overwhelming any manual or horse-driven pumps. Newcomen invented a steam-powered mine drainer: a large horizontal beam, pivoting at the middle, linked to a water pump at one end and a vertical piston and cylinder at the other. Steam entered the cylinder at the bottom and drove the piston upward; at the top of the stroke, cold water sprayed into the cylinder below the piston condensed the vapor back into liquid form, creating a partial vacuum which pulled the piston back down to repeat the cycle. The engine worked—but was bulky, expensive, and inefficient. "It takes an iron mine to build a Newcomen engine," the saying went, "and a coal mine to keep it going."

Skip ahead to a classic moment in the history of modernity. In the winter of 1763–64, a Scottish instrument maker at Glasgow University was asked to repair a model of a Newcomen engine. James Watt, then twenty-eight years old, fixed the model and started pondering the general problem of steam power, especially the obvious waste and inefficiency of the Newcomen design. He tried making the boiler surface larger, and placing the fire in the middle of the water supply, and even using wooden pipes and boilers (because they would conduct and lose less heat than metal components). One Sunday early in 1765, while walking across Glasgow green, Watt finally got it: create a separate condenser so the cylinder could remain at essentially the same temperature throughout the cycle, saving time and fuel because no steam would be lost to condensation from entering a cold cylinder. "I can think of nothing else but this machine," Watt informed a friend. "Write me . . . if any part of what you have to tell me concerns the fire-engine."

For the next crucial step, moving from inspiration to application, Watt needed help. Beset all his life by poor health and severe headaches, timid by nature and easily discouraged, Watt dealt uncertainly with the world outside his workshop. "Jamie is a queer lad," noted the wife of an associate. Matthew Boulton, a Birmingham manufacturer, offered to become "a midwife to ease you of your burthen," as he put it to Watt, "and to introduce your brat into the world." Boulton had more experience than Watt in the metal industry, ready access to money, and many useful contacts. Watt joined Boulton as partners in Birmingham. With a patent obtained in 1769, and later extended, they essentially controlled steam-engine technology for the next three decades. Watt and Boulton formed the first and most important of the many talent-meshing teams of engineer and entrepreneur that later propelled the Industrial Revolution.

With Boulton in the background, prodding and executing, Watt made further improvements, notably a double-acting cylinder whereby steam alternately drove the piston in both directions, yielding two power strokes in each cycle. He also devised linkages and gearings to convert the piston's in-and-out reciprocating action to a rotary motion that could power the machinery of mills and factories. "The people in London, Manchester, and Birmingham, are Steam Mill Mad," Boulton advised Watt, "and therefore let us be wise and take the advantage."

Amid his great success, Watt never stopped fretting about competitors and potential patent infringers. To protect himself and his inventions from the onrushing progress of modernity, he grew defensive and started resisting improvements. He quashed innovations in his own shop (especially efforts to raise boiler pressures and efficiencies beyond a modest four pounds per square inch), refused to license others to use his refinements, and hounded anybody else who dared to build a steam engine. The exploding genie of constant, rapid technological change—which his steam engine had midwifed—finally turned and overwhelmed him. "I do not think that we are safe a day to an end in this enterprizing age," he warned Boulton in 1782. "Ones thoughts seem to be stolen before one speaks them. . . . It is with the utmost difficulty I can hatch anything new." Beset by this immobilizing difficulty, losing his fragile nerve, he stopped trying. But his engine and its revolutionary impacts steamed ahead, gathering speed.

From the 1780s on, various lone inventors in France, Great Britain, and the United States tried to create a steamboat. For the propelling device, some of these pioneers used an application of the familiar waterpower wheel, which converted a stream of water into rotary motion to run a mill or factory: instead of water moving the wheel, the process was reversed so the engine-driven paddle wheel moved the surrounding water and thus the boat. But a paddle wheel was only one of several unsatisfactory early alternatives. Other propelling mechanisms given trials included a set of vertical oars that imitated manual rowing action (by the American John Fitch, in 1786), a jet of water forcefully expelled at the stern (by another American, James Rumsey, in 1787), and palmipedes, or duck-footed paddles (by the earl of Stanhope, in London in 1790). None of these early attempts worked very well or led to any ongoing commercial success. Their inventors tinkered in general isolation from each other, without knowing about or profiting from what their predecessors had done. Steamboats as yet lacked a chain of invention.

William Symington started such a chain through his own inventions and by their later impact on others. He was another Scotsman, born in 1764 in Lanarkshire, south of Glasgow. Educated for the ministry, he was instead caught up in the inventive currents then starting to swirl around southern Scotland. "My natural turn for mechanical philosophy led me to change my object," he recalled, "and to direct my studies to the exercise of the profession of a civil engineer." He made some improvements in the steam engine—earning the suspicion of James Watt—and crafted a model of a steam carriage for road travel. This model brought him to the attention of Patrick Miller, a retired Edinburgh banker who had devised a manually powered paddleboat.

In 1788 Miller hired Symington to build and install a steam engine in this vessel. Symington used his own design, an engine with two cylinders of four-inch diameter and eighteen-inch stroke. A second version with a larger engine had a successful trial a year later, carrying seven passengers at five miles an hour. But this success drew potential legal action by the ever-vigilant Watt for alleged patent infringement. After Miller lost interest in the experiments and withdrew his financial support, Symington dropped his steamboat efforts for a decade and made a living by building mining machinery.

The expiration of Watt's patent in 1800 released a flood of pent-up inventive energy. Thomas, Lord Dundas of Kerse, a large shareholder in the Forth and Clyde Canal, remembered Symington's experiments of the late 1780s. The canal, completed in 1790, stretched thirty-five miles from the River Forth near Edinburgh to the River Clyde near Glasgow, providing a water link across Scotland between the Atlantic Ocean and the North Sea. The canal's average width of about fifty-six feet left little room for a sailing vessel to tack back and forth, so most of the barge traffic was drawn by horses along a tow path. Lord Dundas provided Symington initial seed money for a canal steamer.

In June 1801, Symington's first new prototype ran successfully for two or three miles on the River Carron to Grangemouth. "The nice and effectual manner in which the machinery is applied," a Glasgow newspaper commented, "is an additional proof of the merit of Mr Symington, the engineer, and the whole plan is highly honourable to Lord Dundas." That fall Symington patented his novel arrangement of a connecting rod and crank between the engine and paddle wheel shaft.

A second prototype, larger and more powerful, was named the *Charlotte Dundas* after the sponsor's wife and daughter, who shared the name. The vessel was a broad-beamed towboat, fifty-six feet long by eighteen feet wide, powered by a one-cylinder engine driving a paddle wheel in a recess at the stern. The engine was built at a local foundry, the Carron Works, with a piston twenty-two inches in diameter and a four-foot stroke: an enormous increase over Symington's first steamboat engine of 1788. His solution to the besetting early problem of paddle wheels—the dilemma that drove other pioneers to water jets and palmipedes—was to elevate the wheel quite high above the water. When a wheel was submerged to its midpoint, half in and half out of the water, much of its driving motion was wasted. A paddle entered the water in a horizontal position, slapping downward, and did no useful propelling work until it had run through almost forty-five degrees of its rotation. Only at the bottom of the cycle was it actually propelling the boat forward. On the back stroke, the process was reversed, as for the final forty-five degrees the paddle pushed largely upward until it cleared the surface. About half its energy simply roiled the water up and down to no purpose. To avoid this thrashing waste, Symington placed the eight-bladed wheel of the *Charlotte Dundas* so high in the hull that only three of the paddles reached the water at once, at the bottom of the cycle, all of them working together to move the boat forward.

As Symington later told the story, in March 1802 the *Charlotte Dundas* took on board Lord Dundas, his son Captain George H. L. Dundas of the Royal Navy, and others, and towed two loaded vessels of seventy tons each a distance of nineteen and a half miles along the canal in six hours, against a strong head wind. "This experiment not only satisfied me, but every person who witnessed it, of the utility of steam navigation," Symington later wrote. But the canal proprietors worried that the steamboat's agitation and wake would harm the banks of the canal, and so rejected the plan. Lord Dundas then arranged for Symington to meet the duke of Bridgewater, the leading canal entrepreneur in England. The duke at once ordered eight of Symington's vessels—but he soon died, canceling the deal. This double rejection after apparent successes left Symington too disheartened to persist. "This so affected me," he recalled, "that probably I did not use the energy I otherwise might have done to introduce my invention to public notice."

This version of events has become the standard historical account, but it is wrong in certain particulars. Drawing from memory some twenty-five years later, Symington compressed two separate trials into a single event. On January 4, 1803, the *Charlotte Dundas,* with the two Dundases and others on board, towed a 100-ton boat from Stockingfield to Port Dundas at three miles an hour "amidst a very large concourse of people," according to a newspaper report, "who were exceedingly well pleased with the performance." On March 28, 1803, the steamboat also towed two loaded vessels, a combined 130 tons, from Lock 20 on the canal to Port Dundas, eighteen and a half miles in nine hours and fifteen minutes—a speed about 40 percent slower than Symington later remembered. For this trial he had incorporated suggestions by Captain George Dundas for how to manage the tow lines around sharp bends in the canal. The *Glasgow Herald and Advertiser* praised "the very appropriate mode in which the machinery is constructed, and the simple yet effectual manner its power is applied in giving motion to the vessel." The newspaper also credited Lord Dundas for his generous financial support and perseverance in the "costly experiments."

A few days later, the *Herald and Advertiser* published a testy letter from a Forth and Clyde Canal proprietor which fleshes out Symington's later explanation of why his steamboat was banned from the canal. The letter writer pointed out that a vessel passing through one of the canal's thirty-nine locks used a lockful of water, so a towboat plus barge consumed twice as much water (and the canal had recently been closed by low water); that the *Charlotte Dundas,* contrary to another report, would save no money over tow horses given her initial expense, the cost of coal, her crew, and general wear and tear; and that Symington's earlier steam-boat of 1801 could not run with any ice in the canal, and this problem had perhaps not yet been solved. After all these objections, the proprietor added, "It will be observed too, that the motion of the boat raises such an agitation in the water, as to injure the banks." In conclusion—and this probably clinched the matter—the writer regretted that Lord Dundas had been given all the public credit for funding Symington's efforts. "It should have been added, that the Proprietors of the Forth and Clyde Navigation have already paid about £1700 for these experiments of this ingenious mechanic, without reaping any benefit from them, and with-out even getting any credit for their liberality."

Given this bristling mixture of unmet criticisms and wounded, unap-preciated generosity, and (one may assume) competitive resistance by the local owners of horses and stables, it is not surprising that Symington got no farther with the canal proprietors. Hoping for other wisps of interest in the *Charlotte Dundas* from somebody else, he laid her up near the canal at Bainsford. There she lingered on for almost sixty years, rotting and rusting away, a waning curiosity of the early steam age. Like James Watt, Symington was a gifted inventor saddled with a fainthearted personality, too easily deflected from his purposes. His singular misfortune was that—unlike Watt—he never found his Matthew Boulton.

*R*obert Fulton, the American painter and inventor, knew all the precedents in steam navigation. During twenty years spent abroad, in England and France, he studied the efforts of other steamboat pioneers and tried out his own improvements. In contrast with most of the other pioneers, he was blessed with an overpowering confidence and persistence which, along with good looks and a gift for friendship, brought him the continuing support of rich, powerful patrons. Ultimately he returned to America to build and run the first commer-cially successful steamboat. Today most Americans consider him the prin-cipal originator of steam power on water. The process by which he achieved this reputation—and thus the reputation itself—demands a renewed examination.

For most of his two decades abroad, Fulton was preoccupied with other inventions than a steamboat. Living in France from 1797 to 1804, he devoted himself to an elaborate, quixotic, finally unworkable scheme for submarines and explosive mines, intended to revolutionize naval warfare.

His intermittent interest in steamboats was revived when Robert R. Livingston arrived in Paris late in 1801 as the U.S. minister to France. A man of enormous wealth and political influence in New York, Livingston hoped to develop a steamboat service for the Hudson River back home. Fulton had found his final, most significant patron.

During the summer of 1802, Fulton conducted a series of trials with a model boat powered by a clock spring. After considering all the propelling devices used by his predecessors, he settled on an endless chain with paddles or buckets attached to it. Resembling the tread of a modern tank or bulldozer, the chain was draped over two wheels across the side of the model, dipping into and seizing the water at the bottom of its cycle. Livingston, drawing from his own previous sallies at steamboat invention, preferred paddle wheels; but after Fulton reported on his trials with the model, arguing his case quite vehemently, Livingston was converted to the endless chain. In October 1802 the two men signed an agreement to build a large steamboat in New York, designed for the Hudson River traffic to Albany.

Now came a surprising, puzzling twist in the story. At some point that fall, after insisting so aggressively on the superiority of his endless chain, Fulton decided to adopt paddle wheels as his propelling device. His biographers have guessed that Fulton switched to avoid infringing a French patent, granted earlier that year to an inventor named Desblancs, for a similar steamboat with an endless chain. But Fulton had learned of this patent in June, and as late as September he was nonetheless still urging his own version of an endless chain. Something else must have persuaded him to change this crucial aspect of his design.

A possible explanation was later provided by William Symington. As he told the story in the 1820s, Fulton had come to Scotland to see one of Symington's vessels, explaining that he intended to return to America to build a steamboat, and that his project could lead to a rewarding business for Symington as the inventor. Flattered and intrigued, Symington ordered steam up in his paddle wheeler and took Fulton and others for a ride. From Lock 18 on the Forth and Clyde Canal, they went four miles west and back in one hour and twenty minutes, at an average speed of six miles an hour—"to the great astonishment of Mr. Fulton and the other gentlemen present," according to Symington. Fulton asked questions, took notes, and made sketches of the steamboat. After this single encounter, Symington recalled, he never saw or heard from Fulton again.

The dating of Fulton's visit presents problems. Symington placed it in July 1801 or July 1802. In 1801, however, France and England were at war, severely limiting travel between the two countries. Fulton would have had great difficulty in making his way from France to Scotland; at the time he was also still quite focused on his submarine and mines, to the exclusion of other interests. The Peace of Amiens in March 1802 allowed a brief lull in hostilities, easing travel restrictions. By then Fulton, with Livingston's beckoning patronage, had turned his attention nearly full-time to inventing a steamboat. He spent the summer of 1802 at a resort in the Vosges Mountains of northeast France, too far from the English Channel for a convenient trip to Scotland. That fall he was back in Paris, intent on his steamboat. The most probable date of Fulton's encounter with Symington is thus the fall of 1802, when the *Charlotte Dundas* was almost ready for her first major trial of January 1803. The journey from Paris took three days to London, then about sixty hours by mail coach to Glasgow. He could have made the round-trip in two weeks.

With travel again flowing between France and England, Paris was full of British tourists from whom Fulton or the widely acquainted Livingston might have heard about Symington's boat. A trip to England was clearly on Fulton's mind that fall. His friend Joel Barlow, also interested in promoting a joint steamboat scheme, had recently urged Fulton to go to England "*silent and steady* . . . quiet and quick" to obtain a steam engine. His formal agreement with Livingston in October also bound Fulton to go "immediately" to England for the same purpose. Fulton left no surviving record of such a trip that fall. But he could have gone secretly—silent and steady, quiet and quick—on steamboat business, especially to examine the *Charlotte Dundas,* the most promising such experiment in the world at that time. In late September, he was conspicuously absent from a dinner party given in Paris by the painter Benjamin West. Fulton was a close friend to West, his main mentor in painting. Joel Barlow and his wife, with whom Fulton lived in a ménage à trois, did attend the dinner. If Fulton had been in Paris, he surely would have joined the party. Perhaps he was then quietly off to Scotland.

This mystery turns on hard questions about Fulton's character. Could he have made a clandestine trip to Scotland, borrowed from Symington's work, and later hidden the entire episode? His subsequent history of lies and deceit suggests that he might have. In 1806, for example, he claimed in writing that he had held an American steamboat patent for fourteen years, and that some $280,000 had been subscribed to build twenty of his vessels for service on the Mississippi River—none of which was even remotely true. Later, when embroiled in patent controversies, he forged a "copy" of a drawing he had supposedly made in June 1802 of a Hudson

River steamboat with paddle wheels, at a time when he was actually still committed to an endless chain for propulsion. He also forged a letter, which he dated to 1793, about his supposed interest in paddle wheels at that time. In 1815, shortly before his death, he was caught committing perjury with this letter. All these manipulations were intentional, self-serving lies on Fulton's part.

Symington's later recollections, by contrast, erred in some details, but the essence of his account of the *Charlotte Dundas* is verifiably true. His version of the Fulton story was also corroborated by Symington's engine man, Robert Weir. In 1824, after the matter had become controversial, Weir signed a sworn affidavit that he had fired up the boiler of the *Charlotte Dundas* on the occasion of Fulton's visit and had heard Fulton identify himself by name and nationality. After their brisk eight-mile demonstration, according to Weir, Symington had lamented the difficulty of running his steamboat through the narrow Forth and Clyde Canal, and Fulton had replied that the broad rivers of America would present no such problem. The details and certainty of Weir's affidavit seem authentic.

Fulton's own explanation of how he converted to paddle wheels, later given under duress, must be weighed carefully. In 1811 he asked Joel Barlow to endorse his version of certain events for a potential patent lawsuit. "I want your deposition as follows," he instructed: that in the fall of 1802, while living at Barlow's home in Paris, he had conducted experiments with various propelling devices, which by Christmastime had convinced him to adopt paddle wheels. "You will have this copied on foolscap," Fulton told Barlow, "and sware to it." Barlow apparently complied. It was at about this time that Fulton also forged other documents to bolster his claims of steamboat originality.

The smoking gun in this mystery is the vessels that Symington and Fulton actually produced. In January 1803 Fulton drew up the plans for his first steamboat. Overtly she did not much resemble the *Charlotte Dundas:* long and lean instead of short and stubby, with a different arrangement of the machinery and a distinct means of converting the engine's reciprocating action to rotary motion. But in four crucial respects the boats may be linked. In both cases the engine's cylinder was put in the exact center of the hull, with the boiler behind it. Like the *Charlotte Dundas,* and unlike the vessel recently proposed in his agreement with Livingston, Fulton's first steamboat was a towboat, with room on board just for the machinery, fuel, and crew. Both vessels were propelled by paddle wheels: Symington's by a single wheel at the stern, Fulton's by two wheels attached to the sides.

And—the most telling detail—Fulton's paddle wheels were placed quite high in the boat, as in the *Charlotte Dundas,* so that only three paddles were under water at once, avoiding the wasted up-and-down motions of a more deeply immersed wheel.

It seems more than probable that Fulton did see the *Charlotte Dundas* and borrow from her design without ever acknowledging the debt. His first steamboat, built to the plans of January 1803, underwent a successful trial on the Seine later that year. Fulton eventually returned to the United States and, with Livingston's support and a Boulton and Watt engine imported from England, made the paddle wheel steamboat later known to history as the *Clermont.* Her machinery and paddles closely resembled those of Fulton's first steamboat of 1803—and therefore may also be linked to the *Charlotte Dundas.* With the *Clermont* and her successors, Fulton ran a profitable steamboat service between Albany and New York City, marking the first sustained commercial use of steam navigation. The unfortunate Symington faded into obscurity and died penniless in 1831.

From this point on, geography largely determined the separate development of steamboats in America and Great Britain. In the United States, with its vast internal networks of inland lakes and long, broad, navigable rivers, steam navigation typically took the form of riverboats: large, fragile craft of shallow draft, driven at top speed by high-pressure boilers prone to explosion and disaster. In Britain, the characteristic steamboats were smaller and slower but safer, with low-pressure boilers, and sturdy hulls and high bulwarks designed to survive the heavier seas of coastal and ocean traffic. The future of Atlantic Ocean steamships would unfold mostly in the British Isles.

William Symington's many frustrations had an apparent chilling effect on steamboat building in Great Britain. After he finally laid up his unwanted creation at Bainsford, nine years passed before another British steamboat was launched. The *Comet,* completed in the summer of 1812, became the first passenger steamer in Europe. Her planner and owner, Henry Bell, had been interested in steam navigation for over two decades. But his mercurial nature—his "restless volatile genius," as a friendly biographer put it, "flying from one daring scheme to another"—kept Bell pushing on to the next experiment before finishing his last one. It took him a long time to settle down and produce his first actual steamboat.

Like Watt and Symington, Bell was a Scotsman, born in 1767 near Linlithgow, west of Edinburgh. He came from a family of millwrights and was trained as a mason, millwright, and shipbuilder, with early stints in Glasgow and London. ("I was not a self-taught engineer, as some of my friends have supposed," he later insisted.) Settled in Glasgow, he built houses and public works and started to focus intermittently on steamboats around 1800, after Watt's patent expired. Bell tried to interest various patrons and governments but got no favorable responses. He hung around the Carron Works when the engine and machinery of the *Charlotte Dundas* were being constructed, to the point even of making himself a nuisance to the workmen. Later he repeatedly inspected Symington's boat at Bainsford.

When Bell became the owner of the Baths Hotel in the resort town of Helensburgh, on the Clyde some twenty miles west of Glasgow, he acquired the necessary practical goad that pushed him finally to build a steamboat—for bringing Glaswegian customers out to his hotel. The Clyde, as yet undredged, was then a winding, shallow stream, often filled with sandbanks. Sailing boats drawing only five feet still might be grounded for an hour or two; passengers would be obliged to run on deck from side to side, rocking the hull and loosening the keel from the sand. To reach Glasgow, at the river's eastern and narrowest point, Bell's steamboat for the Clyde had to be small.

In the fall of 1811 he contracted with John Wood, a shipbuilder in Port Glasgow, for a hull forty-two and a half feet long, eleven and a half feet wide, and five and a half feet deep, and a total capacity of only twenty-five tons. John Robertson of Glasgow, a builder of textile-mill machinery, made the engine: a cylinder eleven inches in diameter, stroke of sixteen inches, and four horsepower. Four small paddle wheels hung on the boat's sides. Her smokestack at the bow doubled as the mast for a single square sail (as on an old Viking ship). The *Comet* was named not to suggest her speed but in tribute to Halley's Comet, recently visible in the night sky. Launched in July 1812, she began her Glasgow to Helensburgh to Greenock service a month later. As she puffed along the river, local boys would run down to the water's edge, expecting or hoping to see her blow up. She made the trip three times a week in each direction, covering the twenty-six miles to Greenock reliably in four hours, sometimes under three and a half—as fast as horse-drawn travel by land, and cheaper and much more comfortable than heavy, unsprung vehicles on bad roads. Within a year, four road coaches that had been taking passengers to Greenock stopped running for lack of business.

This quick success provoked a productive steamboat competition. For some years before the *Comet,* Bell had worked on steam navigation designs with John Thomson, a Glasgow engineer. Thomson had made sketches of a boiler and machinery, and he expected to help Bell produce his steamboat. But Bell instead went ahead on his own, leaving Thomson angry and disappointed. He took his revenge by building a bigger, faster boat, the *Elizabeth.* Also constructed by John Wood, she was fifty-nine feet long by twelve feet wide, and forty tons, with a nine-horsepower engine. Her cabin included such touches of luxury as carpets and a sofa, windows with tasseled curtains and velvet cornices, and even a small shelf of books. The *Elizabeth* ran from Glasgow to Greenock and back every day, instead of only thrice weekly, carrying as many as one hundred passengers at speeds up to nine miles an hour, cutting steadily into Henry Bell's business.

Over the next few years, steamboats appeared on most of the major rivers of Great Britain. Just before the first railroads, they started to speed and discipline the pace of life, ratcheting up to the predictable, rationalized clock time of the Industrial Revolution. Steamboats ran at man's pleasure, plowing along through adverse winds and waves, coming and going as ordered. A clock soon became a necessary instrument for doing business. "The merchant, knowing the time of the tide, can count to an hour, in ordinary weather, when his goods will arrive; and will not be disappointed in one case out of thirty," Henry Bell asserted. "I expect in a short time to see all our ferries, and our coasting trade carried on by the aid of steam-vessels."

In May 1815, the first long ocean passage by a steamboat in Europe tested steam's potential for that coasting trade. The *Glasgow* (later renamed the *Thames*) had been built by John Wood a year earlier. She showed steady progress in size and power: seventy-two feet long by fifteen feet wide, sixteen horsepower, and seventy-four tons. Sold to London interests for service on the Thames, she put to sea just for delivery to her new owners, not to start a regular ocean service between Scotland and England. Under the command of George Dodd, a young architect and civil engineer, she set forth from Glasgow with an eight-man crew of a master, four sailors, and a cabin boy—and a smith and fireman for the engine.

The *Glasgow* ran easily down the Firth of Clyde into the narrow channel between Scotland and Ireland. Here she encountered more difficult sailing than anything normally seen on the Clyde, as the ebb tide collided with strong swells sweeping in from the North Atlantic. Unable to make progress, Captain Dodd had to seek shelter in Loch Ryan. The *Glasgow* ventured out again, was tossed around, and nearly wrecked on the rocky Irish coast. She stopped at Dublin for several days of rest and repairs. Naval officers came to see her, agreeing that she would probably not survive a true stormy sea and had better hug the shore. Watched by thousands of spectators ranged along her way, she left Ireland with just two brave passengers for London.

Away from the coast in the Irish Sea, she again met heavy swells. "The movement of the vessel differed entirely from that of one pushed by sails or oars," noted Isaac Weld, one of the passengers. "The action of the wheels upon the water on both sides, prevented rolling; the vessel floated on the summit of the waves, like a sea-bird. The most disagreeable movement took place when the waves struck the ship crossways; but here too its particular construction gave it a great advantage; for the cages which contained the wheels acted like so many buoys." As water flooded into the paddle box on the windward side, the compressed air exploded in an alarming report whose percussive force made the whole boat tremble. This noise exploded again, by reaction, on the other side of the *Glasgow,* then again, much diminished, on the first side. At this point she at least stopped rolling for a while. "During the rest of the voyage," according to Weld, "the vessel made what the sailors call, a dry way, that is, it danced so lightly over the waves, that it never took in one; and in all the passage we were not once wet . . . which could not be expected in any common ship."

As they neared Wexford, at the southeastern corner of Ireland, the *Glasgow*'s thick coal smoke convinced local pilots that the approaching boat was on fire. They scrambled out to sea, expecting to save lives and perhaps seize some profitable salvage—and were surprised and disappointed that the *Glasgow* was just steaming along in safety. She crossed St. George's Channel to England, near Cape St. David, and was again greeted by a flotilla of would-be rescuers not anticipating a smoking steamboat in those waters. Heavy seas tossed up waves so high that at times the crew of the *Glasgow* could not see the coast. Captain Dodd picked his way through, leaving far behind a fleet of sailing vessels trying to keep pace. They stopped for two days at Milford Haven for inspections and to scrape the saltwater scale from the boiler, a problem not encountered when sailing freshwater rivers.

Rounding Cape Cornwall into the English Channel, they encountered their highest swells yet. "It seemed impossible to pass," Weld recalled. "The vessel appeared to suffer. . . . Night approached, and no harbour presented itself, except that which we had quitted, and which was already too distant." Captain Dodd hoisted sail, which helped steady her, and struggled against the waves for hours until reaching calmer waters. The rest of the trip was smooth and easy. At Portsmouth, tens of thousands of people came out to stand back and be amazed. The *Glasgow* reached the mouth of the Thames on June 11, intact and in good order. She had covered 760 miles in a bit more than 121 hours of actual sailing time, spread over almost three weeks.

The voyage showed that a long ocean passage by steamboat was in fact feasible—though not as yet on a routine basis. The apparently insoluble limitation remained the fuel supply. The *Glasgow* burned two tons of coal every twenty-four hours. Coal was expensive and bulky, requiring inordinate storage space aboard ship and, therefore, frequent landfalls for refueling. An extended ocean voyage across open water with no coaling stops was still impossible, awaiting bigger ships and the invention of better engines and boilers. It would be more than two decades before a steam vessel could cross the North Atlantic under sustained power.

Scotland produced the first British steamboats and then dominated that field ever after. By 1822, forty-eight steamers had been launched from the Clyde, more than from any other part of the country. Shipbuilders and marine engineers along the Clyde drew from well-entrenched west-of-Scotland traditions of millwrighting, iron smelting and founding, and engineering. Glasgow also lay at the western end of the geologic formation known as the Clyde Basin, rich in coal and iron deposits. All the necessary human and mineral resources were at hand. The river itself was periodically diked and deepened, allowing access all the way to Glasgow for even the newest, biggest steamships. In these burgeoning circumstances, the Napier and Elder families established durable steam shipbuilding dynasties. With an uncanny (and canny) consistency that came to resemble an orderly series of monarchical successions, these two families, their associates, and their lineal descendants in other firms would build and engine most of the notable Atlantic steamships of the nineteenth century.

David Napier, the first of this line, was born in 1790 in Dumbarton, on the Clyde about halfway between Glasgow and Greenock. The men in his family worked as blacksmiths and iron founders. He attended school briefly, acquiring a little Latin and French, but was inevitably bound for his father's workshop. In 1803 he glimpsed his future when he saw the *Charlotte Dundas* at Port Dundas, near Glasgow. "Although then only twelve years of age," he recalled a half century later, "having been reared among engines and machinery, I took particular notice of it." David went along when his father moved the family business to a foundry on Howard Street in Glasgow. At age twenty, after his father's death, he took over. In another brush with British steamboat history, he built the boiler for Henry Bell's *Comet.* "Not having been accustomed to make boilers with internal flues," he noted, "we made them first of cast iron but finding that would not do we tried our hand with malleable iron and ultimately succeeded, with the aid of a liberal supply of horse dung, in getting the boiler filled." (Napier never forgot that Bell had neglected to pay him for it.)

After the *Glasgow's* pioneering voyage from the Clyde to the Thames, Napier set out to build a steamboat designed for regular ocean service. He studied the sailing packets that took up to a week to run from Glasgow to Belfast, the shapes of their bows and how they moved through the high swells of the Irish Sea. Under sail, the masts acted like tall levers, pushing down the forward part of the hull and demanding extra buoyancy there. Did steam propulsion therefore call for a different kind of hull? Napier tried various models in a tank of water. Eventually he decided to slice the full, rounded bow of the sailing packets into a sharper, finer wedge shape for his steamboats. The *Rob Roy,* the first vessel so designed, was built by his kinsman William Denny of Dumbarton in 1818. She was eighty feet long and eighty-eight tons, with a thirty-horsepower engine by Napier. Under Napier's own command—he would try his hand at anything—she ran from Dublin to Greenock in an unprecedented twenty-six hours. For two years the *Rob Roy* gave reliable service between Greenock and Belfast, then was transferred to the English Channel to run between Dover and Calais.

Over the next few years Napier built progressively larger vessels, up to the 240-ton, 70-horsepower *Superb* and the 350-ton, 100-horsepower *Majestic,* for other packet lines to Dublin and Liverpool. These ocean steamers were bigger, stronger, and more powerful than anything else yet built in Great Britain. Their success meant that steamboats were starting to evolve into steamships—though still, for the time being, with the old masts and sails and wooden hulls. "I was the first that successfully established steam packets in the open sea," Napier claimed in 1822, when obliged to brag by competing claims on behalf of Boulton and Watt. "The Superb is now plying the third year between Greenock and Liverpool, and not a single article of her machinery has ever given way, although she has been out in the worst of weather. . . . The truth is, I have made nearly double the number of engines for boats going to sea that Mr.

Watt has, and their machinery has not in a single instance been so far deranged as to prevent them from making their passage in a reasonable time."

As engineer, shipowner, packet entrepreneur, and sometime ship captain, Napier was forever popping with ideas and inventions. He pushed the evolving steamship forms to their limits, skirting and sometimes exceeding those vague boundaries at which novelty became dangerous. For all his mechanical brilliance, he lacked a sense of due restraint and proportion. He charged ahead like a dashing cavalry regiment, leaving to humbler foot soldiers the grubby tasks of mopping up and administering details. In time he yielded the leadership of Clyde steamship engineering to his cousin Robert, who was less inventive and dazzling but more patient and meticulous and, ultimately, more sound and substantial.

Robert Napier was born in Dumbarton in 1791 with, as he liked to say, a hammer in his hand, the son and grandson of blacksmiths. Of Robert and his three brothers, one became a minister while the others followed family tradition into smithing and millwrighting. At the Dumbarton grammar school, Robert received a liberal education, supplemented by outside lessons in mechanical drawing which gave him a lifelong taste for fine paintings and beautiful objects. His father groomed him for college, but Robert preferred to apprentice in the family workshop. He excelled at ornamental ironwork, fashioning metal into art. In his spare time he made tools and guns, and practiced drawing. At twenty he took off for Edinburgh, armed with an allowance of five pounds from his father and a certificate of good character from the minister of his parish. Soon he was back to work briefly for his father and then left home for good, this time to Glasgow. His artistic side may have craved the heady intellectual ferment of Edinburgh and the Scottish Enlightenment, but he was an engineer at heart, at home on the Clyde.

Bankrolled by fifty pounds from his father, in 1815 he bought the tools and goodwill of a small blacksmith shop. By making millwheels and tools for tinsmiths, he prospered enough to marry his first cousin Isabella Napier three years later. The marriage brought him into closer contact with her brother, cousin David. Restless as ever, in 1821 David let Robert take over his business at Camlachie Foundry, at the east end of Gallowgate. Robert made iron pipes for the Glasgow Water Company, which had just started pumping from the Clyde, and then his first steam engine, for a spinning factory in Dundee.

In 1823, thirty-two years old, Robert Napier found his metier by making his first marine engine. It was installed in the *Leven*, built by

James Lang of Dumbarton for the river traffic between that town and Glasgow. Napier was crucially assisted, with the *Leven* and for the next four decades, by his recently hired works manager, David Elder, who had come from a family of millwrights near Edinburgh. For the *Leven*'s engine, Elder made various refinements in the air pump, condenser, and slide valves. He was using the rudimentary machine tools of the day, which were powered by a central steam engine linked to overhead belts and pulleys. At Camlachie Foundry these devices ran just a few turning lathes (the small pulleys and belts were forever slipping), a horizontal boring mill, and a smaller vertical boring machine. From these modest beginnings, Elder gradually improved his tools, products, and men. The veteran millwrights of the time would not work to the tolerances he demanded, so he preferred to hire cartwrights and house carpenters instead, transferring their fine woodworking skills to the new problems of metal fabrication. "He was a man of great natural force of character," it was said of David Elder, "and maintained his opinions with considerable vigour."

The *Leven*'s steadfast performance brought the firm other marine contracts. For the *United Kingdom* of 1826—the biggest, fastest British steam vessel yet at 175 feet and 560 tons—they put an engine of 200 horsepower in the ship built by Robert Steele of Greenock. In 1828 they moved to a larger site in Glasgow, the soon-famous Vulcan Foundry on Washington Street, near the river. They added heavy new machine tools for making even more powerful engines. Robert Napier and David Elder became, by general reputation, the best engineers on the Clyde.

Any new steam-powered shipping company would routinely seek Napier's advice and active participation; his approval could mark the difference between success and failure. In the workshop, Elder continued his ongoing technical improvements and trained several generations of the top Clyde engineers, including his distinguished son John. Eventually Napier acquired his own shipbuilding yard as well, at Govan on the south bank of the Clyde, and applied the firm's exacting standards to every aspect of producing a steamship. One of his most loyal and long-term customers would be Samuel Cunard.

As Henry Bell had insisted about himself, these pioneers of Clyde steamboat building—from William Symington to Robert Napier—were not just self-taught engineers who worked simply by untutored intuition. They typically had mentors and family backgrounds in their fields. But most of their education did take place outside school, and the best of them then engaged in a continuous process of self-education all through their working lives. Immersed in such a bold new undertaking, they had to contrive their own patterns. They "read Nature's laws in their own fashion," the Scottish naval architect Robert Mansel remarked after the younger Robert Steele's death in 1879. "Admittedly they knew little or no Latin or Greek, and, on the whole, were decidedly averse to talking and talkers." Diligent and laconic in the Scots manner, they left terse, incomplete surviving records of what they did, and nothing whatever about their private thoughts and feelings. Any curiosity about such intimacies would have puzzled them. They poured themselves into their steamboats and steam engines—which also have not survived, except for a few stray shards. Entering their world now requires an act of imagination, with casual leaps over yawning gaps in the historical evidence.

So wedded to the progressive nineteenth century, their work helped change the world within their lifetimes. Whatever they may have thought about this grand transformation has been lost to history, except for off-hand hints. Robert Napier's fine mansion at Shandon on the Gareloch preserved a lingering trace of the old world within its opulent outer walls. The house was built in successive additions around the original modest cottage. A visitor in 1855 marveled at the many beautiful paintings and art objects in the plush outer rooms. David Elder, a music lover, had made his boss a waterpowered pump for the pipe organ in the main gallery. Napier, sixty-four years old in 1855, liked to show the treasures from his lifetime of collecting. At the core of the mansion, happy to remain behind in one of the old cottage's small rooms, sat his wife, Isabella Napier Napier. "A very simple and unaffected Scotch woman," the visitor surmised. The mother of seven children, five still living, she sat spinning by the fireplace, moving steadily to a rhythm older than steam on water. The great Steam Age roared on, around and past her.

PART TWO

The Era of
Cunard Domination,
1840–1870

3

Ships as Enterprise:
Samuel Cunard of Halifax

*T*he Samuel Cunard who appeared so mysteriously in Boston on his *Britannia* in 1840 had come from a tumultuous family history of upheaval and dislocation, of religious and political persecution, and then of neglect and alcoholism in his parents' generation. In the absence of much given structure, he had attained a preternatural early maturity on his own. He essentially invented himself and then took on necessary paternal roles for his younger siblings. Having emerged from such an uncertain background, he might reasonably have wanted a safe future based on some dependable job that provided a secure living. Instead he dealt in ocean ships and shipments, with all their endemic risks and uncertainties. Cunard would spend his working life worrying about cargoes and profits, captains and crews, and an occasional overdue vessel plying the pitiless North Atlantic Ocean. "Those who have the charge of ships," he wrote in old age, "are never free from anxiety."

*S*am Cunard was descended from a group of German Quakers who came to America in 1683. His great-great-grandfather, Thones Kunders, lived in the German town of Crefeld, on the lower Rhine River near the Dutch border. Kunders and his family were religious dissenters, first as Mennonites, then Quakers, at odds with local established church authorities. William Penn granted the Crefeld Friends about 18,000 acres in his Quaker haven of Pennsylvania. They sailed away from intolerance in July 1683, thirteen men with their families, thirty-three people in all: among the minority of immigrants to America who came not for economic opportunity but for reasons of conscience, to worship as they wished. The pilgrims from Crefeld landed in Philadelphia after a voyage of seventy-four days. .

They settled an area to be known as Germantown, later incorporated into greater Philadelphia. For three years, until they put up a meeting-house, the Crefeld Friends worshiped at the home of Thones Kunders. He worked as a textile dyer, his trade in the old country, and was appointed one of the local burgesses by William Penn. Kunders died in 1729, "an hospitable, well-disposed man, of an inoffensive life and good character." At some point he had Americanized his name to Dennis Conrad. In 1710 his sixth child, Henry, married the daughter of another Crefeld colonist. They bought a farm of 220 acres in Montgomery County and had six sons, who later spelled their last name four ways. (With the trail thus obscured, the family line has puzzled genealogists.) Henry's son Samuel took the name Cunrad. In turn, Samuel's son Abraham, born in 1754, later switched two letters and came up with Cunard, where the matter finally rested.

According to an oral tradition passed down within the family, Thones Kunders and his sons were plowing a field one day when they turned up a bag of gold coins—perhaps a pirate's loot, brought ashore and buried but never recovered. This windfall helped establish the family in America. The story, if true, marks the first hint of what afterward was called "Cunard luck" or the "luck of Cunards." Four generations later, Sam Cunard's good fortune in his ever-dangerous shipping business was sometimes ascribed—especially by frustrated competitors—not to alertness or hard work but to his unfair, unearned, uncanny luck.

The American Revolution, however, brought the family nothing but bad luck. As Quakers, the descendants of Thones Kunders could not support the revolutionary cause. The Quaker peace testimony prohibited any violent opposition to governments. Pennsylvania Friends felt no great loyalty to British authority; their pacifism simply made all wars untenable. The local rebels, mainly Presbyterians, took their opportunity to cut into the power of the more established Quakers. This complex internecine conflict, fueled by both religion and politics, became quite bitter. The rebels would place candles in their windows at night to celebrate American victories; Quaker windows without candles might be broken. Friends who declined to join public fast days might have their businesses attached or lose blankets and horses to rebel army requisitions. Soldiers could be billeted in Quaker homes. Quakers could be fined for refusing muster duty or an oath of allegiance to the rebels. Some, like Abraham Cunard's cousin Robert Cunard, were convicted of treason and had their property confiscated.

During the war, and especially after the final American victory, many Quakers (including Abraham Cunard) left for the Loyalist stronghold of New York. When New York in turn fell to the rebels, much of its swelling Loyalist community was banished to the British outpost of Nova Scotia. An elite group of Loyalists petitioned British colonial authorities for land grants and other privileges in their new Canadian home. Abraham Cunard joined a less well connected group of nine hundred others bound for Nova Scotia in asking for their own considerations. "Chagrined as your Memorialists are at the manner in which the late Contest has been terminated," they declared, "and disappointed as they find themselves in being left to the lenity of their Enemys . . . your Memorialists humbly implore redress from your Excellency and that enquiry may be made into their respective Losses Services Situations and Sufferings." Cunard sailed to Nova Scotia with a flotilla of Loyalists in the spring of 1783. Exactly one hundred years after his ancestors had come to America for religious freedom, political and religious strife now forced him to leave home for another new land in a wilderness.

Perched at the southeastern edge of Canada, technically a peninsula but actually more like an island, Nova Scotia in 1783 was a raw frontier territory. It had been lightly settled by immigrants from England, Scotland, and Germany, and by Americans from nearby New England states. During the war it became a bristling garrison for British army and navy forces, who dominated the principal town of Halifax. The newly arrived Loyalists, lured by favorable reports, were generally disheartened by what they found. "All our golden promises have vanished," said one Loyalist. "We were taught to believe this place was not barren and foggy, as had been represented, but we find it ten times worse. . . . It is the most inhospitable climate that ever mortal set foot on. The winter is of insupportable length and coldness, only a few spots fit to cultivate, and the land is covered with a cold, spongy moss, instead of grass, and the entire country is wrapt in the gloom of perpetual fog." Yet Nova Scotia was the most accessible place from New York still under British rule, closer than the West Indies or the Canadian interior, with rich fishery and timber resources and, at Halifax, one of the finest natural harbors in North

America. By the end of 1783 some 20,000 Loyalist refugees had arrived, more than doubling the local population.

In these circumstances of widespread chaos and hardship, of over-crowding, high prices, and temporary shacks, Abraham Cunard found and married a wife. Margaret Murphy was the daughter of Irish immigrants who had settled in South Carolina just before the war; her father joined the British forces and saw action in Georgia. The Murphys fled to Nova Scotia after the evacuation of Charleston in 1782. Abraham and Margaret were married on June 22, 1783; he was twenty-nine, she twenty-five. It was an odd match. The Murphys were Irish Catholics, had owned slaves in South Carolina, and did not share the pacifism, anti-slavery convictions, or abstemious habits of Quakers. This difficult marriage produced ten children over the next two decades. Samuel, the second child and oldest boy, was born on November 21, 1787, and named for his paternal grandfather.

Most of the Pennsylvania Loyalists settled in the new town of Shelburne, at the southern tip of Nova Scotia. Abraham Cunard—perhaps because of his rather heterodox marriage—instead went up to Halifax. The harbor town, less than forty years old, had been laid out on the slope of a steep hill that offered some protection from the northwest winter wind. Cunard found work as a foreman carpenter in the army's lumberyard at the docks. The Cunards lived near the water on Brunswick Street in the north end, a German section known as Dutchtown. Abraham and Margaret compromised their ancestral religious differences by joining an Anglican church. On his own time, Abraham bought vacant property and built houses for sale, turning good profits. He prospered enough to pay eight hundred pounds in cash, a substantial sum, for two waterfront land parcels in 1796 and 1798. He was also granted 1,000 acres of timberland in northern Nova Scotia. At the lumberyard he was promoted to master carpenter, earning nine shillings a day. As far as most outsiders could tell, the Cunards were doing well.

In private, the family was contending with an ongoing crisis caused by Margaret's uncontrolled drinking. Years later, people told stories of her lying in the streets of Halifax, dead drunk, while her children went barefoot and sold produce from the family garden for a few coins. Abraham's response is not known; his extended working hours, between his lumberyard job and the houses he was building to sell, might have functioned as a refuge from his wife's alcoholism—or perhaps a contributing factor to it. What seems clear is that Sam, as the oldest boy, had to assume early responsibilities. After a few years of grammar school, he started working for pay, wasting no time. Driving the cows home at night, he walked along knitting a bag to hold his money. He ran errands, picked dandelions and sold them at market, and purchased fish, potatoes, and other goods at the wharves to sell door to door. At age fourteen he proudly bought a broadcloth suit, his first, with his own money.

Children from an alcoholic home may respond in wildly varying ways. In Sam Cunard's case, he clamped a lifelong tight discipline on his emotions and pleasures. For a family of partly Quaker heritage, trying to make its way in a new and strange place, Margaret's drinking was a shameful secret. But it could not really be kept hidden in a small town isolated by geography and circumstances. Gossips knew and talked about it. From this background, it seems, Sam developed his enduring habit of keeping himself under cover, of not giving public speeches or revealing much even in private letters. His own habits were notably ascetic; he associated heavy drinking with failure and embarrassment. When he later made such bald statements as "I have never known an industrious sober man who has not succeeded," he was referring obliquely to his mother's losing struggle with rum.

From boyhood Sam toiled as a merchant, buying goods and selling them at a profit. He lacked the education for a professional career, like law or medicine, and had no taste for government or military positions, the other main avenues available to ambitious boys in Halifax. Living in a harbor town dominated by its waterfront commerce, he naturally turned to ships and shipping as the main medium for his business activities. On a typical working day he was up early and down to the docks, looking for deals, and finding them often enough to believe that his chosen field would reward hard effort and concentration. "'Tis true that the merchant does not always succeed," Cunard later reflected, "—but with patient industry he generally does—there is one thing certain that no one succeeds without application and close attention to the business he is intended for."

He worked under his father, then with him, and quickly moved beyond him. Abraham got him his first real job, as a clerk in the naval dockyard's engineer department, and next arranged for him to spend a few years down in Boston, working in a shipbroker's office and learning that business. By age twenty-one, in 1809, Sam had returned to Halifax and talked his father into founding the firm of A. Cunard & Son, ship agents and general merchants in the West Indian trade. On his own he also bought two parcels of wilderness land in the lightly settled northern reaches of Nova Scotia, a total of 5,000 acres—the first of many distant land speculations he would try for their potential rents, timber, or minerals.

The prolonged Napoleonic wars and, in particular, the War of 1812 between Great Britain and the United States brought flush times to Halifax. It became the main staging area and supply depot for British army and navy forces in America. At the same time, it continued doing business with those New England states that opposed the conflict. Within three months of the American declaration of war, Sam Cunard was granted a license to import certain goods from the states: flour, meal, corn, pitch, tar, and turpentine, all of them in turn useful for the naval war *against* the United States. Halifax seized the fortunate (if unprincipled) opportunity to supply both belligerents against each other. Privateering and smuggling also flourished, as captured ships and cargoes were auctioned off at low prices. "As all around me are smuggling," one Nova Scotian decided, "I am beginning to smuggle too." Tobacco, soap, and candles could be hidden in hogsheads and puncheons of codfish and then unpacked in a back room, out of sight. The Cunards probably joined in this lucrative, illicit, barely policed trade; the profits were hard to resist. By the end of the war they were buying and selling not just cargoes but the ships themselves.

The windfalls of war made Sam rich enough to take a wife. On February 4, 1815, at age twenty-seven he married Susan Duffus, seven years his junior. She was the daughter of a dry goods merchant and tailor who had come to Halifax from Scotland as a young man. Sam settled his bride in a fine new house at 21 Brunswick Street, adjacent to the home of his parents. His changed circumstances, upwardly striving and soon to include children, pushed Sam into taking a definite step, both merciful and ruthless, about his poor, sodden mother.

In late June 1815, with Susan three months pregnant, he bought farmland out in Hants County, at Pleasant Valley. He had a house built (a better home than most in that area, including a butler's pantry and a central chimney with four fireplaces) and sent his mother to live there, near some of her Murphy relatives. Perhaps, with their first child soon to arrive, Sam and Susan did not want the addled grandmother right next door, visiting and possibly endangering the baby. Sam added more land to the property a few weeks after his son Edward was born. According to the local folklore that persisted in Hants County even into the 1950s, Margaret Cunard was dispatched in order to control her rum supply and to limit social embarrassments in Halifax. She was, significantly, banished without her husband—but with, it seems, his willing assent. Abraham kept working as a master carpenter at the lumberyard, too far from Pleasant Valley for commuting fifty miles a day on horseback. After Margaret died in 1821, he finally retired in his late sixties and went out to live alone in her house.

As this crisp transaction made clear, Sam in his late twenties was functioning as the head of the entire family. The firm of A. Cunard & Son really consisted of the son. Sam also assumed responsibility for the education of his three youngest brothers. He sent Henry and Thomas, eleven and ten years old, to a private school in Pictou, Nova Scotia. "If you think it best, I have no objection to Henry & Thomas learning Latin," he wrote to the schoolmaster, Thomas McCulloch, a Presbyterian minister from Scotland. "The only reason I have for not requesting you to teach them Latin, namely that they are intended for business and that a plain English education answers the purpose. You will say that I have very contracted ideas and I must allow it. I shall feel much obliged if you will have the kindness to supply the little wants of the boys from time to time, they will require as the winter approaches worsted socks, and strong shoes which can be had at Pictou better than here."

Sam was not satisfied with the educational progress of his brother John, fifteen years old, so McCulloch was given another charge. "The masters under whose care he has been heretofore," Sam explained, "have paid but little attention to his improvement and what he learnt at school he has forgot within the last year. . . . I wish him taught what I requested you to teach the other boys, and I hope within one year (the time I propose leaving him with you) that he will have made considerable improvement." Sam's uncertain grasp of grammar and punctuation at times revealed the limits of his own education; he often neglected to end sentences with periods and to start sentences with capital letters, and his spelling could be erratic. But he wanted more schooling for his brothers, despite his businesslike doubts about the real value of learning Latin, and was willing to pay for a privilege he himself had not enjoyed.

Halifax in these years had a population of about 15,000. Seen from the water, it looked like a giant rectangle laid sideways on the slope of a hill: six major streets running parallel to the harbor, intersected at right angles by ten smaller cross streets. Two miles long by a half mile wide, Halifax was capped by a fortress called Citadel Hill and a prominent tower displaying the town clock. Ships, docks, and warehouses were ranged thickly along the waterfront. Only Water Street, closest to the harbor, was paved; the other streets were often muddy or dusty, and buried in deep snow from

December through March. These conditions, along with the steepness of the hill, made carriages impractical. People got about on foot or horseback. The houses, built to no particular pattern, were mostly wooden, of one story, and unpainted.

High society was divided between a small gentry class and a massive military presence. The old settlers and the Loyalists, initially at odds, by now had intermarried and merged their interests. The Loyalists had brought money, energy, and a new assertiveness to the small town. The oligarchy that dominated Halifax consisted essentially of the children of those Loyalists. Allegiances to the mother country still ran deep, in both politics and culture. Newcomers were struck by the pervading Englishness of the place. "Nova Scotia approaches nearer, in most respects, to the customs and ideas most approved in England, than any other part of America," one British visitor noted. "The style of living, hours of entertainment, fashions, manners, are all English. Dress is fully as much attended to as in London."

This Anglophilia was reinforced by British military power. Halifax was both a naval station and a garrison town, its streets filled with soldiers and sailors. Three regiments lived in barracks on the north and south sides of Citadel Hill. Brunswick Street, running between the barracks, was littered with well-patronized grog shops, gambling dens, and whorehouses. (The Cunards lived in a better section of Brunswick Street.) Returning to barracks at night, drunk and frisky, the soldiers and sailors would pick fights with each other and commit small vandalisms. Native Haligonians prudently stayed indoors, out of their way, at such times.

Military officers and the local oligarchy mingled at the Ionic-columned Province Building in the center of town, in the middle of a square enclosed by an iron railing. It was easily the most impressive structure in Halifax, 140 feet long by 70 feet wide by 45 feet high, built to last of locally quarried ironstone. Here met the meshed institutions of Nova Scotian government. The English monarch appointed a governor for the province, who appointed a Council which could amend or reject any bill passed by the Assembly, which was elected by male Nova Scotians who owned houses or land. The Council also designated sheriffs, coroners, and school commissioners and could review some judicial decisions. Occasional democratic pressures from below were, as yet, easily stifled.

In sum: Halifax was a small but quite diverse place, from the deliberating chambers of the Province Building to the nearby dives along Brunswick Street. For a young man on the make like Sam Cunard, it had some of the fluidity of a frontier town, unformed and open to enterprising newcomers. But political power was mostly appointive, beyond any popular control; individual leaders of the oligarchy, in general, came from more privileged backgrounds than Cunard's. Henry H. Cogswell, Richard J. Uniacke, and Thomas Chandler Haliburton were college-educated lawyers. Joseph Howe, the crusading editor of the Halifax *Novascotian,* was the son of a postmaster general and king's printer. The father of the three Bliss brothers was a Harvard graduate who served as attorney general and chief justice of New Brunswick; the Blisses would pepper their letters with French aphorisms and Greek quotations, in the original Greek. The two Young brothers were literate, well-educated lawyers and politicians from Scotland. Moving in such civilized, professional company, Cunard must at times have felt intimidated and culturally inadequate.

His career nonetheless flourished. Just after the war, he obtained his first royal mail contract: a foreshadowing of his later transatlantic steamship line. Given his command of ships and shipping, and useful contacts among the military authorities in Halifax, he was chosen to carry the mail between Boston, Halifax, and St. John's, Newfoundland; occasionally his sailing packets also took letters all the way to Bermuda. The steady performance of his mail ships, year after year, gave him a reliable reputation with British authorities that would help clinch subsequent dealings. "I have always found the Government very liberal and reasonable," he said later, "where the Contractor has endeavoured to fulfill his engagement I have never met with the least difficulty."

After Abraham Cunard's death early in 1824, Sam changed the name of his firm to Samuel Cunard & Company. The new name reflected the long-established reality. Sam took his brothers Edward and Joseph into the firm; his oldest brother, William, had recently died in a shipwreck. On formal occasions Sam now called himself Samuel Cunard, Esquire. The Cunard brothers built an imposing office and warehouse on Water Street, on one of the waterfront parcels their father had bought back in the 1790s: a four-story stone fortress that stretched 110 feet along the street, with a large arched doorway in the middle giving secured access to the wharves. From this solid base, the family engaged in shipping, shipowning, shipbuilding, whaling, timber, iron and coal mining, landowning, property management, and banking. Most of these enterprises succeeded, but ships were always at the base of everything else.

In 1825 Samuel Cunard obtained another crucial connection to British imperial power. The quite English province of Nova Scotia thirsted for real tea, which at the time was produced only in China. Lacking a consistent supply of the genuine article, people had to resort to peppermint, cloves, or aniseed—all deemed poor substitutes. Cunard sailed to London to petition the controlling East India Company for his own tea agency. "Our pretensions are grounded upon our long residence in the Provinces," he wrote in one of his run-on sentences, "and a thorough knowledge of the Trade and People, we possess every convenience in Fireproof Warehouses and means to effect the intended object, we are ready to give such security in London . . . and should you think proper to appoint us to the Agency and management of the proposed Consignments and future business of the Hon. Company you may rely upon our zeal and attention thereto and we shall be happy to give such information"—and so on. The besieged company granted him the agency.

The first tea ship arrived in Halifax a year later, smelling like a gigantic teapot, with 6,517 chests from Canton. Customers snapped them up in a public sale at the Cunard warehouse. For the next thirty-five years, quarterly tea auctions were held there, typically with Sam as the auctioneer. The East India commissions became his most reliable source of income; at times he would use the gross revenues, in the short term, to finance other enterprises and then later remit the balances due. Cunard's coveted tea shipments also strengthened his ascending position in the Halifax oligarchy.

The principal merchants in town persuaded him, briefly, to stand for political office. In the spring of 1826 he agreed to run for an Assembly seat. On the day appointed for the candidates to declare their intentions at an Assembly session, he met with his committee in the morning. Everything seemed in order. He would even make a rare public address. In the Assembly chamber, he stood up, faced the audience, and took a piece of paper from his pocket. "I also had intended to have said a few words from the Hustings," he said, reading, "but recent considerations have induced me to alter my views. . . . I did not come forward to offer myself at the present Election of my own accord, but at the written request of the Merchants, and other respectable inhabitants. I had no ambitious views to gratify, no objects to attain, the good of the country was the sole consideration which induced me to assent to their request." And with that he withdrew his candidacy and sat down, having told no one in advance of his change of mind.

However startling to his supporters, this performance was quite in character. He invariably kept his own counsel, trusting and confiding in nobody outside his family. ("I have always been in the habit," he once said, "of looking after my own business.") A politician would have to make regular speeches, a prospect that quite terrified him. Aside from his inherent shyness in front of an audience, public speaking and thinking on his feet could expose the awkward gaps in his education and the real limits of his verbal powers. "His conduct is strange and has done him no good," noted the Halifax attorney William Blowers Bliss, as flabbergasted as anyone. Though Cunard had offered an implausible explanation, Bliss astutely guessed his actual reason for pulling out: "I believe the real cause to have been that he grew nervous and frightened and timidity got the better of his judgment."

Cunard let his guard down, and relaxed, only at home. After the difficulties of his own childhood, and the inevitable uncertainties of a career in ships, and the watchful complexities of picking his way through the Halifax elite, he found a safe haven within his own expanding family circle. Sam and Susan had nine children in thirteen years. But Susan died in February 1828, at the age of thirty-two, a few days after the birth of her last child. A newspaper notice of her death—the main fragment of historical evidence about her—offered more than the usual conventional pieties: "Those only who witnessed how intimately blended, in her Character, were the mild unassuming virtues of domestic life, with an amiable disposition, sound judgement and religious principle, can appreciate the loss that has been sustained by an attached husband, a numerous family of young children, and a large connection of relatives and friends."

Sam was forty years old when she died. At the time, it was not uncommon for a woman to die from the complications of childbirth, often after having had many babies. The widower then usually married again and produced more children, especially if he commanded the wealth to support a second family. Sam never remarried; he remained permanently "attached" to his dead wife. Susan's mother took over the raising of her grandchildren. Sam sent his two sons, Edward and William, to King's College in Windsor, Nova Scotia, the favored school for scions of the Halifax oligarchy. Eventually Edward, called Ned, became his father's closest business confidant, the only associate he truly trusted with his private plans and ambitions. He traveled with his older daughters, who presided over his homes. His family circle maintained a high wall between himself and the outside world. (On his own deathbed, thirty-seven years after Susan's death, Sam would doze and wake up to speak about many things. At one point, with his sons on hand, he awoke with tears in his eyes. "I have been dreaming about your dear mother," he said. "And a good woman she was.")

*J*n the early 1830s, Cunard took part in his first steamship venture, the *Royal William,* which became the first steam-powered vessel to cross the Atlantic from Canada to England. By its limited success and ultimate failure, this undertaking helped prepare him for his transatlantic steamship line.

The principal coastal ship traffic in eastern Canada ran northeasterly from Halifax, around the tip of Cape Breton Island, northwesterly through the Gulf of St. Lawrence (perhaps stopping at Prince Edward Island and New Brunswick), and then southwesterly down the St. Lawrence River to Quebec City and Montreal. It was an exceptionally dangerous course: whipped by strong, fickle winds and currents, studded with islands and land protrusions, and littered with ice and fog for much of the year. The river froze solid during the long winter. The run between Halifax and Quebec exacted a terrible annual toll in lost ships and men. Under sail, depending on the weather, the trip could take up to five weeks.

The Assembly of Lower Canada, which included Quebec, took steps toward adding steam power to the St. Lawrence in the spring of 1825. For expertise they inevitably turned to the River Clyde in Scotland. Charles Wood of Port Glasgow (the son of John Wood, who had built Henry Bell's *Comet* and other notable early steamboats) suggested a vessel of 500 tons and 100 horsepower, to cost between £10,000 and £12,000, and capable of running from Halifax to Quebec in a week or less. The Lower Canadian legislature offered a subsidy of £1,500; the Assembly of Nova Scotia added another £750. In London, an ambitious prospectus was issued to raise £50,000 for the Halifax and Quebec Steam Boat Company. But the scheme attracted no additional support on either side of the Atlantic and went nowhere. "It does seem a stain upon our enterprise," said the *Novascotian* newspaper of Halifax, "that upon the harbours or estuaries of this Province we have yet received no advantage from the most gigantic improvement of modern times—navigation by steam."

Sam Cunard was always cautious about new ship technologies. In his trips to England and down the American coast, he had seen steamboats and acquired some sense of the current state of the art. He was characteristically waiting for others to make the initial mistakes. In the fall of 1829 some men in Pictou, Nova Scotia, tried to interest him in a steamboat scheme. "We are entirely unacquainted with the cost of a Steam Boat," Cunard told them, "& should not like to embark in a business of which we are quite ignorant & must therefore decline taking any part in the one you propose getting up." But just a few months later, a local steam ferry of thirty horsepower started running from Halifax across the bay to Dartmouth. After overcoming initial problems caused by salt water in her boilers, the ferry gave quick, reliable service. Cunard could watch her puffing back and forth every day, through any wind and weather, and count the additional paying customers she attracted. He began to see possibilities in steam on water.

Early in 1830, the Assembly of Lower Canada doubled its steam offer to £3,000, and the Nova Scotian legislature again added its £750. In Halifax, Cunard formed a committee to solicit stockholders in the renewed steamboat company. At a meeting in March he adroitly maneuvered himself into local leadership of the undertaking. Flourishing a list of 169 people who had promised to buy shares, he proposed a resolution that each subscriber—whether for £500 or £25—would have just one vote in the proceedings, "thus depriving the intelligent and enterprising merchant," one high roller later objected, "of the proper control over his large advances and placing it at the disposal of a number of small shareholders, in most instances entirely unacquainted with the nature of the business." After his resolution passed, the seventy-six subscribers on hand, mostly small investors, elected Cunard as Halifax agent for the steamboat company, granting him the power of general management and control of funds.

Awkwardly balanced between directors in Halifax and Quebec, the company proceeded to build a steamship. The contract went to George Black, a shipbuilder in Quebec City, and his merchant associate John Saxton Campbell. The designer and construction foreman was James Goudie, a local boy who had been sent to Scotland in his midteens to apprentice under a Clyde shipbuilder, William Simmons of Greenock. As an assistant foreman to Simmons, Goudie had worked on four steamboats similar to the one he now laid out in Quebec. He had brought the plans back from Scotland in the summer of 1830. "As I had the drawings and the form of the ship, at that time a novelty in construction," Goudie later recalled, "it devolved upon me to lay off and expand the draft to its full dimensions on the floor of the loft, where I made several alterations in the lines as improvements. Mr. Black, though the builder and contractor, was in duty bound to follow my instructions, as I understood it." When the keel was laid in September, young Goudie was still three months shy of his twenty-first birthday.

The *Royal William*, named after the reigning king of England, was a large steamship for the time, 160 feet long and 44 feet wide overall, with three masts in a schooner rig. The upper strakes of the hull were flared out to contain and protect the paddle wheels, perhaps with the St. Lawrence River's ice in mind; this bulging gave the vessel an inflated gross capacity of 1,370 tons. After being launched in the spring of 1831, she was towed down to Montreal and fitted with a two-cylinder engine of 200 horsepower by Bennet and Henderson. (John Bennet, that firm's senior partner, had apprenticed at Boulton & Watt in Birmingham.) The crank-shafts were forged by Robert Napier at his Camlachie works in Glasgow. Goudie, Black, Campbell, and Bennet were all of Scots background. The boat's designs came from Scotland, as did her crankshafts. Previous accounts have neglected this point: the *Royal William* was actually a Scottish steamship, built and financed in Canada.

In August she left Quebec on her maiden voyage, carrying twenty cabin passengers (who paid six pounds, five shillings apiece, including meals and a berth), seventy in steerage, some freight, and 120 tons of coal. After stopping in New Brunswick, she reached Halifax in six and a half days from Quebec. "Her beautiful fast sailing appearance," noted the *Acadian Recorder,* "the powerful and graceful manner in which her paddles served to pace along, and the admirable command which her helmsman had over her, afforded a triumphant specimen of what steam ships are." Sam Cunard visited her repeatedly, and no doubt proudly, asking questions and taking notes about her speed, coal consumption, and sailing qualities. The *Royal William* made two more round-trips that year before ice closed the river. The proprietors thought about sending her to England for the winter, to ply a coastal route there and earn back more of their investments, but instead she was laid up at Quebec.

She finished her first season amid anemic receipts, and complaints about excessive charges for passengers and freight that scared business away. "While at this port thousands of barrels, and scores of passengers, have been landing from Quebec and Halifax," a New Brunswick newspaper asked, "why has the Royal William been passing our wharves in want of both: as if by the splashing of her paddles, and the smoke of her furnace, she could forever bedim the vigilant eye of an interested public." At the start of the 1832 season, the *Royal William* offered sharply reduced rates in order to draw more customers—but then ran into a cholera epidemic. She made only one trip to Halifax that year, was quarantined, and returned to Quebec after almost two months.

Over the winter, her disappointed owners fell to angry squabbling among themselves. The Quebec stockholders accused Sam Cunard of claiming too large a fee for his services and of not working in harmony with the company. Cunard in turn charged the Quebec authorities with mistreating the *Royal William* during the previous season. "She was neglected in the Winter," he maintained, "and the frost burst the Pipes & otherwise injured the Machinery by which means a great expense was incurred and the sailing of the Boat delayed until the 15th June whereas she should have made two or three trips before that period—this might have been guarded against by a little care on the part of the committee and having an agent in pay they can have no excuse for the neglect." The company was foundering in red ink and feuding leadership. The cholera epidemic of 1832, blamed ever since for the collapse of the enterprise, had merely delivered the final, mortal blow.

In the spring of 1833 the *Royal William* was sold at a sheriff's auction in Montreal for £5,000—some £11,000 less than her initial cost only two years earlier. Her new owners tried a coastal voyage down to Boston and back, and then sent her off to England to be sold again. No steamship had ever tried to cross the North Atlantic from Canada to Europe; it was a voyage now conceived in financial desperation. She left Nova Scotia on August 18, 1833, with just seven bold passengers, 324 tons of coal, and a cargo of six spars, one box, one trunk, some produce, household furniture, a box of stuffed birds, and a harp.

It was a perilous trip. "We were very deeply laden with coal," the captain, John McDougall, said later, "deeper in fact than I would ever attempt crossing the Atlantic with her again." On the Grand Banks of Newfoundland, a gale knocked off the top of the foremast and disabled one of the engine's cylinders. For a time they seemed to be sinking. But they plowed ahead on the remaining cylinder, stopping the engine every fourth day to spend twenty-four hours cleaning seawater deposits from the leaky boilers. They proceeded under sail when the engine was down. After nineteen days they limped into Cowes, on the Isle of Wight in the English Channel, for repairs and a cosmetic paint job. They went on to London, where the *Royal William* was sold for £10,000 to the Portuguese government.

From this whole unlucky episode, Sam Cunard could draw two conclusions. Steamship technology did not, as yet, allow for routine, safe, profitable passages across the Atlantic—or even, for that matter, between Halifax and Quebec. And if he ever got involved in another steamship venture, he would need to run his own show, without having to clear his decisions through ranks of meddling associates. The *Royal William* experience ultimately reinforced his carefully guarded, self-contained ways.

*T*he failure of his first steamship did Cunard no immediate harm in Halifax; the blame could be shifted elsewhere. Now entering middle age, he was reaching the peak of his local career. In the fall of 1830, the governor of Nova Scotia had appointed him to the Council, the twelve-man body that served as the upper chamber of the Assembly. His appointment symbolized inclusion at the highest level of the Halifax elite. "We sincerely hope that the same liberal and expansive views which have distinguished Mr. Cunard as a merchant," Joe Howe declared in his *Novascotian,* "may be observable in his legislative character. He is wealthy and influential—he need fear no man, nor follow blindly any body of men; and we trust that he will not disappoint the hopes which many entertain." He served on the Council for ten years, often displeasing the reformers.

On a social and cultural level, the entrenched Halifax oligarchs still saw him as slightly alien, not quite a peer. In 1831 the lawyer Lewis Bliss urged his brother Henry, who lived in London, to welcome Cunard on his next trip to England. Lewis admitted he did not know Cunard intimately, having dined at his home only once. "I think he may be called a gentlemanly man," Bliss ventured, "—very polished he cannot be expected to be having I believe received rather a scanty education, and moved for the early part of his life not so much in the higher circles now thrown open to him." Yet Bliss guessed that Cunard owned, in whole or part, more than thirty ships, and probably cleared £2,000 a year from his East India Company tea agency alone: the kind of wealth and imperial connections that could almost compensate for an ungentlemanly background. "He is the most liberal as well as the most extensively engaged in business of all our Merchants," wrote Bliss. "He certainly is mild & pleasant in his manners—of an apparently equal temper, and possesses a gentle and not inharmonious voice—in short I look on him as a very good kind of man, and if not very pleasant & agreeable very far from the reverse." Furthermore, Cunard's steady rise from humble origins to heady eminence had not caused any rude behaviors. "He may be said to be modest—free from pride & affectation, and I think ambition, or if ambitious, not manifesting it in his conduct at all turns & on all occasions."

Though his careful manners concealed it, he in fact remained ferociously ambitious. During the 1830s he became a resident director of the Bank of British North America, served as the local agent of the London-based General Mining Association (in charge of coal mines in Nova Scotia and Cape Breton Island), and bought up hundreds of thousands of acres of timber and rental land on Prince Edward Island—all before the most ambitious act of his life. The *Royal William* did not entirely kill his interest in steam navigation, as he ran more modest steamboats in the local coastal traffic. At times he took extravagant risks, skirting financial ruin by moving fluid capital from one enterprise to launch yet another. Most of his undertakings, though, were apparently protected by the famous Cunard luck. On one occasion late in 1832, Haligonians waited anxiously for another overdue vessel to arrive. "As it is one of Cunard's ships," William Blowers Bliss mused, "I suppose she will get in at last, he is too lucky to lose her unless she be well insured."

It was not only luck, of course. Nor, as far as Cunard was concerned, was it the guiding hand of Providence. In his letters he would make pass-ing religious references. "If it should please God that we should all live to see the next year," he might write, ". . . if I should be spared I hope I may yet be useful to our concern." But this was just obeisance to an expected form, perhaps inserted simply to please a pious correspondent. Cunard had no real religious convictions. On his deathbed, when his son Ned suggested the attention of a clergyman, Sam declared "that he did not feel and admit and believe"—a dying confession that told the stark if unwanted truth.

What he really believed in was himself, the hard, driving, ruthless, tireless engine at the core of his being. Over his lifetime, he lived out the story so beloved by dime novelists of the nineteenth century: the poor boy from the provinces who worked hard, curbed his vices, hoped for the best and took optimistic chances, came to the big city and made his deserving way, and finally seized the most coveted material rewards his society offered. He bridged several distinct eras, from a late-eighteenth-century colonial frontier to high Victorian London. Across these steadily more progressive times, Cunard was a quite modern personality, focused intensely and narrowly on the ongoing prosperity of his enterprises. He prudently adopted new technologies when they seemed useful, measuring his success by profits and numbers that he could see and weigh and count. He trusted nothing but his immediate family and his own unquenchable ambitions.

4 ·

Ships as Engineering:
Isambard Kingdom Brunel

*T*he dream of starting a transatlantic steamship line depended in equal measure on enterprise and engineering, or money and machinery. Engineering had to come first. Once it seemed that engines, boilers, and ships had been improved enough to bear the overpowering demands of the North Atlantic Ocean, moneyed investors might come forth to launch the enterprise. Almost nothing in history is truly inevitable; any major event or turning point could have turned out quite differently if shaded by other twists of luck or contingency. But given the ongoing progress in steamship technology, the swelling commercial and political pressures for faster, surer links between Europe and America, and many interested parties on both sides of the ocean ready to invest in any plausible scheme, transatlantic steam seemed virtually certain. The only lingering questions were how soon and by whom. "Indeed, all things considered," said the *Mechanics' Magazine* of London in 1837, "the strangest thing about the matter is, that the object should not have been effected many years ago."

*O*cean steamships became the largest, most complicated machines yet devised. As such, they drew on engineering developments in many different fields. British engineering in general was now approaching its nineteenth-century zenith, a dazzling peak moment of practical imagination, commercial success, and global impact. British engineers were, for the time being, the best in the world. They had started the first Industrial Revolution and then provided the models for its cloning in Europe and North America. Great Britain was producing far more coal, iron, machinery, and technological optimism than any other country. The earliest successful Atlantic steamships could not have come from anywhere else.

Engineering as an exact science was barely a century old. It had originated in France, before the advent of the steam engine, as a real-world application of the Age of Reason. The term "engineer" traditionally meant someone who only built war machines and fortifications; "civil engineering" thus came to mean similar pursuits carried out in peacetime. Influenced by then-current philosophical emphases on rationalist modes of thought, French engineers adapted new ideals of mathematical precision, measurability, and experimentation to their practical building tasks. Such pioneers as Pierre Bouguer and Charles Auguste Coulomb invented the fields of structural analysis, applied mechanics, and hydraulics. The first engineering schools appeared in eighteenth-century France and long remained the most exacting such institutions in the world.

In Great Britain, engineering at the outset was more intuitive and direct, neither assisted nor impeded by much conscious philosophical baggage. The first British civil engineers—John Smeaton, Thomas Telford, John Rennie—mainly worked with the traditional materials of wood, stone, and masonry to build improved roads, bridges, and harbors. In particular, they constructed canals, the prevailing transportation fad during the decades around the turn of the nineteenth century. A canal, a water medium, had to remain as level as possible throughout its course. That meant rearranging the natural environment to an unprecedented degree: building up embankments, running high viaducts across valleys, bridging rivers, cutting down the smaller hills, and tunneling through larger ones. The Sapperton Tunnel on the Thames and Severn Canal, finished in 1789, was over two miles long—an amazing feat at the time. Humans were imposing their will on nature as never before, for all to see, and by their success were encouraged to entertain yet more Promethean ambitions for themselves. As civil engineering matured, it shed its original honest-workman's aura, became a more socially acceptable career, and professionalized itself. The Institution of Civil Engineers was founded in 1818, mainly by canal men. "Civil Engineering is the art of directing the great sources of power in Nature for the use and convenience of man," explained an ICE leader. "The most important object of Civil Engineering is to improve the means of production and of traffic."

The next generation of British engineers typically adopted newer building materials and power, especially iron, coal, and steam engines. The line between the two groups was not quite that stark; Telford and Rennie, from the first generation, had used cast iron in their bridges as early as the 1790s. The real demarcation came down to function. The founding civil engineers built objects that did not move. The later mechanical engineers, as they were called, built machines that snorted and clanked across the landscape. One was best known for canals and bridges, the other for railroads and steam power. Many individuals continued to work at every type of engineering. But with the narrowing of newer specializations, and the relentless deepening of requisite knowledge in any given field, the civils and mechanicals diverged ever more sharply, sometimes feuding with each other. The Institution of Mechanical Engineers, started in 1846 by railroad men, gave this hardening division an organized boundary.

Most of the early British engineers, both civil and mechanical, came from Scotland and northern England. Telford and Rennie were Scotsmen who learned their crafts in Edinburgh, then migrated south to find work. Henry Maudslay, a noted steam engine builder and inventor of machine tools, was from a Lancashire family outside Liverpool. He grew up in his father's carpentry shop but preferred working with iron, so he switched to blacksmithing. Wielding his hammer, file, and chisel, he was a true artist, deft and inventive, utterly in his element. He moved to London and opened a workshop that became famous for its marine steam engines and general excellence. James Nasmyth, one of his many apprentices who went on to notable engineering careers, fondly recalled his first impression of Maudslay's shop at Lambeth in 1829: "the beautiful machine tools, the silent smooth whirl of the machinery, the active movements of the men, the excellent quality of the work in progress, and the admirable order and management that pervaded the whole establishment." Maudslay stressed simplicity and economy to his assistants, demonstrating the lesson by turning a rough piece of metal into a smooth, plane surface with just a few precise strokes of his file.

Civil and mechanical engineers jointly created their most significant early achievement, the steam railway. Mining operations had already produced the first small steam locomotives and had demonstrated the unmatchable rolling efficiency of iron wheels on iron tracks. Because the earliest railroad locomotives lacked much pulling or braking power, the right-of-way had to avoid steep hills; that meant borrowing from the canal builders' leveling techniques for tunnels, viaducts, embankments, and cuttings. Mail coaches and coastal steamboat lines had shown the advantages of providing public transportation on set timetables at fixed fees. All these separate strands came together in tracks and trains. Because the Industrial Revolution had arrived so early in Britain, it happened there long before the railroad—a sequence not repeated anywhere else. The iron horse then exploded on a society already well industrialized, quickly transforming everyday life in ways that steam-powered mines, mills, and factories had not touched.

George Stephenson, the seminal British railway pioneer, was an illiterate engine mechanic born near Newcastle. He always spoke with a thick Northumbrian accent barely intelligible to southerners. After a delayed education, Stephenson built the initial two railroads in England, the Stockton and Darlington (1825) and the Liverpool and Manchester (1830), designing the locomotives and rolling stock as well as laying down the track and its associated structures. The Liverpool and Manchester, the first line to run between major cities, was expected mainly to carry freight such as coal, cotton, and timber between the port on the Mersey and the booming inland factory city. But passengers came forth in surprising numbers, so Stephenson started offering them fast trains on a regular schedule.

What the customers were buying was speed, achieved with a smoothness and consistency previously unknown. It seemed extraordinary that a businessman could leave Liverpool in the morning, travel thirty-three miles and spend his long workday in Manchester, and still return home by that night in reasonable fettle. A mail coach might average only about ten quite jostling miles an hour. A fast horse and rider at full gallop could reach up to forty miles an hour, but only in brief spurts, and with an exhausting clatter and commotion. Railway engines would match a galloping horse and maintain that speed serenely for hours, chuffing along in a steady rhythm with no apparent strain.

In the summer of 1830, the actress Fanny Kemble—fresh from her first great triumphs on the London stage—took an excited ride on a Liverpool and Manchester locomotive, with Stephenson himself driving. She felt inclined to pat the small iron horse, which consisted of just a boiler, stove, engine and gleaming steel pistons, a platform, bench, coals, and a barrel of water. "How strange it seemed," she noted, "to be journeying on thus, without any visible cause of progress other than the magical machine, with its flying white breath and rhythmical, unvarying pace."

No horse, no sail; how did it move? They glided easily through cuttings, across bridges and a viaduct, along raised embankments, and over a swamp. Stephenson described the construction of his locomotive, which Kemble thought she understood ("His way of explaining himself is peculiar, but very striking"). After taking on more water, he let out the throttle, pushing the engine to a giddy thirty-five miles an hour. Sensing the dramatic moment, Kemble stood up, took off her bonnet, and drank it in. The onrushing air pushed against her, forcing her eyelids down. It felt like flying, so fast and yet so smooth and free. "When I closed my eyes this sensation of flying was quite delightful, and strange beyond description; yet, strange as it was, I had a perfect sense of security, and not the slightest fear."

Fanny Kemble's joyful initiation into railbound flight symbolized a turning point in material history. The triumphs of engineering now hooked the nineteenth century on an ongoing expectation of constant, unsatisfied acceleration: speed and progress, reaching into every area of life, ever faster, and regardless of the dangers. "Verily is ours the age for invention," said the *Illustrated London News* in 1842. It was in many ways a Faustian contract, balanced uncertainly between gains and losses. Critics of modernity such as Thomas Carlyle, John Ruskin, and William Morris played a steady minor-keyed threnody in the background as Victorian progress boomed inexorably along. A few dissenting engineers did express timid misgivings about such headlong haste, and about the harrowing, infernal landscape of the Black Country of coal and iron mines in the industrial Midlands. But most practitioners, civils and mechanicals alike, shrugged off such criticisms. "If we would credit these imbecile philosophers, the introduction of every machine is an injury rather than a benefit," one engineer bristled. "There can be no greater fallacy than this."

Most engineers apparently believed their work would improve humankind—lightening its labor, speeding and easing travel, making life more comfortable and abundant. In any case, they devoted themselves to engineering for the more basic reason that they so enjoyed their craft. Engineers worked very hard, to the point in many cases of wearing themselves out at a premature age. R. A. Buchanan, the eminent historian of Victorian engineering, has suggested that they toiled such long hours mainly because they preferred it to any other possible activity. They didn't socialize much, avoided religious and political strifes, and lived simply and quietly. In 1838 a young railroad engineer, Daniel Gooch, made an expected appearance at a dinner party thrown by his boss's family—but quickly escaped. "I believe I did succeed in getting as far as the staircase," he scolded himself in his diary, "and left it disgusted with London parties, making a note in my memorandum-book never to go to another."

Nestled into their workshops, pondering some engineering puzzle of agreeable difficulty, they found their truest happiness in making up a brand-new world. Henry Maudslay took obvious, extravagant pleasure in manipulating his tools, loving the work for its own sake as much as for its applied uses. It called on all the keenest faculties of mind, eye, and hand. To plan their projects, engineers made careful drawings and crafted detailed models. "Drawing is the Education of the Eye. It is more interesting than words," James Nasmyth insisted. "The language of the tongue is often used to disguise our thoughts, whereas the language of the pencil is clear and explicit." Fondling their raw materials on a workbench, shaping and pounding and drilling, the engineers absorbed cues and knowledge directly through their fingertips. Inspiration flowed from the head and eyes out through the hands to the work, and then back again, in a seamless, tactile circuit of material creation. At their peaks, they felt the exultation of artists.

Isambard Kingdom Brunel was a prime inventive force behind the three most innovative ocean steamships built before 1870. Yet he spent most of his career on other projects ashore; he was not a naval architect or shipbuilder or any sort of marine engineer. As a landlubber, prone to seasickness, he never even took a major ocean voyage until the last year of his life. His steamships seem still more imposing as the offhand products of a very busy engineer usually focused in other directions. During his lifetime of great fame and achievement, Brunel was often called a genius for the crunching power, range, and originality of his mind. More successfully than any of his contemporaries, he straddled the widening split between civil and mechanical engineering, resisting the modernist specializing trend. He deplored "the benumbing effect of rules laid down by authority," as he put it, "this tendency to legislate and to rule, which is the 'fashion' of the day." No strict categories or conventions could ever contain him.

He made his first reputation as the engineer to the Great Western Railway. Brunel surveyed its route—a winding course that ran 117 miles west from London to the port city of Bristol—and then planned every detail of its construction, from the locomotives and rolling stock down to the lampposts and stations. "No one can fill up the details," he explained.

"I am obliged to do all myself." He made lavish use of all the canal builders' methods for remaking a resistant landscape, so leveling the grade that the line was known as "Brunel's billiard table." The Box Tunnel east of Bath ran for 1.8 miles through an insurmountable hill, much of it solid rock. The digging and blasting on this single project engaged up to 4,000 workmen and 300 horses at a time, consumed a weekly ton of gunpowder and ton of candles, and killed nearly 100 men in five years. On completion it was the longest railroad tunnel in the world. (The rising sun is said to shine clear through the tunnel on one day of the year, April 9, Brunel's birthday. Given the usual spring weather in southwest England, this intriguing legend can seldom be tested, which may explain its survival.)

Queen Victoria chose the Great Western for her first trip by railway. In June 1842, returning to London from a sojourn at Windsor Castle, she and Prince Albert boarded a special train at Slough. The royal party, in six carriages, was greeted at the station by the Great Western's top brass, and Brunel personally took charge of the locomotive. The train reached Paddington Station in twenty-five fast minutes. Victoria and Albert alighted on a crimson carpet that stretched across the platform, and were cheered by crowds at the station and along the avenue outside. "Free from dust and crowd and heat," the queen noted of her railway baptism, "and I am quite charmed with it." A year later, Albert flew from Bristol to London in just over two hours, averaging a breathtaking fifty-seven miles an hour. Nothing could have better advertised the Great Western Railway—and its chief engineer.

Brunel became a celebrity, an engineering superstar at a time when the public works of engineers were remaking everyday life in large, visible ways and sparking the popular imagination as never before. "Even to shake hands with one so remarkable," an acquaintance later wrote of meeting Brunel, "was a thing to be remembered for a lifetime." He loved any spotlight, courting it and capering in it, presenting himself in dramatic ways. He was a small man, about five feet four inches tall, with an olive complexion and blazing dark eyes under a strong brow. He moved about quickly under clouds of cigar smoke, vital and vigorous, gesturing expansively with his hands as he spoke. Brunel worked killing hours, even by engineering standards, but maintained a boyishly playful disposition, fond of jokes and pranks. Regardless of any contrary fashions, he wore a tall, cylindrical silk hat everywhere, even in his own traveling carriage. He explained, perhaps seriously, that it would protect his head from any blow by collapsing before the skull was struck. "It is at once warm and airy," he elaborated, "and you cannot improve upon it." (It also made him look taller.)

The extent of his fame was revealed in the spring of 1843 when, performing a coin trick for the children of a friend, he accidentally swallowed a half sovereign. It settled in his windpipe, causing pain in the chest and fits of coughing, and could not be dislodged. Brunel designed an apparatus for holding himself upside down, hoping that gravity would help expel the coin. He was inverted and tapped on the back, causing such convulsive coughing that the experiment was abandoned. Sir Benjamin Brodie, a prominent physiologist and surgeon, was summoned. He performed a tracheotomy and poked around with his forceps, but without success. Newspapers issued regular bulletins. Even the august *London Times,* which liked to define serious news coverage, kept its readers well informed. "Mr. Brunel passed a quiet night," the *Times* reported. Four days later: "He was able on Thursday to take a small quantity of fish." And three days more: "Mr. Brunel is going on favourably." At last, after almost six weeks, he was again turned upside down, with the incision in his windpipe kept open. Hit gently between the shoulder blades, Brunel coughed twice, and the coin dropped from his mouth. The *Times* published a detailed final report ("And thus, under Providence, a most valuable life has been preserved").

Over his career, Brunel contrived great triumphs and equally great failures. Everything about him was exaggerated; he vividly displayed both the strengths and deficiencies of genius. He reasonably believed that he knew more, across a wider range of engineering fields, than almost anybody he encountered. Among railroad men, only Robert Stephenson, the accomplished son of George Stephenson, was greeted as a peer. "Stephenson is decidedly the only man in the profession that I feel disposed to meet as my equal, or superior, perhaps," Brunel noted. "He has a truly mechanical head." Anyone else was expected to defer to Brunel's authority. His unorthodox mind and dead-sure tenacity pushed him through any obstacles into bold, original achievements—and also made him a quite difficult associate and boss. It was generally best not to resist or disagree with him. "Admit him to be absolute," one colleague noticed, "and he was not only reasonable, but kind. Hint to him that you had rights, and he was inexorable."

As an engineer, he most valued "usefulness," he insisted, "that characteristic of which we are most proud, and for which we have the vanity to think we are peculiarly distinguished." But "usefulness" to Brunel meant deploying the newest, strongest materials and methods, as called for by the most extravagant engineering standards available. The Great Western was the fastest, most solidly built railroad of its time, but also the most expensive at £6.5 million, well over twice Brunel's initial estimate. He characteristically would brush aside budgets and spiraling expenses, preferring not to think about money, wanting only to be left free do his finest work—thereby distressing his helpless financial associates, endangering and sometimes wrecking the whole enterprise. "He was the very Napoleon of engineers, thinking more of glory than of profit, and of victory than of dividends," a harsh contemporary estimate in the *Quarterly Review* concluded. "He seemed to love difficulties so much that he not infrequently chose the most difficult manner of overcoming them. Whatever was fullest of engineering perils had the greatest charms for him. That which was easy was comparatively uninteresting." Despite his declared focus on usefulness, he was actually the purest of engineers: a demanding, relentless artist intent on finding the most elegant solution regardless of costs or circumstances.

None of his debacles ever impeded his uncanny ability to get jobs and attract new investors. He caught and embodied the relentless engineering optimism of his time. "The most useful and valuable experience is that derived from failures and not from successes," he once wrote. "But what cannot be done?" When testifying before a board of directors or a committee of Parliament, he was a formidable advocate: overflowing with esoteric knowledge, diplomatic yet seemingly candid, speaking tersely to the point, and charming and witty when that seemed appropriate. He could usually persuade even the most skeptical listeners. He disliked writing and thought he had no talent for it, but his memoranda piled up compelling arguments by steady accretion. Brunel was also a facile, accurate draftsman, decorating his workbooks with fine small drawings tossed off for the apparent fun of it, and if necessary he could go to his workshop and make a skillful model of a design in wood or iron. With his command of speaking, writing, drawing, and modeling, he had the rare capacity to explain himself with clarity and eloquence in four modes and three dimensions—a key to his overwhelming powers of persuasion.

Today Brunel remains the only British engineer of his era with an enduring popular reputation. In Great Britain he is virtually a folk hero. Some of his notable engineering works have survived as reminders of his wide-ranging inventiveness. The Great Western Railway still runs across many of his bridges and through the Box Tunnel. At one end of the line, his station at Bristol Temple Meads still stands, though now reduced to a humble parking lot. At the other end, his Paddington Station in London encloses tracks and platforms in a space 700 feet long and 240 feet wide, under a vaulting roof of wrought-iron arched ribs covered with glass and corrugated iron. The Royal Albert Bridge, his greatest feat of bridge building, crosses the River Tamar near Plymouth in two spans of 455 feet each, an artful blend of arch and suspension techniques. With its approaches added, the Royal Albert traverses a total of almost 2,200 feet. The Clifton Suspension Bridge in Bristol, over the dramatically deep Avon gorge, was finished to his designs as a posthumous memorial. The *Great Britain,* his second ocean steamship, was improbably salvaged after a long, checkered career and was brought home to Bristol to be reconstructed and opened to the public.

Other Brunel traces help keep his name alive. The reputations of historical figures often depend on the written footprints they happened to leave behind; Brunel's private papers and manuscripts, amounting to at least twenty-seven thick letterbooks and many other files, are housed at the University of Bristol and at the Public Record Office in Kew. Other Brunel letters are scattered in a dozen archives across Great Britain. One of the fullest research troves available for any Victorian engineer, these materials allow historians an uncommonly rich record of his life. At Westminster Abbey, a Brunel window in the south aisle memorializes him. A Brunel statue stands on the Thames Embankment in London, looking upriver toward the Charing Cross site of his Hungerford pedestrian bridge, now long gone. At Paddington Station, another statue has him sitting down, looking thoughtful, holding his tall silk hat in one hand and a notebook in the other. In Bristol, a third statue presents him standing up, a jaunty hand in his waistband, gazing off toward the river and his preserved *Great Britain* steamship.

*B*runel's biography recapitulates the history of engineering in his time, from its French origins to its ultimate mid-Victorian feats in iron and steam. His father, Marc Isambard Brunel, came from a family of tenant farmers in northern France, halfway between Paris and Rouen. Over *his* father's opposition, Marc decided to be an engineer and spent six years in the French navy. Came the Revolution, and his Royalist sympathies exiled him to America, then to England, where he married an Englishwoman and settled into a picaresque engineering career. He always dressed and carried himself like a gentleman from the ancien régime, with its antiquated manners and costume. Once, in a British court proceeding, he was asked if he was a foreigner. "Yes, I am a Norman," he replied, "and Normandy is a country from whence your oldest nobility derive their titles."

Marc Brunel met Henry Maudslay in 1799, two years after Maudslay had opened his own machinist's workshop. Their complementary skills meshed well: the French-trained engineer explaining his concepts, the skilled British mechanic bringing them down to ground and to practical execution. Brunel and Maudslay worked on projects together for the next twenty years. At the Portsmouth Royal Dockyard, under the supervision of the naval architect Sir Samuel Bentham they devised steam-powered machinery for making the wooden blocks (pulleys) used in great numbers by sailing ships, turning out a cheaper, more consistent product than by the old hand methods. From this first success, Brunel went on to inventions for sawing and bending wood, making shoes and boots, and improving marine steam engines and steamboat paddle wheels. He never quite regained the early heights of his novel blockmaking machinery. Abstracted and absentminded, he would lose umbrellas and take the wrong coach, winding up somewhere out in the country. A financial innocent, at one point he spent three months in debtors' prison.

Marc's greatest work was his son, Isambard, born at Portsmouth in 1806. The boy resembled his father in appearance—small, a large head, dark complexion and eyes—and in his apparently innate knack for drawing and machinery. Isambard grew up in the Chelsea section of London, swimming in the Thames and meeting a stream of famous visitors at home. He found his métier at the Maudslay workshop: "your firm," as he later wrote the Maudslays, "with which all my early recollections of engineering are so closely connected and in whose manufactory I probably acquired all my early knowledge of mechanics." Sent off to school near Brighton, he wrote home that he had been making boats, thus injuring his hands, and asked for his father's eighty-foot tape measure. He spent two years in Paris, studying math and the French language, and apprenticing under a famous maker of chronometers and scientific instruments. Denied entrance to the elite Ecole Polytechnique because of his foreign birth, he returned to England in 1822 and went to work for his father.

Still a teenager, he had already accumulated a range of education and experience—from Marc, Henry Maudslay, and in France—that few British engineers of his generation could match. Bilingual, bicultural, he displayed a precocious sense of engineering theory and practice. His intellectual gifts were obvious. Marc fully recognized them and pushed his son onward. As Isambard's career took flight, his immersion in real engineering projects eventually crowded out his more theoretical French background. "One sadly loses the habit of mathematical reasoning," he noted. He became very much an Englishman, speaking with no French accent, and ever wary of continental tendencies. Later he advised a young man to spurn any writings by French engineers. "Take them for abstract science," he suggested, "and study their statics dynamics geometry etc etc to your heart's content—but never even read any of their works on mechanics any more than you would search their modern authors for religious principles. A few hours spent in a blacksmith's and wheelwright's shop will teach you more practical mechanics—read <u>English</u> books for practice. There is little enough to learn in them but you will not have to unlearn that little."

In 1825 the Brunels embarked on a daring, unprecedented project to build a 1,200-foot carriage tunnel under the Thames. Nobody had ever run a tunnel beneath a navigable, tidal river. The watery riverbed overhead consisted of unpredictable mixtures of clay, sand, gravel, and mud, and was constantly disrupted by tides and river traffic. For these daunting conditions, Marc invented a novel construction shield. It resembled a giant bookshelf, three men high and twelve men across. Each man stood in a separate compartment, digging with pick and shovel; as the ground was excavated, the shield was screwed forward; bricklayers came in behind and shored up the tunnel. The work inched along, beset by leaking water and lighting and ventilation problems. At times the men stood in black water up to their knees.

After a year of difficulties, Marc took sick and told Isambard, twenty years old, to take over. The response of the brawny workmen to their new boss—so young, small, and French-educated to boot—may be imagined. Given all the circumstances, he managed well enough. At one point, with water leaking into the tunnel again, he did not get to bed for five straight nights. "No one has stood out like him!" Marc wrote in his diary. Two hard years into the project, the river broke through from overhead in a gushing flood. Isambard descended on a rope to rescue a workman. For three weeks he could not plug the holes. Marc was harshly criticized by the authoritative *Mechanics' Magazine* of London for not accepting advice or taking responsibility for his crucial mistakes. The leaks were finally sealed and work resumed, but in a changed climate of watchful outside skepticism.

Isambard sought refuge in an extraordinary private journal, the most candid and searching self-appraisals he ever committed to paper. He recorded the details of his daily life, the tunnel work, sleeping five hours a night, and stray thoughts about girls. At twenty-one, despite his adult responsibilities in the tunnel, he was between adolescence and grownupness. Still under construction, he took an unsparing look at himself. "My self-conceit and love of glory or rather approbation vie with each other which shall govern me," he wrote. "I often do the most silly, useless things to appear to advantage. . . . My self-conceit renders me domineering, intolerant, nay, even quarrelsome with those who do not flatter. . . . I am always building castles in the air, what time I waste." Yet that self-conceit had quite adequate cause; he fully appreciated his own special talents and sought fame and reputation. "My ambition, or whatever it may be called (it is not the mere wish to be rich) is rather extensive." So probably he should never marry. "For one whose ambition is to distinguish himself in the eye of the public, such freedom is almost indispensable." Or maybe he should. "Yet, in sickness and disappointment, how delightful to have a companion whose sympathy one is sure of possessing." In this journal, he is less the engineering wunderkind, more any young man in baffled turmoil about his future.

In January 1828 water again broke into the tunnel, more seriously this time. Six men were killed. Isambard was knocked down, suffered internal injuries, and barely escaped alive. It took him over three months to heal. The *Mechanics' Magazine,* no fan of the Brunels, praised his coolness under pressure and brave concern for his men. But investors had lost confidence in the project, still only half completed. Work was stopped and the tunnel sealed. "Tunnel is now, I think, <u>dead</u>," Isambard later wrote in his diary. "This is the first time I have felt able to cry. . . . However, nil desperandum [never despair] has always been my motto—we may succeed yet."

At the time, he felt crushed by such a public defeat. The halting of the Thames Tunnel project did, however, free Brunel from an endless, risky, dreadful burden—and from his father's orbit—to pursue other work on his own. In Bristol, his designs for docks and the Clifton Suspension Bridge brought him to the attention of men involved in starting the Great Western Railway. Elected a fellow of the Royal Society at the early age of twenty-six, he was entering the most successful decade of his career. (The Thames Tunnel project was later resumed, but without Brunel *fils*. It opened in 1843 just for pedestrians, not carriages, and ultimately became part of the London subway system.)

In step with the general progression from civil to mechanical engineering, Brunel's attention moved from tunnels to railways. He took his first railroad trip late in 1831, on the Liverpool and Manchester. The car shook too much for easy writing. "The time is not far off," he decided, "when we shall be able to take our coffee and write while going noiselessly and smoothly at 45 miles an hour—let me try." He got his chance with the Great Western, the longest railroad yet conceived in Great Britain. Appointed its engineer in the spring of 1833, he threw himself into this new work with all the energy of a good engineer at play. He spent long days on horseback surveying and plotting its route, placating resistant landowners along the way, and stayed up late writing letters and reports. Against the advice of most railroad men, he convinced his board to accept a broad gauge of seven feet, more than two feet wider than the tracks of existing lines: a bold departure, ultimately proven wrongheaded, but early evidence of Brunel's forceful persuasive gifts.

For two years he was too busy even to scribble in his diary. The day after Christmas in 1835, he finally sat down and took stock. "The most eventful part of my life . . . emerging from obscurity," he wrote. "What a change—The Railway now is in progress. I am thus Engineer to the finest work in England . . . and it's not this alone but everything I have been engaged in has been successful." (He was perhaps repressing any memories of the Thames Tunnel.) "And this at the age of 29—Faith not so young as I always fancy tho' really can hardly believe it when I think of it. . . . I don't like it—it can't last—bad weather must soon come." He moved into plusher quarters at 18 Duke Street in the Westminster section of London, with easy access to the corridors of influence at Parliament and Whitehall. It remained Brunel's home and office for the rest of his life. Resolving his earlier doubts about possible marital intrusions on those boundless ambitions, in July 1836 he took a trophy wife, a fabled beauty named Mary Horsley whom he had known and intermittently courted for five years.

Marriage and, later, fatherhood did not affect his usual work habits. During the first four months after his wedding, he made decisions about the brick-arched Maidenhead Bridge over the Thames, the Box Tunnel, the tile drains along the track, the heating and welding of iron bars, the sinking of bridge arches and the proper way of laying bricks, the ordering of four locomotives, the size of engine valves relative to piston area, the question of allowing Great Western work on Sundays, and the cheapest wood for posts. It was Brunel's line, all down the line. He installed his own methods for putting down the roadbed and securing the rails, served as architect for every station along the way, and even picked the names for the first locomotives. "It is an understood thing," he wrote one of his men, "that all under me are subject to immediate dismissal at my pleasure."

Brunel's control of every aspect of the Great Western made him the culprit when anything went wrong. As construction took longer and longer, and costs more than doubled, directors in London and Liverpool started having doubts about their young engineer. "The Box Tunnel is operating a good deal against the Great Western," noted George H. Gibbs, a London director. "Connecting it with the name of Brunel, the difficulties of the Thames Tunnel are not unlikely to come into people's mind." The first section of the line, from London to Maidenhead, was opened to passengers in June 1838. When trains did not run as fast or as smoothly as expected, Brunel recommended reballasting the roadbed, replacing springs in the cars, and improving the locomotives. As the trading price of Great Western stock kept falling, shareholders in Liverpool moved to fire Brunel. Even George Gibbs, who usually defended him, felt torn. "With all his talent," Gibbs wrote of Brunel, "he has shown himself deficient . . . in arranging his work in his own mind so as to enable him to proceed with it rapidly, economically and surely. There have been too many mistakes, too much of doing and undoing."

Under fire, for a brief time Brunel felt shattered, even unable to work. His creation, so subject to costly revisions, was mockingly called the Great Experimental Railway. Gibbs had a blunt conversation with him; Brunel promised to cooperate and retained the support of Gibbs and his faction. At a tense showdown during a meeting of the directors, Brunel was again persuasive, defending himself with an even temper and compelling effect. The Liverpool contingent was outvoted, and Brunel proceeded to finish the Great Western. Upon completion, it was acclaimed as the fastest, most strongly built railroad in the world, and its engineer's characteristic problems along the way were forgotten.

*B*runel's first steamship began with a famous jest in October 1835. At a Great Western directors meeting in London, someone objected to the unprecedented length of the line, planned to run all the way to Bristol through many expensive tunnels at the western end. Rising to the challenge and topping it, Brunel replied with what he apparently meant as a joke: "Why not make it longer, and have a steamboat to go from Bristol to New York?" A director from Bristol, an engineer turned sugar refiner named Thomas R. Guppy, took the riposte seriously. He and Brunel talked it over that night. Brunel had almost no prior experience with steam on water, but he recognized few boundaries to his engineering skills. He brought in an acquaintance, a semiretired Royal Navy officer named Christopher Claxton, whom he knew from his earlier work for the Bristol docks. The three men started an informal steamship committee.

Claxton and William Patterson, a local shipbuilder, toured the main steam ports of Great Britain and sailed on every coastal and channel steamboat line. "Great improvements are being gradually introduced," they reported in January, "more particularly observable in the Clyde than elsewhere." For crossing the ocean, they recommended a much larger steamship than any yet built. They invoked a common principle, well known to shipbuilders of the time: that a vessel's resistance as it moved through the water did not increase in direct proportion to its tonnage. As a measure of interior space, tonnage was computed from three dimensions. Resistance was then estimated from just two dimensions, the width and depth of the hull. Thus tonnage increased as the cube of the dimensions, resistance only as the square of them. A much larger ocean ship could therefore include the necessary space for coals and machinery, well beyond the capacity of a conventional ship, without requiring intolerable increases in power and fuel consumption to maintain adequate speed. Claxton and Patterson estimated that a steamship of 1,200 tons and 300 horsepower, loaded with 580 tons of coal, would average between six and nine knots and cross the Atlantic in less than twenty days to the west, and just thirteen days to the east: roughly half the average voyages by sail.

The Great Western Steam Ship Company first planned to build two ships of that size, then decided on a single larger vessel of 1,400 tons and 400 horsepower. Patterson—"known as a man open to conviction," according to Claxton, "and not prejudiced in favour of either quaint or old-fashioned notions in ship-building"—would build her in Bristol. As managing director of the new company, Claxton looked after day-to-day operations. The building committee of Brunel, Guppy, Claxton, and Patterson met about once a week, whenever railroad business brought Brunel to Bristol. In general, on this committee Patterson took charge of the ship, Brunel of the engine. "Mr. Patterson drew the lines," Claxton later recalled; "Mr. Brunel, Mr. Guppy, and myself, often sat over them; Mr. Patterson got instructions and made his own calculations accurately; Mr. Brunel made his also often by my side." Over the next two years, they planned and built the largest steamship yet, the first designed for regular crossings of the North Atlantic.

They were racing against a competing group in London organized by Junius Smith, an expatriate American businessman. His British and American Steam Navigation Company drew investors from both sides of the ocean. This final sprint to steam across the Atlantic came down to three separate but overlapping rivalries: Britain against America, Bristol against London, and the Clyde against the Thames (or the North against the South). Subtly complicated and multiply crosshatched, the contest was played out amid fierce regional loyalties for rich stakes of prestige and fortune.

The crucial technical questions involved engines and boilers. The two leading British builders of marine steam engines were Robert Napier of Glasgow and Maudslay, Sons and Field of London. (Clyde and Thames.) After Marc Brunel's old friend Henry Maudslay died in 1831, the firm had passed on to his sons, Thomas and Joseph, and in particular to Joshua Field, a skilled engineer and manager. "No vessel ever had a sufficient power yet," Field had declared in 1822. "There is a limit, but that limit has never yet reached its fullest extent." As horsepowers kept on growing, the upper border was continuously extended. Progress already seemed infinite. By the 1830s, both Napier and Field were intrigued by the potential honor of powering the first true Atlantic steamship. "I have not the smallest doubt upon my own mind," Napier wrote in 1833, "but that in a very short time it will be one of the best and most lucrative businesses in the country." "The distance is limited," Field added a few years later, "only by the quantity of coal she can carry."

By then both Scottish and English engineers had settled on the side-lever engine as the best mechanism for an ocean steamship. Derived from Watt's old overhead beam engine, it placed the main weight of the power source at the bottom of the ship, lowering its center of gravity to limit rolling and pitching in heavy seas. A vertical engine cylinder drove a horizontal beam pivoted in the middle, with tandem connecting rods at its ends running downward to side levers, which drove a crank on the paddle shaft to turn the paddles. It was complicated and inefficient, moving massive weights up and down, with each stroke coming to a dead stop and then reversing. The bulky rods and levers added weight and took up precious cargo space. But the various parts were easily accessible and well balanced, minimizing friction and strain and needing less lubrication than other engine types. The piston's long stroke made full use of steam in the cylinder. Confined to the ship's closed hold, it was protected from foul weather and did not interfere with sailors moving about on the deck. Napier changed the framing from cast to wrought iron, making it lighter and stronger. The side-lever engine was considered exceptionally rugged and reliable, important qualities for crossing 3,000 miles of ocean.

The earliest marine boilers were kettle types, simply a drum of water heated by an external fire. Around 1830, Maudslay and others introduced a variation on the locomotive boiler, featuring an internal furnace that expelled its exhaust gases through long, narrow flues, making fuller use of the heat to produce more steam. But steamship boilers remained primitive and inconsistent, box-shaped and riddled with fragile seams. Each engine builder made his own boilers, using construction methods and metals of unpredictable quality. No one as yet dared push a seagoing boiler beyond a modest pressure of about five pounds per square inch. Lower pressure held down horsepower and made the engine use more coal, which limited the ship's range and cargo capacity. More than any other technical factor, the state of boiler technology was keeping steamships off the Atlantic.

Several steamers had already crossed the ocean, but not under continuous power or as part of a regularly scheduled service. The American vessel *Savannah* went from the United States to England in 1819, steaming only for about eighty-five hours of the twenty-seven-day passage. Over the next fourteen years, at least five other steamships made an Atlantic crossing, down to the Scottish-Canadian *Royal William* in 1833, Samuel Cunard's first venture into steam navigation. None of these ships could carry enough fuel to steam all the way. In any case, the salt water's scaly deposits in the boilers had to be blown off or laboriously chipped out with hammer and chisel at frequent intervals; that meant stopping the engine for up to a day and proceeding under sail until the puny boilers could be cleaned, refilled, and get up steam again.

One possible solution to these fuel and boiler limitations was to reduce the route across the ocean. The shortest great circle course between Europe and North America ran just 1,900 miles between Valentia, at the southwestern tip of Ireland, to St. John's, Newfoundland. The run from there to Halifax brought the total to 2,400 miles. In 1824 a group in London under Maurice Fitzgerald, the knight of Kerry—an Irish statesman and member of Parliament—launched the Atlantic Steam Navigation Company to carry traffic from London to Valentia to Halifax to New York. The company included Alexander Nimmo, a government civil engineer who was building piers and harbors along the Irish coast, and other men of influence. They planned steamships of 1,000 tons, almost twice the size of any vessel then afloat. In the fall of 1825, American newspapers declared it "almost certain" that the service would start in the following spring. But it never did. In these years, just before the railroad, the journey by coach and steamboat from London to Valentia took fifty hours, and forty hours from Liverpool. The whole trip would have demanded at least four changes of conveyance, with the usual uncertainties of baggage and schedules, to reach New York. It was much easier just to take one of the swift sailing packet lines the whole way from

London or Liverpool. The Valentia company disappeared, though the general idea was periodically revived.

Junius Smith's protracted crusade for an Atlantic steamship line began with an interminable fifty-four-day sailing voyage from England to New York in the fall of 1832. He was then fifty-two years old, a Connecticut Yankee and Yale graduate who had lived in London since 1805, prospering as a merchant. Dawdling across the Atlantic for almost two months, Smith had plenty of time to ponder an alternative means of ocean travel. In his innocence of steamship technology, he conceived a line of four steaming packets, to cost £30,000 each and make the hard westbound run from Portsmouth to New York in just twelve or thirteen days. "I shall not relinquish this project," he vowed, "unless I find it absolutely impracticable." For two years, nobody he contacted in London or New York expressed any flicker of interest. Smith kept trying, virtually alone. "The patience and labor of forming a company in London is beyond all that you can imagine," he wrote an associate in New York. "It is the worst place in the whole world to bring out a new thing, the best when it is done. . . . All the old sailing interest of course is against me."

The project became credible when Smith extended his search for supporters to northern England. Macgregor Laird, from a Scottish shipbuilding family that had moved down to Liverpool, became the secretary of Smith's projected company. He brought acutely needed technical expertise and steamship contacts to the enterprise. They planned four ships of 1,200 tons and 300 horsepower, to make the trip from London (and now Liverpool as well) to New York in fifteen days on average. After the *London Times* announced the scheme in November 1835, over £1 million in stock was subscribed—though not actually paid up—in a few weeks. Many English investors, scenting any reasonable plan, were ready for transatlantic steam. "Job's patience is much celebrated," Smith remarked, "but I don't think that he ever undertook . . . to establish a steam company."

Meantime Brunel and his Bristol associates were getting set to build their own Atlantic steamship, the *Great Western,* starting out a few months behind Smith and Laird. Brunel decided on the size of the engine and picked its builder. He went through the motions of a careful search, taking tenders from three firms. But given his family's intimate ties to Maudslay over four decades, the five years that Thomas Guppy had spent there as a young engineer, the company's long experience in making powerful marine engines, and Joshua Field's eminence as an engineer, the contract could have gone nowhere else. The Maudslay firm agreed to build a two-cylinder side-lever engine of 400 horsepower. For this special project, with its high and public stakes, Field designed a system of engine cams that improved steam economy—useful for the broad Atlantic—and invented new, more efficient double-story boilers. Brunel occasionally dropped by the Maudslay works at Lambeth, checking on the engine's progress, and he mediated squabbles between Claxton and the firm's management about bills and payments. "There are but few good Engine builders," he reminded Claxton, "and it will not be prudent to quarrel with the principal one."

Brunel's role in the ship herself has been exaggerated. He came to the project with no shipbuilding experience and throughout its construction was intensely preoccupied with the Great Western Railway. Claxton and Patterson, the real ship men, had already recommended a very large vessel, and Patterson designed the hull and fittings with just occasional advice from the others. Brunel, drawing on his knowledge of bridge stresses and bracings, did recommend adding to the ship's longitudinal strength with extra iron bolts and trusses. (A longer ship, so the logic went, would need additional strength when suspended from the bow and stern between two high ocean waves.) Brunel wanted to make the ship 400 to 500 tons larger, but Patterson doubted her stability at that size, so she was kept to 1,320 tons. Brunel and Guppy, his fellow landlubber, urged fitting larger cabin windows at the stern for more light and air, as in drawing-room windows ashore. Claxton "took the liberty of reminding them," he recalled, "that there was water outside which was sometimes very uneven in its surface, and unlike the generality of lawns; and strange as it may appear, Mr. Patterson, their builder, agreed." Brunel played a vital part in creating the *Great Western,* but no more than Claxton, Patterson, and Field did. Most accounts ever since have slighted their contributions in favor of their more famous colleague.

In London and Liverpool, the rival steamship project of Junius Smith was splintering into sniping factions. Instead of four ships at once, they had pulled back and decided to build a single enormous vessel of 1,800 tons. As an apparent compromise, the construction of the *British Queen* was split between the Thames and the Clyde. The shipbuilding contract went to the London firm of Curling and Young. Macgregor Laird wanted his friend and fellow Scotsman Robert Napier to make the engine, but Napier's bid of £20,000 was rejected, presumably by the London faction—a fatal error. Another Clyde engine builder, Claude Girdwood of Greenock, got the job instead at a lesser price. "The steamer is going forward in all its branches," Smith noted in March 1837. "I look back with amazement and see how I was guided by Providence in this thing." A few months later, though, Girdwood went bankrupt, and no other builder would complete his unfinished engine. In August, Napier—no doubt with a certain grim satisfaction—agreed to build another engine, of 420 horsepower, for £21,000: more than his spurned offer of a year earlier. This long delay let the Great Western company pull ahead in the race to steam across the Atlantic.

$\mathcal{B}$y the spring of 1838, after more than two years of planning and building, the *Great Western* was ready. At that moment she was, as designed, the biggest and most up-to-date steamship in the world. Most of the technical improvements came from Joshua Field. His version of a spray condenser, which converted some of the engine's used steam back into fresh water, limited scale deposits enough to let the boilers fire continuously across the ocean. Along with his innovative engine cams and double-story boilers, Field had addressed the besetting inefficiency of paddle wheels: the pointless thrashings up and down as the paddles entered and left the water. Borrowing from the geometric figure of the cycloid (the curve traced by a point on a circle as it rolls along a straight line), Field added three staggered boards to each paddle, stepped in from the circumference toward the hub so that each section entered the water at the same place in immediate succession. This "cycloidal wheel" was supposed to reduce both the initial downward slap of a paddle on the water and the heaving motion at the other end of the immersion, as it allowed the paddle to clear and shed the water more smoothly. From boilers to paddles, the *Great Western* was engineered specifically for service on the North Atlantic.

The overall dimensions of her wooden hull, 236 feet long and 35 feet wide, didn't make her look much different from other large ships of the day. "Her size, when seen by herself, does not appear so great as it really is," one visitor noticed, "and it is only when on board, or seen alongside other vessels, whose size is known, that her magnitude is appreciated." The black-painted hull of the *Great Western* presented a flaring clipper bow with a figurehead of Neptune holding a gilded trident. The deck was dominated by four low masts, one looming black smokestack, and two elevated bridges that linked the paddle boxes. A double wheel on a circular platform at the stern allowed two (or more) sailors to muscle her rudder and steer the vessel. Three structures on the deck enclosed a forecabin 46 feet long, the top of the engine room at midship, and the 75-foot main saloon at the rear, the showplace of the ship.

The saloon offered a seagoing opulence and high-ceilinged airiness matched, at the time, only by Edward Knight Collins's Dramatic Line of American sailing packets. The ornamental work was contracted out to Frederick Crace of Wigmore Street and the Jacksons of Rathbone Place, two noted London decorating firms. In early Victorian style, they festooned the saloon with columns that imitated palm trees and large pier glasses that suggested Dresden china, and they painted the walls and ceiling in warm, delicate colors with gold highlights. Edward Thomas Parris, the historical painter to the queen, contributed door panels five feet high that presented vignettes across a carnival of cultures: rural scenery and farming, music, interior views and landscapes, sports and amusements, and the arts and sciences, all in the rococo manner of Louis XV. The main staircase, to the cabins below, had a bronzed and gilded ornamental railing, with woodwork painted in imitation oak. The small cabins accommodated up to 128 passengers and twenty servants. Regardless of Brunel's relative share in her creation, the *Great Western* had emerged as a recognizable Brunel product: made of the finest materials and newest engineering, extravagant and original, truly the Great Western Railway at sea. (One impressed observer inevitably called her a "floating palace.")

Engined and finished in London, on March 31 she left Blackwall for Bristol, whence she would embark on her maiden voyage to New York. She steamed in large majesty down the Thames to the English Channel, the engine pumping easily with contained power, black coal smoke pouring from the stack. Brunel and Claxton were aboard, watching and approving. Everything seemed in fine order on this shakedown cruise— until a serious fire broke out in the engine room. The felt insulation around the boilers, installed to improve steam efficiencies and keep the room temperature tolerable for the stokers, had ignited from the heat of the pipes, and the fire was quickly spread by oil paint and gas to the wooden beams and deck overhead. The flames licked as high as the top of the smokestack, holding back attempts to reach into the engine room. Claxton took a leather hose down to the forehatch and from there poured water on the fire. Brunel started down to help him, lost his footing on the burned rung of a ladder, fell heavily on top of Claxton, and lay unconscious, facedown in a puddle of water. (It recalled his accident in the Thames Tunnel ten years earlier: nearly killed by his own engineering project.) Claxton saved his life by breaking his fall, pulling him out of the puddle, and calling for a rope. Brunel was hauled up on deck, suffering from a dislocated shoulder and a broken leg.

The fire burned on. The commander, Lieutenant James Hosken of the Royal Navy, thought about running out the lifeboats and taking off passengers, but he instead beached the *Great Western* on a flat riverbank, where she sat upright on her bottom. Men finally broke through the deck into the engine room and put out the flames. Refloated on the next high tide, surprisingly undamaged except for the burned felt and some charred wood, the ship proceeded to Bristol. After three days, Brunel felt well enough to dictate a long letter to Claxton about the generally satisfactory performance of the ship and her engine, hardly mentioning his injuries. "I hope the Vessel will be a long way on her Voyage to New York," he wrote, "before I could be in a state to go onboard again."

The fire must have pleased Junius Smith. Nursing an exalted opinion of his own historical significance, he liked to call himself "the father of Atlantic steam navigation." Smith believed that he alone owned the very concept of a transatlantic steamship. He had started his company before the Bristol group got under way, and now, with the long-delayed *British Queen* not even launched yet, he could not bear in frustration and defeat to let the *Great Western* beat him across the ocean. So Smith and Macgregor Laird chartered the *Sirius,* a well-regarded channel steamer of only 700 tons, loaded her down with fuel, and sent her on a risky, shortened passage to New York from Cork, on the southern coast of Ireland. (Starting from Cork knocked a day's sailing off the course to America from Bristol.) The voyage of the *Sirius* was just a heedless, dangerous publicity stunt, a desperate gambit by sore losers, and hardly worth the historical attention it has received ever since.

The *Great Western* left Bristol for New York as scheduled on April 8, four days behind the *Sirius.* The first Atlantic steamship race, contrived and unequal, was under way. Brunel had provided Hosken an engraved Mercator-projection chart of his great circle route, marked with bearings and soundings, to help keep the ship on the fastest course. The *Great Western* carried only seven passengers at thirty–five guineas (about thirty-seven pounds) apiece; fifty additional passengers had intended to go but were scared away by the fire. These seven brave pioneers in transatlantic steaming were amply serviced by fifty-seven crew members, including twenty-four seamen above deck and fifteen sweating coal stokers below. During the entire voyage, the engine was only stopped three times, briefly, for minor adjustments and to take soundings on the Newfoundland Banks. That meant little rest for the stokers. They struggled to bring coal by basket and wheelbarrow from holds at the bow and stern, where the ship's pitching and rolling motions were exaggerated. The stokers balked; Captain Hosken warned them to obey the chief engineer. Without enough coal, the boilers were barely maintaining adequate steam pressure. The stokers were pushed harder and promised extra pay. One exhausted stoker named Crooks got drunk and ornery, which inspired him to try to throw the captain overboard. For this egregious lapse of discipline, Crooks was restrained and tied up. The other stokers stopped working until he was released: not a near-mutiny by real sailors but a hint of proletarian industrial unrest transferred from land to the unfamiliar regimen of a seaborne boiler room.

Up on deck, the more experienced passengers noticed differences from life on a sailing packet. Morning conversations brought fewer fretful speculations about the wind and weather; the wind hardly mattered now. Instead the novel, somewhat frightening steam engine dominated everything on board. Nobody had ever crossed the ocean in the relentless presence of so audible, tangible a machine. All day and night it hissed and clanked, smoked and steamed, heating the deck from below so that tar bubbled up between the seams, sticky and persistent. The smoke and smuts blew around unpredictably, blackening clothes and alighting on hair. The engine lubricants, derived from animal fats with low combustion points, burned and smelled pervasively like a constant, enormous kitchen fire, a bilious irritant to anyone fighting seasickness or confined to a cabin below. The sea atmosphere, usually clean and bracing, felt cooked and greasy. Some people worried about being blown up by the overworked boilers or getting suddenly forced out of bed with no time to get dressed.

The machine was working, though, with the strong and steady rhythm of a heartbeat, practically without stopping, all the way across the ocean. A new era was finally at hand. "How this glorious steamer wallops, and gallops, and flounders along!" wrote a passenger on the *Great Western*. "She goes it like mad. Its motion is unlike that of any living thing I know; puffing like a porpoise, breasting the waves like a sea-horse, and at times skimming the surface like a bird. It possesses the joint powers of the tenants of the air, land, and water, and is superior to them all." As the days passed, the ship clicked off daily runs never before achieved on a westward crossing. A new speed record seemed easily in reach.

The *Sirius*, nearly out of fuel, reached New York on April 23. The *Great Western* came in only twelve hours later after a voyage of fifteen and a half days, the fastest crossing ever from England to America. Of her initial 660 tons of coal, she had 203 tons left in her bunkers—a reassuring margin of safety, for prospective customers, in the most doubted, uncertain aspect of transatlantic steam. The *Great Western* took sixty-six passen-

gers on her trip back to Bristol. After losing almost £4,000 on her first
passage out and home, she made four more round-trips that year, turning
small profits and lowering her own records in both directions. Already, at
this early point, ocean travelers had begun to accept the modernist bargain of steam dangers and discomforts in exchange for consistent,
unprecedented speed. In September the *Great Western* brought 131 passengers to New York and had to refuse 30 more for lack of space. This passage took sixteen days, nine hours—almost one day slower than her
maiden, but still *two weeks* faster than a crack sailing packet.

*T*he *Great Western* puffed back and forth across the ocean while the
British Queen inched along toward completion. It was a surprising reversal of expected form, Bristol over London, the fading western
port over the burgeoning urban colossus; so London, seeking an explanation, blamed Glasgow. During the spring and summer of 1839, partisans
of the Thames and the Clyde engaged in a ferocious public debate about
the practical wisdom of sending the *British Queen* north for her machinery. "Here we have a magnificent vessel dragged from the Thames to
Glasgow, at great risk and expense, in search of engines," wrote a man
from Cheapside in London. "All the world, except the sapient gentlemen
connected with the 'British Queen,' are perfectly aware that London-
made steam-engines (like most London-made goods) are decidedly the
best. . . . Our good friends, the Scotch, proverbially know how to pass off
certain inferior birds *'as swans.'*" In rebuttal, Robert Napier's friends
pointed out that he had been obliged to replace and reinforce much of
the carpentry work installed by the *British Queen*'s southern shipwrights.
Napier also had to lower the keelsons (heavy bracing timbers that ran parallel to the keel) by about six inches just to fit his taller machinery into
the engine room. "The people in the North," said one rebutter, ". . . consider the London-built ships very light and flimsy; in proof of which,
amongst many other improvements made in the *British Queen* at Port
Glasgow, it was deemed absolutely necessary to strengthen her with several additional iron knees. It is notorious that steam and other ships can
be built and fitted out, decorated, and finished, as expeditiously on the
Clyde as on any other river in the world."

These arguments drew on ancient, bitter rivalries between Glasgow
and London, Scotland and England, North and South. British political
and commercial power was centered in London, to the continuous irritation of the provinces. But Scotland could still claim a better educational

system and an older, more eminent engineering tradition than the Thames—and the famous Scottish thrift. Clydeside builders paid lower wages and enjoyed closer, cheaper access to coal and iron than Londoners, which meant they could build steam engines less expensively. That introduced another element to the public debates: "the avarice or parsimoniousness of steam-boat companies," as one Clyde defender put it, "who, finding that their orders can generally be more cheaply executed by Scotch engineers than London ones, run to them, and instead of being liberal in their dealings, screw them down to contracts, not consistent either with good materials or workmanship." Such sharp practices, so this explanation ran, left Clyde engineers the unhappy choice of losing high-standard business or producing shoddy work at cut rates that harmed their reputations as engine builders.

The contest between the *Great Western* and the *British Queen,* overtly a race to dominate Atlantic steam, became an acrid showdown between the two main centers of British shipbuilding and marine engineering. With a lucrative market for transatlantic steamers just opening up, the outcome could have decisive impact on the steam futures of London and Glasgow. "If our Scotch friends would puff their work less, and perform more, it would be more creditable to them," a shipping official in London suggested. After all, shipboard explosions of Glasgow boilers had caused far more deaths than accidents on London vessels; but perhaps—came the reply—that was just because the Clyde had produced so many more steamships than the Thames. "Let any one travel by the Thames river steamers," wrote a Scots enthusiast, "and then go and take a trip by the fleet, strong, and beautifully-built Clyde boats, and then say without prejudice which he prefers. . . . No engineers in the world are more ably qualified for the just, cautious, and accurate execution or manufacture of marine steam-engines, than are the Scotch."

Provincial rhetorics aside, the real-world proof of the matter lay in the ships themselves. The *British Queen* did replace her rival as the biggest, highest-powered steamship in the world: 275 feet long, 1,863 tons, and an engine jacked up beyond its contracted size to 500 horse-power. Robert Napier sent the ship down to the Thames for final fittings before her maiden voyage; his cousin David Napier, who had moved to London, gave her a suspicious inspection. "They unfortunately let one of the boilers get dry while coming round, either carelessly or willingly," David informed Robert, hinting at possible Thamesian sabotage, "which has given the Cockneys another handle against Scotch engineers." The *British Queen* at last left for New York on July 12, 1839 (fifteen months after the *Great Western*'s maiden). Junius Smith and Macgregor Laird went along as her most interested passengers. Also aboard, and quite interested himself, was Samuel Cunard of Halifax, returning home from business in England, and by this time quite intent on developing his own steamers across the Atlantic.

Laird's unpublished diary of the voyage, recently discovered and donated to the Merseyside Maritime Museum in Liverpool, is a doleful litany of worries and discomforts. As the ship's main designer, he felt the burden of responsibility for her performance. At midsummer the weather should have been as favorable as the westward run ever allowed. Instead the *British Queen* fought unusually strong opposing winds and currents, and Laird—famous for an earlier African river expedition, but normally an armchair sailor—spent most of the trip miserably seasick. For days he could eat nothing but brown biscuit; he envied the nine or ten women who lay supine in the ladies' cabin, quaffing six expensive bottles of champagne a day to relieve their queasiness. Hopefully overestimating the ship's speed, a proud father ever blind to his offspring's limitations, Laird kept losing bets on the daily run. "Summer passage indeed!" he exclaimed. "It's hard, very hard upon me—there she goes, pitch and toss! Talk of her being large! She is a plaything on the ocean."

Seven days out, they were still not halfway across. "I'll get nervous if we don't go faster homewards, the only comfort I have is that the ship answers [her rudder] beautifully and is as easy as any slipper, all on board are loud in her praise." Even the large complement of paying passengers did not please him. Laird rather disapproved of the ship's diverse company, which included Englishmen, Americans, Frenchmen, Germans, Spaniards, Portuguese, Russians, and Poles, about 120 people in all, eating and jabbering loudly in strange tongues at dinner. Miserable, lonesome, and ever worried about his ship, Laird longed for home and dry land. "It being my duty I came, but certainly if I could get my living in any other way, than being connected with these passenger steamers, I would most thankfully do it. I am never well, thoroughly well on board ship—I don't care for the talk and society of people I care nothing about and who care as little for me." In the final days, a more favorable wind helped the ship cruise at eleven knots. The *British Queen* reached New York after fourteen and a half days, excellent time by sailing standards but twenty-four hours worse than the *Great Western*'s latest record. "The public will look at the time only," Laird knew, "and not to all the circumstances of the voyage."

A fairer test came at once, as both Atlantic steamships left New York for home on the first of August. The *Great Western* carried 59 passengers, the *British Queen* 103. Sailing at the same time, by similar routes, they encountered essentially identical winds and currents; no differing "circumstances" would console the loser. This first true transatlantic steamship race, between the only two vessels yet designed and built for the North Atlantic trade, was keenly followed on both sides of the ocean. After a head start of forty-five minutes, the *Great Western* steadily lengthened her lead for most of the voyage. On the last two days, though, the *British Queen*—still breaking in her machinery—closed the gap rapidly and reached Portsmouth only about two hours after the *Great Western* came into Bristol on August 14. The *British Queen* did set a new elapsed round-trip record of thirty-two days, twelve hours. Engineers from both the Thames and Clyde could find reasons to preen themselves.

Later voyages, however, proved that Smith and Laird had built a larger but slower vessel. "The *British Queen* was a fine ship," noted Sam Cunard, who was paying close attention, "but she had not power sufficient." During the 1839 season, in three round-trips she averaged seventeen days, eight hours to New York and sixteen days, fourteen hours home. (The latter figure was skewed by an extended December voyage, hobbled by machinery breakdowns, of twenty-two days, twelve hours.) The *Great Western* in six round-trips beat her rival's averages by twenty hours out and three days, five hours home. With a higher ratio of horsepower to tonnage, she showed more effective power against the wind, better sailing qualities with it, and the durability necessary for regular ocean crossing. "Is it not reasonable to conclude," offered a Londoner, "that the engineers of the Thames must be *vastly* superior to those of the Clyde?"

In the entrepreneurial contest over building and managing an Atlantic steamship, Brunel and Bristol had beaten Smith and London. In the engineering battle of the rivers for transatlantic engine-building supremacy, the Thames had won the first round. Across this combined arena of enterprise and engineering, Glasgow—the founder and still center of British steam navigation—had not much to brag about, as yet.

5

The Cunard Line

*S*amuel Cunard had a plan. He characteristically discussed it with nobody outside his family. Only his son Ned, now twenty-three years old and an active partner in the Cunard enterprises, knew what his father was up to. During his annual trips to England, Sam had observed the first efforts at transatlantic steam. Taking their measure, he thought he could do better. In January 1839 he boarded a sailing packet to England and embarked on his greatest gamble. As an outsider, he felt no hobbling allegiance to Bristol or London or Glasgow. Well connected to some British power brokers, yet with no loyalties or commitments to any steam builders, Cunard moved around quietly, asking questions and making judgments. "Altho I am a colonist," he later explained, "I have many friends in this country." The silent colonial attracted little attention; the real transatlantic action seemed to rest in other, more famous hands. Stealthy and independent, he found the right men for his ships and cut the deal of his life. Cunard got his boats built and running—and stole the game away from its earlier players. "The plan was entirely my own," he said later, "and the public have had the advantage of it."

*C*unard went to England early in 1839 because the British government had declared its intention to subsidize steam navigation between England and America, not to carry passengers or cargo (except incidentally) but mainly to transport the mail, that most essential tool of commerce and empire. The *Great Western,* by her five routine round-trip voyages in 1838, had shown the possibilities of regular transatlantic steam. Her performance highlighted the many inadequacies of the British Admiralty's sail-powered mail boats, the ten-gun brigs that ran monthly and unpredictably between Falmouth and North America. In November 1838, rushing to catch up, the Admiralty invited hasty bids from British contractors to provide a monthly steamer mail service between England and Halifax, stipulating ships of at least 300 horsepower. The bids were due in only a month, and the service was to start by April—a schedule so tight that it restricted the field to existing vessels already built for other purposes.

At the time, the Great Western Steam Ship Company maintained a virtual monopoly on transatlantic steam. It had far outclassed the few competing ships, and Junius Smith's overdue *British Queen* would not finally be ready until the following summer. Holding all the cards, the Bristol company proceeded to overplay its hand with the government. On December 13, two days before the deadline, Christopher Claxton wrote Charles Wood of the Admiralty that his company was interested but needed much more time. The last voyage of the *Great Western,* with a slow winter passage of nineteen days to New York, had shown (said Claxton) the need for specially constructed mail ships of 1,200 tons and 450 horsepower, slightly smaller but more powerful than the *Great Western.* Claxton offered to build three such ships within eighteen to twenty-four months, and then to carry the mail once a month in each direction, for £45,000 annually under a seven-year contract.

The government wanted action in four months, not two years. The Great Western company ignored that urgency and rewrote the terms of the tender: in effect, instructing the Admiralty about the realities of steam on the North Atlantic. The company's correcting tone may have annoyed the Admiralty (which believed it had some knowledge of steamships and oceans), and it compounded an earlier, related offense by Thomas Guppy. In the fall of 1837, at a scientific meeting in Liverpool, Guppy—a founder and director of the Great Western company—had sharply criticized the Royal Navy's steamship designs. "Many of the government vessels are of very bad forms; their power and size greatly disproportioned," Guppy declared. "Whoever had seen the fine private steamers belonging to the ports of London, Liverpool, Glasgow, and Bristol, and had then gone to view the government ones in Woolwich basin, must have been astonished at the extraordinary forms there collected; it would be well if a glass case could be constructed over the basin, to procure those curiosities of practical science, as exercised in our naval building yards." (The audience laughed at this sally.) Guppy did not just doubt the government ships; he mocked and made jokes of them. His remarks were published in the *Nautical Magazine* of London, which was carefully read at the Admiralty. And now his company had doubled the insult by turning the government's urgent mail tender inside out. The Admiralty, no surprise, rejected the Great Western proposal on January 10.

On that same day, across the ocean, the Halifax *Novascotian* printed its first announcement of the original Admiralty tender. (This delay of two months itself argued for adding steam power to the Atlantic mail.) In its headlong rush toward steam, the Admiralty was not allowing enough time for any proposals from the colonies. Sam Cunard nonetheless caught the next sailing packet from Halifax to England, unaware of recent developments. He did not yet know about the Great Western offer, its rejection by the government, or the expired deadline for other bids. He had only his own secret plan.

It was not, as it happened, the best time for Cunard to embark on a grand, risky new venture. The British economic crash of 1837 was still lingering over the country, tightening money markets, headed toward a major industrial depression. Cunard was himself overextended at home. George Renny Young, an influential lawyer and politician in Halifax, had drawn Cunard into a grandiose scheme to buy up hundreds of thousands of acres on Prince Edward Island. It brought him a touchy alliance with the Young family, leaders among the local oligarchs. (Agnes Renny Young, the family matriarch, warned her son George about Cunard's "immense" power: "People fear him so much that they keep quiet and submit. He never was friendly to our family and will give you a blow where he can.") Sam took a one-third interest in the land company and soon fought with George Young over the appointment of a company attorney. Young wanted his brother Charles, and Sam favored his son-in-law James Horsfield Peters, who had married his eldest daughter, Mary. Family loyalties, always a tugging allegiance for Cunard, quickly poisoned the venture. "You seem afraid that I intend making a family party of this," Cunard wrote Young in August 1838. "I trust I am sufficiently well known in this community to be believed when I assert that I had no intention of taking advantage. . . . I must now decline any further correspondence on the subject." After Cunard pulled rank, James Peters got the job as attorney, but Young nursed the grievance. The quarrel was only resolved when Cunard bought out Young's shares in the spring of 1839. That left Sam, at this crucial moment, heavily mortgaged and cash-poor for investing in any major new enterprise. (Young and his family became enemies with long memories and later found a damaging chance to strike back at the Cunards.)

What, then, was nonetheless drawing Cunard toward transatlantic steam? Fifty-one years old, still in his prime, he remained as active and ambitious as ever. After his long career in ships and shipping, and (since the *Royal William*) in steam vessels as well, he had in some ways outgrown Halifax and craved a larger arena. An Atlantic steamship line was a logical extension of his lifework. He knew from weary experience the limitations of the British government's mail packets; he had been enduring their slow, precarious service for almost two decades. (On a trip to England in 1831, Cunard had fallen on a packet's deck and broken his arm.) To compete with the fast New York packet lines, the government brigs had been redesigned for more speed. But that made the ships less stable and too prone to foundering at sea. "Almost every year, two hundred or three hundred people were lost in the mail packets, and at last they got the designation of 'coffins,' " Cunard said later. "I came home in those ships very frequently, and of course felt the danger and discomfort of coming in them, and I have lost a very great many friends in them." In January 1839, even as Cunard was crossing to England, that month's westbound Falmouth packet sailed away and just disappeared. "I lost five or six intimate friends," Cunard recalled of that vanished ship. "They were never heard of." This latest packet disaster invested his mission with an immediate, sharper edge and made its own human case for adding steam's protecting power to the ocean mail.

Along with these personal motives, in a larger sense Cunard was acting from patriotic or nationalist incentives. Given his American parentage and his years of business dealings down the coast of New England to New York, he had strong ties to some Yankee ports and individuals. Yet his family had, after all, been forced into unwanted exile by the American victors in the Revolution. He typically regarded his American commercial associates with a goading mixture of fear and respect; over his entire career, nothing else so motivated him as competing with Americans and striking back at them. The New York packet lines, faster and more reliable, had taken most of the transatlantic mail away from the Admiralty's ships. At a time of strained relations between the two countries, British mail depended largely on American vessels: a galling vulnerability that in part explained the Admiralty's sudden sprint toward steam. As a Canadian, British subject, and descendant of American Loyalists, Cunard inevitably savored the prospect of taking the mail back from the aggressive Yankees.

Arrived in London, he took lodgings at a Piccadilly hotel and worked from a desk at the General Mining Association's office in nearby Ludgate Hill. His GMA connections, his two decades of carrying the mail between Boston, Halifax, and St. John's, Newfoundland, and his thriving agency for the East India Company's tea trade in eastern Canada all eased his way through Whitehall and financial offices in the City. He also carried a useful letter of introduction from Sir Colin Campbell, the royally appointed lieutenant governor of Nova Scotia. "I have always found him one of the firmest supporters of the measures of the Government," Campbell had written, "and his being one of the principal Bankers and Merchants and Agent of the General Mining Association, and also Commissioner of Light Houses, gives him a great deal of influence in this community."

Cunard's plan was simple and audacious: instead of a monthly mail service, he intended to run enough ships to maintain a weekly service across the ocean and thus cut more deeply, directly into the frequent sailings of the New York packets. He had taken the Admiralty's tender and quadrupled it. Within days of his arrival in London, he was meeting with Charles Wood at the Admiralty and Francis T. Baring at the Treasury. "I submitted that by going once a week the whole of the letters would be taken by our steamers, and the American packet ships that had previously carried the letters would cease to carry them," Cunard explained later. Wood and Baring "entertained my plan; and they took a great deal of pains . . . spent many hours at different times in going through the calculations and routes with me." The three parties eventually split the difference between Cunard's plan and the Admiralty tender by settling on a mail service to run twice a month at the outset.

Cunard's formal proposal on February 11 committed to paper what they had already thrashed out in conversation. "I hereby offer to furnish Steam Boats of not less than three hundred Horse power," he wrote, "to convey the mails from a port in England to Halifax and back twice in each month." In addition, he would provide steamboats of half that power for carrying the mail between Halifax and Boston, connecting his service to the United States but saving the extra two hundred miles to continue to New York. "Should any improvements in Steam Navigation be made," he added, with a nod to the onrushing pace of technical progress, ". . . which the Lords of the Admiralty may consider as essential to the Service, I do bind myself to make such alterations and improvements as their Lords may direct." For these forty-eight annual transatlantic voyages he asked £55,000 a year. (The Great Western company had wanted £45,000 for twenty-four trips.) The Admiralty and Treasury moved quickly. Within two weeks of the formal proposal, long before any public announcement, word was passing around London's political and financial circles that Samuel Cunard of Halifax had the contract.

Not quite; Cunard had skipped the thorny guesswork, which had so undone Junius Smith and Macgregor Laird, of predicting how soon his vessels would be built and available. To strengthen his case with the government, Cunard needed signed contracts for constructing his ships and engines. Still very much on his own, he appraised the feuding centers of British marine engineering. Because the Great Western company was already operating out of Bristol, and Smith and Laird out of London and Liverpool, Cunard's search naturally drifted north to Glasgow. James C. Melvill, secretary to the East India Company, recommended the Glaswegians Robert Napier and John Wood, who had recently built the swift steamship *Berenice* for his company's trade with India. In late February, two weeks after his successful proposal to the Admiralty, Cunard asked an intermediary in Glasgow to see what Napier and Wood would charge for one or two steamships of 800 tons and 300 horsepower, to be built and ready for sea in only twelve months. "I shall want these vessels to be of the very best description," he emphasized, "and to pass a thorough inspection and examination of the Admiralty. I want a plain and comfortable boat, but not the least unnecessary expense for show. I prefer plain work in the cabin, and it saves a large amount in the cost." Napier was at once quite interested, so Cunard went up to Glasgow to see him.

It is no hyperbole to say that their meeting in early March 1839 set the course of the Cunard Line for at least its first quarter century. Napier had just finished his enormous engine for the *British Queen,* after embarrassing delays and relentless criticism from engineers on the Thames. He therefore welcomed another shot at building an Atlantic steamship engine. Cunard, still so unknown to most British commercial circles, needed Napier's technical expertise and his reputation along the Clyde for shrewd business dealings. The two men were about the same age—Napier was four years younger—and of similar personalities: terse, contained, not given to public displays or extravagant statements, immersed in work, and sheltered by their families to unusual degrees. Each could recognize and (mostly) trust the other. Napier even brought Cunard home to meet his wife and children. Entrepreneur and engineer, the two formed a variation on those symbiotic partnerships that had driven the Industrial Revolution: a kind of Boulton and Watt for ocean steamships.

In Glasgow, Napier took Cunard to see his famous Vulcan Foundry and its redoubtable works manager, David Elder. The foundry sprawled across a large quadrangle on Washington Street, near the river. A sign at the gate advised, "No admittance except on business," and the din and pace of work inside showed that Napier and Elder meant it. Operations were broken down into four specialized areas. In the casting house, furnaces melted raw metal to be poured into sand molds in a pit. Some of the castings were quite large, up to a twenty-four-ton bedplate for a marine engine. This sector was relatively quiet, unlike the open area where boilers and funnels were pounded together. The steady, arrhythmic jangling of hammers on rivets, iron meeting iron, pealed forth the raggedy music of the Industrial Age. It took 10,000 rivets to make an average boiler, each driven home by repeated metallic blows, all day long. The smithery joined the heat of the casting house to the hammered cacophony of the boilermakers: sweaty, muscular blacksmiths toiling over their anvils and forge fires, turning rough metal into finer pieces, with a small steam engine puffing away to force air into the forge fires. The engineering shops, the largest department, held various specialized lathes and boring and planing machines, all driven by steam-powered beltings overhead, to shape and finish to exact tolerances the cylinders, pistons, wheels, and smaller parts of a steam engine. Seven hundred men worked long days at the Vulcan, six days a week. When the noise stopped at closing time, the silence itself was deafening. Sam Cunard could only have been impressed.

As Napier and Cunard got down to the details, the size of the ships kept increasing, a process that would continue through months of revised contracts. They at first agreed on three ships, 200 feet long and 960 tons, of 375 horsepower, to cost £32,000 each. (A ship this size was still only half the tonnage of the *British Queen*.) Napier would build the engines, and his shipbuilding associate John Wood would provide the vessels, all by the spring of 1840. "He appears from the little I have seen of him to be a straight-forward business man," Napier noted of Cunard. "From the frank off-hand manner in which he contracted with me, I have given him the vessels cheap, and I am certain they will be good and very strong ships."

Cunard returned to London, brought his first Napier contract over to the Admiralty and Treasury, and found them "highly pleased." Reporting this news to Napier, Cunard invoked the regional engineering rivalries, then raging, to warn and inspire Napier to his best efforts and punctuality. "You have no idea of the prejudice of some of our English Builders," Cunard wrote, guileless in his guile. "I have had several offers from

Liverpool and this place and when I have replied that I have contracted in Scotland they invariably say 'You will neither have substantial work or completed in time.' The Admiralty agree with me in opinion the Boats will be as good as if built in this Country and I have assured them that you will keep to time." (An oblique reference to the *British Queen* and her delays.) Someone else had told him that Thames engines would use less coal per horsepower, but Cunard assured the man he was mistaken and pointedly reported the exchange to Napier: "The Admiralty cautioned me on this head therefore take good care that you bear me out in my assertions." Cunard also pressed Napier and Wood to start building his ships ("How is Mr Wood progressing—tell him I will be down upon him some of these mornings when he may not expect me").

"I am sorry that some of the British tradesmen should indulge in speaking ill of their competitors in Scotland," replied Napier, not surprised. "I shall not say more than court comparison of my work with any other in the kingdom." The two men, so new to each other, were still forming crucial first impressions by poking around and testing the other guy, gradually settling into what became the most important working relationship of their lives. Cunard sent along reports of the latest patented steam innovations; Napier, playing his expected engineering role, passed skeptical judgments and reassured the entrepreneur. "I was quite prepared for your being beset with all the schemers of every description in the country," he advised Cunard. "Every solid and known improvement that I am made acquainted with shall be adopted by me, but no patent plans."

For his part, Cunard was having serious money problems. He still had no signed contract with the government. The first Napier deal was just with Cunard: he alone had to come up with a binder of £5,000 and the first installment of £5,000 more. At the same time, precisely the wrong time, his falling out with George Renny Young forced him to find about £14,000 to complete his purchase of the Young family's interest in the Prince Edward Island land company. Juggling the two enterprises, Cunard was caught without enough cash in England. When Napier took his first note from Cunard to the bank, it was accepted only after a suspicious delay and objections. Napier was surprised and dismayed. "The truth is," Napier warned Cunard, "had I not been completely satisfied beforehand from other trustworthy sources of your undoubted respectability and highly honourable character, my confidence in you would have been shaken."

If this early crisis had turned a certain way, it could have killed the whole enterprise at the outset. Instead it turned another way and gave the scheme a wider, more secure base, transformed it, and set up the leadership structure that would run the Cunard Line for decades. Napier, now less sure of his new partner, felt obliged to protect his own local reputation for sound dealings. Stepping back, he went around Cunard and confided in Robert Rodger, the Glasgow banker who had doubted Cunard's credit. "I have no wish to put you to the least trouble or inconvenience on my account," he assured Rodger. "The transaction with Mr Cunard is of such a magnitude that I must not have the least risk of trouble or anxiety about the money part of it." The two Glaswegians, acting in their respective self-interests, came up with a protective local solution. They invited other investors into the scheme to provide the cash cushion that was manifestly beyond Cunard's overstretched resources.

As part of this maneuver, they also brought in local experts at running a steamship line, George Burns and the brothers David and Charles Mac Iver. Former rivals, for almost a decade these three men had together been operating profitable coastal steam packets between Glasgow and Liverpool. If Burns and the Mac Ivers were involved, Napier urged Cunard, "the vessels would be well and honestly managed, and save much trouble to all concerned and make money." Furthermore, though contrary to his usual practice, given their participation Napier might himself take "a small interest" in the venture: a rippling expression of confidence sure to lure other investors in Glasgow. "I have several offers but am bound to no one," Cunard replied to Napier. "I should much like to have you and your friends with me." Within a few days, Napier and Burns found their partners (mainly interested businessmen in Glasgow), thus delivering the venture from the uncertainties of Cunard's own money and—most significantly—binding its future to the Clyde. "I want to shew the Americans what can be done in Glasgow," Cunard reminded Napier, turning several screws, "and that neither Bristol or London boats can beat them."

From that point, the pieces all fell into place quickly. In June 1839 Cunard, Burns, and David Mac Iver signed the final revised contract with the Admiralty for the mail service. It called for four steamships of 400 horsepower, 206 feet long, and 1,120 tons, sailing twice each month between Liverpool and Halifax, and then Halifax and Boston, for the government's payment of £60,000 a year for seven years. (The designation of Liverpool, headquarters of the Mac Ivers' coastal packets, reflected the transatlantic enterprise's new ownership structure.) The first ships were to be ready for sea by May 1840. Napier would build the engines, and to meet the soon-looming deadline the ships' construction was parceled out to four different Clyde shipbuilders. ("I dare say I get a good deal of credit for it, but I am not entitled to it," Cunard later said of his first steamships. "Any credit that there may be in fixing upon the vessels of proper size and proper power is entirely due to Mr. Napier, for I have not the science myself; he gave me the dimensions.")

In the final partnership agreement, Cunard took the largest portion at £55,000. James Donaldson, the leading cotton broker in Glasgow, took the next biggest piece at £16,000, followed by eight others at £11,600. Napier threw in £6,100, Burns £5,500, and the Mac Ivers £4,000 each. In all, thirty-three investors from Glasgow, Liverpool, and Manchester provided a total working capital of £270,000. Napier had brokered the whole deal and henceforth was deeply involved, in effect working for himself as both the engine builder and a part owner. Cunard remained the central figure, the founder and ultimate boss. "I had the whole interest for some time in the original contract," he later explained. "But circumstances turned up which made it necessary that I should part with some portion, and I did; but I have still the management." The official name was the British and North American Royal Mail Steam Packet Company. From the start, it was known more simply as Cunard's line or the Cunard Line.

*T*he first four ships of the line were, as Cunard had directed, plain and comfortable, with no unnecessary expense for show. The keels were laid almost simultaneously within a small circle of shipbuilders along the Clyde. John Wood crafted and built the *Acadia,* the pattern for the others. His brother Charles Wood built the *Caledonia*; Robert Steele, the *Columbia*; and Robert Duncan, the *Britannia.* (The names of the ships made careful, portioned reference to Nova Scotia, Scotland, the United States, and Great Britain, the four places that had launched and would then sustain the Cunard Line.) The ships had essentially identical dimensions, varying only in slight particulars: 207 feet long, 34 feet wide, and about 1,150 tons. John Wood was well known for the grace and comely proportions of his wooden hulls. "Remarkable for the great refinement of his taste," the naval architect John Scott Russell, a noted contemporary, later said of Wood. "He was a consummate artist in shipbuilding, and every line was as studied and beautiful as fine art could make it." The Cunard ships were austere beauties, sleek and black, with just a few ornamental touches of gold and red in the paddle boxes and smokestack.

As expressions of steamship technology, they started a durable Cunard tradition of summarizing recent progress in the field and adding only small, careful improvements: advancing the art but not by any risky grand leaps. They embodied the habitual technological caution—ships as enterprise, not as engineering—of their two main creators. Sam Cunard had crossed on both the *Great Western* and the *British Queen,* and Robert Napier knew the latter ship well from providing her engine. Traces of these transatlantic predecessors showed up in the Cunard vessels. They were lavishly trussed and bolted, like the *Great Western,* with (for example) "two strong bilge-pieces in the engine room," as Napier's contract with Cunard made explicit, "similar to what is in the 'British Queen' steam ship and well bolted." To avoid the *Great Western*'s initial difficulties over retrieving coal from distant holds at the bow and stern, the Cunard ships carried their fuel in midship compartments lining the sides of the vessels, from which the coal simply descended by gravity to trapdoors near the furnaces. Ambient heat from the *Great Western*'s boilers had made nearby areas feel and smell uncomfortably cooked; so the Cunard ships included a thick, coarse woolen cloth underneath the cabin floors and on bulkheads around the engine rooms "secured by beams and knees," the contract specified, "so arranged that a space can be left for air courses to ventilate and carry off the heated air and gases." The cycloidal *Great Western* paddle wheels devised by Joshua Field had not worked well; Napier gave the Cunard ships conventional paddles.

Designed to carry mail, not cargo, they were smaller than the *Great Western* and *British Queen*—but a bit faster, with slightly higher ratios of horsepower to tonnage. Napier's newest engines squeezed more power from less fuel by almost doubling the *Great Western*'s boiler steam pressure, from five to nine pounds per square inch, which helped reduce average coal consumption from forty-four to thirty-eight tons a day. Within a year of the *Britannia*'s maiden voyage to Boston in July 1840, the Cunard ships had beaten the *Great Western*'s Atlantic records in both directions, achieving peak average speeds of almost ten knots out and eleven knots home, and cutting the best eastbound passage to just under ten days. Over the first two years, as a fleet they averaged thirteen days, six hours to Halifax and eleven days, five hours to Liverpool. The overmatched Great Western company, competing on its own with just one steamship against four newer, faster vessels backed by the authority and prestige of the government's mail contract, began to lose passengers and profits to the Cunard Line.

Charles Dickens, impressed by what he had heard, took the *Britannia* to Boston in January 1842. His description of the voyage, soon published in his travel book *American Notes,* became the most famous—indeed notorious—account of a nineteenth-century transatlantic steamship trip. Dickens, about to turn thirty, had already achieved great success with the *Pickwick Papers, Oliver Twist,* and other novels. He went to tour America as a literary celebrity and was expecting an ocean passage that conformed with his status. At the Cunard agent's office in London he had seen imaginatively embellished lithographs of the *Britannia's* interiors. When he boarded ship, the actual accommodations caused his first disappointment. The main saloon, the grandest room on the ship, turned out to be "a long narrow apartment, not unlike a gigantic hearse with windows in the sides; having at the upper end a melancholy stove, at which three or four chilly stewards were warming their hands." The overhead racks for glassware "hinted dismally at rolling seas and heavy weather." Grimmer surprises awaited Dickens below. The "state-room" specially reserved for Dickens and his wife was, alas, an "utterly impracticable, thoroughly hopeless, and profoundly preposterous box." It inspired another funereal reference: the bunk beds by their narrow dimensions and thin mattresses reminded Dickens of coffins, a most unfortunate association at the start of a long ocean voyage.

Once under way, he retreated during daylight hours to the ladies' cabin, less noisy and smelly than the main saloon. The stewardess dispensed many tactful services and told "piously fraudulent" stories of previous winter passages, always calm and pleasant. Everybody worried about the stability of their stomachs; at dinner, Dickens noticed the most coveted seats were those closest to the door. Afterward he stayed out on deck till midnight, afraid to go below. Despite his frettings, he looked around and sensed the powerful mysteries of an oceangoing ship at night: "The gloom through which the great black mass holds its direct and certain course; the rushing water, plainly heard, but dimly seen; the broad, white, glistening track that follows in the vessel's wake; the men on the look-out forward, who would be scarcely visible against the dark sky, but for their blotting out some score of glistening stars . . . the melancholy sighing of the wind through block, and rope, and chain."

Finally, too cold to avoid it any longer, Dickens took to his dubious berth. With hatches and portholes closed down for the night, he could fully savor "that extraordinary compound of strange smells, which is to be found nowhere but on board ship, and which is such a subtle perfume that it seems to enter at every pore of the skin, and whisper of the hold." All the woodwork creaked. The stateroom rose and fell with the waves. The night eventually passed. For the next two days, through fair winds and good weather, Dickens mostly stayed in bed, ate hard biscuits, and drank cold brandy and water in a resolute, hopeless effort to avoid sliding over from mere seagoing discomfort into full-blown seasickness.

Teetering on this agonizing edge—nauseous or not?—Dickens fell over hard when the third morning brought a winter gale worthy of the North Atlantic. He awoke to his wife's screams. Objects were floating on the seawater that now covered the stateroom floor. The room pitched and tossed, seemingly standing on its head. "Before it is possible to make any arrangement at all compatible with this novel state of things, the ship rights. Before one can say 'Thank Heaven!' she wrongs again." The ship ran on like a creature with broken knees, as it leaped and dove and somersaulted in jarring sequences and combinations. Dickens hailed a passing steward and asked what was the matter. "Rather a heavy sea on, sir," came the reply, understated and unperturbed, "and a head wind." It continued for four days and nights of relentless motion and noise: wind, sea, and rain, howling in concert; the heavy footfalls of sailors rushing about and shouting hoarsely to each other; high waves pounding over the gunwales and gurgling out through the scuppers, after landing on the wooden deck with the deep, ponderous sound of thunder heard within a confined space; blank, endless nights as the ship rolled to one side, dipping her masts, and then to the other side, and even seemed to stop dead in the water, staggering as though stunned, before plowing onward. "All is grand, and all appalling and horrible in the last degree. . . . Only a dream can call it up again in all its fury, rage, and passion."

The storm blew out, but the weather remained dark and cold. Settling into a determined daily routine, Dickens and his party would gather in the ladies' cabin shortly before noon. Captain Hewitt, recently transferred to the *Britannia* and always in good humor, would drop by and predict better weather. ("The weather is always going to improve tomorrow, at sea.") At one o'clock a bell rang and the stewardess brought baked potatoes, roasted apples, and plates of cold ham, pig's face, and salted beef. At last free of seasickness, and seeking any possible diversion, they ate with hearty appetites and dawdled over the task as long as they could. They read, dozed, and chatted away the afternoon, passing around and chewing over the few available wisps of shipboard gossip: one passenger has lost heavily at gambling, fourteen pounds in fact, and drinks a bottle of champagne a day though he is only a clerk; the head engineer has never seen such awful weather; the cook was found drunk and severely punished; all the stewards have fallen downstairs, and some are sorely injured; the cabins are all leaking. The dinner bell rang at five, announcing more potatoes (boiled this time), various meats (perhaps roast pig if one of the ship's swine had been butchered), flowing wine and brandy, and rather moldy apples, grapes, and oranges for dessert. Then a game of whist, with the tricks placed securely in pockets instead of on the ever-agitated table, and an insistently cheerful good night from the captain.

Approaching Halifax on the fifteenth night, with a bright moon and calm sea, the local pilot—who was supposed to know the harbor so well—managed to run the *Britannia* aground on a mud bank. Everybody rushed up on deck. The engine, "which had been clanking and blasting in our ears incessantly for so many days," stopped suddenly, unexpectedly, leaving a dead stillness. Some of the sailors took off their shoes and jackets and made ready to jump overboard and swim ashore. (This did not inspire confidence among the passengers.) Distress rockets were fired into the night sky, to no point. In the general confusion and near-panic, Captain Hewitt remained calm and in command. The next high tide floated them free. After briefly stopping in Halifax, the *Britannia* took Dickens on to Boston. Wary of enduring another steamship voyage, when his American tour was over he caught a New York sailing packet home to England.

It was a gripping story, slightly exaggerated to improve the telling. The fame of its author and the popularity of *American Notes* ensured the Dickens account a wide, enduring audience—and a cautionary influence on uncounted potential Atlantic travelers. As a piece of historical evidence, though, it remains tendentious and limited, not trustworthy as a generalization about Cunard ships. Dickens described a westward passage in January: the more difficult direction for crossing the Atlantic, at the coldest, wildest time of year. The *Britannia* then steamed through an especially ferocious winter storm, the roughest on the North Atlantic in a long time. Its attendant miseries properly belonged to the indifferent ocean and could not be blamed on the particular ship and crew. The *Britannia* ran aground near Halifax because of mistakes by the local pilot, not the Cunard Line. Many of the problems on the voyage were not the steamship line's fault. And, in general, Dickens had nothing to compare the voyage to because he had never been at sea before. As a transatlantic innocent, he had too readily believed—or so he implied—those alluring promotional lithographs at the Cunard agent's office. His unrealistic expectations collided hard with the actual experience; he felt betrayed and vengeful, perhaps even embarrassed by his initial naïveté, and then took his overstated public revenge in print. (The book sold well but drew generally displeased, unconvinced reviews. "Sneers, vituperations, caustic sarcasms . . . a spirit of entire bad taste," said the *Illustrated London News*. This reviewer doubted in particular Dickens's account of his trials on the *Britannia*: "Of course this is the mere nonsense of book-making exaggeration, written to kill time and tickle the reader.")

A more balanced report on the first Cunard steamships required testimony from someone who had already crossed the North Atlantic by sailing ship and so could compare the passages. Fanny Appleton of Boston took the *Columbia* to Liverpool in May 1841, five and a half years after her tedious voyage to Le Havre by sailing packet (see pages 8–9). "Tho' I miss of course the beautiful shiftings and exhilaration of a sailing vessel," she noted soon after the *Columbia* left Boston, "yet we bound over the waves with no little dignity and grace." Constant shipboard noises on the packet had kept her from reading and writing; her first surprise on the Cunard ship was the prevailing near-silences compared to what she had found on various sailing ships and American steamboats. "One thing excites my unbounded admiration—the marvellous quiet. There is none of the constant bawling of orders (the Capt's are given *sotto voce* to a certain little 'Mr. Finley' who, like a familiar, is ever at his elbow) nor racket of ropes, nor rushing about of sailors, nor even some creaking of masts as in a packet; neither the monotonous plunge of the engine as in our steamers. There is a slight trembling of course but not a sound from the machinery. . . . The only sounds are the bells every hour, the bugle to summon us to meals, the slight sighing of the valves." Napier's machinery was barely audible—and invisible as well, unlike the engines on American steamers, concealed belowdecks and propelling the ship into an "easy majestic motion."

Eight days later, five days from Liverpool, Appleton was still surprised and pleased. "This steaming is all play-sailing compared to packet experience," she decided, "and the big Atlantic itself seems vastly shrunken and dwarfed to me now that we are rushing across it so comfortably, so independant of its head winds, so little wrenched from our equilibrium by its uneasy tossings and tumblings. I little thought I could prefer a steamer but so it is. We have no excuse for grumbling at anything." The passengers behaved well, not too numerous or drunk or talkative. Captain Charles H. E. Judkins, meticulously attentive to every detail, took a southerly course to avoid icebergs. (They still passed a group of ten icebergs about a mile away; the largest—eighty feet high and a quarter-mile long, greenish in the crevices and snowy at the top—resembled a ghostly steamship at that distance.) The *Columbia*'s food was bountiful and varied; the shipboard games, dances, and concerts were all diverting. "Instead of finding a steam-ship a floating Pandemonium as I expected, it certainly puts packets to shame for comfort and luxury and slides me over the Atlantic." Appleton was not even bored; "I feel almost sorry to quit the ship we are having such amusing times." They reached Liverpool in just half the duration of her packet voyage to Le Havre in 1835.

Appleton's account carried its own bias. Instead of westbound in January, like Dickens, she was eastbound in May, through gentle weather. Her father and uncle, Nathan and William Appleton, were business associates of Sam Cunard's in Boston, and when she went ashore in Halifax she was lavishly entertained by the Cunard family. Still, her description seems more believable than Dickens's because their motives for writing were so different. He, a famous fiction writer, wanted to spin a good yarn for publication and to sell books, with the more dramatically harrowing details the better. She wrote her version in private, unpublished letters to her father and to her best friend, with no evident purpose except to report honestly on what she had found. (In general, the more confidential the historical source, the more truthful.) Both had brought unmet expectations to the voyage; he was then disappointed and terrified, she unexpectedly pleased. Revising themselves in opposite directions, they gave their accounts quite different tones. Where they disagreed on specific matters of fact—such as the engine's constant noise level—Appleton had no reason to exaggerate, and Dickens did. So Fanny Appleton probably got it right.

The steamship *President* was the last chance for Junius Smith and Macgregor Laird. Designed by Laird with an almost desperate audacity, built by Curling and Young in London, and engined by Fawcett and Preston in Liverpool, she marked new transatlantic extremes in size, power, and luxury. At 2,360 tons she was a third larger than the *British Queen* and twice the size of the Cunard ships. The first liner with three decks, she afforded passengers an open-air promenade on the spar deck in fine weather and a sheltered turn on the main deck below. An elaborate carved figurehead of George Washington thrust forward from the bow; the projecting stern included other carvings and large plate-glass windows. The engine room, with its ornamented pillars and arches and polished iron and brasswork, reminded visitors of a handsome Gothic chapel: a modernist shrine to steam power. The hull was divided into watertight compartments, like earlier ships designed by Laird, "so that the springing of a leak would be attended with comparatively little danger," it was explained, "and would be readily overcome."

The opulent interior furnishings hardly hinted at the Smith line's ongoing financial straits. The main saloon, some eighty feet long and thirty-four feet wide, was finished in a Tudor Gothic style of delicate colors and grained oak. Its four tables and embossed crimson velvet sofas could accommodate up to one hundred diners at once. A wide corridor with even plusher decorations ran from the saloon to state-rooms at the stern. Ten oil paintings, executed on canvas to imitate old tapestries, depicted scenes from the life of Christopher Columbus. The corridor was said to resemble a picture gallery, or the upper story of a first-class hotel. In flashing these historic references and touches of landed luxury, the *President*'s living quarters were intended to mask a passenger's sense of crossing the ocean on the latest, yet-unproven steamship.

The money lavished on decorations might better have been applied to the engine. At 540 horsepower it was the most powerful on the ocean; but only a bit more so than the engine of the much smaller *British Queen,* which was itself underhorsed compared to the *Great Western* and the Cunard ships. That left the *President* a much lower ratio of horsepower to tonnage than any of the competing vessels. Some skeptics doubted the *President*'s stability, suspecting that the third deck might make her top-heavy and too prone to rolling on the North Atlantic's mountainous swells. She was also shorter and broader than the *British Queen,* not a likely design for great speed. An undeniably impressive sight, the *President* was more notable for her cosmetic features than her basic engineering.

She turned out to be a big, gaudy turkey, and a steady disappointment to her owners. Her first two captains were blamed for slow passages and quickly relieved of duty. The *President*'s lumbering maiden run in the summer of 1840 took sixteen and a half days in each direction. Her second trip home was worse: fighting a heavy gale from the east, in four days from New York she managed only three hundred miles. At that rate she would run out of coal before reaching Liverpool, so she turned around and limped back to New York. Amply coaled up, and with fair winds, she finally arrived at the Mersey on November 28, ten days overdue: a cause of great relief and rejoicing, since nobody in England knew that she had earlier been forced back to New York. ("What on earth or water," Isambard Brunel had asked in passing, "is the President about.") Laid up for two months through the worst winter weather, and refitted at

Plymouth after just two voyages across the ocean, she took three weeks to reach New York in February, her slowest time yet.

The *President* left New York for home on March 11, 1841, with about 110 passengers and crew. She was sailing low in the water because of heavy cargo and full coal bunkers. Stormy seas on previous voyages had weakened and twisted the wooden hull, perhaps fracturing the engine frame. (According to an oral tradition later passed down in the Laird family, shortly after her departure two people—the Smith line's New York agent and the brother of her latest captain, Richard Roberts—had the same dream on the same night. They saw the *President* with a confused crowd on her deck and Captain Roberts on the bridge, giving orders. Then the ship suddenly, unaccountably disappeared.)

She steamed into a screeching winter storm. Toward the end of the second day, another ship's captain sighted the *President* laboring through a dangerous area between the Nantucket Shoals and George's Bank, where the Gulf Stream collides with shallow soundings, sometimes generating starkly vertical waves as high as a five-story house. The other ship's captain saw the *President* rising on top of an enormous swell, pitching and struggling violently. Then he lost sight of her. The captain later guessed that she was shipping heavy seas, perhaps to the point of snuffing her boiler fires and leaving her powerless in the storm.

Once again the *President* was late to Liverpool. Her belated but eventual appearance in the previous November restrained, for a time, the usual worries about an overdue ship. After a month the *London Times* noted her tardiness amid stray rumors that she had been sighted here or there, or hit an iceberg, or come into Bermuda, or run ashore in Newfoundland, or suffered a mutiny, or fallen in with pirates. The line's London office was besieged by friends and relatives of the missing, hoping for any trace of encouraging news. The late-ship deathwatch stretched on for another month, tighter and more hopeless as time passed to no definite resolution. Queen Victoria, on leaving Buckingham Palace for Windsor Castle, asked that a special messenger be sent to her with any news of the overdue ship. When Fanny Appleton sighted a large iceberg from the *Columbia* in May, she declared it the ghost of the *President*. "I do not altogether give up the *President* as lost," Junius Smith wrote on May 14, trying to convince himself, "and yet I fear there is but slight ground for hope."

The *President* had simply disappeared into the North Atlantic, taking 110 people down with her. It was the first transatlantic steamship disaster—and not the last time that the newest, biggest ship on the ocean would steam into a catastrophe.

*T*he competition was dropping away, yet the Cunard Line was going broke. To get his contract, Sam Cunard had offered the government much better terms than the Great Western company's— impossible terms, it turned out. Even before his ships started running, Cunard asked the Admiralty to halve his winter sailings to once a month. "At the time I entered into the Contract," he explained, "steam boats had not crossed during the Winter and I was therefore quite ignorant of the risk and danger I had to encounter indeed I was very hasty in making the arrangement. . . . In the Winter Season there is not much commercial interest, and no passengers and once a month may, I hope be considered by their Lordships sufficient to cross during the inclement Season of the year—I have not only to contend with the storms on this Coast and on the Atlantic but with the severe Winter weather on the coast of Halifax and Boston. . . . These vessels have cost me nearly double the sum I originally expected and I find my Contract is by no means a favorable one, but I am determined to fulfill my engagements." The Admiralty agreed to a monthly schedule from November through February but cut £4,000 from Cunard's already-inadequate annual subsidy.

Back in the optimistic spring of 1839, Cunard when dickering with Napier had guessed that his line would run an annual profit of almost £41,000. No doubt this figure was inflated to entice Napier and his Glasgow investors; but Cunard surely did not expect to operate at a loss. Over its first nine months, the Cunard Line ran £15,355 in the red. When added to his money problems at home, these losses threatened Cunard with financial ruin. He thought about asking the government for a more generous contract but then decided not even to make the request and perhaps to give up the whole, leaky enterprise.

At this bleak moment in the spring of 1841, his partner David Mac Iver replied to a gloomy letter from Cunard with a reassuring blend of caution and optimism. "The day must have been cold, I think, or the subject has had a chilling effect over your spirits," Mac Iver wrote. "It is incumbent on us as shrewd Merchants to have our eyes ever open to the dark side of our doings—there is so much of show, and of the imaginary, in the money matters of steamers. . . . Our own doings, up till now, I rate the experimental; and not sound evidence of our true position or prospects. We have paid, like all beginners in new trades of magnitude, thro costly experience, and are now arrived at that point where we must turn this experience to profit." Their ships could not carry much cargo,

Mac Iver noted, and only attracted such freight in the summertime. So their only way of increasing revenues was to raise fares; Mac Iver suggested new rates of thirty pounds from Boston and forty pounds, nineteen shillings from Liverpool. Drawing from his years of running a steamship line, and knowing the implacable cycles of that business, he counseled patience: wait for better times.

Cunard, having operated ships for three decades, got the point. He, Mac Iver, and George Burns met in London to gather their ammunition. Cunard took the grim statistics of their initial operations over to Whitehall, urging that the contract be doubled to £120,000 a year. The Treasury, not persuaded, asked to see the line's books; Cunard assented "with pleasure." After the Treasury's own inquiry and report, the government granted a new contract raising the annual subsidy to £80,000, with the added stipulation that the Cunard Line build a fifth steamship. "It would have been wrong to let so important a line drop for want of the necessary support," a Glasgow newspaper reported. "We have, moreover, reason to know that the Government complimented the contractors, not only for having acted up to the terms of the previous contract, but also for having far exceeded them."

The new contract of 1841, much less than he wanted, did not ease Sam Cunard's own financial crisis. He had risked his fortune to start the transatlantic line on terms soon revealed to be disastrous. He was the founder and leader, the man who handled those delicate negotiations with the government. In a humbling act that must have embarrassed him, he had to ask his Cunard Line partners for a personal loan. In September 1841 he borrowed £15,000 from them, offering some of his own shares as security, and promising repayment with interest in two years. By the end of the year, Cunard had spent the entire loan on various debts.

Aside from a depressed general cycle in shipping and shipbuilding, Cunard was saddled with wayward investments at home, especially the Prince Edward Island land company. The Young family, his associates-turned-enemies in that venture, were not friendly to Cunard. (At a London party in 1839, William Young and Cunard were dismayed to run into each other. "Sam and I exchanged very distant bows," Young noted in his diary.) As Cunard's money crisis got tighter and more desperate, William and George Young—by a suspicious coincidence—turned up as the attorneys for his more anxious creditors. "They have been at different times employed by persons in England," Sam objected, "and they have resorted to every means in their power to injure me by arresting me and heaping costs upon me. You cannot imagine anything more unfeeling

than their proceedings—they hesitate at no act if it will put a few pounds
into their pocket."

The Youngs were only seeking revenge. Cunard, fighting for his very
survival, was equally ruthless, scratching for money anywhere he could
find it. In Nova Scotia, he sold the house and land in Hants County
where he had sent his late mother to spend her last days. He mortgaged
his wharves and warehouses in Halifax. The Bank of Nova Scotia suspended a rule to loan him £45,000. All his enterprises were squeezed
hard. As the largest landlord on Prince Edward Island, he had his agent
(his son-in-law James Peters) extract every possible penny of rent from
the immigrant tenants. One of them, a bard from the Isle of Raasay
named MacLean, took refuge in a mournful song. Translated from the
Gaelic:

> We left there
> and came out here
> thinking we would receive consideration,
> and that the rent would not be so exacting.
> But Peters is oppressing us,
> and, if he doesn't die,
> we must leave this place
> and Cunard, himself a beast.

The beast had become the quarry. In the spring of 1842, he admitted
to debts of £130,000 and mortgaged property worth £47,000, against
claimed (and probably exaggerated) assets of £257,000. A Liverpool bank,
trying to present a writ for £2,000 against him, sent a sheriff's officer to
the chambers of Cunard's lawyer in London; the lawyer stashed Cunard
in an adjoining room, and he escaped the process. A short while later,
Cunard quietly scurried up to Liverpool and arranged another flight. He
hid overnight in a cottage on the river below Eastham. Several writ
servers, suspecting his intentions, waited for Cunard on one of his
steamships until the last minute before departure. When they were finally
sent ashore, the steamer weighed anchor and started out to sea. Cunard
came alongside in a small boat; his ship slowed down and picked him up.
The eminent founder and leader of the British and North American
Royal Mail Steam Packet Company stole furtively home to Halifax.

From that nadir, Cunard slowly recovered his fortunes. In 1843 he
and his partners again asked the Admiralty for more money, claiming a
loss in the previous year of almost £26,400. The Admiralty allowed them

an additional £10,000 annually. Through all of Cunard's own troubles, his ships steamed across the ocean "with regularity almost unexpected and wholly unsurpassed," as a New York newspaper grudgingly admitted. As competing steamships receded from the Atlantic, the Cunard Line gained an essential monopoly and could raise its fares with impunity, up to forty-one pounds by 1846 for passage from Liverpool. His stock dividends and commissions gradually paid off Cunard's loan to his partners. The shipping business in general cycled upward again, and the Cunard Line started turning profits and paying more consistent dividends to stockholders. Sam Cunard's financial crisis lifted.

*T*he Cunard Line steamed toward solvency on its unmatched reputation for safety and order. That reputation—so coveted by passengers venturing forth on the dreaded North Atlantic—began with the first sailings in 1840. John Quincy Adams, a former president of the United States, and a man whose life was typically an exercise in rigid discipline and organization, took the *Acadia* from Boston to Halifax in September 1840. Adams approved the captain, the food, the crockery and glassware, the cleanliness, and the neat finish to the iron, brass, and woodwork. The *Acadia* was an uncommonly tight ship, Adams confided to his diary: "There is great order and discipline." Those qualities were associated with the Cunard Line ever after. Over the years, other Atlantic steamship lines would run ships that were bigger, faster, more luxurious, or with better service. Cunard ships retained their own unique, dominating cachet: they got you there alive.

No steamship line could entirely escape the North Atlantic's rigors; Cunard always had its share of accidents. In the first eight years, Cunard ships ran aground at least nine times in dense fogs off Ireland and the Canadian and American coasts, and at least twice collided with smaller vessels, sinking them and killing eight of their crewmen. (Other collisions may have happened at night and gone unnoticed, at least officially, by the Cunarder.) Charles Dickens's frightening experience on the *Britannia* was not so unusual. The most serious grounding was by the *Columbia* in July 1843. She had left Boston in a thick fog, picking her way carefully along the New England coast, and was pulled off course by an unusually swift current. Given the fog, the captain could not shoot the sun and take his bearings. Early in the afternoon, while running nearly full speed at ten knots, the *Columbia* struck a notorious reef called the Devil's Limb, 1½ miles offshore and about 150 miles southwest of Halifax. Despite desperate efforts, she could not be budged. A boat came out from a nearby island and plucked off the crew and all eighty-five passengers before the ship, buffeted by chopping waves, broke up and sank—the first Cunard steamship lost at sea.

The most remarkable aspect of these first eleven Cunard accidents is that nobody on the Cunard ship involved was killed. In fact, for the first seventy-five years of the line's history no passenger in its North Atlantic traffic ever died from a shipwreck. Meantime all the other transatlantic steamship lines suffered terrible disasters, one after the other, typically causing hundreds of deaths each time. Only the Cunard Line escaped. Perhaps this was just Cunard luck again, over and over. Or perhaps it was how the line was run.

In the division of labor among the three founding partners, Sam Cunard with his sons Ned and William supervised the line's offices in Canada and America, and Sam on his frequent trips to England represented the enterprise to the government. In Glasgow, George Burns looked after the construction and repair of the Cunard vessels by Robert Napier and various Clydeside shipbuilders. In Liverpool, David Mac Iver (and, after his death in 1845, his brother Charles) saw to the day-to-day management of the ships, keeping the captains and crews up to rigorous standards, making sure the vessels were well supplied and repaired and precisely on schedule both going and coming. These lines of authority at times became mingled; any one of the three founders might briefly take up any role, and major decisions emerged from polite, muted exchanges among all of them, with Cunard usually functioning as the first among equals.

If the safety record has a single overriding explanation, it was the Mac Ivers. Only twenty-eight years old in 1840, Charles stood at the top of the Cunard Line leadership for over forty years, even longer than Sam Cunard or George Burns. Nobody ever labored harder, more persistently or effectively, to keep the line in its usual position of transatlantic supremacy. Almost every day, he went down to the Cunard wharves on the Mersey, watching and measuring, taking notes, and giving orders. Quick and decisive, at times imperious, he had absolute confidence in his own judgments and opinions, seemingly never retreating from them once expressed. "Up to the last moment," a longtime Liverpool associate said later, "he will persist with all the energy of his nature in a course which his reason is gradually convincing him against his will to be erroneous . . . though even then he by some ingenious process satisfies himself and thinks he convinces others that he is not giving way at all."

The Mac Iver brothers were sons of a Greenock sea captain who was washed overboard in 1812 in the Bay of Biscay. As a young man, Charles spent some time in the American port of Charleston, South Carolina, but he soon came home and joined David in their coastal steamship enterprises. In about 1833 he had an experience that, as he told the story, confirmed his core insistence on relentless standards and inspections for oceangoing vessels. He booked passage on a particular sailing ship because he knew the captain. Off the Azores they ran into a powerful gale that lasted twelve hours. The ship, nearly sunk, was damaged and obviously unsound. "When you go home you had better throw up command of this vessel," Mac Iver told the captain, "or you will lose your life." The captain did leave her, but her owners—in their ignorance or greed—sent her out again under another commander. She disappeared with all hands. Any reasonable inspection, Mac Iver concluded, "would have prevented that ship from going to sea."

When the Cunard Line got under way, the Mac Ivers issued and enforced stringent regulations for their captains and crews. The two primary goals for the enterprise, speed and safety, were to some degree contradictory; to favor one could undermine the other. The Mac Ivers impartially emphasized both. "It will be obvious to you," they instructed the commanders of the Cunard ships in 1840, "that it is of first importance . . . that she attains a Character for speed and safety. We trust to your vigilance for this—good steering, good lookouts, taking advantage of every slant of wind, and precautions against fire, are principal elements." At the start of the voyage, the ship was heavily loaded with coal, and the paddles were deeply immersed, not working at peak efficiency. As coal was consumed and the hull rose in the water, the boilers could be fully fired up and the engine pushed harder. But coal use was to be carefully monitored and recorded, along with other supplies in the engine room. Only stokers and trimmers were to carry coal from bunkers to boilers; the sailors up on deck, a separate breed, were not expected to help out.

The Mac Ivers established other rules for the passengers and stewards. "A cheerful acquiescence is expected in the following Regulations and Suggestions," they explained in 1840, "which, if in any instance at variance with the opinions, habits, or inclinations of the few, are framed with a regard to the comfort of the whole." The staterooms were to be swept and carpets taken out and shaken every morning after breakfast. As soon as passengers left their rooms in the morning, their bedding was turned over, beds were made, wash basins cleaned, and slops emptied. Bed linen was changed on the eighth day, and boots and shoes cleaned overnight and returned to the rooms every morning at eight o'clock. Two towels were provided for each passenger and changed every other day, or more often if requested. The wine and spirits bar, always a favorite part of any ocean voyage, closed for the night at eleven but reopened quite early, at six the next morning. Lights went out in the saloons at half past eleven, in the staterooms at midnight, with no exceptions allowed for late readers.

Other Atlantic steamship lines had similar rules. The difference, it seems, is that the Cunard Line extracted routine obedience to its regulations. In January 1847, Charles Mac Iver found that the officers mess in a recent transatlantic voyage had committed "wanton and extravagant waste of the Company's victualling stores," as he put it, and had subverted specific rules "which have for their object general comfort and good order." Unable to discover the particular culprits, Mac Iver sent his prescribed bill of fare to all officers of the mess. "I shall be very happy to receive the resignation of any one who is not satisfied with it," he warned. "No man in this concern has had it in his power to say in truth that he has been otherwise than well treated, but wherever I find a set of men rating themselves only by what they can stow away in their bellies, I have prima facie evidence that they are not the men for the British & North American Royal Mail Service. . . . Specific and known orders shall never be infringed with impunity or trampled upon." Was that sufficiently clear? ("Mac Iver's letters quite discompose one," an associate explained; "you must talk over the matter with him to understand what a fine fellow he is.")

As time passed, Mac Iver's directives became even more definite and specific. His orders to commanders in 1848 ran to eighteen handwritten pages. Each ship was to leave port with enough food and water for thirty days in summer and forty days in winter (though a typical passage took less than half that time). If a very long trip depleted the coal supply, the captain was to put aside enough fuel to run the engine for twenty-four hours and then proceed under sail alone until land was sighted; then the boilers would be fired so the ship could reach port under power. Furthermore: Keep the ship clean to control vermin. "Ventilate, Ventilate." Only safely locked lanterns, no open flames, were allowed in the spirits room. No tobacco smoking was permitted anywhere below decks. "Want of cleanliness in the water closets is a constant cause of complaint, we shall be glad if you can take any measures to remedy this." Every Sunday at sea, the captain must limit unnecessary work and read aloud the Church of England service in the main saloon. Invite a passenger to read from a book of short sermons provided by the line. "If this does not meet with a favorable response, do not press it. Let your crew retire." Don't be too friendly with any particular set of passengers, or make or permit generalizations about any national tendencies; Americans and perhaps Englishmen can be quite sensitive about such matters. "Card Playing on the part of the Captains on board ship has been the cause of so much dissatisfaction and trouble, that it cannot be longer tolerated." Nor should the captain allow card playing or gambling in his private quarters, or in the officers mess, or by any officers.

Year after year, voyage after voyage, life on a Cunard ship expressed the flinty personality of Charles Mac Iver. He resisted luxuries and any fancy touches in the food or furnishings or decorations. He most valued safe predictability: always the same procedures and standards, unchanged unless for a very compelling reason. When accidents at sea happened, the captain and crew followed a precise, well-practiced drill, and so maintained order. The line's amazing safety record was no coincidence. Behind it, sustaining it, was Mac Iver standing at the dock every day, getting ready to inspect another Cunard ship and her personnel, missing and excusing nothing. "The highest court to which you could bring me," he once told a parliamentary commission, "would be my own conscience."

For Sam Cunard himself, the responsibility and success of his transatlantic steamship line made him a transatlantic citizen. He and his children were spending longer periods in England, bringing traces of English culture home to Halifax and setting themselves apart from their neighbors. When Fanny Appleton visited the Cunards in 1841, she noticed "the luxury of Mr. Cunard's house contrasting with the shabbiness of the town." One of Cunard's six daughters presided at the meals, "a very elegant girl of pale complexion, regular features, very black hair and a fine figure who has been to London and did the honors of lunch and Dinner with quite a distingué air." Though Appleton was dressed only in her sailing costume, she was persuaded to stay for dinner, at which the Cunard daughters, "à l'Anglaise, arrayed themselves in full dress." At table the Cunard girls talked about parties and balls with a determined gaiety; it reminded Appleton of scenes from Jane Austen's *Pride and Prejudice*. "It all has the strangest Anglo-American aspect," she decided. "A tedious provincial life they must have of it." Eventually all the Cunard daughters but one married English army officers and left Halifax permanently.

For their father, the break was more difficult. "I have been backwards and forwards occasionally to England," he said in 1847, "but Halifax is my home." His wife and one of his children were buried there. His extensive business operations were still based in Nova Scotia, and—apparently for sentimental reasons—he held on to his local post as a commissioner of lighthouses despite his frequent absences. But his hometown was not keeping pace with the nineteenth century. "What a slow place it is!" exclaimed an American visitor in the 1850s. An English tourist at about the same time chided "the lassitude and want of enterprise of the Nova-Scotians," who seemed to be lagging behind other Canadians; "the Nova-Scotians appear to have expunged the word *progress* from their dictionary." Cunard was a modern personality, permanently moored only to his family. He wanted the snap and zoom of modernity. His steamship line, steadily reducing and taming the unruly North Atlantic, imposing human will on nature, embodied the progressive faith of the age. As he shuttled back and forth across the ocean on his ships, he was increasingly drawn away from his fading colonial homeland and toward the ever-beckoning, up-to-date future that he found in commercial England.

He leased a country estate in Edmonton, eight miles north of London Bridge. Its seventy-four acres included a park, gardens, stables, and a greenhouse with grape plants and lemon trees. The house had plenty of room for visiting children and grandchildren. Playing a country squire, he developed interests in horticulture and joined the Royal Geographic Society. The prestige of his ties to the government and—in particular—the burgeoning fame of his eponymous steamship line eased his way into the higher reaches of London society, despite his usual reticent behavior in public. The actress Fanny Kemble later remembered him as a "shy, silent, rather rustic gentleman" at a party given by the celebrated hostess and writer Caroline Sheridan Norton. A major merchant prince, the man who had bridged the ocean, Sam Cunard still presented himself as an untalkative, unpolished colonial.

The success of his Cunard Line drove out its predecessors. Junius Smith's enterprise went under soon after the *President* disappeared. The Great Western company's profits peaked in 1839, before the maiden voyage of the *Britannia,* then declined in each of the next three years of competing with Cunard. In a last effort to save his company, in 1846 Christopher Claxton tried to pick off part of the Cunard mail contract. "We are quite aware of the excellent way in which Mr. Cunard performs the work," Claxton testified to a committee of Parliament; "no one knows it better than we do; we do not want to injure Mr. Cunard, but we want something to be done for ourselves." The government, well satisfied with the mail service, stuck with Cunard. A year later the *Great Western,* the first Atlantic steamship, was sold into service in the West Indies and South American trade. For the time being, the Cunard Line had transatlantic steam to itself.

6

The Collins Line

*T*he Cunard monopoly was soon challenged. Other nations with transatlantic commercial and political interests wanted their own steamships, without having to depend on the British mail line—no matter how excellent its service. "So far, we have been most fortunate in not having any formidable opposition," Sam Cunard wrote his partner Charles Mac Iver in May 1847, "but for the future we must expect it. I do not apprehend any serious injury from the French—their Ships were built for Men of war and will be strong and heavy and not fast. The American ships will be different. They will introduce all our improvements, together with their own. We shall also have national prejudices to contend with, so that every attention will be required to meet them." The son of American exiles took special notice, as always, of any American competition. "It will behove us," he urged Mac Iver, "to think of any measures that may be considered improvements as the Americans will be alive to every thing."

*C*unard truly feared the American threat, but it also handed him a forceful argument for persuading the British government to subsidize new steamships, bigger and faster, for the Cunard Line. Since the original foursome of the *Britannia* and her sisters, two ships had already been added: the *Hibernia* in 1843 (to satisfy the revised mail contract of 1841) and the *Cambria* in late 1844 (to replace the *Columbia* after she ran onto a rock and sank in a Nova Scotia fog). The two new ships—220 feet long, 1,354 tons, and 472 horsepower—were slightly larger, more power-ful versions of the original four. After the usual shakedown crossings, to let their engines and bearings settle into working order, they set new transatlantic records of nine days, twenty hours, thirty minutes out to Boston and eight days, twenty-two hours, forty-four minutes back to Liverpool, reaching an average speed home of almost twelve knots.

As passenger environments, the *Hibernia* and *Cambria* continued the Cunard practice of austere comfort, emphasis on austere. The main saloon, again placed toward the rear of the main deck, was longer and wider than on the earlier ships, with room for two long tables that could seat up to one hundred diners. The oak beams in the white ceiling included gilt moldings, an atypical touch of plush. Adorning the walls, landscapes painted on slate offered vistas of Glasgow harbor, Liverpool, Halifax, Boston, and—suggesting the line's future—New York. Sofas along the outer walls of the saloon passed daylight through their open bases, from windows overlooking the main deck, down to cabins on the deck below. The roof of the saloon allowed passengers a promenade, pro-tected by a strong brass rail, that was safely out of the sailors' way as they moved around the deck, working the billowy spread of canvas. Below the saloon, the gentlemen's sitting room was painted imitation marble and wainscot; the counterpart for women had large sofas covered by thick, silky velvet. Toward the bow, other quarters could accommodate male passengers and servants in less comfort. (The servants' steerage had a sep-arate entrance.) Fully loaded, the ships could each carry 155 passengers, 130 tons of light goods, and the all-important, enabling mail.

Like the lost *Columbia,* the *Hibernia* and *Cambria* were engined by Robert Napier and built by Robert Steele. Napier, an original investor in the Cunard Line and one of its most trusted advisers, got its engine con-tracts automatically so long as he wanted them. The continued confi-dence in Steele, on the other hand, suggests that Cunard and his manag-ing partners did not blame the loss of the *Columbia* on any defect in the ship herself. In fact, Steele would later go on to build six of the next seven Cunard Atlantic liners as well. His shipbuilding skills and standards must be granted major credit, along with the watchful Mac Ivers, for establish-ing the Cunard Line's enduring reputation of safe reliability.

The two Robert Steeles, father and son, had been building wooden steam vessels since the early 1820s. Their shipyard at Greenock, near the confluence of the river with the Firth of Clyde, gave them easy access to the sea and wider, deeper water for launching large ships than places farther upriver, toward Glasgow. The senior Steele first collaborated with David Napier in 1821 to produce the *Eclipse*, of 140 tons and sixty horsepower, for the trade between Glasgow and Belfast. Only five years later, Steele and Robert Napier built the *United Kingdom*, four times the size of the *Eclipse* and the largest British steamboat yet. The father then retired, and his son, thirty-five years old, took over and continued to work with Napier. These ties exemplify the stability and family connections that became defining hallmarks of Clydeside shipbuilding. Such a situation might easily have led to complacency, technological inbreeding, and shoddy, dangerous work. But it didn't; instead, "Steele-built" became a synonym for high levels of shipbuilding design, materials, and workmanship. The younger Robert Steele, "in his silent, unobtrusive way," as a colleague put it, continued to improve his craft—and kept the exacting Cunard partners satisfied.

Early in 1846, Sam Cunard, at home in Halifax, heard that the U.S. Congress had granted a group of New Yorkers an annual subsidy of $400,000 to start a steamship mail service between New York and England. The news struck him in two sensitive spots: American competition, and his long-nurtured plans to start running his own ships to New York, the booming center of maritime commerce in the United States. Cunard rushed over to London and started dickering with men at the Admiralty and Treasury. "I came to England to point out to the Government that an American line was about being got up, and I wished to prevent that," he explained. "I saw that the American Government were giving encouragement to a mail line which would interfere very much with me, and would interfere equally with the [British] Government; I was satisfied that it would deprive the Government of half the postage, and deprive me of half the passengers." As Cunard made his case, he melded his self-interest with the national interest, presenting both as equally jeopardized by the American thrust.

The government agreed, up to a point. Cunard was already getting £85,000 a year for running five ships on the Liverpool-Halifax-Boston route. He wanted £85,000 more to build four new ships to go to New York. The Lords of the Admiralty were agreeable, but Henry Goulburn, chancellor of the Exchequer, would assent only to an additional £60,000. "I pressed Mr. Goulburn very much," Cunard said later, "and I could not persuade him." Looming over the discussions, crowding out everything else for Cunard, was the threat of the new American line. "I would have taken anything almost in order to prevent it," he said; "therefore I took the sum that Mr. Goulburn offered me, though it was very unjust. . . . He did it to save the money to the country, but he took 25,000 £. a year from me for the good of the country." (In losing the argument, Cunard no longer quite saw his interest as identical to Great Britain's.) The new contract allowed Cunard £145,000 a year for weekly sailings to America, going on alternate Saturdays through Halifax to Boston and to New York; the service was reduced to biweekly during four winter months. The contract restored Cunard's original vision, from back in 1839, for his transatlantic line: a ship leaving every week for the United States, competing directly with the American ships out of New York. Though he might have groused about the money, it was otherwise a benchmark for Cunard.

The new ships—the *America, Niagara, Europa,* and *Canada*—were major advances, about 30 percent larger and more powerful than the last two Cunard steamers. (Even so, they were still shorter and less roomy than Junius Smith's overwrought *British Queen* and vanished *President.* Cunard caution allowed no such reckless leaps.) Robert Napier inevitably provided the machinery and other advice as well, giving suggestions about how to achieve more deadweight without deepening the draft, and greater displacement but with finer hull lines. John Wood constructed the *Europa;* Robert Steele built the other three at a brisk pace, producing vessels of 5,271 total tons in just two years. The new contract paid only a fraction of the final costs of about £90,000 per ship. Most of the capital for these supposedly subsidized steamers had to come from private sources. (Sam Cunard's complaints about Goulburn's stinginess were not unreasonable.)

Once again, little money was lavished on decorations for the new foursome. "Although chastely elegant," a visitor in Liverpool wrote of the saloon, it "is evidently fitted up more for use than ornament." Paintings in the saloon were executed on papier-mâché instead of slate. The new ships tried a novel solution to the endemic problem of providing safe lights below decks: an oil lamp enclosed in a triangular glass fixture illuminated two adjacent cabins and the hallway in front of them. Passengers could not touch or move the lamp, reducing the danger of fire. At the stern, the helmsman had gained a small wheelhouse for protection from the elements. On the poop deck above the saloon, just ahead of the wheelhouse, an officer stood at the binnacle, watching the ship's compass and shouting orders back to the helmsman.

The machinery, the most vulnerable and telling part of a steamer, was by far the most expensive item on board, costing about £50,000 per ship. Steam pressure in the boiler, slowly inching upward since 1840, topped out at eighteen pounds per square inch, double the capacity in the

Britannia and her sisters. Despite that improved efficiency, the new ships burned sixty tons of coal a day, 50 percent more than the first Cunarders; fuel costs inevitably rose much faster than a vessel's top speed. The *Canada* and *Europa* were given slightly more powerful engines of 648 horsepower and were a bit faster than the other two at 630 horses. The *Canada* soon knocked ten hours off the eastbound transatlantic record, and the *Europa* reduced the *Cambria*'s best westbound time by almost a full day. The four ships shared a reputation for swiftness, which always attracted paying customers. In the fall of 1849 the *Niagara* sailed with more than 160 passengers for Halifax and New York, a new record for an Atlantic steamer. Sam Cunard had his ships, it seemed, to meet the American challenge.

*E*dward Knight Collins was the American of Cunard's nightmares. The robust success of his Dramatic Line of sailing packets (see p. 6) had made him the leading figure in New York shipping circles. The unprecedented size, speed, and luxury of his ships marked Collins as a seagoing entrepreneur who embraced new technologies and took grand risks—and, more important, made them work. His lust for the new was grounded in practical realities; he seemed to understand both ships and men. A showman with an intuitive flair for public relations, he treated the press to lavish food and drink when celebrating ship launchings and maiden voyages, and thereby received generous coverage of his feats. In all his ventures he recognized few restraints, ethical or financial. He sighted an objective and charged toward it. And now he was turning toward transatlantic steam.

Like many other rising princes in New York, he was a New England boy who had moved down to the big city to make his fortune. Salt water ran through his entire life and ancestry. He was born at the tip of Cape Cod, Massachusetts, in 1802. His father, a shipowner, sea captain, and merchant trader, followed the maritime drift toward New York, where the son joined him as a teenager. Edward went to sea occasionally on the family's ships—not as a common sailor but as the supercargo (business agent) on voyages to the Caribbean. He even commanded, briefly, a ship in that region. This period may have included brushes with pirates and shipwrecks in Cuba and Florida. He soon came ashore for good and managed the Collins office on South Street, near the East River docks, at the hub of New York shipping. In 1826 he married well, and in a characteristic rush, taking a thirteen-year-old bride named Mary Anne Woodruff, the daughter of a prominent building contractor and active Democratic politician. Coastal packets to New Orleans, and then the Dramatic Line to Liverpool, made Edward Collins rich. He built a country estate of three hundred acres called Larchmont Manor, on the Long Island Sound north of town, and cut a substantial profile around Manhattan.

His dealings with a fellow buccaneer, James Gordon Bennett of the *New York Herald*, suggest a certain tone. The first modern newspaper, the *Herald* cost only a penny and set circulation records by offering a caterwauling pastiche of racy, sensational news about crimes, tragedies, accidents, and seductions. "The *Herald* is producing," Bennett solemnly declared, "and will produce, as complete a revolution in the intellectual habit of daily life as steam power is doing in the material." The paper also quickly became the journal of record for the shipping business, with long columns of news and commentary. At first not friendly to the Collins enterprises, Bennett gave them more space after Collins raised his advertising budget for the *Herald.* (Tit for tat.) In 1840 the city's moral guardians—and the other newspapers in town—tried to kill off the *Herald* through imposing an offended boycott by readers and advertisers. A delegation of boycotting politicians and merchants came to enlist Collins. He asked a clerk how many ads he had in that day's *Herald.* Three, he was told. "Put in three more tomorrow morning," said Collins. "That is my answer, gentlemen. Good morning."

Collins and Bennett shared an early interest in transatlantic steam. As Collins worked toward his own Atlantic steamships, he had the biggest, least inhibited American newspaper in his corner, leading cheers and emitting patriotic screams. In February 1841, eight months after the first Cunard liners started running, Collins submitted a plan to the U.S. government for four steamships of 2,500 tons each (more than twice the size of the initial Cunarders). For this service he asked a mail subsidy of $15,000 a voyage, or twenty-five cents a letter. His steamers would also, he noted, double as warships in the event of war with Great Britain, a constant worry in these years of boundary disputes and diplomatic crises with the mother country. But here Collins indulged in rhetorical overkill; his headlong personality included no hint of the delicate, modulated touch needed for successful Washington lobbying. "Can it be possible," he demanded of the postmaster general, "that our country so much in want of war steamers cannot get its government to accept a proposal unequalled for liberality and importance. The rejection of this proposition by the government may be attended with serious consequences as other nations through our public journals must and will know that the offer has been made." He was acting, said Collins, not for "personal gain" but for "national benefit." The government, not persuaded by this heavy-handed attempt at blackmail by publicity, turned him down. (Over in Liverpool, the ever-vigilant David Mac Iver sent the happy news along to Sam Cunard.)

Five years later, after the Cunard ships had reduced the North Atlantic almost to a regular ferry ride and taken the transatlantic mail away from the New York sailing packets, the U.S. government was more amenable. Its first ocean mail contract went not to Collins but to a hith-erto (and henceforth) obscure New Yorker named Edward Mills. News of this arrangement pushed Cunard and the British government into build-ing the *America* and her sisters for the New York route. The two Mills steamships, the *Washington* and *Hermann,* were almost as large as the new Cunard steamers, but—it turned out—much slower, so Cunard still car-ried most of the mail and the cream of the paying traffic. In response to this national failure, late in 1847 the administration of President James K. Polk pushed through Congress a new mail contract of $385,000 a year for Collins, with a pointed mandate to beat Cunard. Collins sold his interests in sailing ships and, with every resource he could command and all his blunt, rushing energy, switched over to ocean steamships.

Some American shipbuilders were moving in the same direction, but not many. For most naval designers and builders in the United States, ocean navigation still meant sailing ships, now reaching their fleeting peak in the legendary clippers: fast and beautiful, but with limited cargo space, and soon obsolete. American steam continued to take its characteristic form in river and lake steamboats, also fast and beautiful, well suited to their intended purposes, but too fragile and dangerous for long ocean routes. Nobody in the United States could approach the saltwater steam experience of Robert Napier, Robert Steele, and a dozen others along the Clyde and elsewhere in Great Britain. Yet Collins was somehow supposed to produce, from patriotically native sources, better transatlantic steamships than the Cunarders.

The most important early designer and builder of American ocean-going steamships was a mysterious figure named William H. Brown. Unlike other New York pioneers in this field, he had no background of producing sailing packets. Instead he had started with canal boats, per-haps at a yard up the Hudson River in Hyde Park, and later made steam-ers for the Long Island Sound. In the late 1840s he began building ocean steamships at his shipyard on the East River, at the foot of Twelfth Street. He lived in a house nearby on Second Avenue, made a comfortable fortune, and ranked among the largest shipbuilders in the city. His competitors did not like him. After less than a decade of concentrated, significant activity in ocean steam, he dropped from the historical record, leaving no further trace that anybody has been able to discover.

American steam vessels were typically powered by tall steeple or walking beam engines that sat high in the hull, even protruding above the main deck, where they were vulnerable to weather and noisy, smelly annoyances to passengers. Ocean vessels, pitching and rolling on heavy seas, required a lower center of gravity and more protection for their engines. British engineers had therefore adopted the side-lever engine, placed at the bottom of the hull, for ocean steamships. In 1846 Brown built the first American commercial saltwater steamer to be powered by a side-lever engine, the *Southerner,* for the coastal traffic to Charleston, South Carolina. She was small, only 191 feet long and 785 tons, with a low, wide transom at the stern that looked, said one critic, "like the side of a house." But she performed well and set a pattern for the immediate future of American ocean steam. In the next three years Brown built three more ships of similar design but larger dimensions, up to the *Empire City* of 239 feet and 1,751 tons for the trade between New York and New Orleans.

Brown became a key member of the design and building team for the new Collins Line. The mail contract specified that the ships should be strong enough for quick conversion to military purposes; so three engineers from the U.S. Navy, John Faron, Charles Copeland, and William Sewell, offered crucial advice, and the navy had to approve the final plans. In a revealing bow to the current state of the art, Collins sent Faron over to Great Britain in order to appraise (and perhaps borrow) the newest refinements. The Americans believed that the latest Cunard ships ran on boiler pressures of not over ten pounds per square inch. But on the trip home, on the new Cunarder *Niagara,* Faron snooped around the engine room and noticed that the safety valves had thirteen one-pound weights on them. When the ship plunged sharply into the trough of a wave, the valves would open slightly. The boilers were obviously producing more then ten pounds of pressure. To squeeze out more power, Collins therefore made his engine cylinders five inches wider, to a diameter of ninety-five inches, more than six inches fatter than the *Niagara's* cylinders.

William H. Brown built two of the Collins ships, the *Atlantic* and *Arctic.* Jacob Bell, of Brown and Bell (eminent builders of packets and steamboats), made the other two, the *Pacific* and *Baltic.* Engines and boilers came from the Novelty and Allaire ironworks, the two most experienced manufacturers of marine steam machinery in New York. The new line was drawn from the best talent in the city. Collins stood firmly at the center of the entire effort, taking advice from all hands—Washington politicians, naval officers, local engineers—but making most of the final decisions himself. He was "well known in this community as a gentleman of the most enterprising character," the ever-cooperative *New York Herald* reminded its readers. "The advantages of having but a single controlling mind in a project of this nature are obvious."

The four ships manifestly expressed his personality. The contract called for vessels of at least 2,000 tons; Collins made them average almost 2,800 tons and 814 horsepower, huge increases over the *Niagara* and her sisters. The Collins liners rode unusually high in the water, which made them appear even bigger and gave strolling passengers more freeboard protection from rogue waves. The American ships were strikingly, daringly innovative in many designs and details. ("They will introduce all our improvements," Sam Cunard had fretted, "together with their own.") In place of the slanting, flaring bow of sailing ships and all the Cunard liners, the Collins ships had a straight, vertical stem, slicing the water at almost a right angle to the surface and flowing aft into a relatively narrow wedge. They carried less sail than Cunarders: these were *steam*ships. The engine valves and the boilers, designed mainly by John Faron, all departed from standard British practices. The furnaces were arranged in two rows, one above the other, saving precious lateral space in the engine room. A new system of telegraphed signals sent the captain's instructions from the bridge down to the chief engineer in the engine room.

The passenger rooms were decorated and equipped with all the splendid extravagance that Edward Collins had lavished on his Dramatic Line—in inviting contrast to Cunard austerity. Stained-glass windows in the stern offered views of New York, Philadelphia, and Boston; no British cities in sight. Thick glass trunks brought light and air from the spar deck skylights down to the public rooms on two decks below. The saloons set new transatlantic standards of overstuffed luxury, with tabletops of Italian marble, gleaming inlaid wood paneling, ornate mirrors and bronzework, a ceiling of carvings and gildings, and plush Brussels carpets. Men especially enjoyed the novelties of a barbershop, with glass cases of toiletries and a reclining chair for the morning shave, and a smoking room at the stern, a masculine preserve of cigars and spittoons. The finest staterooms, ringing the main saloon, included two washbasins each, a sofa, satinwood berths, and damask curtains. Instead of shouting for the steward, a passenger simply pulled a bell cord. A few "bridal suites" even provided double beds for romantic couples (a feature of the toniest American steamboats, but new to the ocean). Steam heat, another innovation, kept the public rooms comfortable through North Atlantic winters without the smoke and danger of stove fires. No barnyard animals intruded their sounds and smells from the top deck; a cold house with up to thirty-five tons of ice kept meat and other perishable food fresh for the voyage.

The Americans had rethought the whole concept of a transatlantic steamship, pushing it ahead by giant bounds in size, power, luxury, and sophistication. All this came at equally extravagant costs. The ships were delayed by unexpected problems of design and construction. Some of the heavy tools and machinery required were not available in the United States and had to be specially made or imported from Great Britain, which took more time and money. Collins had said he could build his ships for $500,000 each; that bill eventually rose by nearly 50 percent, to a total of almost $3 million for the four steamers. As with the Cunard Line, the Collins mail subsidy only provided seed money and the prestige of a government contract. Most of the initial capital had to be scrounged from private investors.

Collins found his angel in James Brown, of the international banking house of Brown Brothers. The Brown family, from Ballymena, north of Belfast, had come to America in 1800. William, the eldest of four brothers, returned to the United Kingdom in 1809 and became a noted Liverpool banker, merchant, and member of Parliament, wielding immense, rippling influence in his adopted city. James, the youngest brother, ran the firm's flourishing New York office at 59 Wall Street. Both Browns had been involved with the maritime enterprises of Edward Collins since the late 1830s. (This Brown family was not related to the shipbuilders William H. Brown or Brown and Bell.) When Collins got the steam bug, he naturally turned to his associate James Brown in New York. But William Brown had already invested heavily in the Cunard Line; of its first stockholders, only Sam Cunard and James Donaldson had contributed more than Brown's share of £11,600. As the Cunard Line lurched uncertainly toward solvency in the mid-1840s, William Brown wanted no more such dubious plunges in transatlantic steam—especially from an American competitor.

This family showdown became a complicated sibling rivalry. The prudent oldest brother, always the dominant Brown, already had his steamship line. The striving youngest brother wanted his own counterpart. Rebuffed in Liverpool, James tried to recruit other members of the family. Cousin Stewart, of the New York office, was not personally charmed by Collins ("certainly a strange being . . . he has a low unrefined mind with an overbearing disposition"). William Brown dismissed the whole scheme. "On no question is Mr. W. B. so fixed and determined," noted his Liverpool partner, Joseph Shipley. "Nothing will induce him to think of embarking any money in it—and I entirely agree with him. The Agency of the steamers here can be no desirable object."

William remained adamant, but he could not stop his youngest brother. James supervised a stock offering in New York that raised a bit more than $1 million for the Collins Line; he then arranged additional loans of about $2 million, thereby scraping together the $3 million needed to pay for the four ships. Of the original investors, James Brown took the biggest chunk with $200,000 of his own money. Collins stood second at $180,000, with $20,000 more from his father-in-law, Thomas Woodruff, which essentially set up the two founding partners as equal players. Brown assumed the office of president while Collins took aggressive charge of the line's everyday operations.

The double engines at the head of the Collins Line made a rather odd couple. Brown was a devout, active Presbyterian layman, meticulously dressed in the black uniform of a New York merchant, physically frail, and by nature quite contained, punctual, disciplined, modest, and reticent. Collins was an utterly secular figure, fleshy, expansive, unbuttoned, heedless, noisy, disorderly. One felt most at home in the quiet inland countinghouses of Wall Street, the other along the bawling, brawling West Side steamship wharves on the North River. It is difficult to imagine just the two of them in a room, talking business and puzzling over the other man. Yet through their peculiar symbiosis, each contributing what he did best, they had after heavy labors produced the most substantial challenge yet to Cunard's domination of Atlantic steam.

In Liverpool, still not reconciled to what he deemed his brother's folly, William Brown took steps to protect the family's investment in both steamship lines. He dreaded an undercutting fare war that could ruin both enterprises. In the spring of 1849, a year before the first Collins ship set forth, he approached Charles Mac Iver to suggest "such arrangements as would prevent injury to either concern." He also asked his brother to see if Sam Cunard was interested in controlling competition. They struck a deal: a secret agreement that fixed minimum rates for passengers and cargo, and guaranteed a split of gross revenues, with two thirds to Cunard and one third to Collins. This pact, signed by Mac Iver and William Brown in May 1850, was the first ocean steamship cartel in history. Not illegal at the time, but potentially awkward in the irascible climate of Anglo-American tensions, it was carefully concealed. As far as the public knew, Collins and Cunard were intending to drive each other off the Atlantic.

*V*irtually from their start in 1850, the Collins liners were preferred by most passengers, especially the Americans. The Cunard Line had drawn its first commanders from the Royal Navy, and its service retained residues of military abruptness and austerity. A typical Collins Line commander came from the New York sailing packets, with their traditions of opulent food and comforts. When James West became captain of the *Atlantic,* the first Collins liner, he brought along his own shipboard chef of the previous decade, an African American known by the generic nickname of Blackie. "A better sea-cook I never wish to find," an English traveler, William H. G. Kingston, declared after comfortably crossing the Atlantic on the *Atlantic.* "How can I sufficiently praise his super-excellent Irish stew, a dish fragrant to the nose and delicate to the taste, unlike any Irish stew I ever before eat . . . his delicious puddings, his exquisite *petits plats?*" Blackie (and his assistant cooks) turned out such confections all day long, from breakfast at eight till late supper thirteen hours later. "The advent of every meal was announced by a gong," Kingston noted, "and as people were always feeding, the gong had little rest." All the meals were generous, and marked by the American custom of offering ice to cool the water, milk, and wine.

Dinner at four was the grand occasion of the day. A typical Collins menu offered two soups, six types of boiled meat, three fish dishes, nine kinds of roasted meat, nine entrées with elaborate French names, two vegetables, two salads, eight choices for dessert, fruits and nuts, and coffee and frozen lemonade. (Wines and spirits cost extra.) At the outset, as diners mulled over their many choices, the head steward looked around the dining saloon to see that everybody was seated, and then waved his hand. The waiters filed in with trays of fish and soup. For the second course, the waiters marched in bearing covered dishes and halted smartly, facing right and left. At a signal they put the dishes on the tables; at another signal they removed the covers. So it went through the meal, everything hot or cold, as desired, and crisply executed, orderly, and luxurious.

The public rooms and fittings also contrasted favorably with Cunard ways. The public rooms, especially when brilliantly lit at night, were dazzling. "Rich enough for a nobleman's villa," noted Benjamin Silliman, a Yale geology professor, on the *Baltic* in 1851. He counted about fifty mirrors among the various saloons. At one point he looked up from his reading and beheld six duplications of himself. As the editor of one of the leading American scientific journals, and a veteran of prior crossings by sail, Silliman took a professional interest in the *Baltic*'s steaming and sailing qualities. The hull timbers were tightly fitted, well caulked, and braced by iron plates and bars. "The ships of this line are as strong as wood, iron and copper can make them, and they hardly leak at all," Silliman decided. "The movement of the machinery, and the stroke of the waves produce scarcely a perceptible tremor. . . . It travels onward with the most admirable ease and regularity; even with a heavy head wind and opposing waves, it moves like clockwork, without apparent labor."

The Collins service and furnishings were powerful inducements. But what the paying customers really noticed was speed. After some initial problems with coaling and stoking, the Collins ships clicked off successive records for the route between New York and Liverpool. In May 1851 the *Pacific* made the first run from New York in under ten days, averaging a bit more than thirteen knots. (The Collins Line awarded a silver vase to her engineers.) Three months later, against the prevailing winds and currents, the *Baltic* came home in less than ten days. These records were soon beaten—by the *Arctic* and *Baltic*. Overall, during 1851 the Collins liners averaged seven hours faster than the Cunarders to Liverpool and twenty-two hours faster to New York. The American triumph seemed complete, as measured by any standard: bigger, better, faster. "The United States have never yet done any thing which has contributed so much to their honor in Europe," exulted an American preacher, John S. C. Abbott, after crossing on the *Arctic*. "We have made a step in advance of the whole world. Nothing ever before floated equal to these ships."

The eagle was screaming, but the evidence looked irrefutable. Captain Lauchlan B. Mackinnon of the Royal Navy took a Cunarder to New York, then the Collins Line home. "They are, beyond any competition, the finest, the fastest, and the best sea boats in the world," he declared of the American steamers. "I am sorry to be obliged to say this; but, as a naval officer, I feel bound in candor to admit their great superiority. Their extraordinary easiness in a sea can not fail to excite the admiration of a sailor; I never beheld any thing like it. There was none of that violent plunging—that sudden check usually attending a large ship in a heavy head sea. . . . The whole fore-length of the vessel appeared to sink gently down until almost level with the water, and as gradually to rise again after passing." Mackinnon believed that the Collins liners plowed through adverse waves so smoothly because their designers had eliminated the boltsplit (or bowsprit), the spar that extended forward from the bow of a sailing ship to secure thick ropes from the foremast. "The vast and heavy boltsplit of the Cunard line, therefore, is an absolute excrescence—a bow-plunging, speed-stopping, money-spending, and absurd acquiescence in old-fashioned prejudices about appearance."

Behind the public scenes, beneath the design innovations and glossy surface appearances, were certain troubling realities. The heavy engines and pistons, pounding at top speed through a typical voyage, strained and twisted the wooden hulls. Paddles were sometimes broken and lost. In January 1851, the *Atlantic* on a westward passage fractured her main shaft in midocean; with her engine disabled, she tried to head west under sail, was beaten back, and finally retreated to Ireland after being out for six weeks. When a Collins liner came into New York, mechanics from the Novelty Iron Works might have to toil night and day to get the machinery ready for the next trip. Once, in Liverpool, a curious newspaperman was denied access to a Collins ship's engine room for several days. (The chief engineer said that it was "not yet in the exact order he would wish.") Collins liners would be quietly withdrawn from service for three or four months at a time, with no announcement or explanation. It was unusual for all four ships to be running at once; schedules had to be adjusted. "In power of engines, durability of works, and strength of frame," the *New York Times* concluded early in 1852, "the English sea-steamers as a general rule, are superior to our own. They can be what is technically termed *pushed.*" American ships were faster, the *Times* allowed, but the Cunarders were still stronger and surer.

The triumphant speed and service exacted enormous costs. "The proprietors of this American Line," Sam Cunard happily informed Francis T. Baring, now the first lord of the Admiralty, "have discovered that the working expenses amount to double the sum they anticipated or calculated on. . . . Messrs. Brown have a large interest in it, but the Stock is also at a heavy discount." By its own reckoning, the Collins Line was running an average loss of almost $17,000 per round-trip. All the supplies, coal in particular, piled up excessive bills. A typical run, out and home, cost $8,600 for fuel, $9,215 for repairs, $8,900 for insurance, and $21,600 for wages, provisions, and general expenses. Constant overhauls and heavy interest payments on the initial $2-million loan kept the line steaming through red ink. It never paid any dividends to stockholders. By the end of 1851, even as its ships were declared the unrivalled queens of the North Atlantic, the Collins Line was in fact teetering toward bankruptcy.

After two years of technical success and commercial failure, Collins went back to Congress for a raise in his mail subsidy of $385,000 a year. President Millard Fillmore and his fellow Whigs, the Republicans of their day, were quite friendly to American business interests. But Congress, controlled by the Democrats, had to approve a larger appropriation for the Collins Line. The congressional debates, and intense lobbying behind them, stretched on for months. The factions were roughly arrayed along party lines—and, more sharply, across sectional schisms. In general, the North supported the Collins proposal, and the South opposed it. Only nine years before the Civil War finally erupted, the bitter political and cultural divides of the impending crisis spilled over into ocean steamships. (During the Senate debates about the new subsidy, William H. Seward of New York led the Collins forces, and Robert Hunter of Virginia and Solon Borland of Arkansas led the opposition. When the war came, Seward served as Abraham Lincoln's secretary of state, Hunter as the first Confederate secretary of state, and Borland as a brigadier general in the Army of the Confederacy.)

In the fight of his life, bailing hard to keep his ships afloat, Edward Collins seized every weapon at hand. An attorney and journalist named John O. Sargent was then editing a Whig daily newspaper in Washington, the *Republic,* as a paid mouthpiece for the Fillmore administration. Sargent was a skilled writer, well connected, and discreet. Collins sent him an obliquely worded proposal, marked "<u>Strictly confidential,</u>" in January 1852. "What I now write, must not pass beyond your own thoughts," Collins told him. "Our memorial to Congress will no doubt be presented to-morrow—in it I think I see your finger." If it passed, Collins hinted, he would have more money at his disposal. "You are the only person to whom I dare make this strictly confidential communication—and you must consider yourself <u>my</u> employee." PS: "Do not fail to write <u>private</u>, on the address of your letters to me." Sargent and his newspaper had just been added to the Collins payroll.

The best argument for a fatter subsidy was those magnificent steamships. "Had not the ships <u>outstripped the World</u>," Collins pointed out to Sargent, "any attempt at further increased pay, would be idle." They decided to send the *Baltic,* freshly painted and polished, down to Washington to perform her own kind of lobbying. Preparing for this gambit, Sargent measured Collins and decided his presence in Washington would not help the cause. He urged Collins to make no comments or comparisons that might agitate opponents: "<u>Keep quiet and keep away.</u>" Collins, hurt and surprised, wanted to sail down to Washington on his ship and bask in her fame. "<u>Others</u> may manage my affairs better than I can," he shot back, "but up to this time my experience does not tell me so—yet from you I expect the most candid advice. Let me have your <u>reasons frankly</u> for me not to visit Washington. I may at this moment be silencing the very opposition." Sargent, having annoyed his employer, but now backing and filling, prudently changed his mind. "I was apprehensive," he explained, picking his words delicately, "that in your go-ahead, impulsive, free-spoken style, you might possibly say something that would be perverted or employed to your disadvantage. . . . Your visit of a few days with the Baltic might be beneficial." (Of a few days.) "When I visit Washington," Collins replied, still offended but mollified enough, "I shall endeavor to succeed in <u>diplomacy</u> as well as I have in Steam."

Collins and the *Baltic* steamed up the Potomac to Washington and prepared a fancy reception on board for any interested politicians. "These persons," Senator Borland complained, "are acting upon a saying we sometimes hear throughout the country—a proposition which is an insult in itself—that the nearest way to the hearts and understanding of Senators is down their throats." On March 2, a Tuesday, Congress adjourned for the day, and Collins stood proudly on the deck of the *Baltic,* greeting the president, his cabinet, diplomats, members of Congress, and over 2,000 others attracted by the free food and drink. After touring the ship and taking a short cruise, they crowded boisterously into the refreshments. "Many were the salutes fired by inoffensive corks, in all directions," the newspapers reported. "Every body was happy, joyous, hilarious." (And drunk.)

Safely back on land, Congress argued into the summer. The Collins advocates invoked history, the irresistible steam revolution, and the need to smite the British. "In this, *the third war with England* which we are now waging, I, for one, mean to be on the side of my country," said Representative Gilbert Dean of New York. "The power of the British Government is, and for years has been, devoted to the maintenance of this Cunard line." The opponents spoke of "corrupting influences," of rumors (denied) that certain "foreign capitalists" held stock in both the Collins and the Cunard lines, and of undue favoritism to New York over other parts of the country. Even the opulence of the Collins liners raised some suspicions about their military potential; "a bare inspection of them," said Senator Benjamin Wade of Ohio, "will show that they are better calculated for eastern seraglios than for ships of war." Finally, by narrow votes in both houses, Congress more than doubled the subsidy, from $385,000 for forty voyages a year to $858,000 for fifty-two annual voyages. The Collins Line would survive—perhaps despite its founder's headlong style. "I hope Collins will keep quiet," one of his exasperated lobbyists wrote to Brown Brothers in New York, "and not crow too loudly."

*T*he *Arctic* left Liverpool for home on September 20, 1854. It was the end of the summer tourist season in Europe, so the ship went out overloaded with 281 passengers, including a higher proportion of families—women and children—than usual, and more people of wealth and social prominence than the mercantile class of travelers that made up a typical transatlantic ship's list. Among the notables on this trip were five members of the Edward Collins family and seven of James Brown's relatives. The *Arctic,* four years old, was one of the faster vessels in the Collins fleet. For two and a half years she had held the current eastbound record of New York to Liverpool in nine days, seventeen hours, and fifteen minutes. Her captain was James C. Luce, forty-eight years old, a veteran of the Red Star sailing packets who had joined the Collins Line at its inception. He was, as ordered, pushing his ship to her limit.

Nothing remarkable happened until seven days out, when the *Arctic* entered the notorious graveyard of the Atlantic on the Grand Banks of Newfoundland. (Exactly three years earlier, in the same ever-dangerous area, the *Arctic* had charged ahead despite a thick fog and nearly collided with a large sailing ship on an opposite course.) During the morning of September 27, she steamed through a dense, clumpy fog; at times the white smoke would blow away, allowing bright sunshine to break through, but then it would close in again, to the point that a man standing near the bow of the *Arctic* could not see her superstructure at midship. The Collins liner kept plowing onward at her full speed of thirteen knots. Just after noon, as passengers were sitting down to lunch in the dining saloon, a smaller steamship suddenly emerged from the fog, under full sail, bearing down on the starboard bow of the *Arctic.* A lookout rang a warning bell and somebody shouted, "Hard to starboard." The helmsman spun his wheel, to no avail. The other vessel—the *Vesta,* a French steamer headed for Europe—slammed bow-first into the wooden hull of the *Arctic,* about twenty feet aft of the stem on the starboard side, and stuck there, her crumpled snout wedged into holes below the waterline.

At the moment of impact, a man standing in the *Arctic's* main saloon felt a slight tremor, nothing more, and suggested to a friend that they might have struck a rock. "Hit him again," said the friend, brushing the matter off lightly. Down in the engine room, the telegraph from the bridge struck one bell, the signal to stop the engine: an ominous command with the ship so far out at sea. "What's up now?" barked J. W. Rogers, the chief engineer. A second signal told him to reverse the engine, as the *Arctic* was trying to detach herself from the *Vesta*. After ten minutes of backing and churning, she was free—and the holes in her hull were now wide-open. At the bottom of the ship, the men in the engine room saw the water slowly spreading in from the bow. Rogers started the regular steam-powered pumps and, with a wrench, opened the bilge injectors, so the main engine would pull water for the boilers from the bilge instead of from the ocean, thus transforming itself into an enormous pump.

The *Arctic* was already listing forward and to starboard. The heavy anchors and chains at the bow were shifted to the port side or thrown overboard. Captain Luce told the passengers who were rushing up on deck, puzzled and curious, to move toward the rear of the ship, in order to raise the bow and perhaps expose the holes there. The *Vesta*, more visibly wounded, looked doomed. Luce sent a lifeboat under the command of his first mate, Robert Gourlay to assist her, and ordered his second mate, William Baalham, to drop over the bow and examine the holes in the *Arctic*. They were worse than expected: three distinct gashes, the largest five and a half feet long by two feet wide, together admitting perhaps 1,000 gallons a second. The crew tried to patch the holes with sails, pillows, and mattresses; but the openings were too jagged, and bristling with protruding remnants of the *Vesta's* bow, to be covered effectively.

Under ferocious, swirling pressures, beset by bad news, Luce had to make a crucial decision. His lifeboats could accommodate only about 40 percent of the 434 people on board. He decided to make a risky run for Cape Race, some sixty miles to the northwest. He left the *Vesta* (which would survive) and the boat with his first mate (which would not), and ordered full steam ahead. In the fog he ran over a lifeboat from the *Vesta*, drowning all its occupants but one. As the *Arctic* cut through the sea, the rushing forward motion brought even more water in through the holes at the bow. Luce badly needed his first mate; Gourlay was a robust adjutant, feared and obeyed by the crew. Luce evidently lacked his force of personality. (The American consul in Liverpool, the writer Nathaniel Hawthorne, had briefly dealt with Luce in the course of his duties. "He was a sensitive man," Hawthorne noted in his journal; "—a gentleman, courteous, quiet, with something almost melancholy in his address and aspect.") Luce was not the right man for the tightening crisis on the

Arctic. He seemed almost paralyzed at times, barely able to speak, and then commanding only in lurching, feckless spurts.

In the engine room, meanwhile, a mutiny was brewing. The engineers, oilers, firemen, and trimmers could see, more clearly than anyone else on board, that the ship was sinking. They carried the coal from the holds, shoveled it into the furnaces, tended and raked the fires, nursed and pushed the boilers at maximum pressures to keep the paddle wheels turning and the pumps working as hard as possible. And still the water crept steadily up their legs, inch by inch; nothing could stop it. After thirty minutes, the rising water snuffed out the lower row of furnaces, with a great hissing of steam and exploding clouds of acrid black smoke. The struggle was already lost at full steam power. At half power, the end could only come sooner. Suddenly, most of the engine room men threw down their tools and shovels and rushed up on deck, desperate and wild-looking, some with knives drawn.

"From this time," one passenger said later, "all order and discipline ceased on board." Most of the passengers stood bewildered on the quarterdeck at the stern, watching, praying. Some of the men joined the sailors at the manual pumps on deck. Others tried to fill and launch the lifeboats; the choppy sea and the steadily downward, sideways tilt of the deck complicated the process. One boat full of women and children slipped its rope and tackle, spilling everyone into the sea and drowning them. An apprentice engineer, Stewart Holland, fired the *Arctic*'s signal gun at close intervals, hoping another ship would hear it and come on through the fog. He stayed at his post, firing until the end.

Chief engineer Rogers and about nine others from the engine room seized the small boat from the roof of the fore deckhouse, stocked it with food, water, and cigars, and launched it off the liner's bow. The boat had room for more people, but they were discouraged at gunpoint. Second mate Baalham filled another boat with sailors and deserted the ship. Some of the firemen and trimmers pushed passengers aside and took their places in lifeboats, or jumped into boats as they were being lowered, landing heavily on top of others. Captain Luce, flailing around and brandishing an ax, tore the shirt off one fireman, Patrick Tobin, and ordered him to let the passengers take the boats. "But it was every one for himself," Tobin said later, not apologizing, "and no more attention was paid to the captain than to any other man on board. Life was as sweet to us as to others."

Third mate Francis Dorian (one of the scarce heroes on the *Arctic*) and a few other men scavenged materials for a raft—spars and yards from the masts, some gratings, water casks, anything that would float—threw these pieces over the side, and started lashing them together. No ship's carpenter or sailors were left on board to contribute their emergency expertise. To assist the construction, Dorian put the last boat out just beyond the raft, after removing its oars so the raft makers in the boat could not leave yet. People still clambered across the unfinished raft, thereby interfering with the work, and piled into the boat. "A perfect mania seemed to seize all on board," Dorian said later, "and a universal rush was made for the boat, so that no possible entreaty or threat could stop it."

Four and a half hours after the collision, with passengers still crowding the quarterdeck, the bow rose, the heavily weighted stern dipped, and the *Arctic* sank amid a great welling, sighing sound from her boilers and funnel. It seemed oddly human, a final death rattle. The fragile, overloaded raft was struck by the sinking ship, broke in two, and wound up with seventy-six people clinging to the remaining piece. The sea was briefly dotted with scattered thrashers, swimming and floating in metal life preservers. The water, at a lethal forty-five degrees, soon subdued the scene. The clammy air, still fogged in, felt chilly and penetrating. A final act of selfish cowardice remained: the lifeboats did not rush back to rescue survivors. "Our principal endeavor was to keep at a distance from the ship," one man later explained, "for the purpose of avoiding the eddy, should she sink suddenly."

Two of the six lifeboats eventually reached Newfoundland. Passing ships picked up a third boat and a few hardy souls from pieces of wreckage. By a grim justice, the mutinous boat commandeered by chief engineer Rogers disappeared into the fog and was lost. The raft drifted around; within four hours, as the sky darkened, seventy-five people succumbed to exhaustion or exposure and slipped into the ocean. After twenty-six hours, a ship happened by and plucked off the last survivor, a stubborn waiter named Peter McCabe. Captain Luce, standing by the starboard paddle box, intended to go down with his ship but instead was tossed up as she sank and then was improbably saved.

Of the 281 passengers, only 23 survived. Of the 109 women and children aboard, every one died. But 61 of the 153 crew members were rescued, including 4 of the 5 top officers and half of the firemen and trimmers. The *Arctic* disaster shattered high Victorian notions of how men were supposed to respond under duress. In this desperate crisis, the muffling, protecting layers of civilization slipped away too easily, and humans behaved like any other wild, self-preserving animals. It broke the social contract in alarming, revealing ways. Worse than shocking or tragic, it was shameful. "Oh, what a manly spectacle that must have been!" keened a New York newspaper. "It is enough to make us all ashamed of humanity, and envy the better nature of the beasts of the field."

The private purgatories of Edward Collins and James Brown may only be imagined. Their steamship line, the proud, laborious product of their own best efforts, had caused the deaths of 350 people on one of their celebrated vessels—and in a manner that could hardly be contemplated. As though in penance, Collins had lost his wife, two of his three children, his wife's brother, and that brother's wife. Brown had lost his favorite daughter, another daughter, the son considered his heir apparent, that son's wife, and two grandchildren. When James Brown heard the news, he fell to his knees and cried, "My children, my children!" His wife, Eliza Coe Brown, became mute. She fled into an impenetrable seclusion and did not speak for two years.

*T*ransatlantic travel always carried the risk of a terrible disaster. But this was different, and worse, not just another ship gone to grief in the graveyard of the North Atlantic. The Collins Line was hauling such a heavy freight of patriotic hope and brag. The *Arctic* was one of the four finest ocean steamships yet built. And the death toll was so high, full of so many prominent names, and so embarrassingly distributed among passengers and crew, and men, women, and children. The brutal stories of those dreadful, bullying scenes on the sinking ship hung in the air like nightmares that would not end. In the history of transatlantic steam, no other event had such thudding, mournful impact in both America and Britain until fifty-eight years later, when the *Titanic* sank.

The *Arctic* disaster delivered a staggering blow to the Collins Line, but not a mortal one. The line received an insurance settlement of $540,000, which it needed for its continued operating deficits. For the time being, it put off replacing the lost ship and maintained schedules with the remaining three vessels. Watertight hull compartments and more lifeboats were added to the ships, for extra safety and to reassure the public. The line also benefited from Great Britain's involvement in the Crimean War. The Cunarders, like the Collins fleet, were—as part of their mail contract—available for military uses. During 1854 and 1855, eight Cunard ships were borrowed for varied stints of transporting troops and horses to the Crimea, forcing Cunard to reduce its Atlantic service. That gave Collins a most welcome opening. In 1854 Cunard took in 70 percent of the gross transatlantic mail revenues, a business not covered by the secret cartel between the two lines; a year later, Cunard and Collins split those revenues evenly.

Since the advent of the Collins Line, Cunard had added three major new steamers to its Atlantic route: the *Asia* and *Africa* in 1850, and the *Arabia* in early 1853. All three were built by Robert Steele and engined by Robert Napier, again with paddlewheels and wooden hulls, and they repeated the line's usual careful, gradual growth over their Cunard predecessors in length, power, and interior space. The *Arabia* blatantly imitated the Collins liners, copying such features as steam heat, stained glass windows in the stern, a smoking room, and other rather un-Cunard-like touches of luxury. But the three new ships were still smaller and slower than the Collins liners, and they retained the unnecessary (and perhaps hobbling) bowsprit that so agitated Captain Mackinnon. For whatever reason, they could not match the Collins sailing qualities. The *Asia* "goes bowling down the Mersey, carrying a sea before her enough to swamp a revenue cruiser," an English newspaper remarked, while the Collins ships "slip down the Mersey with scarce a ripple at the bow, dividing the water like a Gravesend steamer." The traditional Cunard caution against too many technical improvements, too quickly, received sharper scrutiny when these new ships still could not outrun the older Americans.

The secret cartel arrangement between Cunard and Collins satisfactorily restrained their competition by keeping cargo and passenger fees artificially high, and was renewed each year despite occasional mutterings from both parties about violations of the terms. "We do not attempt to do any thing in opposition to them," Sam Cunard confided to Francis Baring, "we go on very quietly with each other." The revenue-sharing aspect of the cartel involved relatively small sums; Collins received £1,427 from Cunard in 1850, and, when the Americans started carrying more passengers and freight, Collins paid Cunard a total of £21,973 over the next four years. The cartel's real impact lay in preventing a fare war, a potential boon to the paying public but a dreaded danger for steamship lines operating on narrow profit margins.

The concealed ties between the two lines, linked by the Brown Brothers stock holdings in both enterprises, made a rare—if unexplained and unnoticed—public appearance in the summer of 1851. As part of the celebration of that year's Great Exhibition at the Crystal Palace, William Brown arranged a splashy reception in Liverpool for the exhibition's sponsors. Brown first led a tour of the Cunarder *America,* just before she embarked for New York, and then escorted his party on to the Collins liner *Atlantic,* anchored in the Mersey, for another impressing visit and a plush banquet on board. "The *Atlantic* is fitted up with a splendour which is truly astonishing," reported the *Illustrated London News,* which spread a drawing of the banquet across three full columns of a page. It did not seem odd, apparently, that William Brown would undertake such visible displays for both competing lines.

Within the framework of the cartel, the two parties still fought at will for reputation and business. Cunard got permission from the Admiralty to run weekly ships all year long, including the winter months, and to steam directly to New York, without stopping at Halifax: both in order to match Collins. The Collins Line made sure the public fully appreciated its faster times across the ocean. "Nothing seems to touch the English nearer than this question of nautical superiority," Nathaniel Hawthorne noticed; "and if we wish to hit them on the <u>raw</u>, we must hit them there." From his hotel window in Liverpool, Hawthorne watched the Collins and Cunard ships passing in the Mersey and firing their guns in polite salutes, "the two vessels paying each other the more ceremonious respect, because they are inimical and jealous of each other."

Until it produced a faster steamship, the Cunard Line would not beat the Collins speed records across the ocean. But it came close and never stopped trying with the ships on hand. The two lines could not stage a matched race because they left port on different days of the week (from Liverpool, Cunard on Saturday and Collins on Wednesday). They were nonetheless sometimes charged with "racing" each other, to the heedless peril of arriving safely—an accusation they would piously deny. "It is our duty," Sam Cunard declared in a letter to the *London Times,* "to convey the mails with as much despatch as possible, consistent with safety, beyond which we have never gone, and no consideration would tempt us to exceed the bounds of perfect security. It is, therefore, not right to accuse us of racing, which may give unnecessary alarm to passengers." It was noted, after the fact, that the *Arctic* had steamed at full speed through that fatal Newfoundland fog. The shock from the disaster finally abated, though, along with any persuasive misgivings over reckless speed. Cunard and Collins continued to charge across the Atlantic, at times regardless of ice or weather.

Granted the boon of the Crimean War and its diversion of Cunard ships, Collins reached new peaks during 1855. It maintained consistent schedules month after month, without lowering its own speed marks but more steadily than before, carrying 138 passengers on an average trip and setting a new Atlantic record of 7,176 passengers for the year. The Collins ships, lengthening their lead over the helpfully reduced competition, averaged twenty-eight hours faster than the Cunarders to Liverpool, and over thirty-one hours faster to New York. After six years of operations, of recurring financial crises, political maneuvers, and one horrible catastrophe, the line finally seemed reasonably well established and stable.

*T*he *Pacific* departed Liverpool for New York on January 23, 1856. She carried 45 passengers (a typical reduced complement at midwinter), 700 tons of fine goods, and 141 officers and crew. Overhauled and refitted the previous spring, the *Pacific* was considered fully shipshape. She had recently taken on a new captain, Asa Eldridge, and a new first mate, Hugh Lyle. Neither man had much prior experience in transatlantic steam; Eldridge came from the crack clipper ship *Red Jacket,* and Lyle had been working as a pilot on coastal steamers between Halifax and Boston. The *Pacific's* chief engineer, who joined the crew just after Captain Eldridge, was also new to his job and machinery. The great circle route to New York was especially dangerous that winter. Icebergs and field ice were already appearing on the Grand Banks, earlier and more thickly than usual, and at points farther south than their normal pattern.

The *Pacific* did not appear in New York after her usual passage time of ten to twelve days. No other ships in the transatlantic lanes had sighted her in passing. On February 7, a passenger on a steamer heading home to Glasgow thought he saw broken ice and some floating cabin furniture on the Grand Banks. The Collins agent in Liverpool issued a statement: "The handles of the doors of the *Pacific* are similar to those described; but so we believe are they in almost every large vessel sailing out of New-York. We do not attach much importance to the report, as we think there would have been a much larger quantity of wreck about had it been the *Pacific.* Still we are not in a position to deny that it may have belonged to her." A month went by, and still no sign of her anywhere.

Anxious observers reminded themselves that, five years earlier, the *Atlantic* had broken her shaft and gone missing before news of her arrival in Ireland at last reached New York—after seven weeks. The famous commodore of the Cunard Line, Captain Charles H. E. Judkins, said he felt certain the *Pacific* was safe, that he knew of many cases where ships had become ice-locked for a month, unable to move but in no great danger. Officially the Collins Line agreed and refused to give up. On March 17, an acquaintance in New York ran into Edward Collins, "who was quite exuberant in commendation of the fine weather," the man wrote in his diary. "Wanted, but hadn't the heart, to ask him about the *Pacific.*" On that same day, dropping his public face of determined good cheer, Collins quietly applied to the Navy Department for permission to substitute another steamer to take the *Pacific*'s place in carrying the mail.

"No good reason exists for the popular belief that this unfortunate ship has perished with her passengers and crew," a steadfast Englishman insisted in late March. Her engine was probably disabled, he supposed, and recent prevailing winds had been keeping many sailing ships off the coasts of Great Britain and the Continent, unable to make port, and out of touch. "She can probably be found by a searching squadron. . . . Urgent reasons exist for sending relief to her without delay." The British Admiralty sent out two steamships to look for her, the *Tartar* and the appropriately named *Desperate*. After due searching, they couldn't find anything. "A little longer," said the *New York Times* after almost three months, "and the agony of contending hopes and despondency will be over."

By then it was indeed over, no matter what anyone said in public. As with the vanished *President* in 1841, people slowly came to accept that the *Pacific* would never show up, and they would probably never know what had happened. For those closest to the situation, with friends or family on board, the cruel denial of final certainty kept hopes flickering for too long, clinging to the slightest wisps of rumor or possibility. Everyone felt a need for some plausible explanation. At the time it seemed most likely that the *Pacific* had foundered off Newfoundland, after another foggy collision or—the prevailing theory—after hitting one of those plentiful icebergs. But these were just guesses, attempts to impose order on a tugging riddle.

(There the mystery remained for the next 135 years. In 1991, divers in the Irish Sea, about twelve miles northwest of the Welsh island of Anglesey, found the bow section of the *Pacific*. Contrary to most informed surmises in 1856, she had made only about sixty miles from Liverpool when something went wrong, suddenly and completely. A boiler explosion or a fire, perhaps; the new chief engineer, in his first hours at sea with the *Pacific*, might have made a fatal mistake in the unfamiliar engine room. Or maybe a collision with another ship that sank both vessels so quickly that it left no time for lifeboats, or even floating bits of wreckage, to tell the story. It was only a dozen miles from land, but too cold in January for anybody to swim to safety. Yet why did nothing identifiable—not even a life preserver—wash ashore? To find out, at last, where the *Pacific* went down still left the most tantalizing questions unresolved.)

A second Collins Line disaster in sixteen months doomed the enterprise. The deaths, now, of 536 people under its care sent the traveling public streaming back to the slower but safer Cunard ships. The storied Cunard reliability had never seemed more appealing. With half its fleet gone, the American line could no longer maintain its mail contract; a new Collins ship, the *Adriatic* (long promised but long delayed), was still far from finished. Political support for the line in Washington, already precarious, drifted away in the administration of President James Buchanan. Congress cut back the subsidy to its original level and then sliced it further. "I think Collins is a dead lion," said Thomas Rusk of Texas, a former Collins advocate, on the floor of the Senate in March 1857. "His contract is drawing to a close, and I understand that he is about to be bankrupted under it."

For William Brown and his associates in Liverpool, the sudden collapse of the Collins Line vindicated their original (and sustained) doubts about the entire scheme. "The wise course will be to throw up the contract and sell the boats for the most they will bring," urged one of the Liverpool partners. "So long as they run there will be nothing but trouble to Mr. James Brown and loss to all concerned. . . . It would be the best news I could hear, that Mr. James Brown had decided to run the boats no more." James Brown finally agreed. Still mourning his lost children and grandchildren, depleted by years of deficits and aggravations, weary of fending off critics in Liverpool and even among his own friends in New York, late in 1857 he scuttled what was left of the Collins Line. The three remaining ships were soon auctioned off for a pittance. After a few final gestures of stubborn undefeat, Edward Collins retired from maritime enterprises. (He eventually married again, tried but failed to recoup his fortune, and died in respectable poverty in 1878. Captain James Luce, who never went to sea again after the *Arctic* disaster, came to his funeral.)

After a decade of such promise and heady achievements, the Collins Line ended poorly, in a puddle of tragedy, unpaid debts, and ill feelings. "An odor of corruption and roguery hung round the line to the day of its death," *Harper's Weekly* declared in 1860. "It was badly managed, badly officered, and badly engineered in every way." This was too broadly harsh; the naval engineering of the Collins ships established new standards of performance and expectation, and the novel passenger comforts were validated at once by Cunard with the *Arabia's* imitations. But the line's managers, especially Edward Collins himself, and his shipboard officers were another matter. The chaotic denouement on the *Arctic* would probably not have happened on a more disciplined Cunard ship, no matter the circumstances. It was murderous incompetence, not just bad luck.

For Sam Cunard, on the other hand, the two Collins disasters could be taken as an especially gruesome instance of Cunard luck. His ships generally could not beat the Collins liners in design, performance, or service. But the Cunarders could outlast them and then welcome passengers back after the double catastrophes. "The whole capital of the company," Cunard said of his beaten rivals in 1860, "was sunk, I suppose at least 700,000£. [about $3.4 million], in the time that they were running, and the only thing that saved them from bankruptcy earlier was the loss of two ships, for which they recovered a quarter of a million of money [$1.2 million] from the underwriters here." Their ships might have been very expensive to operate as well, he allowed. "I do not know; but," twisting his knife, "there was something wrong about them." Even so, Cunard still worried, as ever, about new competition from the United States. ("I knew that the Americans would never be satisfied.")

Not satisfied, perhaps, but clearly defeated, and at dreadful costs in political prestige and human lives. In the messy aftermath, British commentators happily discerned national tendencies that explained the outcome. "America is weeping for her steamers, and will not be comforted," the *London Times* offered. "Like other young people, the United States have had their visions and castles in the air, but the illusions have vanished, and they are now steadily settling down to the realities of actual existence." The Americans, as youthful parvenus, had embraced the showy and ornamental, and mistaken an early lead for a final victory. But the British were more mature, solid, careful. "The Cunard Company has always avoided alluring novelties," the *Times* explained, "and relying solely on the safe teachings of experience, has uniformly sacrificed show to substance and speed to safety. They have gradually increased the size and power of their ships, but in each case only so much as experience showed to be advantageous." Furthermore, their officers and crews had always displayed "that confidence in each other, that discipline and courage, which such a system cannot fail to produce." The *Times* was, after its usual fashion, smug and condescending toward upstart Americans—but, in this case, too close to the hard truth for Yankee comfort.

Other American challengers would continue to run steamships across the North Atlantic. The end of the Collins Line, though, marked a historic turning in the struggle for transatlantic steam supremacy. The Civil

War, foreshadowed in the congressional maneuvers over the Collins Line, also shifted American attention away from ocean steam. The United States turned inward, to the war and its tangled aftermath, and then to westward expansion and the transcontinental railroad. For the rest of the nineteenth century, and well beyond it, American ship designers and builders would not again achieve even a brief moment of beating the Cunarders. American challengers generally had to contend with higher prices of labor and supplies for their shipyards, costlier insurance premiums, unfriendly navigation laws, and political indifference in Washington. Behind all these specific limitations, though, was a larger and simpler obstacle, impossible to overcome. Facing toward the south and west, preoccupied with its own mighty internal convulsions, and expanding essentially just within its own spacious borders, the American national will in effect conceded the Atlantic Ocean to the British. The contest was over.

7

Distinguished Failures

$\mathcal{L}$ike war and politics, the history of technology is usually written by and about the winners. The terms of the contest, as well, are defined by the winners. A triumphal perspective points out that ocean steam was inherently a commercial enterprise. Ships therefore must finally be judged by whether they turned profits for their owners. Everything else—speed, size, beauty, technical innovations, legends, and romance—must give way to the cold, remorseless verdicts of the ledgers. And yet, the two most intriguing transatlantic steamers of this era, the *Great Britain* and the *Great Eastern,* were plump commercial failures on the Atlantic. Their red-inked histories still achieved a grandeur and impact not matched by most of the safely profitable steamers of the age. Outsize in every sense—indeed great—daring and unprecedented, dazzling but impractical, these two ships were (of course) created by the ambiguously brilliant Isambard Kingdom Brunel.

$\mathcal{T}$he Collins Line ships were the final, finest expressions of a soon-obsolete technology: paddle wheel ocean steamers built of wood. They only half expressed the nineteenth-century revolution in iron and steam. Smoking and clanking, true machines at sea, they still used wooden paddles and wooden hulls, and carried wooden masts and canvas sails for prudence and extra speed. They were like a factory powered by steam, but with wooden machinery and laboring waterwheels on the side;

or a railroad locomotive running on wooden wheels and rails, and carrying a team of oxen to pull the train in case of emergencies. Long after factories and railroads had fully joined the revolution, ocean steamships held
back, not quite convinced, still retaining odd traces of the old ways. A
breakdown in midocean, after all, presented sharper dangers than a balky
loom or locomotive on land. The North Atlantic mandated its own
imperative level of caution.

Iron and steam were irresistible. One invention kited into the next,
iron and steam provoking each other, into tight chains of invention that
finally girdled all of modern life. The notion of making a boat of iron came
not from shipbuilders still wedded to wood, but from ironworkers and
steam men accustomed to metal. Late in the nineteenth century, Robert
Duncan of Greenock looked back on the multiple transformations he had
seen in his lifetime. Descended from a family of traditional shipwrights, he
remembered the initial attempts on the Clyde, in the early 1800s, to craft a
boat from iron. "In the marine boiler shop of that period we have the origin
of the iron ship," Duncan recalled. "The boiler maker began to think that if
the shell of a boiler would keep water in it would keep it out." As a small
boy, Duncan for the first time beheld—"with great surprise, not unmixed
with contempt"—an iron vessel sent out from Glasgow to have wooden
beams and decks installed by his father. Along with most other Clyde shipbuilders, Duncan long remained skeptical of iron; in 1840 he and his father
built the wooden *Britannia,* the first Cunarder on the Atlantic.

The idea of an iron boat did defy common sense. A piece of wood
thrown into water floated, and a piece of iron did not. Yet when shaped
into a watertight vessel, it was surprisingly buoyant. In its traditional
form of cast iron, with a carbon content of 2 to 4 percent, the metal was
hard but brittle, readily broken, and too easily pulled apart because it
lacked tensile strength. A wooden ship's heavy timbers, by contrast, could
flex and creak and twist in heavy seas without breaking. The invention of
a new puddling process in the late 1700s let ironworkers readily remove the
carbon to produce wrought iron, softer and more ductile than cast iron, yet
with four times its tensile strength. That made possible iron bridges, buildings, railroads—and ships. The new material also matched the historical
moment. Great Britain was running out of forests, but it soon found abundant deposits of iron ore in South Wales, Staffordshire, North Yorkshire,
Durham, Northumberland, and Scotland. Whether cast or wrought,
depending on its particular application, the metal became what the railroad engineer Robert Stephenson gratefully called "that all-civilizing
instrument, Iron."

The earliest iron vessels were unpowered barges, to be towed on rivers or canals. In 1787 John Wilkinson, a famous ironmaster near Birmingham, launched an iron barge, seventy feet long and eight tons in capacity, for local traffic on the River Severn. "It answers all my expectations," Wilkinson noted, in mingled pride and doubt, "and has convinced the unbelievers, who were 999 in 1,000. It will be a nine days' wonder." Over the next three decades, iron vessels appeared on Scotland's Forth and Clyde Canal, on the River Clyde—to Robert Duncan's boyish contempt—and elsewhere in Britain. But they remained more expensive than comparable wooden boats, more difficult to repair, subject to rust and corrosion, and unfamiliar to traditionalist crews who spurned such radical innovations.

The crucial next step was to reunite the two main elements of the Industrial Revolution, iron and steam power. When applied to shipbuilding, wrought iron was both stronger and lighter than the hardest of hardwoods. It could better endure the pounding, eccentric forces of a marine steam engine and paddles. The first iron steamboat, the *Aaron Manby*, came from no shipyard. Instead she was built in 1821 for Aaron Manby himself, an inventor, at the Horsley Iron Works in Staffordshire, in the smoky industrial Midlands. She was then taken apart, sent to London, and reassembled on a dock. With a cargo of linseed and iron castings, the *Aaron Manby* crossed the Channel (the first iron vessel at sea) and proceeded up the Seine to Paris (supposedly the first vessel of any kind to travel directly from London to Paris). For twenty years she gave reliable service on French rivers, along with three other iron steamboats of similar design by Manby.

British ship men still resisted iron hulls. Their indifference left the field to pioneers like William Fairbairn, who came from inland roots in Roxburghshire, Scotland, near the English border. The son of a farmer, he was barely educated. ("To the present day," he said much later, "I am unable to determine whether I write or speak correctly.") After apprenticing to an engine mechanic at a colliery, and much diligent self-directed library work, he built mill machinery and waterwheels, and started his own ironworking firm in Manchester. When the Forth and Clyde Canal Company asked him to improve the speed of its iron barges, Fairbairn turned his engineering attention to naval architecture. His experiments with different kinds of hulls convinced him of the advantages of iron.

In 1831 Fairbairn built his first iron steamboat, the *Lord Dundas,* for the Forth and Clyde Canal. She was sixty-eight feet long and seven tons, with a hull of iron plates, and powered by a locomotive engine. (The canal and family connection recalled an earlier steamboat of note, the *Charlotte Dundas* by William Symington.) Fairbairn followed her with a much larger iron steamer, the 100-ton *Manchester* for the coastal trade. In 1836 he established his Millwall Iron Works on the Thames, the first iron shipbuilding yard in London. (Maudslay, Sons and Field had already made an iron steamer for the East India Company, but most of its business was steam engines and machinery, not ships.) At Millwall, Fairbairn in thirteen years built over one hundred iron ships, of up to 2,000 tons, but finally had to sell his firm at a loss. Unsuited to enterprise, Fairbairn remained active as a theorist and tireless evangelist, conducting important experiments and publishing esoteric papers on the properties and best construction methods of iron. "It is from this material that we derive the instruments of our civilisation," he later declared; "our progress in useful art depends upon our knowledge of its application . . . it supports our wonderful industry, and is the soul of our commerce. . . . The iron age of the world has come."

John Laird of Birkenhead was the first iron shipbuilder from a marine background—and, perhaps for that reason, the first iron boat man to earn regular profits from his unorthodox vessels. The Laird family had moved down from Greenock, on the Clyde, and settled first in Liverpool, then across the Mersey in Birkenhead. Like their fellow Liverpudlian Scots, the Mac Iver brothers, the Lairds ran steam packets and other shipping enterprises on the Irish Sea. In 1824 the Lairds started making boilers in Birkenhead. Recognizing the ironbound future even before railroads started running, in 1829 they launched their initial iron vessel, a sixty-foot barge of sixty tons hammered together in the boiler works. Spectators at the launch wondered about her metal buoyancy; but she drew only fourteen inches of water, "being a less draught," a Liverpool newspaper explained, "than that of a vessel of equal tonnage built of timber. She is of a beautiful mould, and looked exceedingly well."

During the 1830s, John Laird and his younger brother, Macgregor, built a series of notable iron steamships. The *Garry Owen* (1834) of 300 tons was the first such vessel with transverse bulkheads dividing the hull into separate watertight compartments. Suggested by Charles Wye Williams of the City of Dublin Steam Packet Company, the bulkheads were apparently intended mainly to strengthen the hull with cross bracings to make up for the absence of a traditional massive wooden keel. But the additional advantage, of containing a leak or hull rupture to just part of the vessel, was soon appreciated. Bulkheads became a standard feature of Laird ships, and the practice spread to other builders. The *Rainbow*

(1837) by John Laird had six such compartments and a steam engine of 180 horsepower. At 600 tons and 213 feet, she was the longest iron steamship yet. "It is no difficult matter to see," the authoritative *Nautical Magazine* remarked in November 1837, "that these iron steamers are gradually becoming general."

By 1840 iron steamships were no longer dismissed as novelties. Shipbuilders on the Clyde, Mersey, and Thames, and on the northeast coast around Newcastle and Wallsend, were routinely turning them out. They were undeniably stronger, lighter, and more rigid than wooden ships of the same size. Wooden hulls were inherently limited to a maximum length of about 300 feet; beyond that dimension, they could not be made taut enough to avoid flexing leaks and quick deterioration. Iron hulls could be extended to any theoretical length within the current reach of shipyards and builders. With less bulky internal structures, they had more space for passengers and cargo. They were not threatened by dry rot or wood-eating marine organisms. As British iron became cheaper, and timber dearer, they were no more costly to build.

But they also had unique problems. The vaunted rigidity itself seemed oddly dangerous. "There is no elasticity," said Junius Smith in rejecting Macgregor Laird's suggestion for an iron-hulled transatlantic liner; if the hull were twisted, Smith supposed, rivets might pop and the ship would sink. Iron plates and girders were not yet fabricated to known, consistent standards, and metal weaknesses could not be surely detected by an external examination before construction. Iron would rust and corrode, especially in the bow and stern, in highly aerated water, and at key points like the boottopping, where the side of the hull met the water and metal was exposed to constant, destructive cycles of wetness, wind, and drying. Below the surface, iron hulls attracted barnacles and other marine growths which fouled the bottom, adding lumpy deadweight and impeding a smooth, efficient passage through the water. Unlike wooden hulls, iron could not be sheathed in copper to limit such fouling because iron and copper, when immersed in salt water, together caused a galvanic reaction and electrolytic corrosion. And finally, an iron ship in the course of her hammered, vibrating construction took on her own magnetic field, which then might cause serious, unpredictable compass deviations. Coastal and river steamers could usually navigate by visible landmarks. For seagoing steamers far from land, an errant compass reading might prove lethal. None of these iron-hulled problems had adequate solutions in 1840.

The oldest, most enduring puzzle of inventing a steamboat, the propelling device, also remained unresolved. Paddle wheels, hung at the stern

or on the sides of a vessel, were vulnerable to disruption by waves or floating objects. Most of their motion was wasted: out of water for more than
half the cycle, spinning pointlessly through the air, then pushing down or
up once immersed instead of performing useful lateral work through the
water. Attempts to improve the paddle wheel's wasteful inefficiencies, like
Joshua Field's cycloidal wheels for the *Great Western,* had not succeeded.
Paddles survived as the universal harnessing mechanism for steam on
water only because nobody had yet come up with anything better.

Screws in a stupefying variety of forms had long been applied to other
tasks. Ever since the ancient Greek inventor Archimedes, an extended
screw within a cylinder had raised water for irrigation or to bail a boat.
Familiar windmills and smokejacks (small windmills within chimneys,
turned by the rising smoke and heat) drew from the same basic mechanism in different mediums. In the same way that paddle wheels had been
logically derived from mill waterwheels, the essential principle of a screw
was already widely appreciated. By the time of the Industrial Revolution,
no one had to invent it. From the mid-1700s down to 1836, at least two
dozen unconnected tinkerers in France, Britain, and the United States
came up with their own versions of a screw propeller for a steamboat. A
few of these devices were tried out and seemed to work, after a fashion, but
none functioned well enough to provoke further development.

The screw that finally led somewhere, in Britain, came from a most
unlikely source: not a shipbuilder, or a naval architect, or a steam engine
builder, or an iron engineer, or a boilermaker. Francis P. Smith was just a
grazing farmer in Hendon, Middlesex, northwest of London. At a time of
steady professionalization among mechanical engineers and ship men, he
represented a throwback to the old intuitive, untrained lone inventor,
working in utterly creative isolation because he had no idea what others
in the same field had already done. But Smith was quite persistent, unlike
most of his predecessors, and he was lucky.

Francis Pettit Smith was born in February 1808 in the town of Hythe,
on the southeastern coast of England. His father was the local postmaster.
From boyhood, Smith liked to build model boats, trying out different
modes of propulsion. (When he was ten years old, yet another fruitless
marine screw experiment was essayed, in Hythe, by a Captain Duvernet
of the British Staff Corps.) After brief schooling by a minister in Ashford,
Smith found his livelihood in tending livestock, eventually settling in
Hendon. In 1834 he ran a small boat powered by a screw and spring across
one of the horse ponds on his farm. Its success led to Smith's first patent,
in May 1836, for a steam vessel propelled by a screw at the stern.

Smith and his patent attracted supporters: a banker, a merchant, an admiral, and a versatile engineer named Thomas Pilgrim. Contributing resources and expertise that Smith alone could not command, they built a small thirty-four-foot screw steamer of ten tons and six horsepower. Its propeller, installed slightly ahead of the stern, did not at all resemble the form that such devices later assumed: a long, thin helix that wound around its horizontal axis through two complete turns. It worked well enough, in trials around London and out to sea as far as Dover and Smith's hometown of Hythe. One of these excursions produced a fortunate accident. The propeller hit an object in the water and broke in half; the boat, instead of stopping, at once sped up. Smith therefore fitted a new helix of just a single turn and got still better results, achieving an average speed of eight miles an hour in a voyage of four hundred miles.

The lords of the Admiralty, traditionally cautious about technical changes, requested a demonstration in March 1838. Intrigued but not convinced, they then asked to see a screw applied to a larger vessel of at least 200 tons. Smith and his associates gave the boat contract to Henry Wimshurst, a Thames shipbuilder. The *Archimedes* was launched at Wimshurst's yard in the fall of 1838 and fitted with an engine of ninety horsepower. In her initial trials the following spring, she ran down to Portsmouth at higher speeds than expected, of up to ten knots, but with excessive vibration at the stern and loud, clattering noise from her machinery. Coming up the Thames to London from Portsmouth, Smith—perhaps too aware of the impression he was making on spectators along the river—ordered extra steam. The boiler lacked a steam gauge, and its safety valve had been disabled. The boiler exploded, scalding two men and killing the second engineer. But the accident could be blamed just on Smith's inexperience with steam at sea, not on his propeller itself.

The *Archimedes* was laid up for five months of repairs and refitting. At Wimshurst's suggestion (so he later claimed), a new propeller was installed of just two distinct blades, each winding through only half a turn. This latest reduction in Smith's original concept of a continuous, elongated helix gave the *Archimedes* more speed with less vibration. In tests by the Admiralty in the spring of 1840, she outran sailing mail packets at Dover. She then embarked on a triumphal publicity tour of all the principal ports of Great Britain: a steamship that somehow glided along without sails or any visible sign of mechanical propulsion except her smokestack. "Everywhere the vessel became an object of wonder and admiration," noted one former skeptic, the mechanical engineer and writer John Bourne. "It was impossible to resist facts such as the performance of the Archimedes afforded. Ancient opinions, in many cases negligently taken up, had to be modified or abandoned." Isambard Brunel, never a slave to ancient opinions, took notice.

The Great Western Steam Ship Company was then building a second Atlantic steamer, to join its pioneering *Great Western*. As first planned within weeks of the *Great Western*'s maiden voyage in the spring of 1838, the new ship would break no technological ground: just another wooden paddle wheeler, but faster and larger than her predecessor. With two ships, the Great Western company could become a true transatlantic line, crossing the ocean in both directions at the same time on a more frequent sailing schedule, and becoming an assumed habit for travelers. The driving philosophy was ships as enterprise, not as engineering; commerce above invention. Over the next few years, these initial intentions were turned inside out to produce the *Great Britain*. At a time of rapid, sweeping technological changes for ocean steamships, the Bristol company's planning and building process took many unforeseen twists and whizzed off on its own Brunelian course.

The initial drawings and lumber for a wooden vessel were on hand when, in the fall of 1838, John Laird's channel packet *Rainbow* stopped at Bristol. Only a year old, the largest iron steamer yet built, she presented a powerful argument for the advantages of metal hulls. Brunel sent his steamship associates Christopher Claxton and William Patterson on an investigating voyage to Antwerp and back on the *Rainbow*. They returned as converts; their report, revised by Brunel, persuaded the Great Western directors to switch to iron.

The building committee for the new ship, as for the *Great Western*, again consisted of Brunel, Claxton, Patterson, and Thomas Guppy. But the balance of power within the committee shifted toward Brunel. He turned his chronically overtaxed attention to ship questions as never before, and the *Great Britain* became truly his baby in ways that the *Great Western* had not. He at once asked John Grantham, an authority on iron shipbuilding, for advice about who made the best iron plates and braces for marine purposes, and drew up painstaking details for the engine and boilers—all to produce a vessel ready for sea by the spring of 1841. "If these points are well considered," he confided to Claxton in November 1838, "we will secure the superiority of the Bristol line for the next five years at least."

The hull designs drawn up by Patterson kept being revised. Iron fabrication allowed for a much bigger ship; the material drove its own internal logic, and the process began to redefine itself. The *Great Britain* grew ever greater with each new set of plans. By the fifth take, she had reached 3,400 tons—1,000 tons larger than the biggest Atlantic steamship yet planned, the doomed *President* still under construction by Junius Smith and Macgregor Laird. In July 1839, workmen in a specially built dock in Bristol harbor started to bend, shape, hammer, and rivet the hull of the *Great Britain*. "It is most important to the interests of the proprietors that she should be at sea in the early part of 1841," the company assured its stockholders. "Your Directors . . . have warmly to acknowledge the aid afforded them by the science of Mr. BRUNEL; nothing indeed can exceed the obligation under which the Company is indebted to that gentleman for assistance, which, at every spare moment and at any personal inconvenience to himself, has been at the service of your Directors."

In the spring of 1840 another revolutionary steamship pulled into Bristol: Francis Smith's *Archimedes,* on one of her promotional stops. Brunel had been trying to find a better propelling device, looking into a new type of paddle, and considering experiments to squeeze more speed with less coal consumption from a standard wheel. Nothing seemed of real potential until the *Archimedes* hove into unpaddled view. Alerted by Thomas Guppy, Brunel was intrigued. ("I have long since learnt," he said a few years later, "to be prepared to be convinced of any thing.") Claxton and Patterson resisted the screw, still so unproven, for such a major project as the *Great Britain*. Francis Smith, sensing a significant new customer, lent the *Archimedes* to the Bristol group for a series of trials. Construction of the *Great Britain* was halted pending the outcome. Brunel already knew the conclusion he favored. "The result if as I expect satisfactory would be of great importance to you," he wrote a Smith associate. "The cautious must have the positive proof."

Armed with the experimental results he wanted, Brunel wrote a long report for the Great Western directors. Taking up twenty-three pages of a letterbook, the report shows Brunel at the height of his persuasive powers. His tone was initially impartial, as though he were simply applying one of the finest engineering minds of the time to an interesting scientific question. He considered the problem of "slip," of how much energy a screw or paddle wasted in moving a vessel through water, and described tests of resistance and friction on a hull. On these points screws and paddles were judged comparable, with slight advantages for screws. In apparent candor, Brunel noted that the *Archimedes* was actually a poor model of a screw-powered vessel: limited by a faulty engine and gearing, the screw surface already roughened and made less efficient by corrosion, the hull badly designed, and the ship generally "uninhabitable" for passengers because of noise and vibration.

With proper design and engineering, though, the screw offered compelling advantages. It would save weight, especially at the top of the vessel, where high paddle wheels and their machinery raised a vessel's center of gravity and made her less stable in heavy seas. It allowed for a simpler, more compact ship structure, with more space below for cargo and passengers. By eliminating the bulky paddle boxes, it offered less resistance to head winds and waves. When a ship rolled from side to side, a screw would remain submerged and moving the ship forward, unlike side paddle wheels, which could roll out of the water, alternately racing and overloading the engine, harming machinery. Without paddle boxes, a ship would present less beam for tight maneuvers in rivers and harbors. Paddles might start a voyage too deeply immersed under a full load of coal, then rise too far out of the water as fuel was burned; a screw stayed efficiently under water regardless. Screw machinery cost less initially, and—because it was less vulnerable to passing shocks—would be cheaper and less trouble to maintain. Brunel marched irrefutably to his conclusion: "The wisdom I may almost say the necessity of our adopting the improvement."

In commercial terms, what the Great Western company needed in that fall of 1840 was a second ship, as soon as possible, to compete with the newly established Cunard Line. It most emphatically did not need any more delays or major revisions—or risky new gadgets—in the building of the *Great Britain*. Only an engineer of Brunel's reputation and rhetorical skills could have talked the company directors into switching, so far along in the project, to screw propulsion. The half-finished paddle engine was abandoned. After nine months of suspended activity, building resumed in the spring of 1841—by which time the ship was supposed to have been finished. The half-completed hull was redesigned for the screw machinery. Brunel knew his personal prestige was now on the line. "If all goes well," he wrote Guppy, his fellow advocate for the screw, "we shall all gain credit, but quod scriptum est manet [what has been written will remain]; if the result disappoint anybody, my written report will be remembered by everybody, and I shall have to bear the storm. . . . I feel more anxious about this than about most things I have had to do with." Yet he retained the faith of a pure engineer: "All mechanical difficulties of construction must give way, must in fact be lost sight of in determining the most perfect form."

Guppy, a Bristol resident, took charge of the project as its superintending engineer. Brunel, from his office in London and from other points where his varied projects took him, asserted his ultimate authority. His letterbooks from the early 1840s are packed with queries, opinions, and directives about the *Great Britain*. Any detail whatever might catch his attention; but he focused especially on the interior structure of the iron hull, the engine, and the screw, the three most innovative and challenging aspects of the ship. He conceived the hull as an iron railroad bridge, a structural problem he knew well from his years as engineer to the Great Western Railway. The crucial stresses, he judged, would hit the hull mainly lengthwise from bow to stern. In nautical parlance, which he did not use, he wished to prevent "hogging" (the bow and stern drooping when a wave lifted the vessel at midship) and "sagging" (the midship drooping when the ship was raised at her ends). The *Great Britain's* extraordinary length of 322 feet made her hull uniquely vulnerable to such strains. She was generously trussed and braced, with ten continuous longitudinal girders along her double bottom, heavy keel plates, five watertight compartments, and lapped, double-riveted seams in the hull plates. Taking full advantage of the novel properties of iron shipbuilding, Brunel made his steamship massively, even redundantly, strong.

The engine, of an unprecedented size and design, was built by Maudslay, Sons and Field in London. The ancestral ties between the Brunel family and the Maudslay firm did not inhibit Isambard from instructing Joshua Field about staying alert to the newest developments. A screw engine, unlike that for a paddle ship, had to sit low in the hull to meet the propeller shaft, and the rotating motion was turned laterally across the hull instead of bow-to-stern. "Remember when I urged you some years ago," Brunel reminded Field, "to consider the importance of a compact engine instead of the beam engine—you were rather disposed to think it more prudent to keep to your old form. You have thought better of it since and invented a very beautiful arrangement—now keep pace with these improving times and don't grudge a little trouble to obtain what will be sought after." Brunel provided the basic design, borrowed from an old patent of his father's: two pairs of cylinders inclined upward at sixty-degree angles to each other, producing 1,000 horsepower and driving an overhead crankshaft. "These are moving days," Brunel again warned Field. "Do not run the risk of quarrelling with screws."

The *Great Britain's* prospects for commercial success, for speed and reliability across the Atlantic, came down to that screw. The device was so new, still unproven in a large ship and on the deep ocean. Nobody knew the best form for a screw: how many blades, the most efficient size and shape, how to fabricate it, the right metal to use, how to attach it to the shaft, or how to run the shaft into the stern in a manner that was somehow both leakproof and friction-free. In designing the *Great Britain,* Brunel spent more time on her screw than on all other aspects of the ship together.

In Bristol, the building committee ran tests on eight kinds of screws. The Admiralty in these years was also conducting its own screw experiments. (Screws especially interested the British navy because the vulnerable propelling machinery could be buried below water, protected from gunfire.) Brunel served as a very active consultant on the Admiralty tests, incidentally applying their findings to his own ship. The Bristol trials achieved their top speeds with a screw designed by Francis Smith, of four wide blades that tapered down to narrow attachments at the shaft. "We must stick to four arms," Brunel informed Guppy, "—altho I feel pretty confident that with our pitch Six would be the right number. . . . One half the entire disc may be <u>unn</u>ecessary but a little more can do <u>no harm</u>." The four-bladed Smith screw was announced as the final choice for the *Great Britain* but then was withdrawn in favor of a screw with six narrow blades and much more open space between them, as Brunel wanted. The great engineer still prevailed.

Building the *Great Britain* was a typical Brunel project, riddled with endless delays, revisions, and spiraling costs. As always, he sought "the perfect form," expenses be damned. At the same time, competition from the Cunard Line was cutting into the Great Western company's steamship profits. This diverging curve—of rising costs and declining revenues—dictated an even slower building pace for the *Great Britain,* with work suspended for months at a time. The company's wearied stockholders, annually reassured that the ship's construction was moving forward, that she would soon be ready for sea, understandably became impatient. In the fall of 1842 they were told that the ship would cost not £76,000, as previously divulged, but over £100,000; and she still had not even been launched. One battered stockholder declared himself "sick of the continued demands, and want of money," and of "the system of making, altering, and re-making, which has been so long practiced to our cost, and of advantage to none but our salaried servants." This man had a draconian solution: finish the damn boat, sell her, and scuttle the whole company! "Judging from past experience, I dread to anticipate, that the more we spend the more we shall be required to supply."

So much was at stake. The *Great Britain* was the largest ship yet built, the first iron steamer bound for the North Atlantic, and the first propelled by a screw. Only one of these distinctions would have, by itself, made her notable. Probably no ship had ever received so much attention before even hitting the water. "It must be granted," declared the *London Times,* "the experiment is of vast importance in a national point of view." The *Mechanics' Magazine* devoted an article of sixteen pages to minute descriptions of her construction and machinery. When she was finally launched, in July 1843, Prince Albert came out from London on the Great Western Railway to attend the ceremony. (Brunel supervised the locomotive.) Albert spent almost two hours inspecting the ship, then sat down to a lavish, hang-the-expenses banquet for 520 people.

Her fitting out predictably took longer than expected. At last, in the summer of 1845, the *Great Britain* was ready for her maiden voyage—seven hard years after the Great Western company had decided to build its second steamship. Too large for the Bristol channel to the sea, she departed from Liverpool. Passengers approaching her on the dock beheld a vessel multiply unlike anything ever seen before. Their initial impression was of overpowering size: more than 50 percent longer and three times roomier than the first Cunard liners of 1840. The ship carried six masts, for a full schooner rigging, and a single black funnel at midship, slightly raked. The hull was a looming black hulk relieved by a narrow, encircling band of black-dotted white midway between the upper deck and the water line. She looked her most handsome when seen from the side, where the great length could be appreciated; viewed head-on, the outward bulging of her sides from midship aft—needed to make room for the engine and propeller machinery—left her looking rotund and dumpy.

The entire ship, except for the upper decks and the passenger furnishings, was made of iron. Stronger but less bulky than wood, the metal fittings made the *Great Britain* appear strangely fragile, each part seemingly less substantial than its wooden equivalent would have been. Even the ropes in the rigging were made from iron wire instead of hemp (an innovation borrowed from American sailing packets); the thinner wire rigging was expected to cause less air resistance when the ship was beating to windward. Around the edge of the upper deck, slender iron rails and netting replaced the traditional solid wooden bulwarks. The iron masts looked less sturdy than traditional pine versions. One visitor to the ship just before she sailed was startled by "the impressions of fragility and lightness" and "the light, rakish appearance" when seen up close—such an unexpected contrast to the massive exterior viewed from a distance.

At the fourth mast, stairs descended from the upper deck to the promenade deck, with passenger cabins fore and aft, and then down to the dining saloon. The furnishings were plain and inexpensive, no doubt a reflection of the Great Western company's declining fortunes. ("There is little by way of ornament," noted the *Illustrated London News*.) Officers and seamen bunked in their traditional place, the forecastle at the bow. Engineers, coal stokers, and passengers' servants were housed near midship, along with the lavatories and twenty-six water closets for up to 360 passengers. The boilers and engine at midship stretched from the bottom of the hold all the way to the upper deck. The engine's crankshaft was linked by four enormous chains to the propeller shaft; gearing raised the revolutions, by a ratio of 3 to 1, up to the higher speeds required for screw propulsion. The propeller shaft, forged by the Mersey Iron Works in Liverpool, was the largest single piece of the ship, 130 feet long and weighing thirty-six tons. At sea, running hard, the spinning shaft was cooled by continuous infusions of water.

The *Great Britain* sailed for New York on July 26, 1845, carrying forty-five passengers and an impossible load of expectations. The ship's many novel aspects focused attention on her performance. She made the first crossing in fourteen days, twenty-one hours, at an average speed of 9.25 knots. This was fully a knot and a half slower than the current record westward passage by the Cunarder *Cambria,* but maiden voyages with new machinery were not expected to set speed records. She came home on the faster eastward course in thirteen and a half days, again in quite unremarkable time. The screw, it was reported, produced no perceptible vibration at the bow, "a mere tremulous motion" at midship, and "far less" motion at the stern than on a typical paddle ship. The chains and gearing were judged much quieter than on the *Archimedes.* But the ship rolled excessively, especially in a calm, when the sails were not stiffened by wind to help stabilize her. Her bulging sides at midship, and the absence of paddle boxes acting as outriggers, probably made the rolling worse. On an ocean as stormy and choppy as the North Atlantic, it was a serious defect—in particular for the comfort of passengers fighting seasickness.

Back in Liverpool, in an effort to improve her speed, two additional inches of iron were riveted to each blade of the propeller. The *Great Britain* left for New York on September 27 with 104 passengers. Five days out, she ran into a heavy squall and lost her foremast. Later three of the six arms of the propeller broke off, all on one side, leaving the screw hopelessly unbalanced. Passengers again complained of "intolerable" rolling. The ship made it to New York in eighteen days. After repairs, she carried only twenty-three passengers to England. The propeller again fell apart, piece by piece, until it was down to just nubs of two arms. She reached Liverpool under sail after twenty days and was laid up for the winter. Her first season on the Atlantic, of only two round-trips, had been a technical and financial disaster.

The company had no choice but to sink yet more money in her. The *Great Britain* was installed in the queen's graving dock in Liverpool for extensive renovations. The third mast was removed, perhaps because its iron was suspected of skewing compass readings on the adjacent bridge. To keep the ship from rolling so badly, two bilge keels—flanges of metal 110 feet long and projecting outward 2 feet—were attached to the bottom of the hull. Joshua Field oversaw improvements in the engine and boilers to obtain more power and speed. The iron rigging, which had proved balky, was replaced with conventional hemp ropes, and the ship was rerigged to improve her handling and steering. ("Decidedly more in accordance with nautical notions," observed the *Manchester Guardian*.) The fragile propeller was thrown out and replaced by a four-bladed screw, a single piece of cast iron without the inherent weaknesses of welding and rivets. So many changes after just two voyages amounted to a confession of grave design flaws in the original ship. Public discussion of the ship, once so admiring, began to turn. The *Great Britain*, the formerly enthusiastic *Mechanics' Magazine* now declared, had succumbed to "a leviathanism which was wholly uncalled for," and manifestly "beyond the reach of any engineering skill possessed by those who planned and built her."

Her second season offered a shot at redemption in 1846. The first two round-trips, though hobbled by more mechanical breakdowns, came off in reasonable times, but necessary repairs and unexpected wear to one of the chain drums kept her from maintaining a frequent sailing schedule. She did not leave on her third trip to New York of the season until September 22. Crossing the Irish Sea, Captain James Hosken—formerly of the *Great Western*—committed a series of truly inexplicable navigation errors and ran her hard aground at Dundrum Bay, on the northeast coast of Ireland. Everybody aboard was landed safely. The ship's iron hull and masts made faulty compass readings a plausible explanation, but Hosken insisted the compass was tested afterward and found accurate. The *Great Britain* remained grounded, buffeted by waves and weather, with winter closing in.

In December Brunel went to see her. His report, though written in carefully contained language, seethed with barely suppressed emotions. "I have returned from Dundrum with very mixed feelings of satisfaction and pain," he led off, "almost amounting to anger, with whom I don't know." The ship was only slightly damaged, could easily be recovered, yet lay unprotected on rocks. "It is beautiful to look at," Brunel reported. "It is positively cruel; it would be like talking away the character of a young woman without any grounds whatever. . . . I could hardly help feeling as if her own parents and guardians meant her to die there." Brunel had invested time and reputation so prodigally in this revolutionary ship. His concern now, as he watched her lying there wounded and vulnerable, was paternally outraged but helpless.

Under his directions, she was protected for the winter. The ship's very survival in such circumstances proved the iron strength of her construction. To get her back in the water, though—on that unsettled coast, far from the needed heavy machinery—took months of pulling and hauling. The biggest ship in the world amounted to quite substantial deadweight. Finally, in August 1847, she was floated free. By then the company was already ruined. The *Great Britain* had cost £117,000 to build, plus £53,000 for her own buildings, machinery, and dock in Bristol, plus £34,000 to recover her from Dundrum Bay: a total outlay, even without including repairs and refittings, of over £200,000. In 1849 the last vestiges of the company sold her, after only eight trips across the Atlantic, for just £25,000. Invention over commerce, all for the perfect form; yet the form, it turned out, was quite imperfect.

*F*or a long time after the *Great Britain* grounding, Brunel gave hardly any thought to steamships. Years of heroic effort, of stubborn persistence against mounting odds and public hootings, had come so quickly to so little. The *Great Britain*, refitted again but to more modest expectations, went on to a long, honorable career on other ocean routes, working mostly under sail. But she had failed as a steamer on the North Atlantic, the most visible and lucrative of all sea passages. In the wake of that embarrassing defeat, Brunel for years stayed away from any more marine enterprises. He was so far removed that at one point he had to ask William Patterson for the dimensions of his own ships, the *Great Western* and the *Great Britain*. But then he conceived the *Great Eastern*. She was, for the third time in his engineering life, merely the biggest, most audacious steamship anybody had ever seen. The difference on this occasion was that Brunel built her with a partner of comparable force, ego, and talent: the eminent naval architect and shipbuilder John Scott Russell. The *Great Eastern*—grand leap, white whale, and Waterloo—

emerged from the fecund, testy, fragile relations between these two most willful men.

Alike in too many ways, they came from quite different backgrounds. Russell was a Scotsman, raised between classical scholarship and the calloused shipbuilding traditions of the Clyde. He was born near Edinburgh in 1808, the eldest son of a minister who expected his boy to follow him into the church. An early knack for tools and mechanics sent the son on his own way, a lifelong habit. Like Brunel, he performed equally well in classroom and workshop. At sixteen he graduated from the University of Edinburgh in a traditional curriculum that as yet included no engineering classes. After school and during vacations, though, he worked for mechanics and millwrights to learn practical skills and methods. For a few years Russell taught science and math at various schools around Edinburgh; his popular lectures were clear and notably forceful. At just twenty-four he was called back to his college as an acting professor of physics.

The burgeoning power of steam, the wonder of the age, pulled him along. He lectured and conducted experiments on steam engines and boilers. "Steam was made a water-pumper," Russell wrote in 1832, "but has now become a miner and a mariner, a coal-heaver and a cotton-spinner, a cook and a coffee-mill, a universal agent and jack-of-all-trades." (Its application to railways, he noted, still remained unproven.) He invented a steam road carriage that ran the ten miles between Glasgow and Paisley a few times; a boiler explosion and other breakdowns forced him to return under horse power. A Scottish canal company fortunately took him from land to water by asking him to look into the problem of steam navigation on canals. This job—another echo, three decades later, of William Symington and the *Charlotte Dundas*—led Russell to his most significant early achievement.

With no prior experience in steamboats, he could approach the problem from fresh angles. Instead of the steam engine or the propelling device, he focused on the vessel itself. He set out to discover the hull form of least resistance: the shape that would move the particles of water out of the way, but no farther than necessary, and leave them behind and undisturbed as quickly as possible; thus the boat would move through the water at maximum speed and efficiency. To measure a hull's resistance, Russell decided to study the waves generated by a hull in motion. The waves, examined with systematic care, would provide the visible evidence of how a given hull was doing.

Russell built a long, thin trough, a foot wide, a foot deep, and filled with six inches of water. At one end he added a quantity of water, all at once, from a point above the trough. "The little released heap of water acquired life, and commenced a performance of its own, presenting one of the most beautiful phenomena that I ever saw," he recalled a quarter century later. "The first day I saw it was the happiest day of my life." The added water formed a wave that quickly ran the length of the trough, leaving the original water quiet and undisturbed after its passage. On horseback Russell then tracked this phenomenon, later called a carrier wave or solitary wave of translation, along miles of the canal, noting its consistent speed and size. It was, he concluded, identical to the wave generated by the bow of a vessel as it sliced through the water.

Still limiting his studies to rivers and canals, Russell identified three other kinds of waves: a simple ripple; oscillatory, as generated by a stone dropped into water; and waves with a broken top, called surges. The surges and the carrier waves let Russell test the effectiveness of various hulls. The prevailing cod's-head designs of the time—fairly blunt, rounded bows sweeping back from the cutwater in either straight lines or convex curves—proved the least efficient, generating the greatest disturbance. The best hull, it seemed, was a sharp entrance followed by a concave (hollow) curve into the widest part of the hull, and then tapering narrowly into the stern. The widest hull section was therefore moved from its traditional place toward the bow back to a new position aft of midship. "By these improvements, the form of the old vessel was pretty nearly reversed," Russell noted, "to the great annoyance of the old school." His findings, presented at meetings of learned societies in Scotland and England, established Russell's reputation before his thirtieth birthday.

Switching from theory to practice, Russell became the manager of Robert Duncan's shipyard at Greenock. Here he built four small iron canal steamboats on "wave-line" principles, then four larger vessels on the same lines. After four years he moved on to a prominent Clyde engine builder. Meantime he published scientific papers on such diverse topics as tides, parallelograms, and acoustics. For the seventh edition of the *Encyclopaedia Britannica,* in 1841, he wrote a long treatise on steam navigation. "Naval architecture is scarcely recognized as a science in England," he declared. "The properties and structure of steam vessels is an enquiry which has hardly as yet systematically commenced." Two of every three vessels built failed to achieve their intended purposes, Russell estimated. The same errors were committed again and again. He included his wave-line among the prevailing theories of hull design, but his conclusions were carefully tentative and subject to revision. "Even the elementary principles of hydrodynamics are yet to be learned, before we can apply them to the ends of steam navigation. What seems the law of a fluid to-day, tomorrow shows to be a plausible fiction, of doubtful verisimilitude."

Russell moved down to London in 1844 and took editing jobs on a railroad magazine and a daily newspaper. What he most wanted to do, though, was build ships. Nothing else so engaged both the experimental and the practical aspects of his exceedingly active mind. With partners he started his own shipyard on the Thames in 1847. In five years he built scores of vessels—but none of any real size or novel significance. Russell's headstrong personality, and the daunting range and depth of his knowledge, did not attract major clients. "If we could only get shipowners to advance their views of naval architecture," he said later. "The difficulty is, that the taste and knowledge of the customer do not come up to the ability and the powers of the manufacturing and professional man. The shipbuilder can do anything, provided only an intelligent man will give judicious orders." His contracts did not match the grand designs in his head. "I am afraid," he later confided to a son of Robert Napier, "in building vessels to sell, you must a little, consult the taste of the buyer." His early shipbuilding career was apparently constrained by his artistic reluctance to bend to the wishes of clients.

Russell's most important work during his first decade in London was for the Royal Society of Arts. This private organization, nearly a century old, awarded annual prizes to encourage developments in manufacturing, commerce, and the arts. In decline and debt when Russell joined, the society was revived and enlarged under his leadership as its secretary. He raised the intellectual level of the papers read at its meetings, added a new prize in industrial design, and launched an ambitious series of yearly national exhibitions. Among the influential new members recruited by Russell was Isambard Brunel. The two men had known each other at least since 1839, when Russell was consulted in the planning of the *Great Britain*. In London they saw each other occasionally at meetings of the Institution of Civil Engineers. Russell made Brunel a vice president of the Royal Society of Arts in 1848.

The Great Exhibition of 1851, the first world's fair, was initially planned and organized by Russell and two associates at the society. After Russell persuaded Prince Albert to join the undertaking, it gained decisive support and momentum. As the planning moved forward, though, Russell managed to alienate virtually all his colleagues in the venture. By turns unstable, inattentive, and petty, he was also accused of insolence and lying. When he lost Albert's confidence, Russell was frozen out of the process. The exhibition, held at the specially built Crystal Palace in Hyde Park, attracted over 100,000 exhibits and 6 million visitors. By any criteria, it was a historic success: British engineering and faith in industrial progress at their midcentury zeniths, on display for the Western world to see. But Russell, because of his prickly ways, was denied the public credit he deserved.

As the 1850s began, the careers of Russell and Brunel had gone through remarkably similar arcs. After showing precocious early promise, both men had stalled as they entered middle age. Russell had not matched the youthful feats of his wave-line studies. His shipbuilding yard had turned out no real breakthroughs, and he was not recognized for his founding role in the Great Exhibition. For Brunel, after the successes of the Great Western Railway and the *Great Western* steamship, the 1840s had brought a string of setbacks: the suspension of construction on his Clifton bridge because of money problems, the defeat of his broad gauge for railroads by an act of Parliament, the collapse of his major attempt to power an "atmospheric" railroad in South Devon by vacuum pressure, and the conspicuous transatlantic failure, grounding, and sale of his *Great Britain.*

The curse of precocious brilliance is that it generates nearly impossible expectations for the rest of a person's life. Artists, writers, and musicians often do their best work when young—and then spend frustrated lifetimes trying to match it. Engineers, building their technical knowledge and professional contacts through long careers, ought to produce their finest work in middle age. Russell and Brunel, middle-aged engineers in the 1850s, were perhaps too anxious for another great achievement, something to balance their recent disappointments and equal their early coups. They might have wanted it too much, to the fatal point of carelessly underestimating risks and of overreaching certain implacable realities.

*A*t some time late in 1851, Brunel started thinking about a really big ship. He had been mulling the idea for years. It began with the problem of coal supply on a long voyage to the East. With the Suez Canal as yet undug, a steamer bound for the British Empire's distant outposts in India and Australia, by way of the southern tip of Africa, had to stop several times en route to refuel. Such coaling wasted time in insignificant ports, and the coal supply itself, which had to be shipped out from British sources, cost four times more than at home. To save that time and money, Brunel determined to build a ship large enough to carry sufficient coal for a round-trip from England to the eastern empire. Hence the name of the projected vessel: the *Great Eastern.*

Having exhausted his supporters in Bristol, Brunel sought technical and financial allies in London and the industrial Midlands. His first recruit was John Scott Russell, who snatched this opportunity finally to make a thumping statement as a shipbuilder. Russell's presence in turn helped give the nascent project technological plausibility. "Every body seems to take to the thing," Brunel wrote to prospective investors. "With respect to the form and construction of the vessel itself nobody can in my opinion bring more scientific and practical knowledge to bear than Mr. Scott Russell." Brunel found his two most useful supporters in Thomas Geach, a banker and iron manufacturer from Birmingham who previously had financed Russell's shipyard on the Thames, and Henry Thomas Hope, a wealthy gentleman and conservative member of Parliament who had joined in the planning for the Great Exhibition. Both Geach and Hope came into the new ship project by way of Russell—further tightening his, and their, commitment to the venture.

Brunel and Russell worked out the initial dimensions between themselves. Brunel, as usual, recognized no limits or precedents. "The wisest and safest plan in striking out a new path," he told Russell, "is to go straight in the direction which we believe to be right disregarding the small impediments which may appear to be in our way . . . without yielding in the least to any prejudices now existing." As an example, Brunel noted their disagreeing about the shape of the hull, "in which my fresh ideas had the advantage even of your much greater knowledge—hampered by a little preconceived idea." (Russell presumably thought of his accumulated ship expertise as more than mere prejudices and preconceived ideas.) Brunel kept pushing Russell to make the ship longer and wider, with more room for coal, cargo, and passengers. Russell would refigure his calculations, come back with another proposal, and then Brunel would ask for yet more. "How I wish I could venture to add 100 ft. more to length," he urged, knowing he might; "it would make everything perfectly easy. I suppose I shall do it before long."

The final dimensions were breathtaking: a ship 690 feet long (twice as long as any ship ever built) and of 22,500 tons (*six times* larger than any previous ship). All the other specifications of the *Great Eastern* also exceeded anything ever seen. She would carry up to 4,000 passengers (or 10,000 troops in a wartime pinch) and 18,000 tons of coal and cargo. For moving all that bulk, she would have six masts and 65,000 square yards of canvas, and paddle wheels, and a screw propeller. No other ship had ever combined paddles and a screw. The engine for the paddles would churn out 1,000 horsepower, and the screw engine—the largest steam engine yet built, whether for land or sea—would add another 1,600 horsepower. To provide enough steam, seventy-two furnaces would belch coal smoke through five funnels.

It was boggling, preposterous, hardly believable. And, in part for those reasons, a very hard sell. Brunel, Russell, and Geach spent eighteen months raising enough money to begin actual work on the ship. To a greater degree than in any of his previous projects, Brunel was leading the promotional and business aspects of the *Great Eastern*. As he ruefully acknowledged, "Everybody understands the 'proposed large steamer' to mean 'Brunel's absurd big ship.'" He wrote many letters in the unaccustomed, expansive argot of capitalist enticement, spinning out grand visions of success and profits, entire fleets of giant ships plying the eastern route to riches—and, to broaden his appeal, on the lush North Atlantic route as well. "Of course I am not and never pretended to be a capitalist or a speculator," Brunel noted, "and no man knew that better than my good friend Geach." At a critical point in December 1853, when too many of the promised subscribers had not paid up, Brunel and Geach wrote personal checks for the balance due on the stock offering. They managed to pull together the first £120,000 to start building.

Once the size was determined, the design details emerged from a fractious dance between Brunel and Russell. Each man was so confident of his superior knowledge, so sure of what he wanted in a project that should crown his career. Neither was in the habit of deferring to anybody. Brunel was called the "engineer," Russell the "contractor," leaving the delicate question of ultimate authority implicit but not defined. "I have slaved to get the specifications ready," the engineer wrote to the contractor, "—don't you delay the contracts for want of drawings—Besides the first attempt is sure not to satisfy me." Brunel contributed two key features, the double-skinned hull and the cellular construction of the upper decks, a concept borrowed from the box girders recently used in the *Britannia* railroad bridge by Robert Stephenson and William Fairbairn. Like other ships previously built by Russell, the *Great Eastern* had a flat bottom, longitudinal framing with no keel, and transverse bulkheads for strength and safety. Her bow was a straight vertical stem, as proven efficient by the Collins Line ships, sweeping back into Russell's usual hollow, wave-line hull shape. The total cost was initially estimated at £492,750.

When construction began in the spring of 1854, Russell assumed more direct authority simply because he was there every day, in charge of all details. The ship was supposed to be finished in two years, a tight schedule for such an enormous, unique vessel. Brunel and Russell quickly fell into squabbling over the pace of the work. The *Great Eastern* directors tried to install a buffer between them, a resident engineer who would report only to the company, but Brunel refused to accept the arrangement. The colliding egos bouncing around the ship blew up irreparably in November. The immediate cause was trivial: an article in a London newspaper that slighted Brunel's conceptual role in favor of Russell's. Brunel, wounded and angry, smelled a cabal between Russell and the company to deny him due credit. "I am by no means indifferent to a statement," he made clear, "which would lead the public, and perhaps by degrees our own friends, to forget the origin of our present scheme." No one should be telling a newspaper writer, "while I have the whole responsibility of its success resting on my shoulders . . . that I am a mere passive approver of the project of another which in fact originated solely with me and has been worked out by me at great cost of labor and thought devoted to it now for not less than three years." Russell tried to repair the breach—"you are the father of the great ship and not I"—but Brunel would not excuse it. They now had to complete this project, of massive ambition and difficulty, without trusting each other.

*T*he Cunard Line, which was then losing customers to the Collins Line and facing future competition from the monster rising slowly in the Russell shipyard at Millwall, built its own version of a giant ship. The *Persia* was in some ways another exercise in Cunard technological caution. She still carried paddle wheels, instead of a screw, and was powered by another side-lever engine, while other Atlantic lines were switching to more modern types. But she was also the first iron Cunarder on the Atlantic. Sam Cunard had toyed with building an iron steamer as early as 1846, during the *Great Britain*'s second season. The Admiralty, which had to approve any plans because of Cunard's mail contract, would not yet allow iron construction, believing it offered less resistance to enemy shot than hardwood. By the mid-1850s, though, the Admiralty had finally come around to metal hulls.

At 3,600 tons and 376 feet long, the *Persia* was much larger than any previous Cunarder, with 50 percent more internal capacity than the *Arabia* of 1853. No other Cunard ship had ever made such a daring leap forward in size. She was in fact the biggest ship in the world (until the

Great Eastern was launched). The hull and machinery both came from Robert Napier's new yard at Govan, two miles down the Clyde from Glasgow. Robert Mansel, the resident naval architect for Napier, designed an exceptionally beautiful ship, graceful and lightsome despite her bulk. Her flaring clipper bow may not have moved through the water as cleanly as a straight stem, but it looked undeniably prettier, soaring and dramatic. The *Persia* had seven watertight bulkheads and was especially strong at the bow, buttressed by a new system of diagonal bracings devised by Napier. "The effect is that," it was explained, "in the case of collision with other ships, or with rocks or icebergs, the strain would fall upon the very strongest material within the structure, and the *Persia* would have a good chance of safety."

On her maiden voyage, in January 1856, five days out from Liverpool she reached the graveyard of the North Atlantic and, sure enough, ran straight into an iceberg at eleven knots—and did not sink. The impact stove a gaping hole in the bow, popped rivets for sixteen feet on the starboard side, and twisted the rims of the paddle wheels. The first compartment at once filled with water; but the inrushing sea was contained there and went no farther. Slowly, carefully, down at the bow, the *Persia* made her crippled way to New York. She arrived days late, to general relief and celebration. (Cunard luck, again, reinforced by Napier construction.) Her collision-but-survival reinforced the Cunard reputation for safety and attracted more customers from the faltering Collins Line to the demonstrably hardy *Persia*.

She was pretty, sturdy, and fast as well. The *Persia* took the transatlantic speed records back from the Collins Line. Within eight months of her maiden, she lowered the westbound standard to nine days, sixteen hours, sixteen minutes, and the eastbound to eight days, twenty-three hours, nineteen minutes. Both records lasted for the next seven years. To run so fast, she burned too much coal: up to 176 tons a day. Charles Mac Iver, ever the Scotsman, was aghast at such wanton expense. But for the traveling public, indifferent to coal bills, the *Persia* functioned superbly as an advertisement for the Cunard Line. She became the most popular transatlantic steamer of her time, consistently laden with passengers and cargo.

*W*hile the *Persia* was planned, built, and raced across the Atlantic, the construction of the *Great Eastern* stuttered ahead. Brunel's customary revisions and additions caused expensive delays. Money problems became worse with the death of Thomas Geach, the project's principal investor. In these tightening financial straits, Brunel discovered odd discrepancies in Russell's iron supply: the total of the iron already used in the big ship and that still on hand did not equal the supply that had been delivered to the shipyard. The shortfall amounted to some 900 tons. Brunel suspected Russell of stealing iron intended for the *Great Eastern* and applying it to other ships under construction in his yard. "Look the facts in the face," Brunel urged Russell, "and if there is a problem and there is—let us look at it boldly and meet it—like sensible men." Russell shrugged it off; Brunel snapped back: "Do you disregard every exhortation and entreaty I make." Brunel took his case past Russell to the directors but gained no satisfaction.

Work stopped altogether for almost four months in 1856, then resumed and staggered along with Brunel and Russell barely on speaking terms. The engineer fired off raging, impatient directives; the contractor replied in urbane, unruffled terms, masking whatever he might have actually been feeling about the situation. By the fall of 1857, the ship was finally ready for launching. Built parallel to the river because she was judged too long to launch lengthwise into a narrow section of the Thames, the *Great Eastern* was supposed to be eased sideways down to the water, then float away on a high tide. Nobody had ever tried to move so much inert mass, weighing about 11,000 tons, across dry land. Brunel had no real precedents to guide him. Worried about all that weight, and determined to avoid an unfixable accident, he suddenly decided to replace the customary greased wooden ways with iron-shod cradles under the ship running down to the river on lubricated iron rails. This late change in the launching plan, hasty and fatal, still defies rational analysis.

It is painful to witness the decline of a great mind at such a critical moment, precisely when Brunel needed his faculties at their sharpest. All the pressures of building this unprecedented ship, which he insisted on taking on himself alone, were wearing him down. In September 1857, harried by endless problems, he sent an odd, stilted letter to Henry Thomas Hope, the chairman of the company. "I am anxious to allow nothing to disturb me in the one pursuit of success," he wrote, ". . . of not allowing my feelings to be acted upon so as to disturb me" or "risking the slightest collision with anybody." He wished only the completion of the ship "and the relief from a heavy burden." In this addled state of mind, he decided to launch by sliding iron on iron. The friction between the two metal surfaces would bite, just as the spinning wheels of a locomotive on leaving a station gradually bit and seized the rails. Brunel, as a veteran railroad engineer, should have known better. Six exhausting years into the project, he was losing his usual sharp edge.

Russell, according to his later recollection, told the directors that it could not work, but he was overruled. "It has just been settled to try the launch on Tuesday next about midday," he wrote a friend in Scotland, four days before the event, "and if she don't go, it will be tried again next day. I don't think there will be any spectacle." Brunel warned beforehand that the ship might stop halfway down, or not move at all, and then require two or three tides before the launch was complete. Nobody was prepared for what did happen. The ship moved a few feet, farther at the stern than the bow, and was intentionally slowed; the iron bit into iron; the ship became stuck and could not be budged. It took three months of desperate measures, of trying different techniques and more powerful mechanical and hydraulic forces, to get her into the water at last. The launch alone wound up costing £120,000, money that was supposed to go toward finishing the ship. With that final shove, the company collapsed. "We were ruined by a great folly," Russell later wrote, "—the folly of allowing Brunel to try the insane experiment of launching the ship on iron ways."

For almost a year afterward, work stopped and the *Great Eastern* lay idle in the river, gathering rust. New investors formed another company to finish her. Hoping for more immediate returns, they decided to start the ship on the North Atlantic route. Brunel, dying of Bright's disease, took a long vacation to Egypt, then returned in fragile health for the final stages of his terminal project. "I wish you would answer my letter about the flywheels," he wrote to the screw-engine builder in his last letter, bracing a contractor even to the end. "You have never answered me about the stuffing box." Visiting the nearly completed ship in September 1859, he was photographed standing next to a funnel, looking old and worn out at age fifty-three. He suffered a stroke a few hours later and was taken home paralyzed. A few days afterward, the *Great Eastern* was steaming along on a trial run when the forward funnel casing blew up, killing four men. (This explosion recalled the fire on the *Great Western,* just before her maiden voyage, twenty-one years earlier.) Brunel was told about the accident—just another blow delivered by his big ship—and soon faded away and died.

The disasters of her construction flowed seamlessly into the disasters of her oceangoing career. The *Great Eastern* was not just unlucky but cursed. Her first captain was drowned while rowing ashore off

Southampton. Under steam, the ship wagged excessively at the stem and stern, was several knots slower than hoped, and—like the *Great Britain*—rolled too much in even moderately heavy waves. Her two engines, almost obsolete by the time the ship finally put to sea, lacked the power to move her huge mass rapidly. The *Great Eastern* made a few trips to New York but could not compete against more advanced ships that burned less coal and required smaller crews. Ponderous and ungainly, she kept colliding with other vessels. A howling gale on the Atlantic tore loose her paddle wheels and rudder. She ran aground off Long Island, ripping a hole in her bottom eighty-three feet long and nine feet wide, but she was saved by her double skin. She found humble short-term employment in laying several Atlantic telegraph cables; no other ship could carry all the wire needed to cross the ocean. The *Great Eastern* was later kicked around for years, unused and unusable, and was finally turned into scrap iron in 1889. The gargantuan efforts of her building and launch came to practically nothing.

The *Great Eastern* hung over several generations of British ship-builders and naval architects as a highly charged inspiration/admonition. After Brunel, no other individual was ever granted such unlimited power and money to produce an enormous, radically new kind of steamship. Changes would come gradually, in measured increments, as approved by layers of interested parties. A few of the *Great Eastern's* innovations—such as placing the public rooms and passenger cabins at midship, where the vessel's pitching motions were minimized—later were generally adopted. Most of the others—Brunel's cellular construction, Russell's longitudinal bracing, and the doubled propulsion of paddles and screw—were not. In general, contemporary ship men made clear their final judgment of the *Great Eastern* by not imitating her. It was four decades before anyone dared to build a larger steamship. As one noted shipbuilder and designer, William John, said in 1886, the big ship remained "a warning that the highest flights of constructive genius may prove abortive if not strictly subordinated to the practical conditions and commercial requirements of the times."

*B*runel's reputation as a steamship pioneer needs unsentimental scrutiny. Of his three ships, the most successful in both engineering and commercial terms was the first, the *Great Western*—the one for which he took the least influential role, no more significant than the contributions of Christopher Claxton, William Patterson, and Joshua Field.

Brunel assumed greater responsibility for the *Great Britain* and the *Great Eastern*. Though undeniable technical breakthroughs, they were just not very good ships as judged by their sailing qualities and economies. For the *Great Eastern*, John Scott Russell must share the blame in almost equal measure with Brunel. It is significant that after the big ship, Russell—only in his early fifties when she was finished—for the rest of his life built just two more vessels. Among all her other casualties, the *Great Eastern* in effect ended both Brunel's life and Russell's shipbuilding career.

8

Emigration and
the Inman Line

*T*he summer of 1846 in Ireland was even wetter than usual, with steady light rain and mild breezes—perfect growing conditions for the deadly potato fungus *Phytophthora infestans*. It had appeared, nobody knows how or why, a year earlier in Belgium, then spread through northeastern Europe and southern England before reaching the ideal, verdant fields of Ireland. The spores are delivered by wind, rain, and dew, brought to voracious life by the smallest drop of moisture. Entering any part of the potato—leaf, stem, or tuber—the spore both chokes and consumes the plant, eating it alive. *P. infestans* is one versatile, speedy fungus, reproducing bisexually or asexually as opportunities offer, and zipping through the entire disease cycle from penetration to dispersal in as few as five days. An apparently healthy field on Monday may be gone by Friday, too soon for detection or early harvest. A stricken potato leaf develops black spots on a mottling gray background, withers, and dies. The harvest then may yield starved potatoes the size of walnuts or apparently sound tubers, full-size and firm, that are invisibly infested. In winter storage they ferment and rot away in a few weeks. The fungus spends the winter undetected in slightly affected potatoes which, planted next spring, will renew the cycle and so doom harvests for years to come.

*T*he Great Famine in Ireland provoked a new kind of transatlantic traffic, unprecedented in its size, poverty, and desperation. Over the previous half century, the potato's advantages as crop and nutrient had tripled the Irish population to 8.5 million, more than half of whom then relied on potatoes as their sole or staple food. Given this inflated overdependence on a single crop, the successive failed potato harvests from 1846 on brought starvation and disease, death for at least 1 million people—and flight from Ireland by another 2 million survivors. Of these famine emigrants, almost 1.5 million landed in the United States by 1855.

For all but a few of the fleeing Irish, the ocean passage to America was a long, grim ordeal, recalling the worst shipboard conditions before the advent of steam power on the North Atlantic. Most of the emigrants came from central areas of Ireland (South Ulster, East Connaught, and the Leinster midlands), places that were crippled by famine and tenant evictions yet not so devastated as the western counties, where many people were too poor and broken even to get up and leave. The emigrants typically made their way to the port cities of Cork and Dublin and took small steamboats to Liverpool. Crossing the Irish Sea, for up to thirty-six hours they sat out on open decks, crowded beyond reason by five hundred or even seven hundred passengers, with no shelter from storms and high waves. Naive country folk—mostly farmers, laborers, and servants, many speaking no English, accustomed to premodern attitudes and behaviors—in Liverpool they were preyed upon by brokers and runners for the sailing packets. From an emigrant song of the time:

> To Liverpool I'll take my way.
> Amelia, whar' you bound to?
> To Liverpool that Yankee school,
> Across the western ocean.

> Beware these packet-ships, I pray.
> Amelia, whar' you bound to?
> They steal your stores and clothes away,
> Across the western ocean.

Most of the emigrant ships were old, slow, leaky vessels. The passage usually took about six weeks, sometimes longer, through cold, stormy weather against the prevailing winds and currents. Below decks, the emigrants were crammed into bunks of rough planks, with no separation of the sexes and no supervision. Women lacking escorts had little protection from predatory sailors. Closed down for the night or during storms, the cargo holds lacked ventilation; people might sleep with clothes over their heads to try to mask the overwhelming odors of bilge and body. No bathing or laundry facilities were available. Sometimes the famine-weakened emigrants would bring the plagues of cholera and typhus on board. In such close circumstances, as the weeks slowly passed, an epidemic had time to spread through the steerage. On the *Virginius* in 1847, carrying 476 emigrants in nine weeks from Liverpool to Quebec, 158 people died from typhus at sea and over 100 more were sick on arrival. Another emigrant song:

> Before we were ten days at sea, the fever seized our crew
> And falling like the autumn leaves bid life and friends adieu,
> The raging waves sweep o'er their graves amidst the ocean foam
> Their friends may mourn they'll ne'er return to Erin's lovely home
>
> My loving sisters both took ill, and shortly life gave way,
> And O to me 'twas worse than death to throw them in the sea.
> I breathed a prayer to heaven—alas, that we did roam,
> To end our days far far away from Erin's lovely home

The *Washington,* a large sailing ship of 1,600 tons, left Liverpool for New York in the fall of 1850, overcrowded with more than nine hundred emigrants. The undermanned crew pushed the passengers around, cursing and kicking them. The promised water and food supplies of oatmeal, rice, flour, and biscuits were not issued adequately. At times the crew would demand bribes of whiskey or cash for better fare. One passenger tried to hand a letter of protest to the captain; the first mate knocked him down. The ship's doctor charged for medical supplies that were supposed to be free. A dozen children died of dysentery caused, apparently, by the poor food. When the ship reached New York, a reform-minded English passenger, Vere Foster, tried to start legal action against the shipowner, but he got nowhere.

At their worst, the emigrant ships ran aground, or sank in storms, or caught fire, or just vanished. In 1847 the *Exmouth,* a tiny vessel of only 320 tons, was blown by a storm onto a rocky Scottish coast, killing 240 emigrants; the *Carrick* after six weeks at sea hit a shoal and broke up off

Newfoundland, with about 180 deaths; the *Canton* ran onto rocks near Durness, Scotland, leaving no survivors among the three hundred on board. On the *Ocean Monarch* in August 1848, only a few hours out from Liverpool, someone lit an illegal fire below deck; the ship burned to the water line, killing over four hundred emigrants bound for Boston. In 1849 the *Maria,* heading from Limerick to Quebec, ran into an ice field in a storm, could not get free, collided hard with an iceberg, and went down with 109 people. Together these five shipwrecks caused well over 1,200 deaths; and during the first six years of the famine, in all about fifty ships foundered on their way to North America. So it went, year after year, a steady drumbeat of disasters with no evident remedy.

Yet despite all these dangers and difficulties, the annual totals of Irish emigrants to the United States kept rising: 68,000 in 1846, then 118,000, and 151,000, and 180,000, and over 184,000 by 1850. From America, newly prospering recent Irish arrivals—who had just made the ocean trip and so well knew its hardships—still sent home to relatives in Ireland more than a £1 million a year, "passage money," to bring them over as well. This swelling traffic, against the many known perils and burdens of crossing the North Atlantic, by itself suggests just how truly desperate these emigrants were to escape the famine, and to try out what they had heard about the distant dream and reality of America.

Prodded by alarming reports, both the British Parliament and the U.S. Congress imposed well-intentioned regulations on the emigrant ships. In 1848, for example, Parliament decreed that such vessels should maintain certain standards of ventilation, cleanliness, and order, with no open flames or smoking below decks, and marauding sailors banned from passenger quarters. But these rules could seldom be enforced. The government's medical inspector in Liverpool was paid a pound for every one hundred embarking passengers he inspected. Employed by the shipowner or broker, he wanted only to approve quickly as many emigrants as possible. When a ship arrived in America, no matter what had happened during the passage, the relieved emigrants walked onto the dock and disappeared, without the time, money, or social standing to pursue slow legal remedies for violations. The government regulations provided, by inference, a concise summary of the worst problems in the emigrant trade; but as dead letters, they only described intended, not actual, shipboard reforms.

If the emigrants could have afforded the Cunard or Collins steamers, they would have arrived in America in reasonable health in just ten to twelve days. The cost of passage on the emigrant sailing ships ranged from only three pounds, ten shillings up to five pounds—about one-eighth of cabin fares on the crack Cunard and Collins ships. Those steamers carried few passengers in steerage. For all but a tiny fraction of the Irish emigrants, the pricey mail lines offered no reachable alternative to the horrors of sail-powered steerage to America. "You could not possibly, I think, send them by steamer," Sam Cunard declared in 1847; "it is so very expensive."

As it turned out, the most substantial improvement in the emigrant passages came with the invention of a new kind of steamship: not quite as fast as the mail steamers, but as safe and well ordered, and—the crucial point—much cheaper to run than any steam vessel yet on the ocean. With improved engines and boilers, a more efficient propelling device, and perhaps better hull design, coal consumption and other costs could be reduced enough to bring steamship steerage fares down to a level within reach of most emigrants. Steam power might then keep the ship less vulnerable to being blown defenseless onto rocks or icebergs, and cutting the trip from six weeks to two would limit the danger of typhus and cholera epidemics. The emigrants would gain a markedly improved chance of reaching America alive. Much more effective than any humanitarian reformers like Vere Foster, or those easily flouted government regulations, this technological solution became available while (and because) the potato famine was still emptying Ireland.

*D*avid Tod, fifty-five years old in 1850, was another Clyde shipbuilder in the Napier tradition. After more than three decades of quite varied experience with steam on water, he had the technical skills to create a different type of steamship. In addition, his particular life experience gave him some understanding of the specific needs of the famine emigrants. Over the course of his own lifetime, he had known the entire revolutionary cycle of the nineteenth century, from muscle-powered farm work to waterpower to steam and iron, from the country to the city, and from static premodernity to headlong modern progress. All these vaulting changes found oblique or direct expression in the breakthrough ocean steamship he launched in 1850, the *City of Glasgow.*

Tod was born toward the end of the eighteenth century in the farming village of Colen, Scone Parish, Perthshire, in the east midlands of Scotland. His father worked modestly as a farm servant, with no land of his own. Scone lay on the River Tay, the longest river in Scotland, across from the town of Perth. Along that section of the Tay, rapids and cataracts provided waterpower for a spreading network of flour and textile mills. In his teens, David Tod was apprenticed out to a country millwright, showing his skills in tools and mechanics. Steamboats appeared on the Tay in about 1815, filling the air with unaccustomed noise and smoke, but drawing the young Tod toward the new world beyond Colen. He moved down to Glasgow, the expanding center of British steam navigation, to find work. There his second job in the city sealed his future course: Tod came under the demanding tutelage of David Napier at his Camlachie workshop.

Napier was then in the early phase of the most inventive, productive decades of his career, marked by fearless—at times reckless—innovations in hull, engine, and boiler design. His *Rob Roy* of 1818 had sharper, leaner lines than other Irish Sea steamers of the day. In one of his first assignments for Napier, Tod served as her chief engineer on mail runs between Greenock and Belfast, thereby learning about steam at sea from the vital working perspectives of the engine room and stokehole. At the Napier workshop, Tod met a fellow apprentice, John Macgregor, about five years younger. The son of a clockmaker from Stirlingshire, north of Glasgow, Macgregor also spent time at sea on a Napier boat. In the early 1820s he sailed as chief engineer on the *Belfast,* twice the size of the *Rob Roy,* when she plied between Liverpool and Dublin.

As the manager of Napier's expanded Lancefield works in Glasgow, Tod probably helped his mentor develop the famous steeple engine. Its frame was a high, peaked triangle—hence the name—with a horizontal beam at the top attached by a long connecting rod to a low drive shaft, right over the cylinder. Though a tall structure, often sticking up through a ship's top deck and impeding work there, the steeple engine took up less lateral space than a side-lever engine. It was a light, pretty design, with all its parts easily seen and accessible; Tod used it for the rest of his engineering life. In the early 1830s John Macgregor drew the significant job of seagoing engineer on the *Clyde,* Napier's first steamer with the steeple engine, on her regular runs between Glasgow and Liverpool.

Napier pushed Tod and Macgregor out on their own when, harried by a few lawsuits over boiler explosions, he decided to leave Scotland and move his business down to London. He offered Tod and Macgregor the Lancefield works; but they, wishing to begin more modestly—and perhaps to avoid any association with David Napier's recent reputation for taking dangerous risks—instead started their own engine-building workshop in 1834 at the foot of Carrick Street in Glasgow. Tod, about to turn forty, was the engine man and driving force. Macgregor specialized in hull design and supervised the firm's shipbuilding yard on the Clyde.

They became well known for their iron-hulled paddle wheel steamers in the Liverpool trade. Following the lead of the Lairds of Birkenhead, the most successful pioneers of iron shipbuilding, they strengthened and subdivided their hulls with watertight transverse compartments. The *Royal Sovereign* and *Royal George* of 1839 each had five such compartments. Their owners, the Glasgow and Liverpool Royal Steam Company, were so well pleased that they gave Tod a free hand in designing another ship. Left to tinker at will—any engineer's dream—in 1841 he produced the *Princess Royal,* about 75 percent bigger and more powerful than her immediate predecessors. At 380 horsepower and 800 tons, an unusually high ratio of power to tonnage, she was well designed for quick passages. The *London Times* noted her "extraordinary speed" when she set a record of fifteen hours, twenty-three minutes from Glasgow to Liverpool.

After more than twenty years of working on ships and engines, Tod could appreciate the extraordinary privilege of being given unlimited freedom by a client. Usually the shipbuilder and shipowner worked at cross purposes toward what was supposed to be a common goal. The builder wanted to enhance his reputation for speed, power, and innovation, and to be paid well and on time. The owner wanted a reliable ship, not including any risky unproven innovations, as quickly and cheaply as possible. The owner, as John Scott Russell had ruefully come to appreciate, held all the cards; so enterprise usually trumped engineering. The more inventive builders, like Tod and Russell, strained against the artistic compromises of these commercial restraints.

Robert Napier, who had succeeded cousin David as the leading figure among Clyde engineers and shipbuilders, enjoyed a long, fruitful working connection with a unique client, a rich gentleman named Thomas Assheton Smith. The *Glasgow Mail* declared him "the most princely steamboat fancier of the age." Known mainly for his truly heroic devotion to foxhunting, Smith—whose homes included an estate on the water in Wales—was also a yachtsman who liked to play with different vessel designs. Completely untrained in the field, he intuitively, independently came up with his own version of Russell's hollow, wave-line theories for improving the efficiency of a hull moving through water. Smith, easy to caricature as a riding-to-hounds sportsman with too much money and time on his hands, was quite serious about his steamboat fancies.

Starting in 1830, Robert Napier built eight steam yachts—both paddlers and screws—for Smith over the following two decades. The client planned the hulls, in a winsome variety of shapes and sizes from 110 to 700 tons, and the engineer was granted all but absolute freedom to design and build the engines. For Napier this meant the rare gift of repeatedly letting his engineering imagination run in almost any direction it took him—for a wealthy client who could not be displeased. The artist in Napier had found his patron. "What struck me most," Napier later recalled of his dealings with Smith, ". . . was the complete confidence he placed in me from first to last, to which I responded by doing everything I could to meet his wishes, and on the lowest terms I could, as I knew he did not build his vessels for mercantile purposes, but purely for the improvement of steam navigation. So sensible was Mr. Smith that I wished to serve him in the most liberal manner, that he seldom would look at my accounts beyond a glance at the sum total. This I did not like at first, as I knew he was very particular in his business dealings with others."

In the mid-1840s, Napier built a sequence of three vessels for Smith that all were named *Fire Queen*. For the second of these boats, a screw steamer of eighty horsepower and 230 tons in 1845, Napier devised a special version of the steeple engine. One of the most baffling puzzles of early screw design was how to raise the engine's revolutions, by a factor of two or three to one, from the relatively slow paddle wheel standards of the day up to the higher speeds required by a propeller. In the *Great Britain*, Brunel had resorted to a bank of four enormous chains linked to a massive driving wheel which protruded above the top deck and bulged the ship's sides, contributing to her distressing tendency to roll in heavy seas. For the second *Fire Queen*, Napier contrived a compact, elegant solution to this problem. He placed the engine's two cylinders slightly to one side of the vessel's bow-to-stern midline, with their weight balanced on the other side by a large geared wheel (driven by a steeple-engine system of beam and connecting rod), which meshed with a smaller ring of gears on the propeller shaft. These gearwheels neatly provided the faster revolutions for the screw—but without the immense, destabilizing, fragile chains and sprockets of the *Great Britain*. Napier made the *Fire Queen*'s engine and machinery unusually strong, with a prodigal use of expensive wrought iron, but she still ran fast in the water, up to fourteen miles an hour—a striking blend of strength and speed.

Napier's experimental, uncommercial solution became much more than a rich man's amusement because David Tod then applied his revised version of the *Fire Queen*'s engine and gearing to his own *City of Glasgow* in 1850. This special ship was built under special circumstances. Shipbuilders would sometimes construct a vessel on their own account, trading the usual signed contract and sure payoff for the gift of design independence, and then hope to attract a buyer later. Lacking an indulgent patron like Thomas Assheton Smith, Tod adopted this alternative tack toward engineering freedom for his first Atlantic steamer. He in effect appointed himself his own client.

Like most Clydeside builders, Tod had taken his skeptical time about converting from paddles to screws. His first major screw vessel was an iron steam yacht, the *Vesta*, in 1848. Built for a young Liverpool business-man, William Inman, at 112 feet long, 148 tons, and forty horsepower she was one of the first oceangoing iron screws launched on the Clyde. (A vessel built purely for recreation allowed more artistic floor space for innovations, with lower stakes, than a merchant ship.) Still, Tod found the novel challenges of screw construction hardly worth his trouble. "I hope," he told Inman afterward, "you will not bring us any more screw boats—we prefer paddle steamers." The *Great Britain* had not settled any arguments about screws on the North Atlantic. Ship men still wondered about an oceangoing screw ship's speed, durability, tenacity against a head wind, and stability in cross seas. The screw's blades might break off, it was thought, and the stern bearings could wear out and leak.

Yet in the summer of 1849, only a year after his displeased experience with building the *Vesta*, Tod started construction on his breakthrough screw ship *City of Glasgow*. Something, perhaps the performance of the *Vesta* between the Clyde and Inman's home in Liverpool, had changed Tod's mind about screws. Not just a gentleman's hobby or a pleasure yacht, the new ship was a major steamer, 237 feet long, of 1,600 tons and 350 horsepower. She had room for 52 passengers in first class, 58 in sec-ond, and 400 in steerage. With the famine emigration still flooding into Liverpool and crowding its docks, Tod—perhaps drawing from memories of the farming poverty of his Perthshire childhood—had designed the first Atlantic steamer that was intended to carry emigrant traffic.

Her iron hull, which came under John Macgregor's purview, followed recent trends in both Britain and America toward a longer, leaner, sharper form. In profile the *City of Glasgow* resembled the fast American clipper sailing ships of the 1850s: a steeply raked and flaring concave bow, trim at midship, and a well-rounded stern, undercut like a champagne glass for easier motion in heavy seas. Her proportion of length to breadth stood at about 7 to 1, sharper than the 5.7 of the *Great Britain*, the 6.2 of the Collins liners, and the 6.6 of Cunard's *America* and her sisters. David Tod modified Napier's gearing arrangement for the *Fire Queen* by moving the larger geared wheel from above to beside the smaller wheel, thereby drop-ping the center of gravity in the interest of stability. In her hull, screw, engine machinery, and internal accommodations, the *City of Glasgow* was novel enough to please the purest engineer—and to distress any normally prudent client.

Working on their own time, Tod and Macgregor built her quickly in their yard at Meadowside, on the west bank of the Kelvin by its junction with the Clyde, a few miles below Glasgow Harbor. She was more than 400 tons larger than any previous ship by the firm. Under the general supervision of Archibald Gilchrist, who had started working for the company as a draftsman in 1842, separate work crews toiled on the hull and the machinery. They had steam up within two weeks of the launch, a remarkably short interval. (Regular clients must have wondered why builders couldn't manage that speed for them as well.) The first-class cabins, of two or four berths each, were ranged aft of the single funnel, and the cabins for second class forward of it, with berths for four or eight passengers. The two classes would eat in separate dining saloons. "The internal fittings are tasteful and genteel," one journalist reported, "and got up with a view to comfort and pleasantness, rather than to dazzle with unprofitable brilliancy." Not overladen with luxuries, the ship did include a bathing room, with a pump providing hot or cold seawater. For safety the *City of Glasgow* had five watertight compartments, two lifeboats and four smaller craft, and three masts with a square-sailed bark rigging.

In April 1850, only eight months after the start of construction, the builders sent her to New York with 127 passengers. All in first or second class, they had paid fares of twenty-one pounds or twelve pounds, twelve shillings—about half of what Cunard and Collins charged. "We look upon this enterprise with considerable interest," said the *London Times.* "It is an effort to combine a reasonable degree of speed with certainty and cheapness." The trial run raised both commercial and technical questions. In the checkered wake of the *Great Britain,* could an iron screw really succeed on the North Atlantic? "It is to be hoped," ventured the *Liverpool Chronicle,* "the result will be such as to establish confidence in iron-built ships, as well as in the screw propeller, properly applied."

On her maiden trip westward, fighting severe weather, high seas, and an ice field that induced her captain to shift to a longer, more southerly course, the *City of Glasgow* made New York in sixteen days, twenty-one hours—several days slower than the mail liners but twice as fast as the swiftest sailing ships. She went home with a full complement of passengers, then made two more round-trips, both well patronized, in the next few months. The screw and machinery were standing up well to the rigors of the North Atlantic; and the most impressive, significant aspect of the first three round-trips was her miserly consumption of coal. Unlike paddles, the screw stayed submerged all the time, churning away with less wasted motion to drive the ship. That, along with Tod's efficient engine, gearing, and boilers, meant the *City of Glasgow* burned just twenty tons of coal a day. Cunard's *America* and her sisters, slightly larger wooden paddle wheelers, used sixty tons each day. Such drastic economy by itself suddenly established the iron screw ship—or at least this one—as a serious commercial competitor on the North Atlantic.

In October 1850, as she again set forth well loaded to New York, the *City of Glasgow* was sold for a rewarding price to Richardson Brothers, a linen importing and foodstuff firm in Liverpool. The deal was arranged by a junior Richardson partner, none other than William Inman, the man whose yacht *Vesta* had first induced Tod to try building a screw. Her new owners started the *City of Glasgow* on a brisk regular schedule between Liverpool and Philadelphia—since Cunard and Collins were already running to New York and Boston—and they announced plans for the firm of Tod & Macgregor to build a sister for her, of the same design but larger and faster. David Tod's major engineering gamble had paid off. An important new transatlantic steamship line had been started without any mail contract or government subsidy. (The nascent line was not, as yet, available to poor emigrants. "No steerage passengers taken," explained the first newspaper ads.)

*W*illiam Inman was a different breed of transatlantic steamship mogul: an inland businessman, just lightly touched by salt water, from a comfortable family background. He endured no hardscrabble childhood, like those of Sam Cunard and Edward Collins, nor did he spend his adolescence alongside docks or on the ocean. Instead, a series of business interests (and an amateur's interest in screw propellers) brought him into management of the *City of Glasgow*. He then perceived that his future, and that of transatlantic commerce, lay with iron screw ships; and for the rest of his unflamboyant commercial life, he seldom thought about anything else. "I prefer to be silent and quiet," he once wrote when asked to provide biographical details. "From the time I took to steamers I have devoted all my business time to them, to study all points connected with them, their models, lines, plans, engines, screw, economy, and I am in no other business."

The Inmans were an old merchant family from Yorkshire and Lancashire, near Liverpool. William's father worked for two decades in

Leicester, in the English Midlands; William was born there in April 1825. He later recalled a happy boyhood ("very fond of my pony and my leaping pole, at school strong in mathematics"), with a hint of his future career ("always fond of the water though a bad sailor at sea"). When he was thirteen, the family returned to Liverpool, where his father prospered in banking and commerce. His older brother Thomas became a noted physician and historian of religions. William's headmaster at the Liverpool Royal Institution urged him also to go on to college, but he preferred to try himself in business. He left school at sixteen, clerked around for a few years, and then joined Richardson Brothers in Liverpool. After just three years, at the age of twenty-three, he became a partner.

The Richardsons were Irish Quakers from Belfast—a fact of crucial, shaping consequence in the early history of the Inman Line. Quakers had introduced the linen industry in northern Ireland, and the Richardson family had for years been making and importing the cloth to the United States. John Grubb Richardson and his five younger brothers branched into a general merchant business in Belfast and Liverpool; John sent his brother Thomas to America to run the family's offices in New York and Philadelphia. From provisioning ships and importing linen, the firm naturally started running its own sailing packets on the Atlantic. William Inman took charge of that part of the company, mastering it quickly for his fast ascent to a partnership.

For a long time, as he later told the story, he had been intrigued by screw propellers. As a teenager he had seen a trial of the *Archimedes* in Liverpool and later had "closely watched" the *Great Britain*. The latter ship's many tribulations did not put him off the trail. Even before he made partner with Richardson Brothers, he had persuaded David Tod to build the screw yacht *Vesta*. When the *City of Glasgow* demonstrated its durable economies across the Atlantic, Inman seized his moment. With three Irish partners—John Grubb Richardson, his brother Joseph, and Joseph Treffry, another Belfast businessman—he formed a special steamship group within Richardson Brothers.

They had no mail contract, but they did enjoy access to substantial capital. The major investors in the new line included the Birley family of Lancashire, who owned flax and cotton mills, and the Lepper family of Belfast, also in the linen business. Like the Richardsons, these investors produced fine textile goods in a volatile and fast-moving market that needed swift, sure transportation between Liverpool and America. The Inman steamers built by Tod & Macgregor soon proved themselves the best combination of price and speed crossing the ocean: cheaper than the mail liners, with compact screw machinery and more cargo space, but much faster and more predictable than sail. Inman's principal backers were providing themselves a better transportation system for their own wares.

As with Robert Napier and the Cunard Line, David Tod bought stock in the Inman Line, thus binding his engineering skills to the new enterprise. Over its first five years, Tod & Macgregor built five new ships for the line, with gradual increases in size and power. In design these ships all resembled the *City of Glasgow:* a clipper bow and hull, of similar but progressively narrower proportions, with three or four masts in a bark rig, a single funnel, and Tod's steeple engine and gearing. Given that structural uniformity, they could be drawn, framed, and built more quickly and cheaply. These first five Inman ships cost an average of about £62,000 each. By contrast, Cunard's *America* class cost £90,000 apiece, and the go-for-broke Collins ships topped out at some £150,000 each. The frugality in the Tod & Macgregor shipyard matched the frugality of its ships' coal consumption. As a tight private enterprise with closely watched ledgers and no government cushion, the Inman Line allowed itself no fat or frills.

The initial policy of not carrying emigrants in steerage was revised at the behest, it seems, of John Grubb Richardson. An idealist in business, he sharply felt the service mission of his devout Quaker faith. (From his parents, he recalled, he was "strongly impressed with the duty we owe to God in caring for the welfare of the people round us.") In 1847 he opened a model linen factory and self-contained industrial colony at Bessbrook, County Armagh, southwest of Belfast. It was an exercise in strict capitalist benevolence, patriarchal and forcefully abstemious. Richardson provided housing for his workers and banned alcohol on the premises, "enabling us to control our people," he hoped, "and do them good in every sense."

From the outset of the potato blight, Irish Quakers were especially active in famine relief. They gave out food, clothing, and money, ran soup kitchens, and distributed seeds for turnips and other new crops to wean Irish agriculture from its fatal overdependence on the ruined potato harvest. Still, the awful miseries went on, sending more waves of forlorn people to the cities and seaports. Irish emigration to the United States reached a new high of 219,000 in 1851. For Richardson, making room for emigrants on his steamship line was—like his Bessbrook colony—a canny effort to mingle doing good with doing well. In April 1852 the line announced, "A limited number of Third Class passengers will be taken from Philadelphia and Liverpool, and found in provisions." The initial fares were six pounds, six shillings—about thirty-one dollars—from Liverpool, and twenty dollars from Philadelphia.

William Inman resisted this new policy at its start. Like most Atlantic ships of the time, the Inman liners carried only enough lifeboats for the crew and cabin passengers; the steerage merited no such safety net. Furthermore, "I had great fear at first," Inman later wrote, "of carrying steerage passengers from fear of fire." (The *Ocean Monarch* disaster of 1848, by which over four hundred emigrants had died in a blaze at sea, remained a cautioning memory.) Inman was still only twenty-seven, a junior partner, so John Grubb Richardson as the senior founding partner inevitably had his way about the steerage. Later, when discussing the start of the line's Irish emigrant business, Inman implicitly credited his colleagues from Belfast. "Our company was in fact started by three Irish gentlemen," he explained, "who were my partners (it was an Irish line in the beginning), with some English partners."

To reassure himself about this new traffic, Inman and his wife took passage on the *City of Glasgow* to Philadelphia with some four hundred emigrants. "I then learned to look upon them as the safest cargo," he recalled, choosing a revealing noun. "They made the ship buoyant and fire could not break out without some living person on board immediately knowing it." Instead of endangering the ship, the emigrants might function as a vast company of below-deck sentries. "After that time I drew my own plans for the general crew and passenger accommodation on board and no arrangement has since been made without my 'passing' it." Inman put the single male emigrants at one end of the steerage, the single women at the other, with married couples between them as a moral barrier: another case of social control in the hope of doing them good, and surely a comfort to unescorted women.

From the start, emigrants filled the Inman liners to America, three hundred or more on each trip. The fare was only a few pounds more than steerage on a sailing ship. It was good business for Inman as well. Packed into stacked rows of steerage bunks, the emigrants took up much less ship space per person than cabin passengers, ate the cheapest food obtainable (which the Richardsons as foodstuff purveyors could buy at wholesale), and did not expect—or receive—the constant attention of the stewards scurrying among the first- and second-class cabins on the upper decks. They were indeed safe cargo. And, most important, the Inman steamers cut the ordeal of a North Atlantic emigrant crossing down to about two weeks, in apparent dispatch and security.

*T*he *City of Glasgow* left Liverpool for Philadelphia on March 1, 1854. The flagship and model of the Inman Line, she was finishing her fourth year of steady, uneventful transatlantic service. On this trip she went out fully loaded with 111 cabin passengers, 293 down in steerage, and 76 in the crew: 480 people in all. Deeply laden, she steamed into relatively pleasant late-winter weather, day after day of clear sailing. She was breaking in a new captain, but he had previously served as her first and second officer, and knew the ship well. Iceberg season had not yet reached its height, though some ships were bringing back reports that bergs were already drifting down into the great-circle shipping lanes and clumping more thickly than in recent years.

By the eighth of April, the *City of Glasgow* had been out for five and a half weeks, more than twice the duration of even an unusually slow westward passage for her. The Richardson Brothers office in Liverpool, which had been waiting for good news that did not come, released a statement. "Some anxiety being felt for her safety, we consider it our duty to lay the following particulars before the public." The ship was "in a state of perfect efficiency at starting," and her compasses (still a matter of predictable concern when an iron hull was involved) had been adjusted within five days of sailing. She carried enough fresh water for forty days and could produce more from her condensers. She had forty-six days' worth of fresh and salt meat (at a pound per head per day), fifty-four days of bread and flour, six tons of potatoes and vegetables, and adequate supplies of tea, coffee, and sugar. With economy, these provisions would last sixty to seventy days. The coal on board could power the ship for only about twenty-five days, but she had previously proven capable of beating to westward, under sail alone, against adverse winds. "We believe the vessel to be detained in the ice on the banks of Newfoundland, and unable to make her way out of it," the statement concluded. "We ourselves feel no anxiety for her safety."

On April 21, in midocean, the sailing vessel *Baldaur* sighted a large steamer from two miles away. She was listing sharply to port. No people or smoke were visible. The captain of the *Baldaur* shifted course to investigate, but the other ship disappeared, leaving a bobbing field of biscuits and boxes in the water. The mystery ship was described as having a black hull and funnel, drab paint on the inside, three masts, and yellow paddle boxes. This description roughly fit the *City of Glasgow*, except for the paddle boxes; but she had yellow deckhouses that could have been mistaken, at a distance of two miles, for paddle boxes. The *Baldaur*'s news was telegraphed as soon as possible from Queenstown to Inman and the Richardsons in Liverpool. They scraped the report down for any tiny glint of hope. The absence of visible survivors sparked slight, fleeting hopes that the crew and passengers of the overdue steamer had somehow been saved and soon would send word. Such miracles had happened before, so people told each other.

Over the next few months, the American steamer *Franklin* sighted a disabled ship at sea, but she was not the *City of Glasgow*. The Cunarder *Hibernia* ran across a wrecked vessel with white door handles, which the missing ship did not have. Nautical men hoped she might have changed course and put into the Azores for repairs or coal, but ships kept reaching Britain from those islands bearing no such tidings. Rumors flew around that she had run aground on the Massachusetts coast (not true), and that she had foundered at sea and all aboard had made it safely to the African coast (no). The bark *Mary Morris* from Glasgow fell in with the hull of a large iron vessel about 250 miles west of Ireland. The vessel looked Clyde-built and was painted black with a red bottom, like the *City of Glasgow*. A fire had burned all the woodwork out of her. Thick weather prevented a closer examination, but the *Mary Morris* managed to pick up the hulk's carved wooden figurehead and brought it to New York. It was not from the missing ship.

The deathwatch wound on to no resolution. In May, William Inman privately acknowledged the "probable loss" of his ship. "The most painful apprehensions are entertained as to her fate," admitted a London weekly. After three months, in early June, she was quietly removed from the Inman Line's newspaper advertisements. The ship and her saga gradually faded away, without a climax or a clean ending; just another bottomless mystery of the sea. Observers needing some explanation groped for the usual suspects. Inman concluded that she was probably another iceberg casualty, somewhere in the littered graveyard of the North Atlantic.

The loss of the *City of Glasgow* was by far the worst transatlantic steamship disaster yet. (The grim toll of 480 deaths would not be surpassed for another two decades.) At the time, though, it did not have the screaming impact of the sinking of the Collins liner *Arctic*, which happened a few months later and killed 130 fewer people. That catastrophe had a boundaried, knowable outcome, with survivors to tell horrific stories about bullying crewmen and undefended women and children, tales that filled magazines and newspapers for months afterward. The *City of Glasgow* simply vanished, leaving no survivors and no stories except the periodic reports of hopeful-then-disappointed sightings. Stratified levels of money and fame also made for distinctions between the disasters. The *Arctic*'s victims, all in first or second class, included many of wealth and social standing, the Collins and Brown families and others. On the *City of Glasgow*, over 60 percent of the dead were emigrants from the steerage. It didn't seem to matter as much.

*I*n just its fourth year of operations, and two years into its revised policy of carrying emigrants in steerage, the Inman Line was in crisis. Late that September, the *City of Philadelphia*—the newest Inman steamer by Tod & Macgregor—on its maiden voyage from Liverpool hit a submerged, gouging rock near Cape Race, Newfoundland. It was eleven o'clock at night, in a slanting rain and dense fog. Wounded by a mortal hole, six feet by four feet, in her hull, the ship was backed off the rock and beached in twenty feet of water, about three-fourths of a mile from shore. Her lifeboats safely landed everyone on board, some six hundred people, in a few hours. But the ship was a total loss, the second for the Inman Line in just six months.

A double blow, so close together, inevitably provoked questions about the emigrant policy and—once again—the safety of iron screw ships on the North Atlantic. The *City of Philadelphia* for two days had been running through dark, overcast skies that made solar and star navigation impossible. That meant the ship could only be guided by her compasses; but of the six on board, no two were giving the same reading. The ship and crew had blundered onto that well-charted rock because they were sailing blind, at night, in a fog, in the rain. The failure of the compasses could be blamed, fairly or not, on magnetic distortions by the iron hull. William Inman had bet and built his company on iron screws. It now seemed possible by inference (and in the blank absence of any direct evidence on the matter) that the *City of Glasgow*, too, had been lost because of inherent technical flaws in a still unproven type of ocean steamship.

At this low point, the Inman Line was saved by an accidental sequence of events. The British and allied forces in the Crimean War needed ships to carry troops and supplies some 2,000 miles to the battle theater. (This diversion of Cunard ships had briefly pumped up the Collins Line.) Inman was willing to rent his ships to the war effort, but his Quaker partners—John Grubb Richardson in particular—were not. As an observant Friend, Richardson could not take part in any such killing. His recent ship losses and the 480 deaths on the *City of Glasgow* may also have affected his decision to quit the enterprise. Toward the end of this awful year on the North Atlantic, in November 1854, Richardson, his brother Joseph, and Joseph Treffry dissolved their steamship partnership with Inman, who "by mutual consent" took sole charge of the line. Freed of Quaker pacifism, Inman then rented out his three remaining ships as war transports—not to the British but to their French allies, who offered better terms.

It was fortunate timing. Inman was still only thirty years old, with the weight of the steamship line management now on him alone. Just at this crushing moment of his own ill fortune, he could take his ships off the dangerous North Atlantic route and, for a while, let the French government assume full responsibility for them. He only had to collect the rents. ("We made our chief profits in the war," he said a few years later.) For almost a year and a half, no Inman ships ran to America. By the time they resumed, in the spring of 1856, memories of those double iron-screw disasters of 1854 were fading away, and cargo and passengers came back to the bargain fares and (once again) steady service of the Inman Line.

Inman's luck was shifting, pulled on by events beyond his control that kept breaking his way. The collapse of the Collins Line handed Inman another fortunate opening. The two lost Collins ships were both wooden paddle wheelers; so iron screws were not, it seemed, the only modern steamers vulnerable to the perils of the North Atlantic. The Collins failure left its New York traffic, much larger and richer than Philadelphia's, available to Inman. Having severed the Quaker connection that had first linked his line to Philadelphia (which still had many Friends among its shipping interests), he took over the Collins sailing dates in New York.

That brought him into direct, port-to-port competition with the Cunard Line. "We had the Cunard Company against us with a great name," Inman said later, "and we had to earn a name against them." For years the rivalry was so unequal that it hardly engaged anyone's attention. Sam Cunard spent most of the 1850s preoccupied with the Collins Line and other American rivals, real and projected. The Inman liners, running different routes with novel ships at lower speeds, were not even playing on the same field with Cunard and Collins. During 1852 the Inman ships averaged seven knots across the North Atlantic, more than two knots slower than Cunard's *America* and her sisters, and three knots behind the Collins liners.

Sam Cunard's habitual technological conservatism hardened into stubborn muleheadedness when applied to the issue of screws on the ocean. A few years earlier, the disappointing speed, fragility, and running costs of the *Great Britain* had fixed in Cunard a skepticism toward screws. "They answer very well in coasting, and in making short voyages," he had said in 1849. "For long voyages paddles unquestionably are the best; there can be no doubt upon the subject. I have had the experience of many voyages. . . . The screw propeller is an excellent auxiliary to a sailing ship, but to carry mails with regulated time you must have paddle wheels." The newer Inman ships, built in the mid-1850s, were a bit faster than the earlier ones. When they took over the Collins routes, they sometimes outran Cunard ships (except for the crack *Persia*). Early in 1857, the *City of Baltimore* beat the *Africa* home. Sam Cunard remained unpersuaded. For him, doubting the screws became linked with fighting off the new Inman challenge. The two stances wound together in his mind, reinforcing and nearly interchangeable, braided together into an inflexible policy of some potential risk to the Cunard Line's future.

He did take the new rival seriously enough to try to fix the competition. As his government mail contract came up for renewal in 1857, Cunard retained the support of the Admiralty and Treasury, but not of the Post Office. To strengthen their hand, Cunard and Charles Mac Iver approached Inman with a deal. By an arrangement that recalled the secret Cunard-Collins cartel to distribute profits, the two companies struck agreements to set passage rates and to arrange sailing schedules so they left port on different days. In return for this stabilization of his still-insecure business, Inman—as a virtual dinghy attaching itself to the glittering mother ship—agreed not to bid on the mail contract. The terms were drawn up and never signed, but they were acted upon by both sides. Inman soon came to believe he had been unfairly outmaneuvered. He made the arrangement public and cried foul. "The Cunard Company quieted us by this agreement," said Inman in 1860, "and within three months after this agreement they got their new contract. . . . It took us entirely by surprise, and this agreement kept us quiet."

"There was an agreement made, that we should not interfere with each other more than possible," Sam Cunard replied. "There was only an understanding, not a signed agreement." Roused to the fight, he charged that the Inman ships—as screw vessels carrying many emigrants—were not safe. "Generally throughout the year we have found that the paddle is more regular than the screw," he maintained. "I can hardly state why, but it is so. . . . The shaft frequently gets out of order, and the fans are frequently broken off." As for the emigrant traffic, "I have always considered it inconsistent with the safety of the mails to carry emigrants, and have never, during the twenty years we have been engaged in this service, carried any." He cited the need to consider lifeboat capacity and, in emergencies, the importance of keeping Cunard levels of good order. "Our ships are provided with boats," he explained, "sufficient to secure the safety of the crew and [cabin] passengers in case of accident from fire, or any other cause; and generally we have enough to do to take care of them and the mails. But if we had 400 or 500 emigrants, the emigrants will cause the immediate loss of the whole, for they rush to the boats, and we have no control over them. . . . If you carry emigrants, you run the risk of losing the mails."

Well, he was asked, had the Inman Line—known to all for carrying many emigrants on its screw ships—lost any vessels? Two, Cunard allowed; "I do not like to talk about them." But he had made his point. Had the Cunard line in twenty years lost any ships? "Not one," said Cunard, slipping the memory of the *Columbia* sinking in 1843. Had any passengers lost their lives on Cunard ships? "Not one," said Cunard, in over 1,500 voyages across the ocean—and that was true. On the North Atlantic, Cunard still believed, it was paddles and safety, or screws and take your chances. "We should not, after twenty years' experience, continue to build paddle ships if we had not some good reason derived from our experience."

*A*ctually the Cunard fleet already included three times more screws than paddles. During the 1850s, the enterprise expanded globally, adding to existing schedules around the British Isles and from Halifax to Bermuda, stitching together networks of steamship lines to the Mediterranean and even Australia and South America. These new lines fed into each other and linked up with Cunard's main transatlantic service, as a commercial counterpart to the general spreading of the British Empire across the world. After the *City of Glasgow* had proven her strict economies in ocean steaming, the Cunard feeder lines were stocked with screw vessels: smaller and slower than the mail ships but cheaper to run, with more compact machinery and thus much more room for cargo. "The mail steamers," even Sam Cunard pointed out in 1856, "take but a small quantity of freight—for instance, the *Arabia,* of 2,500 tons, can only carry between 400 and 500 tons of goods. A screw steamer of the same tonnage will carry 1,800 tons, and sail at half the expense." For his secondary lines, which hauled more cargo than passengers, under more relaxed schedules, Cunard thought screws were good enough.

Some of the larger Cunard screws—the *Alps* and *Andes* of 1852, the *Jura* of 1854, and the *Etna* of 1855—joined the transatlantic service, to be sent as auxiliaries to Boston or New York as needed to maintain schedules. Built by William Denny and other Clyde shipbuilders, they lacked the stability and horsepower of the mail paddle wheelers of the time and were sometimes overwhelmed by the stormy North Atlantic. "When we first commenced making screw vessels," Charles Mac Iver said later, "we made them very sharp forward and not quite so sharp aft. We found that the effect of that was that the nose went down and the heels went up." On one especially troubled passage, the *Alps* took fifty-three days from Liverpool to New York. (William Inman suspected an intentional Cunard plot to discredit all screws on the Atlantic. "I really thought," he grumbled, in private, "they had brought her out to throw ridicule on our enterprise.")

These early difficulties confirmed Sam Cunard in his insistence that only paddle ships should carry the Atlantic mail. The Admiralty, which typically followed his lead in such matters, went along with Cunard. But his two principal partners, George Burns and Charles Mac Iver, wanted to build a regular mail liner with screw propulsion. The *Persia* of 1856, the fifteenth Cunard paddle wheeler to be engined by Robert Napier, was a transatlantic favorite, fast and beautiful. But she was also a coal hog, dangerously expensive to run, and therefore with limited resale value to be expected at the end of her Cunard career. Burns and Mac Iver hoped she would be the last of her kind.

Over their two decades of association, the three partners had worked in remarkable harmony, each sticking to his own department. Never intimate friends, they addressed each other by last name and lived in three different cities: Burns in charge of shipbuilding in Glasgow, Mac Iver prowling the docks and maintaining standards in Liverpool, and Cunard making financial arrangements and dickering with government offices in London. In their infrequent arguments, Burns and Mac Iver could occasionally deflect Cunard from an intended course. Always partial to his family, Cunard had wanted to replace the line's well-connected Boston agent, S. S. Lewis, with his son Ned; Burns and Mac Iver objected, so Ned later went to the New York office instead. In general, though, Cunard—as the founder, largest initial investor, and (not least) the essential name on the door—did as he wished and ultimately prevailed.

George Burns started moving toward retirement in 1856. Sixty years old, defeated by the decision to build the *Persia* and then the extravagance of running her, he feared the Cunard Line would still go ahead and build another ship of the same type. He had also quarreled with his older brother James, his first partner in steamship ventures, and the family business was suffering. In the fall of 1856, Burns told Mac Iver he was "most heartily willing to retire from all business." Though concious of his ongoing responsibilities to partners, "I am prepared to conjoin with you and others to work resolutely towards a winding up." A winding up? Mac Iver was only forty-four, many years from retirement. If Burns did intend to leave, the vigorous, aggressive Mac Iver hoped to acquire his shares in the enterprise, with the added power over future company policies they represented. "So long as retiring member's stock can be taken up by continuing partners," Mac Iver confided to Cunard, "there's no winding up I reckon."

Cunard soon turned seventy. He as yet had no intentions of retiring, but the fierce ambitions that had carried him from provincial Halifax into transatlantic steam and fame were finally waning. He now spent more time thinking about his children and grandchildren, and about the mood and balance he wished to achieve for his remaining years. In the irascible faceoff between Burns and Mac Iver, he played the mediator, soothing and placating. "I have the most perfect confidence in the ability and integrity of my partners," he wrote Mac Iver on New Year's Day in 1858. "A kindly interchange of sentiments on all subjects connected with our extensive concerns may be useful. . . . If it should please God that we should all live to see the next year let us hope that no unkind feeling or word shall have passed between us—business may be carried on with equal correctness and advantage, in harmony and peace. Let us cultivate these sentiments and trust that a blessing will attend our undertaking. We have now been nearly twenty years connected in business, a large portion of an ordinary life, and we have been most providentially protected. . . . If I should be spared I hope I may yet be useful to our concern—you may rest assured that I am not idle."

These sweet hopes of the New Year were soon undone. "I regret to say," Cunard told Mac Iver after meeting with Burns, "that I am compelled to agree with you in the conclusion 'that some of our partners are running to seed.'" In grave financial straits, Burns wanted to quit the company but still leave his shares to his two sons. Coveting those shares for himself, Mac Iver asked Cunard for his opinion. Cunard pondered it through a sleepless night. "We should not forget," he wrote Mac Iver the next morning, "that Mr. Burns, our partner and old Friend, is in trouble and that it is our duty to come forward to do all we can to assist him, and we should do it in the way least calculated to hurt his feelings." Burns had been kind and generous to Cunard during his own financial crisis in the mid-1840s; Cunard was returning the favor. Again to Mac Iver, later in 1858: "We should assist him in his distress and not aggravate his misfortunes—do not turn your thoughts to law pleadings—you will in the end thank me for my apparent obstinacy. I am well stricken in years—and when you are as old you will see things differently." Burns retired to his castle in Scotland, where he passed the time reading religious and scientific texts. Mac Iver wound up with a chunk of Burns's shares, but the bulk went to his sons, John and James Cleland Burns.

For their next Atlantic liners, Cunard and Mac Iver split their technological differences. In 1862 they launched one last paddle wheeler: the magnificent *Scotia*, 379 feet long and 3,900 tons, the fastest steamship of her day. Robert Napier had retired, and in the orderly family tradition of Clyde shipbuilders, the business was passed along to his son John. The *Scotia*, John Napier's first major ship, was a larger, faster extension of the *Persia*, again with two masts and two funnels, and strikingly pretty. "In model she is exceedingly beautiful," noted the *Liverpool Albion*, "all her lines being sweet and flowing, while the general harmony of her proportions is such as to suggest the idea of great strength combined with easy and steady motion." The *Scotia* soon sliced the transatlantic records in both directions to eight days and a few hours. But she was technically obsolete from the start. Her side-lever engine was bulkier and more complicated than more contemporary types. As her designer, Robert Mansel, later acknowledged, when the *Scotia* was deeply laden with fuel and cargo, the paddle wheels became quite inefficient. That contributed to her heavy coal use of 164 tons a day. Despite her steady popularity with passengers, the *Scotia* did not earn the Cunard Line acceptable profits.

At the same time he was constructing the *Scotia*, John Napier also built for Cunard its own version of an Inman liner. The *China* was the first Cunard screw ship designed for the Atlantic service, with planned room for emigrants. At 326 feet and 2,500 tons she was smaller and a bit slower than the *Scotia*, not destined for the swiftest mail traffic. But she could carry 160 passengers in first class and 770 more in steerage (against the *Scotia*'s 300 passengers, all in first), and she burned just 82 tons of coal a day. That made her both more versatile and more profitable than the final Cunard paddle wheeler. For its next screw the line went directly to the Inman source: Tod & Macgregor built the *Cuba* of 1864, essentially identical to the *China*. With these two successful ships, the Cunard Line at last stopped building paddles for the Atlantic.

Sam Cunard's time was passing away. Knighted in 1859 for his ships' services during the Crimean War, he had moved into town from his leased country estate in Edmonton. He spent his last years at 26 Prince's Garden, in Kensington, with his unmarried daughter, Elizabeth. After suffering a minor heart attack in 1863, he retired from the steamship business. In the spring of 1865 he started declining. "I feel quite well," he wrote his daughter Jane in one of his last letters, "and I hope a little fine weather will put me all right again." His sons came to sit with him, Ned from New York and William from Halifax. As they hovered by his bed, he would doze and dream about his long-dead wife and many other things, then wake up and tell them about the strange reveries passing through his head. "He has been a kind good father to us all," Ned wrote to Jane. "He is so patient, and thankful, so afraid of giving trouble and so grateful for what we do for him, that I get quite overcome." Sir Samuel Cunard died on April 28, seventy-seven years old.

One year earlier, John Scott Russell had measured the Cunard achievement. Russell noted that he had often been cast as a rival to the Cunard Line and had never had any formal connection to it. His judgment therefore carried no inherent bias in favor of Sam Cunard's lifework. "The Cunard Line stands unrivalled on the Atlantic for speed, regularity, and security," Russell declared; "the brightest example of a wise, useful, and public-spirited, but hazardous enterprise, carried out with an amount of wisdom, prudence, forethought, and conscientious execution, rarely met with in human undertakings."

While the Cunard Line became more Inmanlike, the Inman Line grew more Cunardesque. The two steamship companies were converging. In 1866 Tod & Macgregor delivered its latest steamer, the *City of Paris,* an Inman screw with Cunard ambitions. Long and lean by recent standards, with a length-to-beam ratio of 8.6, she looked unusually tall when seen from an adjacent dock, measuring seventeen feet from the top hurricane deck down to the waterline when fully loaded. She was the first Atlantic liner to be steered from a forward position, like an American steamboat. Her wheelhouse, on the bridge between the first and second masts, was expected to allow the helmsman cleaner peering through fogs and thick weather. At 350 feet, 2,875 tons, and 550 horsepower, the *City of Paris* was notably larger and more horsed than previous ships of the line. The first Inman ship built to compete in every sense with the best Cunard steamers, in 1867 she made a westward crossing of eight days, four hours—only an hour slower than the *Scotia's* standing record.

Transatlantic passengers still argued the respective merits of paddles and screws. Paddles seemed steadier in heavy seas, as the paddle boxes improved lateral stability and dampened rolling. But the screws were perhaps safer, with simpler machinery less prone to breakdowns, and their undeniable coal economies meant cheaper fares for both cabin and steerage. On the *City of Paris* in 1867, a Protestant minister from New York, accomplished in rhetorical flourishes, reached for his command of analogies to describe the sailing experience: "Smooth as a snake, and with a sting in its tail from which it seems fleeing in terror." And: "She rolls like a revolving auger, boring an endless gimlet-hole in the eastern horizon." And: "She glides through the water like a bird through the air, without jerk and without pause." Slithering, drilling, and flying, the ship corkscrewed across the ocean.

In speed and service, Cunard and Inman had become essentially identical. They both sailed twice a week from Liverpool and New York, and they both carried emigrants, stopping at Queenstown (later renamed Cobh) on the southern coast of Ireland to pick up and discharge Irish customers. Henry Morford, an American cabin passenger on Inman's *City of Boston* in 1865, had embarked as a skeptic toward screws—"the dernier resort of poor people," he supposed, "who could afford nothing better." He was disappointed by the size of his shared stateroom, which required a polite minuet between the occupants when dressing and washing, and he disapproved of the "dirty" steerage denizens when they came up on deck to clean their dishes. But the sea air gave him a ravenous appetite. He pounced on the steak, eggs, and biscuits at breakfast and ate pastries three times a day. "I have never breakfasted and dined so well elsewhere as during all that period," Morford later concluded. "England and America seem to be alternately ransacked for the juiciest hams and the most succulent beef and mutton; milk and eggs are carried freshly from Sandy Hook to the chops of St. George's Channel; fruits seem to grow in mid-ocean quite as early as they do in mid-continent." Both the victuals and ship machinery left him well satisfied. As for the ongoing debate about propelling devices, "The Clyde-built screw is my ideal of blended safety and comfort at sea."

While upgrading its cabin services, Inman continued to solicit its traditional steerage trade. Irish emigration to America had fallen from the famine-era high of 219,000 in 1851 to an annual average of about 75,000 in the late 1860s. Inman ships still filled their steerages with emigrants—seven hundred or more on a typical trip—from elsewhere in Great Britain and Europe. On the *City of New York* in 1865, an English traveler making his first trip to America, Samuel Amos, shared his cramped steerage room with nineteen other men of nine nationalities: two Yankee sailors return-ing from China, two Englishmen, four Irishmen, four Frenchmen, two Germans, a Polish priest heading to Mexico, a Spaniard, two Swedes, and a Dutchman. They all behaved in ways that, for Amos, confirmed their national origins. Their bunks ranged in three tiers around a tiny central floor space, six feet long by three feet wide, for standing and dressing. The taller occupants had to hang their legs over the ends of their bunks. The men shared their private food stashes, sometimes without first notifying the owner. The fare provided by the ship's stewards was simpler than up in the main saloon—just hot rolls, butter, and coffee for breakfast—but seemed "good and abundant" to Amos; "the only thing one craves is a lit-tle more variety." (Inman's steerage food was not always so consistent. After the *City of Paris* brought more than 1,000 people in steerage to New York in the summer of 1869, they complained of unbaked bread, no fresh meat, and inedible potatoes, salt meat, and fish.)

The *City of Paris* had other priorities. The spearhead of Inman's push into the higher-stakes Cunard market, she was driven for speed and the needs of cabin passengers. From the start of her service, the steamship-watching public paired the *City of Paris* in ocean races against the top Cunard ships like the *Scotia* and *Cuba*. Whenever they left port at about the same time, an undeclared contest might unfold. The steamship com-panies, as always, denied they would ever endanger the public by such dangerous sport on the deep ocean, certainly not, or run an engine beyond its safe tolerances, or charge through fields of icebergs, or push boiler pressures above prudent limits. "I do not make bets on steamers," said Ned Cunard in 1868, "and have none pending on the *Cuba*." At sea, though, a veteran captain could—with the known if unspoken approval of his bosses—make certain decisions on his own. The scent of a race made the crew snap more sharply to its work and gave passengers a brac-ing diversion from the voyage's progressive tedium.

The *City of Paris* acquired her closest rival, a near replica, when Cunard brought out its *Russia* in the spring of 1867. By an apparent departure from nearly three decades of tradition, she was the first major transatlantic Cunard mail ship not engined by a Napier. In the Cunard style, though, she represented a safe departure hedged by dynastic lines. The *Russia* was designed and built in Glasgow by George Thomson, who had learned his craft as Robert Napier's assistant manager during the building of the first six Cunard mail steamers. When Thomson and his older brother James left Napier to start their own Clydebank foundry in 1846, they laid out an "almost slavish" copy of Napier's Vulcan and

Lancefield works. Among the early managers they hired were Andrew Burns and Thomas Russell, both from working backgrounds with Robert Napier.

The Thomsons first established themselves with fast screws like the small *Iona* (for the Clyde river traffic), the medium-size *Jura* (for Cunard's Atlantic auxiliary and West Indian trade), and the large *Australasian* (for Cunard's new Australian mail line). Coiled into the inbred culture of Glasgow ship men, their brother Robert worked for Cunard as its first superintendent engineer. Given that connection, the Thomsons—like any ambitious Clydeside builder of the time—most wanted the prize of building a Cunard mail steamer for the Atlantic. They bid against Napier, their old mentor, for the contract that became the *Scotia*. "We must confess," they wrote Charles Mac Iver when notified that Napier had won, "to no small disappointment at losing the best chance we have had as yet of making a reputation in our business." Yet coming so close, in such restricted circles, still enhanced the firm's cachet. "The best proof of our standing," they told a potential client, "is the fact of the Cunard Company considering us as the only parties in the Clyde worthy of tendering along with Mr. Napier for their very large and important vessels."

The Thomsons got their chance with the *Russia*. She and the *City of Paris,* the two newest, finest ships on the Atlantic, looked sisterly in profile: a clipper bow, champagne stern, three masts, and a single funnel between the first and second masts. The *Russia* was a bit longer and beamier, the *City of Paris* deeper in the hold. One's engine had wider cylinders, the other a longer piston stroke. One was built by students of Robert Napier, the other by engineering progeny of David Napier. Appearing toward the end of the first three decades of transatlantic steam, both descended from the dominant family of Scottish ocean steamship builders, they summarized all the lasting improvements of their era. They were as closely matched as two spirited racehorses of crosshatched bloodlines.

For almost two years, they were carefully not scheduled against each other. In February 1869, the *City of Baltimore* was supposed to leave New York in tandem with the *Russia*. But the Inman ship needed unexpected repairs after a passage from Liverpool through seas and weather that were brutal even by normal westbound, midwinter standards. So the *City of Paris,* just arrived in New York on the heels of the *Baltimore,* was quickly turned around and loaded for sea in her place. The *Russia* was poised to leave on the same tide. The great race was finally happening. Crowds came down to Castle Garden, at the southern tip of Manhattan, to watch the joint departure and make sporting wagers. Odds slightly favored the

Inman liner. Somebody told one of the captains he was inclined to bet on the other ship. "Well, I'll tell you one thing," the captain replied, "and that is that I'll drive my vessel to hell or win this race." The passengers on board felt a rippling tingle of excitement, eyes on high beam, everyone lit up in curiosity and anticipation.

The Inman ship, under Captain James Kennedy, cast off first, at 1:35 in the afternoon of the tenth. The Cunarder under Captain Theodore Cook followed sixty-five minutes later. For five days they bowled along within sight of each other, black coal smoke pluming from the hard-worked boilers and marking their spots on the ocean like dark, smutty peacock tails. Through a baffling wind that started out fresh and strong but turned unsteady and difficult to judge, the *Russia* puffed up behind the *Paris,* ran even with her—the hardier passengers on both ships out on deck to wave and cheer—and then pulled ahead on the fifteenth, slowly leaving the *Paris* eight miles back near the edge of the western horizon. A brisk westerly gale blew up. The *Paris,* known for fast sailing under canvas, set out her full spread of sails and gained twenty-three miles in two days. Though now on slightly different courses, out of sight from each other, the Inman liner overhauled her rival and retook the lead. For the final three days they raced each other in keen isolation. The *Paris* pulled into Queenstown with a ninety-minute lead. Heading up to Liverpool, Captain Kennedy throttled down, the contest apparently won. The *Russia* kept charging along, through St. George's Channel and around Anglesey to the Mersey.

They came up the river a few hours after midnight, through curtains of fog. For a week, Liverpudlians aware of the race had been declaring their favorites and placing bets. Now the spectators along the banks could identify the ships only by their lights and rockets. "The excitement, nevertheless, was very great," the *Nautical Magazine* reported, "and both Captain Kennedy and Captain Cook were hailed with cheers as their respective vessels steamed slowly up the river. Nothing like it was ever seen on the Mersey at such an early hour." The *City of Paris* finished first, but the *Russia* got there faster by half an hour, arriving just thirty-five minutes behind. After swift trips of about eight days, eighteen hours, settled by so narrow a margin, it was judged a split decision. Cunard and Inman were still running neck and neck.

9

Life on a Steamer

A transatlantic steamship voyage in the middle decades of the nineteenth century lasted long enough to pass, typically, through five
distinct phases. These wildly shifting stages depended—like almost everything else at sea—on the latest moods of wind, weather, and ocean. At
times the North Atlantic simply closed down for long, unrelieved blocks
of dark storminess, days and even weeks at a stretch, allowing travelers no
relieving nuance or variation. On the *Great Western* in the late fall of 1842,
Henry Wadsworth Longfellow hardly left his berth for the first twelve of
the fifteen days to New York. Enduring this especially "boisterous" westward passage, he just lay in bed, seasick and bored, composing poems in
his head through the endless, sleepless nights. "I was in the forward part of
the vessel," he wrote afterward, "where all the great waves struck and broke
with voices of thunder. In the next room to mine, a man died. I was afraid
that they might throw me overboard instead of him in the night." But this
ordeal was not typical in its relentless gray monochromes. On most voyages, the intensely varied outside conditions induced equally fluid moods
inside the charged, confined ocean steamships.

F or new or infrequent voyagers, it was a strange, disorienting
stretch of time with few comforting similarities to any related
experiences on land. Like a railway car or a river steamboat, an ocean voyage took a group of assorted strangers and threw them together, for a

while, in order to move them from one point to another: a jumbled intimacy accepted for the sake of travel. But crossing the Atlantic was so different. A train or steamboat stayed on or near familiar landmarks, seldom out of sight of the assumed safety of human civilization. A transatlantic steamer spent most of a much longer voyage alone on a vast ocean, without landmarks or known human signs, sailing toward a distant, invisible destination. For day after day, the ship remained in the center of an immense, shifting circle of salt water bounded by the sky. The dangers and mysteries were so obvious they came unwilled to mind, needing no imagination or exaggeration. Ocean neophytes left the dock, according to one Atlantic veteran in 1870, with "a certain sense of solemnity, heightened by an element of vague and undefined fear."

The first day or two of steaming might flow quite smoothly, to the general relief of passengers, thus provoking the most cheerful expectations for the rest of the voyage. The initial course—down the Mersey, across the Irish Sea, and along the southeastern coast of Ireland, or the great circle route up from New York or Boston, skirting the northeastern coast of North America—stayed close to the shore, within occasional sight of land and often in mild weather and gentle seas. The ship, still so novel and impressive to unaccustomed senses, pulsed and chuffed with an aura of effortless, contained force: so huge and powerful, obviously an adequate match for whatever the ocean might offer. The machinery actually felt smoother and made less noise than the much smaller engine and propelling device of a river or coastal steamer. Any initial qualms on leaving the dock might subside. Passengers could relax and promenade the deck, look around, and try to feel at home.

Home would be one's cabin, grandiloquently called a stateroom. Over the first three decades of Atlantic steam, this space gradually became larger and more comfortable than the cramped box on the *Britannia* that had so dismayed Charles Dickens. On the *Persia,* the celebrated Cunarder of 1856, a typical cabin was about nine feet long, with two bunk beds across an aisle from a comfortable plush couch. A small porthole admitted light and, in calm seas, some air. Each end of the cabin offered a washstand and mirror. But the oil lamps were still dim and were extinguished for everybody by midnight. The berths remained quite narrow and thinly padded; anyone of large dimensions might spread beyond the edges. ("I turn over with care," noted a man on the *Persia.*)

The worrisome ritual of appraising new cabin mates did not change either. The staterooms in first and second class held berths for two to eight people. Most passengers would find themselves assigned to divide their small quarters with strangers with whom they would share the most intimate, undefended aspects of crossing the ocean. A matter of pure chance, subject to random choices made by the ticket agent or the purser on board, the situation required good manners and much mutual for-bearance—qualities that the North Atlantic might not encourage. An English lady named Isabella Lucy Bird took the *Canada* to Boston in the summer of 1854. For a cabin mate she drew a sodden Englishwoman who had been living in New York, "and who combined in herself the disagree-able qualities of both nations," Bird later exasperated. "She was in a fre-quent state of intoxication, and kept gin, brandy, and beer in her berth. Whether sober or not, she was equally voluble; and as her language was not only inelegant, but replete with coarseness and profanity, the annoy-ance was almost insupportable." While she mocked all religions and declared her militant atheism, Bird sat reading the Bible, determined and pious. "I wish you'd pitch that book overboard," said her companion by way of friendly conversation, "it's enough to sink the ship."

The main dining saloon was placed toward the stern, either in a house on the upper deck or enclosed within the main deck just below. The largest room on the ship—up to sixty-two feet long and twenty feet wide, on the *Scotia* of 1862—it served breakfast, lunch, dinner, tea, and late supper each day and became a social center during the brief intervals between meals. Diners sat crowded together at long tables on padded benches that were bolted to the floor. Overhead racks held bottles and glassware, close at hand but secured against the ocean. Sofas and chairs for smaller groups were placed at the ends or sides of the saloon. Daylight in fair weather came through side windows and skylights in the ceiling. Coal stoves hissed and glowed against the penetrating chill of North Atlantic winters, often losing the battle, until improved by steam heating on the Collins ships. Mirrors lining the walls offered a welcome illusion of multiplied brightness and open space, brilliantly lit up at night.

The first dinner at sea was a defining event. Knowing passengers had already marked their seats with visiting cards; the most coveted spots were near the captain (for social cachet) or close to the doors (for graceful exits in case of emergencies). Less desirable were places at the middle of the benches, which restricted quick departures, or toward the stern, where the ship's queasy pitching motions were exaggerated and—on an Inman ship or the late-model Cunarders—the screw's noise and vibration might impede digestion. The places claimed at the first dinner would, in gen-eral, remain unchanged for the entire voyage. "The five meals make the company near one at table quite important," a passenger noted in 1867. "We have very good neighbours."

Given fair weather and full attendance, the first dinner provided an initial exposure to all of the ship's company at once. A sailing packet, intended mainly to carry cargo, usually included only a few dozen passengers. The mail steamers, by contrast, hauled little cargo but as many as several hundred diverse people, tossed together for up to two weeks amid confined and trying circumstances. In high Victorian society, both British and American, social life ashore was conducted within rigid stratifications of class, ethnicity, nationality, race, and religion. The middle and upper classes might deal with "others" in their working lives but could safely avoid them when dining or in clubs or at home. An ocean steamer broke down these ancient rules of community by exclusion. A prime symbol of modern progress, the ship forced encounters with the looming diversity of modern times. "Steam is the great democratic power of our age," wrote an optimistic American, Samuel Laing, after steaming to Europe in 1848, "annihilating the conventional distinctions, differences, and social distance between man and man."

On the *Great Western* to New York in 1845, a diligent Philadelphia librarian cataloged "ministers of the Gospel, and ministers from John Tyler whose missions were over,—preachers and players,—painters and physicians,—artists and amateurs,—mechanics and musicians,—merchants and merry-andrews,—singers and sewers,—patentees and plaster-venders,—booksellers and basket-makers,—picture-dealers and portrait painters,—barbers and bakers,—Irish, Italians, Germans, Swiss, French, English, and Scotch,—young newly married people and newly married old people,—men older than their wives and wives older than their husbands, and rather remarkably so,—women who would neither walk nor talk, and women who did too much at both,—ladies and gentlemen, and people who were neither." Such variety could be seen on the streets of any major city; but here, at sea, they were all sitting together in one room up to five times a day, talking loudly and gesturing after their respective fashions. Isabella Bird noticed an Irish pork merchant seated next to a Jewish man who ate no swine, a grand dame from a ducal family elbow-to-elbow with a French cook heading to a San Francisco restaurant, and an army officer bound for high command in Halifax next to a California miner wearing buttons of gold nuggets. American and British passengers found Frenchmen quite amusing—their funny-looking clothes and the emotional intensity they invested in the most trifling subjects—but stiffened in the presence of French table manners. On the *Niagara* to Liverpool in

1854, the American journalist George Wilkes was appalled by the return-ing French ambassador to Washington, Bois le Compte, because he scratched his head with a knife handle at dinner, lifted the lids of dishes before the waiters had done so, and used the promenade deck "for pur-poses which are usually consigned to the most private portions of the ship."

Democratizing hopes aside, the clash of cultures aboard ship might confirm prejudices instead of softening them. In 1845 the black American abolitionist Frederick Douglass tried to book cabin passage on the *Cambria* but was told by the Cunard office in Boston that most American passengers would object to such social intimacy across races. Douglass accepted steerage accommodation in the forecastle cabin but did deliver a typically fiery abolitionist speech in the main cabin which caused roaring exchanges and near-fistfights among some passengers. When he returned to America two years later, the Cunard agent in Liverpool made Douglass agree to take a cabin by himself, eat his meals alone, and not mingle with saloon passengers. The British press erupted in patriotic dudgeon over this intrusion of peculiar American racial mores on a Royal Mail ship. "No one can regret more than I do," Samuel Cunard responded, "the unpleasant circumstances respecting Mr. Douglass's passage; but I can assure you that nothing of the kind will again take place in the steam-ships with which I am connected."

A more persistent irritant was the Cunard policy of allowing only a Church of England service in the main saloon on Sundays. The captain was supposed to read the Anglican liturgy, and no minister of any other faith was allowed to speak or hold devotions elsewhere on the ship. To many passengers—prickly Americans in particular—this seemed to deny freedom of religion on the high seas. George Burns, a founding partner of the Cunard Line, was a devout Anglican, and Charles Mac Iver enforced the restrictive policy in the interest of good order. Sam Cunard, who har-bored no religious beliefs, tried to make his ships more ecumenical. "I am under the impression," he urged Mac Iver in 1847, "that you would do well to reconsider the regulation respecting the religious observance of the Sabbath. I think in addition to the Church Service, which should always be performed as at present, that any clergyman on board who may wish to preach should be allowed to do so in the Saloon immediately after the Church Service. I think this would give general satisfaction and could do no harm. . . . This course would prevent the repetition of the charges made against us of bigotry and exclusiveness—talk this over with Mr Burns and do just as you and he may agree upon." (Cunard knew that his partners felt more strongly about the service than he did.) The policy was slightly liberalized, allowing passengers to read aloud from the Bible, and ministers of the Scottish Presbyterian church to preach. But firm sectarian lines were still drawn: the Cunard ships remained floating bastions of established British Protestantism, a continual annoyance to passengers of other creeds.

The feuding cousins, the British and Americans, found themselves most often at odds because—amid all the shipboard variety—they most resembled each other. Together they comprised a clear majority on most voyages; the ships were, after all, shuttling between England and the United States. The touchy Anglo-American diplomatic and political tensions of the time found inevitable expression aboard ship, wound tighter by the trials and hardships of the ocean. Traveling to England in 1847, an American lawyer and author named George S. Hillard felt crowded by "a motley crew of persons of various nations," he confided to his journal. "The English had their usual sub-division: the fastidious and exclusive few and the vulgar many. The natural and unadulterated English is an unrefined animal. It takes two or three generations of favoring culture to beget fine influences." Hillard was flinging the usual criticism of Americans by the British right back at them; in his mind, Columbia maturely led the way toward higher culture while rough-hewn Britannia lagged behind in torpor. Yet when he reached Liverpool, Hillard felt a surprising, uncanny sense of coming home. "I should have been almost ashamed," he admitted, "to let the Englishmen around me know with what emotion I landed upon the soil of their country. It seemed like going back to my father's house."

Amos A. Lawrence, from a prominent Boston textile family, took the *Great Western* to Bristol in 1839. The ship's company did not remind him of Boston. As he walked the deck on the first day at sea, a bold Italian claimed acquaintance. At table, an Englishman laughed at molasses. A Frenchman sang. Another Englishman called blessings on the queen, which created a stir. Sir Lionel Smith, a lieutenant general in the British army, asked Lawrence about the Yankee states. "I told him," Lawrence wrote in his journal, "adding they were the finest country in the world. Afterwards in order to please him I told him they were very proud of their Eng. ancestry." Sir Lionel persisted in teasing good humor: soon there would be a king of the northern American states—"they love royalty and Britain"—and they would return to Great Britain, not as colonies but allies. "No," said Lawrence, "perhaps England will come over to us in time." "Ah, you love titles in Boston." "No. . . . We admire only an aristocracy of merit." You will have carriages and crests, Sir Lionel predicted, with foot-men, and poodles looking out the carriage windows. "Spare the poodles," Lawrence riposted, "they are English, we never have them in America." Sir Lionel urged him to wed an Englishwoman, adding that he must not marry less money than he already had. "I tell him that in America we marry for love." Later on, well lubricated, they toasted the ladies of Great Britain and sang "Yankee Doodle."

*T*he second phase of the voyage began when the ship stopped hug-ging the shore and steamed out to sea. On the deep ocean, the water changed color and the air smelled different, more purely oceanic, with no land vegetation to mix into the aroma. The swells got longer and higher. The ship rose and fell, pitched and rolled, more noticeably. Walking the deck became an adventure, amusing at first, then less so. The bow sliced into oncoming waves, sometimes splitting them apart with small explosions of spray to each side, sometimes burying its nose under water before struggling back up, water running out of the forward scup-pers. The crew cautioned the passengers about heavier weather coming. In the main saloon, waiters set out the table guards: strips of wood three inches high arranged around the edges and across each place setting, to keep plates and glasses from sliding away. The cabin stewards brought out side boards to hold people safely in their berths as they rolled back and forth. "Oh, Stewardess," a woman asked on the *Cuba* in 1874, "you do not think that it is going to be very rough soon, do you?" "All I can say, ma'am," came the reply, "is the sooner you get into that berth the happier you'll be."

Many steeled themselves for a grim struggle not to surrender to the stalking, progressive discomfort they could feel rising within. "A certain regimen may be pursued," the American poet Lydia Sigourney declared after returning home on the *Great Western* in 1841. "One of its principal elements must be an energy of will." After the full attendance and ani-mated faces at earlier meals in calmer seas, each successive seating saw fewer people straggling into the saloon, looking overcast and subdued. A once-festive dinner scene might now resemble a wake, with just a few quiet mourners keeping sadly to themselves. "I sat down at breakfast heroically," noted a man on the *Persia* in 1856, "—but suddenly retreated and had only time to reach my stateroom, before I————." Perhaps fresh air up on the deck would assist self-control; at least that offered escape from the bilious smells below. On a rocking Cunarder in 1867,

Elizabeth Rogers Cabot of Boston spent an unhappy day packed in shawls on deck, "very cold and a little sick." She and a gentleman companion decided to stick it out. "Still Harry and I cling to the deck and have our meals there," she wrote in her journal two days later. Rain and heavy wind drove her indoors. She could not bear the saloon, so she spent an evening at the entryway, on a seat near the door leading up to the deck, "clinging on for dear life." When the moon rose over a lovely, stormy sea, the captain carried her up on deck to savor the spectacle. A day later, the struggle was lost. The date's entire journal entry reads: "Awful life."

Popular remedies abounded. Cyanide of potash and a water compress. Ice enclosed in an India-rubber bag applied to the spinal cord. Hot West Indian pickles mixed with potatoes. Marmalade. Brandy, champagne, rum, port wine, lemonade. Grapes, chicken, oysters, raw celery. Eating no jam whatever. A mustard plaster. Exercise or bed rest, remaining horizontal or vertical, sunshine or darkness, ear plugs or nose plugs. Nothing really worked. A few lucky souls, often those boasting prior ocean experience, sailed along through any seas in serene equilibrium and hearty appetite. They were loathed and envied. Women seemed to suffer more, for longer periods, than men. To men, therefore, the act of resisting became a stern test of manhood, to be met bravely until the final exploding moment.

Most transatlantic travelers did finally have to admit defeat. Sliding into the final stages, from queasiness into misery toward nausea, it became risky to walk or move or even sit upright. The slightest jarring—a flickering candle, or dribbling wax, the flaring light or smoke of an oil lamp, a face in a mirror, any odor of cooking, a wave glimpsed through a porthole—might provide the final nudge. "So long as I was able," wrote Horace Greeley of the *New York Tribune* after crossing on the *Baltic* in 1851, "I walked the deck, and sought to occupy my eyes, my limbs, my brain, with something else than the sea and its perturbations. The attempt, however, proved a signal failure. . . . Naked honesty requires a correction of the prevalent error that this malady is necessarily transient and easily overcome."

Every sense became absurdly alert as sufferers lay in their berths, praying for a delivering shipwreck. The creaking of the ship's woodwork, the steady wash of the waves, the humming and thrumming of the engine, the footfalls of the sailors overhead, and especially the plaintive calls for the steward and those intermittent retchings and barkings—all of it sounded unreasonably noisy and irritating. "There was a queer feeling of having a windmill, instead of a head, upon my shoulders," one woman noted. "You study physiology involuntarily, and doubt if your heart, your lungs, or indeed any of your internal organs, are firmly attached, after all; if you shall not lose them at the next lurch of the ship. Your head is burning with fever, your hands and feet like ice, and you feel dimly, but wretchedly, that this is but the beginning of sorrows."

Writers who put to sea with literary understandings of the ocean's nature soon felt betrayed. The Anglo-American intellectual generation that came of age around the middle of the century had grown up reading the romantic poets and their happy assurance of a natural world that was friendly and beautiful. "And I have loved thee, Ocean!" Byron sang,

> And my joy
> Of youthful sports was on thy breast to be
> Borne, like thy bubbles, onward; from a boy
> I wantoned with thy breakers, . . .
> And trusted to thy billows far and near

Hauling such notions, Harriet Beecher Stowe took the *Canada* to Liverpool in the spring of 1853, just after the astonishing success of her *Uncle Tom's Cabin* in both America and Britain. Dreaming (she said) of poetry and romance, she was expecting a smooth, pleasant fortnight, like a river or coastal cruise, with generous blocks of free time for reading, sewing, sketching, and chatting. Instead Stowe found herself laid out, paralyzed and prostrated. "Let me assure you," she wrote afterward, ". . . that going to sea is not at all the thing that we have taken it to be." Disabled from any of her favorite pastimes, she could barely button a garment. "You lie disconsolate in your berth, only desiring to be let alone to die. . . . I wonder that people who wanted to break the souls of heroes and martyrs never thought of sending them to sea and keeping them a little seasick."

Literary impressions of the deep ocean, it turned out, depended on where authors were reposing at the moment of composition—and the real extent of their seagoing experience. Byron had never crossed the North Atlantic. The ocean blew through countless moods, hitting every imaginable extreme, but on balance was neither friendly nor hostile but indifferent. It just didn't give a damn. "Our most poetic descriptions of ocean life," an American magazine writer decided after steaming to Liverpool on the doomed *Arctic* in 1852, "have been written in the enjoyment of warm and comfortable firesides on the land. Cushioned upon the parlor sofa, the idea is delightful, upon the ocean waves to be 'borne like a bubble onward.' But there is altogether too much prose in the reality."

On, briskly, to the third phase. For most passengers, the worst ordeals of seasickness lifted after a few days. A morning came with an arresting sunrise into a clear cerulean sky over a flat sea: the sun, sky, and ocean, all in sharply etched, shining detail, and no visible human signs anywhere in the surrounding liquid disk or the vaulting inverted bowl overhead. The natural world, in this purest state, again seemed wide-open and gorgeous. After the dark, confined miseries of the previous few days, the welcome was irresistible. "The sunshine and calm brought out the hitherto forlorn and sorrowful passengers," noted an Englishman on the *Atlantic* in 1853, "and those who had been shrunk up like mere grubs in their berths for many a day came forth in gay suits, and skipped and frolicked about like gaudy butterflies; but not like them silent, for louder clatter was never heard from human tongues." Up on deck at last, exulting in the splendid freedom and fair weather, they felt like condemned prisoners miraculously snatched from execution.

With most of the voyage still before them, passengers settled into a daily routine. It included hardly any work or serious activity. Except in an extraordinary calm, the ship never stopped rocking up and down, or side to side, or both motions at once. "As to the question whether 'pitching' or 'rolling' is the less miserable," suggested the Reverend Henry Bellows on the *City of Paris* in 1867, "it must probably be settled by saying that the form not immediately present is the more tolerable of the two." The powerful engine and machinery, throbbing away all through the voyage, caused persistent trembling vibrations, especially at midship on a paddle wheeler and toward the stern on a screw. These natural and mechanical agitations made reading and writing so difficult that the best intentions were easily thrown overboard. Trashy novels taxed even the most earnest attention span.

The passage became an enforced vacation that few travelers cared to resist. "The infinite delight of freedom from all labor," as Richard Henry Dana, author of the sea classic *Two Years Before the Mast,* exulted on the *America* in 1856, "the certainty of nothing to do—the certainty that there is nothing I <u>can</u> do. No matter how many strings you have left lying, no matter what occur to you as things you might do, or ought to do, you banish and forget them all." The driving pace and busyness of life on shore were impossible at sea. Everyone throttled down into simpler, longer rhythms. Clock time barely mattered. Passengers were called to meals, and hardly anything else was scheduled. A whimsical irony: on a steamship, the Industrial Revolution at sea, passengers reverted to pre-industrial notions of time, measured out in vague approximations by the sun and night sky. Urban, middle-class attitudes toward work and discipline also became out of place. "The absence of all care," noted the Reverend Bellows, "and all necessity for exertion of will, intellect, heart, has been a negative pleasure. The sea appears to paralyze the conscience for at least ten days. I feel no reproach in an idleness which on shore would drive me into bitter remorse. Nonsense or listlessness seem innocent and appropriate occupations."

In this middle phase of the voyage, with nowhere to go and nothing they had to do, passengers fell into an amiable, flexible humor, easily amused and diverted by trifles. On deck, people spent lazy hours just sitting and staring at the passing ocean. Almost any novel sight out there— a sail in the distance, a bird, a whale, a flying fish, or the porpoises that liked to play around the ship's bow, leaping and rolling—could bring idlers to the railing, to exclaim and strain for a closer look. In cold weather, the deck contingent stood in the lee of the funnel, huddled up to its warmth. Men puffed on their pipes and cigars there or in the fiddley, an open alleyway above the boilers, until they gained their own smoking room on the Collins liners. Men (and an occasional woman) would play shuffleboard, a deck game borrowed from the American sailing packets which appeared on steamers by the early 1850s. It was easy to pick up but, given the erratic lurches of the ship, difficult to master: an undemanding time killer that resisted too-earnest intentions, perfect for the relaxed mood at sea. The favorite regular pastimes on deck were the simplest, just walking, talking, and dozing fitfully over a book from the ship's library. Serious reading and intellectual discourse were generally impossible. "The dampness seems to strike into the wits as into the lucifer-matches," the American writer James Russell Lowell noticed, "so that one may scratch a thought half a dozen times and get nothing at last but a faint sputter."

The ocean and sky were continual wonders, an ever-shifting slide show of fluid moods and colors. The light changed everything. On a cloudy day, both sea and sky looked gray and lifeless, dull and gloomy, cold and threatening. When the sun came out, some cosmic switch was flipped, and the scene clicked into bright, glittering Technicolor. Both sea and sky seemed bluer than when seen from land, deeper, richer, and infinitely more enveloping. At times, in a choppy sea, the waves looked like small mountains topped by snowy crests; they rose and then broke into cascades of shifting greens and blues, the spray sparkling like diamonds. With the sun positioned right, the tops of the whitecaps became small, fleeting rainbows, disappearing in the time it took an observer to point them out. Passengers might sit for hours in contented exhilaration. Almost hypnotized by the ship's motion and the passing light display, with hair and clothes whipped by the constant wind, travelers could feel an odd, clashing, thumping blend of both invigoration and exhaustion.

Turning back to the ship's company, passengers took inordinate pleasure in the tiniest morsels of shipboard gossip. By this point in the voyage, individuals had sorted themselves into distinct clusters organized, usually, around women of particular charm, beauty, or conversational dexterity. They held court at sheltered areas of the deck or in the ladies' cabin, a haven of feminine taste and decoration that was placed close to midship to minimize the vessel's pitching motions. A man could enter only when approved by a woman. The resident social arbiters would exchange news, such as it was, and pass judgments. The presence of so many men and women so far from home, with nothing to do, sparked diverting rumors and flirtations. Standing at the rail, a man would naturally—just as a matter of common chivalry—put his arm around a woman, to steady her and offer manly protection from the wind and spray. Toward the stern of the ship, sheltered by a deckhouse, a "love lane" or "lover's walk" offered a limited seclusion, just exposed enough to be exciting. Precisely what happened at night, behind the closed doors of staterooms, went unrecorded by Victorian diarists. But anyone could speculate deliciously about couples seen from a distance. "If the gentleman is very attentive to the lady," a Cunard stewardess explained, "then they are *going* to be married. If the lady is very attentive to the gentleman, then they have *just been* married. If they do not seem to care any thing about each other at all, then they have been married some time!"

The favorite, most anticipated shipboard activity was eating and drinking. Freed of intellectual and workaday ambitions, life came down to the elemental basics. After the forced fasts of the seasick days, when food seemed so threatening, appetites quickly returned in hunger and relief, sharpened by the sea air and other agents. On a Cunarder in 1847, tippling spirits founded the "Cock-Tail Club" for the proclaimed purposes of "the prevention of sea-sickness and the promotion of an appetite for dinner." Several ladies were awarded honorary memberships. They all gathered a half hour before dinner to quaff a concoction of equal measures of brandy and water, with sugar nutmeg and a touch of cordial bitters. (The club was dissolved after nine days because, with its noble goals achieved, nobody was still willing to pay for a round.) Alcohol was the universal remedy at sea, applied for all ailments—especially boredom. The drinking water on board, stored in wooden barrels or iron tanks, smelled and tasted foul; better to slake one's thirst with a beverage kept in glass. On Cunard ships, the bar closed for only seven hours a day, from eleven at night till six the next morning. A new Cunard liner in 1848 went out provisioned with assorted spirits—and 600 bottles of port wine, 600 of Madeira, 600 of Rhine wine, 1,200 of sherry, 1,200 of champagne, 2,400 of London porter, and 3,600 of Scotch ale. If the vessel were fully loaded (so to speak) with 140 adults in the saloon, that still worked out to 73 bottles apiece for the two-week voyage: more than five bottles a day.

In the early years, the cuisine on the first Cunard ships left passengers well fed and content. But as competition from Junius Smith's line and the Great Western Company fell away, and Cunard essentially had the steamship business to itself, standards in the galley apparently slipped. On the *Caledonia* in 1848, Pelham Warren of Boston gagged on "wretched" wines and spoiled butter. "The bills of fare are long and showy," Warren wrote in his journal, "filled with French names and all served according to rule, but nothing tastes right. All is exactly alike, and it is difficult to tell whether you are eating boiled chicken or roast beef." The other passengers told him they had never eaten worse fare on the ocean. The *Caledonia* was eight years old, showing her age, only a year from being retired from the North Atlantic trade. Warren found the gentlemen's cabin dull and cold, the mattress in his berth worn thin and hard, the pillow just as bad, and the berth so narrow that he could not bend his knees. "I have never been so disappointed," he decided, "as in the experience of what life is on board a Cunard Steamer."

The Collins Line set new benchmarks with its food and service. The newer Cunard ships (the *America* and her sisters of 1848, and the *Persia* and other vessels of the 1850s) had to meet or at least approach this American competition. Isabella Bird, returning home on the *America* in 1854, was impressed by the ample breakfast spreads of cold tongue, ham, Irish stew, mutton chops, broiled salmon, crimped cod, eggs, tea, coffee, chocolate, toast, and hot rolls. Everybody still came to lunch, only three hours later, but without much appetite. They were saving themselves for dinner. On the *America* to New York, William Chambers of Edinburgh— a noted author and publisher, and a man of sophisticated tastes—could scarcely believe the daily dinner cascades of fine soups, fish, meat, fowls, game, many side dishes, pastries, and fresh and preserved fruits. "The elegance and profusion of these dinners is surprising," he wrote. "How so many things can be cooked, how there can be so much pastry dressed up daily, is a standing wonder to everybody." When the captain entered the saloon and took his seat at the upper end of the first table, on the left-hand side, the blue-jacketed stewards stepped forward smartly to remove the lids of the silver soup tureens on the eight tables. Dishes were cleared, the next courses produced and handed along in crisp, silent efficiency, like a well-drilled military operation. "No matter what be the state of the weather," Chambers noticed, "the dishes are brought in at the appointed time; and I verily believe that if the ship were sinking, the stewards would still be continuing to serve the dinner."

Harriet Beecher Stowe, wondering how it was done, spent hours watching the cook on the *Canada*. He was a tall, thin, sad-faced man who went about his tasks with an apparent expectation of a disaster about to descend. Stuffing a turkey, making a sauce, pinching ripples into the crust of a tart, or just washing pans and dishes, the mournful cook found no comfort or satisfaction, only confirmation of this life as tragedy and defeat. Stowe recognized a kind of artist, doomed to struggle on against impossible odds and conditions. "I can see how these daily trials," she wrote, "this performing of most delicate and complicated gastronomic operations in the midst of such unsteady, unsettled circumstances, have gradually given this poor soul a despair of living, and brought him into this state of philosophic melancholy." He excelled at robust meats and English plum pudding, dishes requiring "grave conviction and steady perseverance, rather than hope and inspiration." His jellies needed more lightness, just like the cook.

After dinner was cleared away, the saloon became an animated parlor well into the evening. Tea and then late supper were laid out for anybody who could still contrive an appetite. New shipboard friends, bound together by their intense, temporary intimacy, moved around the room in shifting, rotating groups. Drinks and party conversation dominated, with occasional music and dancing. Players bent over their chess and backgammon boards. Whist enthusiasts shuffled and dealt their cards. Isabella Bird overheard snatches of talk: "Do you really think me pretty?—Oh flattering man!" "It's your play—You've gammoned me." "Holloa! steward, whisky toddy for four—I totally despise conventionalisms." "Brandy punch for six—You've thrown away all your hearts." William Chambers absorbed the burbling cacophony, looked down the long, low-ceilinged saloon, and imagined himself far from the ocean, at a large evening party in a watering place on shore. With windows shut against the nighttime chill, as the evening wore on the air inside became a stuffy bouquet of pickles, roasted onions, myriad forms of alcohol, and the earthy odors of packed humans at vigorous play.

It was sharply refreshing to take the night air up on deck. In contrast with the saloon scene, the circumstances again became unmistakable: a steamship at sea. On the westbound Inman ships, and then the Cunarders from the late 1860s, the saloon passengers in first and second class might amuse themselves by peering down into the lower gangways at the carousing emigrants from steerage. To the plain music of fiddle, bagpipe, or piper, barefoot Irish girls might be dancing. (The glimpse of an unclad foot and ankle would intrigue a Victorian gentleman.) The laughter seemed louder among "the unwashed but the jovial from all the European lands," Henry Morford noted on the *City of Boston* in 1865, and the fun more boisterous than in the main saloon. "Smoking, chaffing, skylarking, erewhile dancing—while the saloon-passengers look down from the railings above, sometimes not half so happy, and often, I think, aching to have the privilege of joining in the rough amusements."

As the great circle route arced northward, up to a high latitude of fifty-five degrees north, the summer twilight lasted far into the evening. At nine o'clock a passenger could still read on deck. Hours later, a soft light was still visible on the northern horizon, like the first glimmer of morning. At other times of year, in lower latitudes, a clear night flung out uncountable heavenly bodies, from just above the circular horizon to the infinite extremities of sky above and around. The moon and, in its absence, the constellations, planets, and stars all looked unnaturally bright against the black sky at sea. On the *Britannia* in 1846, the British geologist Charles Lyell watched an especially brilliant display of the aurora borealis. "The sky seemed to open and close," he wrote, "emitting, for a short period, silvery streams of light like comets' tails, and then a large space became overspread with a most delicate roseate hue."

Gazing up and around, surrounded and engulfed, an observer might recall the landbound feeling of looking down from a high mountain peak on a wilderness forest spreading out below: that simultaneous sense of both being very small and unimportant, yet comfortingly meshed into colossal, ultimate mysteries. A tiny human on a tiny ship on the endless ocean; yet the ocean that had first spawned life on earth, the ocean that human ancestors had crawled out of so many eons ago. It was primordial, timeless, unchanging, so unlike the progressive, human-centered push on shore. Here, at sea, the waves and occasional fishes tossed up quick, teasing flashes of blue-green phosphorescence on the black water. The light came from microscopic organisms in chemical reactions with an enzyme, another cycle of life at sea, which produced new, glowing molecules. From the stern, the radiant wake stretched out behind, thrashing at first, then trailing off and blending back into the ocean—a brief human presence, leaving no permanent trace on the water. "A night of marvellous moonlight beauty," a man wrote in his diary on the *Russia* in 1869; "the wind fair and the sails set and swelling, and the sea, a moving calm. . . . I am writing this by moonlight at 11:30. A night so beautiful that it is difficult to go to bed."

*D*uring the long middle phase of the voyage, the passengers—always casting about for diversion—watched the crew and toured the ship. They experienced the steamer as a ship and a building; the crew also knew it as a machine, the largest and most complex yet built. The advent of steam had multiplied the tasks on a seagoing vessel, adding ethnic enclaves and blurring responsibilities. On a sailing ship, the captain knew and could command every department from topmast to keel. On an Atlantic steamer, the Scottish engineers and Irish coal handlers came to comprise a separate lower fiefdom, its industrial expertise unknown to the captain. Serving the upper decks, the catering and steward staffs—limited on a sailing packet—kept adding hands as the steamers carried ever more passengers. And, in a residue from the old ways, steamships of this era still needed dozens of sailors to deploy sails in favoring winds or when the machinery broke down. The ever more complicated steamer as machine inevitably needed a more specialized, fragmented workforce to tend it.

While passengers slipped back into preindustrial modes of time, the crew lived by the most rigid clock time, sounded by the ship's bell every thirty minutes, with work split up precisely into four-hour watches. Early in the morning, under the right conditions, travelers woke in their berths to a distant choir of singing sailors. ("Give me some time to knock a man down" or "With rosy cheeks and coal black eyes" or "I hang my banjo on a tree, /And take my love along with me.") To a steady rhythm of call and response, solo and chorus, the seamen pulled on lines to raise sail. Singing kept them on the beat and hauling together—and it asserted their waning presence in an alien place, this giant machine. "A notion appears to have got abroad," a steamship captain warned in 1873, "that steamers can be navigated independently of seamanship, in which I should be very glad to share if fine weather, a smooth sea and steamers' engines could always be depended upon."

The sailors for the time being still claimed their own territories, the spars and rigging overhead, and the forward part of the upper deck, above the dark holes in the forecastle where they slept and ate. An old custom that marked this boundary persisted on steamers: if a cabin passenger came forward of midship, and a sailor chalked the deck behind him without being seen, the passenger had to pay a fine. The fines went into a pot to be split by the sailors, and the payer was then granted freedom of the ship. At slack times the seamen might perform tricks for children: sewed up in a canvas bag with just his head showing, a man would jump across a stick; or a sailor sitting cross-legged on an oar had to extend a heavy stick a certain length without falling over. (Such feats called on the same balance, strength, and agility that sailors needed when skylarking around the rigging.)

In daylight hours, the captain paced the bridge. It loomed over the forward part of the upper deck, at first between the paddle boxes, then as an independent structure on screw ships. The captain scanned the ocean, especially toward the weather side, the direction from which winds and storms were coming. He gave deck orders to the boatswain, who blew his distinctive whistle and told the sailors what to do. The boatswain's whistle would then time the job and signal its finish. On the quarterdeck at the stern, watching the compasses, the third or fourth officer walked the deck athwartship, reporting to and taking orders from the captain, and passing steering instructions back to the helmsmen at the wheel. The captain and his first two officers only appeared on the quarterdeck for all-hands activities such as coming to anchor, getting under way, or taking in or tacking sail. Otherwise they stayed on the bridge—absorbing information through all their senses, barking out commands—and pushing the ship into a constant balancing of speed and safety.

The sailors had to work hardest in the worst conditions. Unlike outdoor workers on shore, who could knock off and stay inside during severe weather, the seamen had to perform their most difficult, dangerous tasks when the weather and ocean turned foul. In a prolonged siege, sleep and dry clothes became impossible luxuries; a crew might spend an entire voyage in cold, wet clothing. On the *Arctic* in 1852, sailors went up to house the topmast late on a stormy afternoon. As the ship rolled in deepening billows, the men clung to ropes high in the mast—a long, exaggerating lever that snapped them around like the crack of a whip. Suddenly a cry was heard from above, followed by long groans. A sailor had caught his arm, and it was being mangled between spars. Other men pulled him out and brought him down to the ship's surgeon. "From this scene," a passenger noted, "so sad, so gloomy, I descended to the ladies' saloon. How great the transition!"

Dealing with such uncontrollable conditions, and always the most vulnerable of all the ship's people to caprices of ocean and weather, sailors took refuge in signs and superstitions to give order to their precarious lives. Luck was so vital; it needed warning and encouragement. Friday was a risky day for starting a voyage, Sunday a fine day. Children on board brought good fortune, as did hanging a rope over the ship's side. Cats and ministers of the gospel were sure to bring bad luck; "always a head wind," the sailors said, "when there's a parson aboard." A flash of St. Elmo's fire meant that Heaven was caring for the ship and crew. Killing a bird would cause a prolonged calm. Whistling would raise a wind. It was dangerous to carry a dead body on board, sneeze to the left, lose a mop, or drop a bucket into the sea while drawing water. Dolphins and porpoises, if bounding around with marked vigor, foretold an approaching gale, and so did the small seabirds called Mother Carey's chickens, or stormy petrels. If a shark followed the ship for several days, someone was going to die on board.

Down at the bottom of the ship, the engineers and coal handlers toiled in a small, industrial space closed to most outside elements. Here the variables were coal, air, fire, water, and steam, all seemingly under their control. Stripped of sailor superstitions, they relied just on themselves and their machines, acting out the progressive dream of the nineteenth century. Trimmers brought coal from the bunkers, in baskets or wheelbarrows, and dumped it on the floor by the boilers. Stoking required skill, speed, and extraordinary endurance. It took years to learn. "No refinement of construction," Joshua Field reminded Isambard Brunel in 1852, "or attempt at economy of fuel will do half so much as good stoking, keeping the water in the boilers in a proper state and keeping the tubes clean." The stoker (or fireman) flung his shovelful into the furnace to form an even layer of coals of consistent density, not so thick that the fire burned too sluggishly, or so thin that cold air could pass up into the furnace through exposed fire bars. Manipulating the shovel demanded both hard-grunting strength and a delicate touch at the finish: a punch with a flick. Clinkers and ashes had to be raked out frequently to keep the fires burning at top efficiency. Shoes and tools could become uncomfortably hot. The engine almost never stopped calling for steam, which meant the trimming and stoking seldom ceased.

The working conditions were awful, like the worst coal mines or factories on shore. The *Russia* carried forty stokers and burned eighty-five tons of coal a day; each man had to lift and throw more than two daily tons. Curious passengers who descended to the boiler room were reminded of other infernal regions: the fire, heat, sulfurous smells, and the constant heavy clanging of metal on metal all building into an overpowering assault, even on idle observers. Sweat poured like steady rain from the torsos of the firemen. Every few minutes, they would have to pause and stand under the wind sails and open gratings that brought some air down from the deck. They hard-earned their daily ration of rum. On the *China* in 1869, an English homeopathic physician and Swedenborgian author named J. J. Garth Wilkinson wondered how men could keep working at that pace and heat. "It is incredible," he wrote in his diary, "to find airy deck and gilded saloon and snug ventilated berth built up around such bowels of furnaces. If the men who suffer down there, and upon whom the whole passage hangs, were properly paid for their own destruction, invention would soon become humane, and take care of their health in the cost. But it seemed to me, that that has had no proper consideration."

The firemen were generally young, strong men; a stray veteran like Walter McCrossan, who stoked on Cunarders for forty-three years until 1884, was quite exceptional. They worked hard, tended to drink and fight for amusement, and left scant written record of themselves. If not actually illiterate, they seldom put their thoughts down on paper. One of the few documents purporting to give voice to the firemen was a statement of thanks sent to Captain James West after the *Atlantic* broke her shaft in midocean in 1851: "You will please deign to receive at the hands of the firemen of your ship their warmest thanks and heartfelt gratitude to you, sir, for your kind, and almost parental, regard for us (whom so few care for, and so many disregarded) during the past trying, and, at times, perilous situation of the Atlantic. . . . In our situation in life, we have neither gold or silver to offer; but such as we have is cheerfully offered, which is our humble services." The language, perhaps refined by the Collins Line office, seems too elegant and truckling for the sweaty denizens of the stokehole. But it may truly express their appreciation for considerate treatment in skirting a disaster.

The chief engineer stood between the firemen and their machinery. Reading his gauges, hearing and feeling the hum of his engine, and always responding to commands from the officers up on the bridge, the engineer gave the firemen their orders. As the central link among all these entities, he held a position of unique power and responsibility. Nobody at sea was more aware of the onward march of steamship technology. The *Britannia* of 1840 had burned 38 tons of coal a day in boilers under maximum pressures of nine pounds to generate 400 horsepower. Only twenty-two years later, the *Scotia* used 164 tons of coal a day at twenty-five pounds to produce 1,000 horsepower. The machine was always becoming larger, stronger, and more dangerous. On board ship, the engineers alone mustered the esoteric knowledge and seagoing experience to control this brimming, ever-growing force. Though less strenuous and risky than the work of sailors and stokers, engineering still took casualties. On the *Arctic* in 1851, an assistant engineer named William Irwin slipped from his post, fell into the engine, and was instantly crushed to death. (At lunchtime, the machinery was briefly stopped while the saloon passengers were told. They raised a subscription of five hundred dollars for Irwin's dependent mother.)

In general, the job required routine maintenance and unblinking vigilance. Before leaving port, the engineer would overhaul his condensers, pumps, and valves, check the tightness of bearings, fill lubricators with grease and oil, blow out the cylinders, and test the engine ahead and astern. Under way, both economy and safety required that steam pressure be kept as stable as possible; that meant maintaining a constant level of water in several boilers and consistent fires in many furnaces. To achieve such plateaus, the engineer scanned an array of gauges and made frequent small adjustments. "Stoke freely when under steam," a marine engineering textbook taught in 1867, "but not too heavily, so that when stopping suddenly, the combustion can be lessened by opening the flue or smoke box door and closing the damper doors." If the engine were halted under a full head of steam, the excess vapor would have to be vented out the steam pipe attached to the main funnel: a piercing eruption of vapor distressed nearby passengers and wasted the company's money. The engineer monitored his daily coal consumption, checking it against the remaining supply in the bunkers, and carefully tracked the salinity and cleanliness of the boiler water. When it became too dirty, water and steam might mix together and boil over into the spewing mess called priming. This could strain the engine, especially the cylinders, and overheat and damage the tubes and plates of the boiler. At its worst, priming would cause a catastrophic explosion.

The machine at sea was still a quite delicate mechanism, engineered to rough tolerances, with its own internal strains tested by unpredictable ocean forces. As a paddle wheeler rolled from side to side, the paddles would be alternately dunked and lifted out of the water, causing the shafts first to stall, then to run free. When a screw steamer pitched heavily, with her nose buried and the tail up, the propeller might leave the water, briefly race, then thump down into abrupt resistance. Such circumstances confounded the engineer's efforts for regular steam pressures and a smooth flow of power. The sudden alternation of free running and forced slowing might fracture shafts and engine parts. When machinery broke down at sea, the officers and stokers could only stand by, waiting for the engineering department to pull off a miracle without the right tools and materials. Shortly after she left Halifax in 1842, the *Britannia*'s larboard steam pipe broke. Her engineer fixed it with canvas and rope yarn, and the ship made it home under lower steam pressure. At other times, engineers at sea would fashion crude propellers from whatever came to hand, repair shafts and bearings, and replace pistons—all done in cramped, oily spaces, under dim lighting, with the ship rearing and plunging.

The deck and engine departments, above and below, remained separate, feuding domains. Engineers typically learned their trades in factories and machine shops on shore and did not grow up accepting the rigid chain of command and obedience expected on ships. They thought of themselves as engineers—one of the favored, even romantic occupations of the nineteenth century—and not as seafaring men. "The atmosphere of the engine shop is entirely unfavourable to the teaching of subordination and discipline," a steamship owner noted in the *Nautical Magazine*. "The engine-room crew are seldom civil, and generally abusive, when spoken to by the deck officers. . . . If engineers were properly brought up to the sea, and learned their duties in the engine-room instead of in the shop, the result would be more satisfactory; they would become accustomed to ship routine, and gradually drop into the position required of them." By birth status and education, the chief engineer and his assistants were social peers of the captain and his top mates. (Indeed, the assistants got higher pay than the mates.) The chief engineer wore a uniform, like the captain, but for years was not allowed to join the captain in the fancy meals and social prestige of dining with passengers in the main saloon. The engineers were generally confined below, just one rung above the firemen and trimmers in the shipboard pecking order. Thrown into working together for the duration of the voyage, the engine and deck men managed a testy, sniping coexistence riven by guarded trade secrets and mutual suspicions.

Up on the bridge, the captain remained the final authority: the only man that everybody on board, passengers included, was supposed to obey. He would delegate deck tasks to his boatswain and mates, and engine and stoking details to his chief engineer. But the captain took sole responsibility for navigation. On the vast, trackless ocean, nothing mattered more than where the ship was located and headed. At high noon every clear day, the captain and a few of his officers brought out the sextant (for shooting the sun) and the graduated quadrant (for indicating the position of the ship on her keel). They took several readings and compared them. The captain oversaw the throwing of the log, a heavy object attached to a line segmented by colored knots. Cast overboard and timed by an hourglass, the log gave a rough sense of the ship's current speed. With these bits of information, and the precise time back on shore provided by the ship's chronometer, the captain could retire to quarters to plot his current bearings and course, and compute the total distance run in the previous twenty-four hours. (This daily run gave bored passengers something new to chew over, always welcome, and settled wagers.)

Bearing so much responsibility, the captain had to know his business. The general level of competence among transatlantic commanders was quite high; gross, fatal mistakes, like those by James Hosken in grounding the *Great Britain* or James Luce on the sinking *Arctic,* were uncommon. Captains on the harsh North Atlantic had typically come up through decades of varied service on different ocean routes. The endemic dangers of the life actually attracted them, imparting a sharp snap and challenge to their jobs. James Price, a commander for several Atlantic lines, was the son of a Welsh ship captain. Raised on rousing stories of seafaring adventure, he never had to choose his calling. "In fact," he said in 1870, "my imagination had been worked up to such a pitch that I thought the greater the danger and the more hazardous the position, the greater the glory of a sailor's life." He took gold miners out to Australia on a sailing packet and circled the world more times than he could recall. Along the way he was caught in floating ice, twice shipwrecked, burned out at sea, and spent six days and nights on the ocean in an open boat. After all that, the North Atlantic's fearsome reputation did not daunt him.

Armored by their air of command, preoccupied with hard, unending duties, captains might not welcome friendly overtures from passengers. The Cunard Line was especially noted for the flinty gruffness of its commanders. Captain Lauchlan Mackinnon of the Royal Navy found a "degree of pompous mystery" on Cunard ships: "If a passenger ventured to ask a question from one of the officers, he met with a sullen reply." (Whereas on the Collins liners, Mackinnon added, he had encountered a "universal and cordial civility.") Captain E. G. Lott of the *Persia* beamed from a jolly-looking countenance that emboldened innocent passengers to approach him. "I tried it once," said an Englishman. "I thought I was drifting into the Gulf Stream, but I found I had struck an iceberg." Theodore Cook, the first commander of the *Russia,* was a small man with a low, quiet voice. He was shooting the sun one day when a strolling traveler ventured, "I'm afraid that cloud prevented you from making your observation." "Yes, sir," Cook replied, "but it did not hinder you from making yours." A few years later, a woman on the *Cuba* concluded that her captain regarded a passenger as "an inferior animal who comes on board with its easy chair and asks silly questions; it must be treated gently, but with firmness, and kept at a distance, and any notion that it may have picked up from observation—about the ship's going too slow—economy of coals—rotten sails, etc.—must be nipped in the bud and treated with the greatest contempt. . . . There is a shake of the head and a turn of the brow that makes us feel keenly what miserable creatures we are."

The captains might have replied, in character, that they had more important tasks than answering stupid, repeated questions. Captain Lott made few attempts to charm; but after his three hundredth crossing, grateful customers from the *Persia* threw him a celebrating dinner at the posh Delmonico's restaurant in New York and gave him a set of silver plate. Captain Cook commanded twenty-eight Cunard ships and retired only because of failing health after forty-seven years at sea. The Cunard captains were steady, faithful reflections of Charles Mac Iver's insular, assertive management style. Like their shore boss, they just did their jobs year after year, safely and predictably, without extra flourishes or grace.

Charles Henry Evans Judkins, the legendary commodore of the Cunard Line, established these standards. Joining the line at its start in 1840 (when he was thirty-one years old), he commanded the major Cunard ships on the Atlantic until his retirement in 1871. Judkins was a blocky man with a strong, handsome face and a prickly presence of unmistakable authority. Reading the Church of England service on Sundays, he seemed almost the Voice of God. ("Few clergymen read as well," Richard Henry Dana noticed.) Anybody on his ships, from Samuel Cunard on down, was expected to snap to his will. "A dove-like sweetness of manner," it was said, "was not the commodore's best point." After so many safe North Atlantic crossings, he had no patience with fretful inquiries. A brave woman once asked him if the dreaded Newfoundland Banks were always fogged in. "I don't live upon them, madam," he explained. Again, in a gale, somebody dared to ask how far they were from land. "Two miles," he said, pointing downward.

Judkins did harbor one vulnerability. As John Burns, the son of a Cunard founder, later observed, the commodore might melt in the presence of a beautiful woman. When Fanny Appleton of Boston took his *Columbia* to Liverpool in 1841, Judkins behaved in ways that would have astonished most of his later passengers. Then twenty-four years old, Appleton was smart, pretty, and ruthlessly unsparing in her judgments; Longfellow had been courting her for years without, as yet, any success whatever. Judkins simply attached himself to her. He walked with her on the upper deck, showed her the machinery below, had her sit at his table and eat the choicest foods, gave her the use of his cabin, played his guitar for her, promenaded with her at sunset and under moonlight. (He predicted, in passing, a great future for ocean steam navigation. "Every future Nelson," he declared, "will be the hero of a paddle box.") At the end of the voyage, Appleton sent quite different reports to her father and to her best friend, Emmeline Austin. "He is a right good fellow," she told her father, "and an admirable Captain, giving all his orders in the quietest manner and devoting himself, heart and soul, to the comfort of his passengers." "Never was Captain so devoted to any she but a ship, as our right worthy one to me," she told her friend Em. "But don't be alarmed, there is a Mrs. J on shore to whom I must speedily resign him. He has just been giving me such a sailor-like account of his courtship from the first penchant to the happy finale."

*T*he fourth phase of the voyage, mercifully brief, brought boredom or terror, or both; too little going on or too much. At a certain point, the limited diversions of the passage no longer looked interesting. The constant annoyances became more intolerable. Everybody seemed to share the same desperation: When is this ever going to end? "One long disgust is the sea," Ralph Waldo Emerson wrote in his journal on the *Europa* in 1848, nine days after leaving Liverpool. "No personal bribe would lure one who loves the present moment. Who am I to be treated in this ignominious manner, tipped up, shoved against the side of the house, rolled over, suffocated with bilge, mephitis and stewing oil. These lacklustre days go whistling over us." (Two days later, as the ship entered Halifax harbor, the fog suddenly pulled back to reveal the shore and town. The entire ship's company applauded. Emerson resolved never to travel again.)

The first three phases had induced successive moods of curiosity, misery, and relief. Now a feeling of restless melancholy prevailed. Each day felt exactly like the last, and the next. The ocean, the sky. Eat, drink.

Walk, talk. The ocean, the sky. Try to read or write, but give it up. Play cards in the saloon, reaching yet harder for still-untried byways of conversation. Perhaps a flirtation, sparked more by boredom than ardor. The ocean, the sky. "There is nothing so desperately monotonous as the sea," James Russell Lowell decided, "and I no longer wonder at the cruelty of pirates." Travelers found bitter amazement at just how wide the North Atlantic was, how remarkably long in duration the voyage had to stretch out between Britain and North America. Despite so many days at sea, and ample time to adjust, few people ever got comfortable with the damp, cramped berths and rocking, rolling motion. Almost everyone felt soggy and hard to light. Even the most determined journal keepers left more days blank; nothing to write about. Tempers grew short, and good humor drifted away. "It's awfully stupid the life aboard," William Makepeace Thackeray wrote after eleven days on the *Canada* in 1852. "I'm weary of guzzling and gorging and bumping in bed all night and being 1/2 sick all day. . . . O it will be comfortable to be in a bed that doesn't jolt, and on a floor that doesn't give way under you. . . . The waves are immense: about 4 of them go to the horizon—but I'm disappointed in the grandeur of the prospect. It looks small somehow—not near so extensive as 1000 landscapes we have seen."

In this mopey, dopey mood of lethargy and tedium, even an approaching storm might seem appealing. It was at least not boring. A serious storm at sea came on gradually, cued by plunging barometer readings and odd, unsettling clouds in a distant slice of the sky. The officers on deck could see it and smell it coming as they started getting ready. Late in the voyage, with so many tons of coal burned, the ship rode higher in the water. She rolled farther over, lingered there, and took longer to come back up. To a passenger, the storm might feel like riding an especially spirited horse that kept going faster and faster, finally slipping the bit and running out of control. At first it was bracing, exhilarating fun. On the *Cuba* in 1867, the Reverend Newman Hall went to the foredeck as the ship began to pitch harder. An officer warned him about a possible rogue wave. "But I could hold on?" Hall asked. "No, you couldn't!" Hall still saw no sign of danger, but he took cover. The horse ran harder. By subtle degrees, the ride shifted from exciting past uncomfortable into terrifying. Unable to stop or get off, a passenger could only hold on and wait it out.

The *Great Western,* nine days out from England in 1846, steamed into a slating, screeching gale. The sea broke over the ship, flooding down into the engine room. The stewards decided not to try to serve breakfast. A high wave washed away two lifeboats from their davits; perhaps a crucial loss, given the rising circumstances. Passengers heard the bad news in a moist, heavy silence. "See," said one, "no one converses, no one reads, all are engaged each with his own thoughts." A wave demolished the port paddle box. As people huddled in the saloon, the sea crashed overhead, broke through the skylights, and poured down into the staterooms. Women screamed and fainted. No one slept much that night. Toward morning, the overcast sky lifted in the east, revealing an amber belt of light just above the horizon. This pretty display augured an ocean tornado. It whirled toward them, pushing and lifting clouds of spray and foam. When it hit the ship, the *Great Western* keeled over, her gunwales buried, and stayed there for an interminable moment. The paddles kept turning. The ship came back to an upright position. She headed into the storm and resumed her course to New York, battered and leaking.

On the *Baltic* in the late fall of 1852, 350 miles west of Ireland on a passage to Liverpool, Lauchlan Mackinnon noticed the swells increasing around noon. Off the starboard bow, toward the southeast, he saw a dense, dark, copper-colored swatch of sky. The barometer, dropping for the last thirty-six hours, bottomed out at 28.40, the lowest reading Mackinnon had ever seen in all his years at sea. The swells from the east rose into heavy waves, with strangely vertical slabs on the sides, and solid, curly tops. The ship throttled back to eight knots. A spindrift of spray broke over the bridge, drenching the captain and his mates. Driving into the storm, the ship faced an oncoming mountain of a wave. The wave rolled toward them, gathering power, bearing down, finally blotting out the horizon before breaking. The *Baltic* rose gently, just splitting the crest of the mountain, her high bulwarks even with the roiling water. It happened repeatedly as she rode out the storm. Mackinnon's high professional opinion of the Collins liners was confirmed.

In the middle of a storm, when it might be abating or still getting worse, and the small question of the ship's survival remained unknowable, passengers endured long, slow hours of terror. The huge ship was being tossed around like a fragile shell, plunging down and coming up trembling. Sinking into the trough of a wave, people could see walls of dark water rising overhead on every side, apparently about to engulf the ship and everyone on board. The officers and stewards tried to act unconcerned and reassuring. "Fine weather this!" boomed the captain during a near-hurricane on the *China* in 1870. "This is what we sailors call a *stormy wind*." The noises built in layers into a howling bedlam: the wind whistling through the rigging, waves breaking against the hull "like a thousand sledge-hammers knocking at once," the woodwork grinding and creaking, dishes breaking, furniture rolling around, water splashing on deck and roaring out the scuppers, random shouts and screams in the distance, and sailors clumping around overhead and yelling to each other. Waiting out a storm on the *China* in 1866, Henry Carey could hardly stay in his berth. He wedged himself between a pillow on one side, an overcoat on the other. He napped and woke up with every bone aching. "Rocked in the cradle of the deep," Carey sang to himself, "I lay me down in crooks to sleep." ("Poetry," he reflected in his diary, "is not always reality.")

About 2,000 miles out from Liverpool, a westward ship entered the Grand Banks of Newfoundland, the graveyard of the North Atlantic. It was by far the most dangerous sector of the voyage. Dense, persistent fog muffled the sea and air, especially in the summer months. The Gulf Stream collided with the Labrador Current, throwing up storms and heavy seas. Hundreds of small fishing boats came out to the teeming cod and mackerel grounds of the Banks and squatted—tiny, slow moving, difficult to see—squarely in the transatlantic shipping lanes, menacing both themselves and the big steamers. Icebergs floated down from Greenland, usually starting in January and peaking from April to July; but a stray berg could appear at any time of year, when least expected. A passenger might hope to see an iceberg, beautiful and thrilling in its massive white danger—but only in daylight, at a safe, harmless distance. Such benign cameo appearances resisted secure scheduling.

Steam power made traversing the graveyard both safer and more hazardous. Crossing the Banks in 1841, the *Great Western* picked her way through a closely packed field of three or four hundred icebergs of every form and size. Lydia Sigourney watched the captain calling out orders to his engineer—half a stroke, a quarter of a stroke, stand still, let her go—as he steered carefully through the icy obstacle course. "It was then," Sigourney noted, "that we were made sensible of the advantages of steam." Freed of the fickle wind, a steamship could keep going at the speed and safe direction chosen by the captain. But steam also let the ship charge through fogs and iceberg fields at full, reckless power, regardless of sea or weather conditions. Such well-respected authorities as Matthew Maury of the U.S. Navy maintained that it was actually safer to run through a fog at top speed because the ship spent less time in danger, thus encountering fewer vessels, and made more splashing noise than at lower speeds, warning others of her approach.

Speed gave the massive ship a headlong momentum, however, leaving her harder to stop or deflect, and more lethal in collision. Steam on the Grand Banks amounted to a typically double-edged exercise in modern progress. In June 1846 the *Britannia* encountered several hundred icebergs up to four hundred feet tall. While the watch at the bow of the ship was tripled, both aloft and below, the ship went on at full speed. One night she came within less than a ship's length of a large berg; a British naval officer on board declared she was going too fast for the situation. A year later, the *Caledonia* charged into the fog and, suddenly, passed within two hundred yards of a New York packet. Early in 1854, a transatlantic steamer racing through a dense fog came upon a huge iceberg, taller than the ship's topgallant yard, just a hundred yards ahead. "The captain looked as if he had been struck in the face," a passenger recalled, "and, though his lips moved, he could not speak to give the necessary orders." The second mate, taking over, shouted commands to the helmsmen; slow seconds ticked by; a paddle box cleared the berg by twelve feet. The captain then resumed full speed, sparking a near-mutiny by passengers.

The *Arctic* disaster of September 1854, caused by a high-speed collision in a Grand Banks fog, shocked the steamship lines into addressing safety issues. In its wake, Samuel Cunard assured the public that his ships, at least, carried enough lifeboats for all passengers. "Captain Luce [of the *Arctic*] is known to be a good seaman and a firm and resolute man," Cunard added. "He did his duty under the trying circumstances in which he was placed. If his crew had stood by him all might have been saved." But what about the practice of running full speed across fogs and iceberg fields? Three months after the catastrophe, *Harper's* magazine of New York reflected gravely on "the national wail which has not yet died away." After a snug general sense in recent years that modern science and skill had subdued the ancient terrors of the ocean, one event had brought them all back again. The problem, so *Harper's* insisted, was the reckless lust for speed, the foolish determination to plow ahead, despite fog, simply to arrive a day sooner. "Speed is not safety," the editors concluded. "This was not an unavoidable accident."

Under scrutiny, put on notice, the steamship lines ordered steam whistles and slower speeds in ice and fog. Passing through a Newfoundland blanket in July 1856, the *America* screamed her steam whistle every few minutes. Two sailors at the forecastle also blew horns. Lookouts stood on each paddle box, at the bow, and in the foretop high in the foremast. The ship proceeded at half speed for three days, peering and making noises. The ship's bell rang every thirty minutes and each lookout cried "All's well" in turn. Richard Henry Dana, one of the passengers, slept in his clothes just in case. Captain Wickman never went to bed or came to his table in the saloon. When the fog lifted and the ship resumed speed, Dana told Wickman that the passengers liked his caution in the fog. "Whether they like it or not," said the captain in his best Cunard manner, "it is my duty to do it." ("This was a little gruff," Dana thought, "but I liked the downright honesty of it.")

As time passed and the *Arctic* disaster receded into the folklore of the sea, steam whistles remained as its lasting memorial to the 350 dead. But the chastened care about speed did not last. Passengers still wanted to arrive as soon as possible. Cunard and Inman kept building faster ships. Competitive pressures across the Atlantic reasserted the mandate for ever shorter passages, and steamers again charged at full bore through the graveyard, regardless of visibility or objects in the water. The modernist bargain, of speed in exchange for safety, proved irresistible.

*T*he final phase of the voyage came none too soon for most travelers. As the ship approached land, the waves usually subsided. The ocean again changed color, and the air smelled comfortably of the shore. Seagulls and land birds flew out to case the ship. The crew might be set to polishing brass and scouring woodwork, wiping out the residues of salt water and fog, returning to shipshape for inspection by the company men waiting on the dock. Passengers came up on deck in their best clothes, reaching for the first sight of land, with everybody in high, relieved spirits. "No one who has not crossed the ocean can conceive of the joyous excitement of the scene," wrote a man on a Collins liner in 1852. "All the discomfort of ocean life was forgotten in the exhilaration of the hour."

The last day—depending, as always, on the ocean and weather— might turn into a playful lark: implausible athletic contests on deck, a mock trial of the captain for allowing contrary winds to prevail, a last chance for courtship or matchmaking. A formal concert in the saloon would display passengers and crew members with claims to musical talent. The final dinner at sea was especially fancy and raucous, with ample champagne provided by the captain and fulsome rounds of toasts and compliments to all hands. "Very flattering things are said of the qualities of the ship and the skill and virtues of the captain," an Englishman wrote after crossing on the *Cambria* in 1845, "of the vast advantages of such speedy communication between the two greatest nations in the world. . . . As soon as these agreeable subjects are exhausted, the passengers find it agreeable to walk on the deck a little and cool their heads, heated with champagne and eloquence."

A betting pool was organized as to the exact date and time of arrival. A local pilot came out to steer the ship through the final legs of coast and harbor. In the high spirits of the moment, passengers would place bets about the pilot's hair color, whether he would wear an overcoat, which leg he would first step on board with, whether he would have a mustache, the color of his hat, and the number—odd or even—on his boat. As the steamer at last eased into her dock, the winners collected their rewards. The grand, life-risking gamble of crossing the North Atlantic came down to smaller, more frivolous gambles at the end. "We are all sorry to part with each other," Thackeray concluded, "and glad that the voyage is over."

PART THREE

The Era of Steamship Competition, 1870–1910

The White Star Line

*W*hen the White Star Line launched its first Atlantic steamer, in 1870, it marked the start of an expansive new era in transatlantic competition. Cunard had dominated the first three decades of Atlantic steam: shrugging off Junius Smith and the Great Western company, outlasting the Collins Line, untouched by Brunel's *Great Britain* and *Great Eastern,* and meeting the recent challenge of the Inman Line's shift into crack mail steamers. White Star would give Cunard rougher, more lasting competition than it had ever encountered—and just when Cunard's internal management was struggling, in the wake of Samuel Cunard's death in 1865 and before the next generation of company leaders had fully taken hold. Given the Cunard Line's vulnerability, White Star moved into this opening with startling speed and command. But its quick ascent, White Star rising, required more than just seizing the strategic moment presented by a weakened competitor.

The story came down, once again, to the ships. White Star simply built better steamers, of revolutionary designs that made them more comfortable and luxurious than anything else on the ocean. The man behind the White Star liners was Edward J. Harland, perhaps the most distinguished British naval architect and shipbuilder of the nineteenth century. In planning his ships, Harland zeroed in on how the oceangoing experience would feel to a typical passenger. "Comfort at sea is of even more importance than speed," he declared. From that insight flowed a corollary: "You cannot have comfort at sea unless you have size, and in that size you must not forget the length." Atlantic steamers were always getting bigger—to carry more coal, cargo, and passengers, and to allow their owners temporary bragging rights for claiming the grandest ship of the moment. Only Harland so insistently connected size and comfort, and he rode that formula to independent success.

Harland's boldness and originality recalled the best work of Isambard Brunel. Like Brunel, Harland was a mechanical engineer before he strayed into shipbuilding; and, like Brunel, he built his ships far from the River Clyde. They were both outsiders, not meshed into Scotland's tight family lineages of steamship builders. It is significant, therefore, that Harland explicitly did not place himself in the Brunel tradition. For Harland and many others in his shipbuilding generation, the *Great Eastern* lingered on as a textbook example of how not to construct an ocean steamship. "The machinery and the propelling power in her," Harland maintained, years after the deaths of Brunel and Russell, "were, to my mind, simply two masses of very miserable failures." The engines, paddles, and screw propeller, he said, all reflected naive, headlong incompetence by her builders. "Experience had hardly warranted their embarking on such an undertaking, and . . . they took a much more gigantic stride than was warranted," said Harland. "We must pass the *Great Eastern* as a melancholy illustration of great ambition but great ignorance."

In historical terms, Harland may be understood as Brunel crossed with Sam Cunard: an unfettered engineering imagination skeptically grounded by practical, commercial realities. Invention at the service of enterprise. "We have the pockets of the shipowners to consider," Harland insisted at a meeting of prominent ship men in 1879. "If the naval architect enters his profession and follows it out on scientific principles simply, he will make a great mistake; and if he follows it out from a shipbuilder's point of view, he will make a mistake. He has to be scientific, and he has to be practical, and he has to be something of a merchant. In other words, the result of the whole thing must be satisfactory to everybody, and particularly to the man who pays for the vessel and owns her." Steering carefully between accounting ledgers on his left and pure science on his right, he saw more danger to starboard. "We must not lose ourselves in science," Harland warned. "I should be the last man in the world to say a word against it, but I think at the present time we are disposed to make too much of what is called education."

*E*dward Harland's own education took place mostly outside classrooms. A Yorkshireman, the son and brother of physicians, he was lightly schooled but deeply engineered on the northeast coast of England—an area not previously noted for its contributions to transatlantic steam. Harland was born in May 1831 in Scarborough, a spa and resort town on the North Sea. "I was slow at my lessons," he recalled, "preferring to watch and assist workmen when I had an opportunity of doing so, even with the certainty of having a thrashing from the schoolmaster for my neglect. Thus I got to know every workshop and every workman in the town." He paid special attention to the wooden-ship building yards that turned out coastal colliers, for bringing coal down from Durham and Northumberland, and sailing ships of 1,000 tons for the East Indian trade. Hanging around the shipyards, young Harland studied each step—molding a timber, bending a plank, lining off a spar, laying down the launching ways—and then crafted his own model yachts, of the fastest, neatest hulls he could imagine.

Already, by himself, he had found his métier. At the same time, both his parents encouraged habits of invention and tinkering. His father, William, had shown original mechanical skills since boyhood. Spurning the usual games with schoolmates, he preferred his workshop, where he made models of water-powered mills and devised new machines. Such pursuits seemed merely recreational to him, however; William went up to Edinburgh for his medical degree and then opened a practice in Scarborough. On the side, he invented a steam road carriage, taking a patent in 1827 for its boiler, condenser, and machinery. Edward Harland's mother, Anne Pierson Harland, "was surprisingly mechanical in her tastes," he said later, "and assisted my father in preparing many of his plans, besides attaining considerable proficiency in drawing, painting, and modelling in wax." Under her tutelage, her eight sons and daughters turned their nursery into a home workshop that produced an inventive stream of toys, dolls, boats, and dollhouses. "It was in a house of such industry and mechanism," Harland recalled, "that I was brought up."

William Harland did not, however, want his sons to make their livings at such workmanlike callings. Engineering was not yet granted the full social status of the respectable professions. After Edward's fitful encounter with grammar school in Scarborough—one imagines him sitting by a classroom window, listening to the sounds of saws and hammers from the shipyards, and yearning to bust out—he was sent to the Edinburgh Academy. The classical curriculum did not engage him, and he came home after two years. His father urged him to study law. He envisioned a life like his own for Edward, with a professional career and mechanical dabblings in his free time. The son wanted to be an engineer. Britain was then exploding with steam and iron, tossing off newly astonishing feats of ships, railroads, factories, and bridges. The son longed to join this beckoning future; so the father finally gave in. Through his friendship with the Stephenson family of railway builders, William arranged for Edward to apprentice at the Stephenson works in Newcastle, some eighty miles up the coast. He started the five-year apprenticeship on his fifteenth birthday in 1846, play transmuted into work.

"I was now in my element," he later wrote. The Stephenson factory, which included brass and iron foundries, specialized in boilers, steam engines, and large locomotives. Ranked at the top of its field in the world, it demanded long hours and ruthless standards. In his teens, Harland toiled from six in the morning till quarter past eight at night—except on Saturdays, when work knocked off at four in the afternoon. He made his way through various departments, learning one trade after another. After three years he was put in full charge of building one side of a locomotive. To keep up with his mate on the other side, a muscular, experienced Scotsman, Harland did nothing but work, eat, and sleep until the job was done. Sent on to the machine shops, he was proud to be assigned to the best screw-cutting and brass-turning lathe on the premises. Finally, finishing up in the drafting office, he refined the drawing skills first taught by his mother to turn out detailed designs for engines and other machines. A good Victorian engineer, he learned to handle a broad range of tools from hammer to sketch pencil.

Robert Stephenson's major project of the late 1840s was the Britannia railroad bridge across the Menai Straits in Wales. With the advice of William Fairbairn, the iron shipbuilder and ferrous advocate, Stephenson came up with a new type of wrought iron bridge: essentially a giant girder, 472 feet long, 15 feet wide, and 30 feet high, braced and enclosed by riveted iron plates on all four sides and by additional cells of iron bars at the top and bottom. The largest iron construction yet attempted, the bridge was so self-reinforcingly strong that initial plans for overhead suspension cables were dropped as unnecessary. Given this daring design, and the mighty puzzles of how to raise the four segments into place, the Britannia Bridge project was widely watched and debated during Edward Harland's apprenticeship. (The box girder structure later showed up in his White Star ships.)

His education stretched beyond the walls of the Stephenson works. On his holidays and long summer evenings, he had time to visit the principal mines and factories around Newcastle, picking up additional trades and techniques—and adding to his lifelong suspicion of book learning. He trusted only what he could see, heft, drill, hammer, file, screw, turn, weigh, measure, and sketch. "The sons of shipbuilders are over-educated," he later contended, "and are apt to lose themselves in figures, because they are very apt to follow the easier course of study, having books and papers before them, and neglect the workshop." Harland never neglected the workshop. Among his peers, by his later testimony, he found no one equally ablaze. About thirty other apprentices were training along with him at the Stephenson factory. "Some were there either through favor or idle fancy," said Harland; "but comparatively few gave their full attention to the work, and I have since heard nothing of them."

While absorbing every aspect of engineering available to him, he had not lost his boyhood affection for boats and ships. In the Stephenson drafting office, he worked on occasional designs for iron vessels. A government contract assigned the factory to build three large floating iron caissons for a dock. Since these caissons resembled ordinary iron ships, the works manager put Harland on the project. On his own time, he invented a radical version of a lifeboat: an enclosed metal cylinder, propelled by a screw at each end turned by sixteen men, with room for another dozen passengers. It was Harland's first new boat design of substantial ambition. Staying up late at night—sometimes never going to bed at all—he built a scale model, thirty-two inches long and eight inches wide, complete with all the internal fittings of tanks, seats, pulleys, and crank handles. He took it to sea and threw it into the water. Launched without ceremony, the craft rolled over but righted itself, riding waves or passing easily through them. Inspected afterward, it had taken on just a trace of water through the joints in the sliding hatches. Harland submitted the craft to a lifeboat competition sponsored by the duke of Northumberland, but it did not win—because, he later supposed, of its unprecedented novelty.

At some point during his apprenticeship, Harland met Gustav Christian Schwabe: the most important contact of his professional life. Schwabe came from a wealthy Hamburg merchant family, originally Jewish but baptised as Lutherans in 1819. Along with other relatives, Schwabe migrated to Britain for its freer, more industrialized business climate. In Liverpool he prospered as a textile merchant and ship manager. It so happened that Schwabe and Dr. Thomas Harland—an uncle of Edward's—had married women who were first cousins, the daughter and niece of John Dugdale of Dovecot, a wealthy merchant in Liverpool and Manchester. By this intricate network of family ties, Edward Harland came to the generous attention of Gustav Schwabe. He approved of the young man, eighteen years his junior. (Both were interested in ships.) Schwabe became Harland's sponsoring angel, dispensing a fruitful abundance of favors, connections, and—ultimately—major financing.

Harland completed his apprenticeship in 1851, on his twentieth birthday. For the next few years he passed through a series of brief engagements. The Stephenson managers hired him as a journeyman for twenty shillings a week. But business was poor, so Harland quit and went down to London to spend two months at the Great Exhibition in Hyde Park, luxuriating in the displays of art and mechanics at the Crystal Palace. It amounted to a daily graduate course in the latest prodigies of engineering from the entire industrialized world. Well inspired, Harland returned home and—with the help of Gustav Schwabe—took a job in Glasgow with the firm of James and George Thomson, builders of steamships and marine engines. Hired for the same wage of twenty shillings a week, he spent two years in the main center of British steam navigation. "I found the banks of the Clyde splendid ground for gaining further mechanical knowledge," he later recalled. On Saturday afternoons, he would walk along the river, visiting the famous shipyards of Robert Napier, Tod & Macgregor, and others, boarding their ocean steamers under construction and paying careful attention. At Thomson's he was promoted to head draftsman, but with no increase in pay. When a shipyard in Newcastle offered him the position of manager, Harland left the Thomsons. "They were first-class, practical men," he said later, linking those qualities, "and had throughout shown me every kindness and consideration. But a managership was not to be had every day." The Newcastle shipyard, however, did not match the engineering standards Harland had learned from Stephenson's and the Thomsons. After just a year, in 1854 he switched to a shipyard in Belfast, where he remained for the rest of his working life.

Like his home ground on the northeast coast, Belfast and northern Ireland had not yet produced any ocean steamships. The city's shipyards, active since the 1790s, lay twelve miles from the sea, on a river more narrow than the Thames. The dredging of the Victoria Channel in 1846 gave Belfast deep-water possibilities. On Queen's Island, which was created by reclaimed land from the dredging, an iron manufacturer named Robert Hickson took over a shipyard. Since he knew nothing about shipbuilding, Hickson allowed his new works manager, Edward Harland, a liberal hand in the enterprise. When Harland cut wages, his men went out on strike. He responded by bringing in experienced shipyard workers from Glasgow and Newcastle, and the strike was broken.

After three years, Harland—still only twenty-six years old—resolved to start his own shipbuilding business. Aside from the difficulty of finding skilled workers in Belfast, he had to import coal from Scotland, iron from England, and timber from Norway or Canada. Accordingly, with the help and advice of Gustav Schwabe he tried to obtain suitable property in Liverpool or, across the Mersey, in Birkenhead. This was the crucial turning point of Harland's life—and, as it turned out, of many thousands of other lives. If he had located his firm on the Mersey, the shipbuilding futures of both Liverpool and Belfast would have been vastly different. But the Liverpool authorities turned him down, citing his youth and lack of experience. They thereby did him a stupendous favor. Schwabe came to the rescue; family bonds again framed the arrangement. Schwabe's young nephew and namesake, Gustav Wilhelm Wolff, had apprenticed as an engineer in Manchester, then worked as Harland's assistant in Belfast. When Schwabe bankrolled Harland to buy out Hickson's interest in the shipyard for £5,000, Wolff became the head of its drafting department and then, a few years later, a full partner. The great oceanic shipbuilding firm of Harland & Wolff was launched on Harland's technical skills and Schwabe's nephew and money.

The firm's best customer for its first decade was the Liverpool steamship line of John Bibby, Sons. (They were introduced by, of course, Gustav Schwabe.) The Bibby company ran steamers down the coast of Europe into the Mediterranean. After Harland built his first Bibby ships to conventional designs, for the next order he suggested certain innovations that had first occurred to him a few years earlier, when he was working at the Thomson shipyard in Glasgow. The Thomsons had engined a steamer, the *Tiber,* that was 235 feet long and 29 feet wide. Such a high ratio of length to beam was then considered dangerous. "Nonetheless, she seemed to my mind a great success," Harland said later. "From that time I began to think and work out the advantages and disadvantages of such a vessel."

As he then understood the matter, the advantages were added carrying capacity at the same speed, but with no commensurate increases in power or coal consumption. The owner got the benefits; the builder got only the puzzle of how to make the longer, thinner hull rigid enough to resist hogging and sagging in heavy seas. The new design therefore needed a stronger system of internal construction. For his solution,

Harland drew—consciously or not—on the Britannia Bridge. By making the bottom of the hull essentially flat, and building the upper deck entirely of iron instead of wood, he made a long, enclosed, self-reinforcing iron box: a near duplication of Robert Stephenson's railroad bridge across the Menai Straits. "In this way," Harland later recalled, "the hull of the ship was converted into a box-girder of immensely increased strength, and was, I believe, the first ocean steamer ever so constructed."

The first two ships of this design—the *Grecian* and *Italian* of 1861—and three more that quickly followed were called "Bibby's coffins" because of their long, boxy shape and supposedly dangerous structure. Skeptics knowingly predicted that the reduced breadth would cause lateral instability and make the ships roll too much. If caught on the beam by a serious Atlantic swell, it was thought, Harland's creations would capsize, and if struck on the end, they would break in two. Not so; the ships made routine fast, prosperous voyages between Liverpool and the Mediterranean. But their sailors did nurse a legitimate gripe: Bibby's coffins were unusually wet ships. "All long vessels," Harland conceded in 1866, "ship water more than short ones." A shorter, beamier vessel stayed drier because it bobbed up and down and rose to every sea. The longer, thinner, sharper-bowed hull would rise just slightly and then slice through the crest of an approaching wave, shipping some water but—as Harland pointed out—taking a shorter, steadier route through the sea.

The *Persian*, which Harland built for Bibby in 1866, was 360 feet long and only 34 feet wide, an unprecedented ratio of length to beam of 10.6 to 1. Because of her length, Harland also gave the *Persian* a different snout. A typical ocean steamer still presented a flaring clipper bow. It looked undeniably dramatic and beautiful; seen from the front, it resembled the rising wings of a giant bird in flight. But, when combined with the usual bowsprit, jib boom, and figurehead, it weighed down the front of the ship. In tight situations, such as negotiating narrow rivers or channels or when docking, this busy forefoot made the ship longer and less maneuverable. The channel in Venice, where the *Persian* was supposed to call, was thought impossible for a ship of 360 feet. So Harland chopped off the conventional bow and rig and left the *Persian* just a straight, vertical stem. Compared to a clipper bow, it looked quite stubby and dumpy; and the banished figurehead seemed a sentimental loss. But the straight stem—as function over form—worked out well, creating no special sailing problems in heavy ocean weather, perhaps even keeping the nose higher and drier. And the *Persian* could still steam up the Venetian channel despite her length. The straight stem became another hallmark of Belfast steamers.

Over a decade, Harland built twenty ships for Bibby. During that time, the vessels grew from 270 to 400 feet; in length-to-beam ratio, from 7.9 to 10.8. These sharpening trends had no apparent inherent limit. Just how far would Harland go? "After a great deal of heavy and constant work," he later wrote of his Bibby ships, "not one of them had exhibited the slightest indication of weakness, all continuing in first-rate working order." Harland took special measures to make his skinny vessels strong. Aside from the overall box girder, he devised a tighter, tougher way of filling in the spaces between the iron frames of the upper deck. Usually this was done with wood. But iron and wood, metal and fiber, responded quite differently to changes in temperature and moisture, which caused deck strains and loosening. So Harland instead filled the spaces with Portland cement and covered the iron plates with cement and tiles, making a more rigid deck in all conditions. At midship, the point of greatest strain, his workers secured the sheer strakes on the sides of the hull with quadruple chain riveting, twice the usual application at the time. "My idea of a ship," he said to a group of naval architects and shipbuilders in 1868, "is that you should be able to kick her about like an old shoe without knocking a hole in her."

In his late thirties, on the eve of the White Star Line, Edward Harland was entering the major phase of his career. He was now well settled in Belfast. He had married a local woman in 1860; they had no children. Harland served on the Belfast Harbour Board, which carried out many improving port and dock projects of real benefit to the city's major shipbuilder. Though he spent his long workdays in the smoke and din of the Queen's Island shipyard, with its six building slips and 2,400 employees, he did not look like a famous shipbuilder. His general appearance and presence were unremarkable. "An unassuming gentleman," said an acquaintance, "who, to the eye, seemed scarcely able to distinguish a liner from an omnibus." For a man in his rough, vigorous trade, he was considered unusually attentive to his wardrobe: sometimes designing his own clothes, returning custom-made suits and shirts if they weren't just right, and harrying his bootmaker without mercy.

This fastidious care simply expressed, in a different medium, the way he ran his shipyard. He allowed no smoking; anyone caught in the act was fired. He was rumored to keep a telescope at home for checking on work in his absence. When he walked through the various departments, he seemed to notice every detail without looking to the left or right. If he spied a defect on a ship in progress, he pulled a chalk from his vest pocket and circled the offense. Odd inspirations could strike him at any moment; he would pick up any scrap of paper and rough out a sketch. After marking his first impression in Belfast as a strike buster, he was not considered an easy boss. Personally he was often quick-tempered and impatient. "He has been described as over-reserved and distant in his manner," according to a contemporary account. "That he was of hasty temperament can hardly be denied, and it was a fact he himself deplored. . . . The heat of the moment passed, his generous spirit prompted him to make instant reparation." Harland's partner, Gustav Wolff, was a tactful soul of burbling wit and good humor. Managing the shipyard in Harland's absences, he helped balance and soften his colleague's relentless driving qualities.

*E*arly in 1869, Gustav Schwabe invited Thomas Henry Ismay to dinner at his home in Liverpool. Ismay was thirty-two years old, an ambitious young shipowner who was then running vessels to Central and South America, Australia, and New Zealand. Schwabe was thinking about conferring yet another favor, of a sound and mutually rewarding business nature, on his nephew Gustav and his distant relative Edward Harland. After dinner, the two men retired to the billiards room for a friendly game. As they circled the table, no doubt in a cloud of cigar smoke, Schwabe got to the point. He wanted to put together a new transatlantic steamship line. If Ismay were willing to run it, and if Harland & Wolff built the ships, then Schwabe would raise the money.

Ismay was more than willing. Coming from several generations of ship men, he seemed to know about ships "by a sort of hereditary knowledge," the *London Times* later remarked, "as some men know about horses." Ismay grew up near the northern border of England, in Cumberland, across the Solway Firth from Scotland. His great-grandfather had owned one of the first shipbuilding yards in Maryport; his father started as a shipwright and became a timber merchant and shipbroker. As a boy, Thomas carved ship models with his first penknife. At school he was known for his love of the sea and for smoking tobacco. (His nickname was "Baccy.") When he was sixteen, an uncle arranged for his apprenticeship with a shipbroking firm in Liverpool. There he learned the business, at least from the landbound perspective of a clerk's stool in an office.

At nineteen, his apprenticeship completed, he went to sea for the first time. It was not a brief voyage: Ismay shipped out as the supercargo—in charge of selling and buying goods—on a sailing vessel bound for Chile. His shipboard diary, now preserved in manuscript at the National

The Cunard office and warehouse on the Halifax waterfront; passengers and cargoes arrived at the dock at the rear of the building and passed through the arched gate at the front. The building was torn down in 1919. (MMA)

Samuel Cunard of Halifax, Canadian shipping entrepreneur and founder of the Cunard Line; 1849 portrait by the American painter Albert G. Hoit. (PANS)

Cunard in his later years. (PANS)

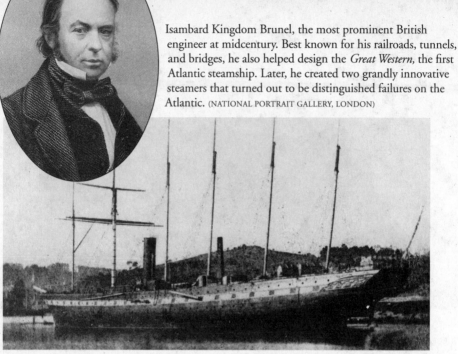

Isambard Kingdom Brunel, the most prominent British engineer at midcentury. Best known for his railroads, tunnels, and bridges, he also helped design the *Great Western*, the first Atlantic steamship. Later, he created two grandly innovative steamers that turned out to be distinguished failures on the Atlantic. (NATIONAL PORTRAIT GALLERY, LONDON)

Brunel's *Great Britain,* the largest ship yet, and the first iron-hulled screw steamer on the Atlantic. This photograph, one of the earliest of a steamship, was taken in 1844, during her fitting-out at the Brandon Wharf near Bristol. (NMM)

Brunel's *Great Eastern* under construction on the Thames in the mid-1850s by John Scott Russell. Six times larger than any previous ship, she was underpowered, unwieldy, and a costly debacle—"a melancholy illustration of great ambition but great ignorance," according to Edward Harland. (NMM)

The Collins liner *Atlantic,* built by William H. Brown. One of four luxurious American-built sister ships, the largest and fastest Atlantic steamers of the early 1850s. (NMM)

Edward Knight Collins of New York, founder of the Collins Line, in the mid-1800s the most serious American challenge to Britain's domination of transatlantic steam. (LIBRARY OF CONGRESS, "AMERICAN MEMORY" ON-LINE COLLECTION OF DAGUERREOTYPES)

Thomas H. Ismay of Liverpool.
In 1869, he co-founded with
Gustav Schwabe the White Star
Line, the most formidable
British competition that Cunard
ever encountered. (NMG)

Launched in 1870, White Star's *Oceanic* was the first modern ocean liner. It was
designed by Edward Harland after ten years of working out his innovations in
steamships for the Bibby Line, and built by Harland & Wolff in Belfast. (NMG)

The remains of White Star's *Atlantic* after she ran aground and sank off Nova Scotia on April Fool's Day, 1873. The ship, broken in two, lies with the forward part on its starboard side, and some masts still above the water. This was considered the worst transatlantic disaster (585 deaths) of the nineteenth century. (MMA)

White Star's *Britannic* (1874), the second generation of Harland-designed flyers built by Harland & Wolff. (NMM)

Guion's *Arizona* (1879), the first of many Atlantic greyhounds designed and built by William Pearce at John Elder's Fairfield works, at Govan on the Clyde. (NMM)

Cunard's *Etruria* (1884), also designed and built by Pearce at Fairfield, was the result of the reorganized line's decision, under John Burns, to reenter the costly race for transatlantic records. (NMM)

Inman and International's *City of Paris* (1889), one of the first two twin-screw steamers on the Atlantic. Designed by John Harvard Biles and built by the Thomsons on the Clyde, it was otherwise an American ship, masterminded by Clement Griscom and financed by the Pennsylvania Railroad and Rockefeller's Standard Oil Company. (NMM)

White Star's *Teutonic* (1889), designed by Alexander Carlisle, built by Harland & Wolff, and adored even by the easily displeased Henry Adams: "The big Atlantic steamer is a whacker." (NMG)

Cunard's *Lucania* (1893), the last notable Atlantic steamship built at Fairfield after the death of William Pearce. (NMM)

The American Line's *St. Louis* (1895), designed by Lewis Nixon and built by the Cramp shipyard in Philadelphia, was the first American-built fast Atlantic steamship in four decades. (NMM)

North German Lloyd's *Kaiser Wilhelm der Grosse* (1897), designed by Robert Zimmermann and built at the Vulcan shipyard near Stettin, was the most innovative Atlantic steamer since the *Oceanic* of 1870, and the first German-built ship to reign as the biggest, fastest liner on the Atlantic. (NMM)

A deck scene on White Star's *Germanic*, in which passengers are well muffled against the usual North Atlantic chill; taken perhaps in the 1880s. (NMM)

Passengers at determined play on the *Germanic;* the lady is trying her hand at quoits. (NMM)

Andrew Laing, the most brilliant marine engine specialist of his time, designer and builder of engines for Fairfield's Cunard and North German Lloyd flyers of the 1880s and 1890s. Dismissed by Fairfield in 1895, he later got his revenge with the *Mauretania*. (*Engineering Magazine* [New York], June 1893)

One of Laing's triple-expansion engines for Cunard's *Campania*, 1893. At 15,000 horsepower, it was one of the most powerful marine engines yet built. (*Engineering* [London], April 21, 1893)

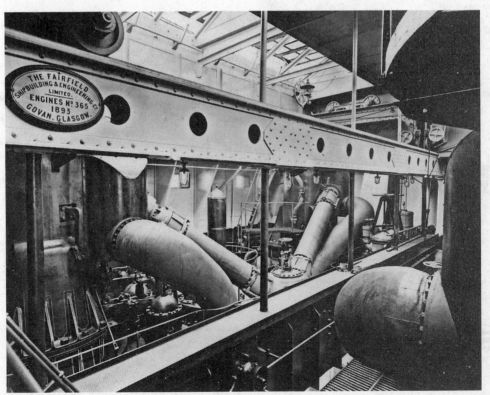

The engine room of Laing's machinery for the *Campania*'s sister, *Lucania*, seen from above; overhead, the engine-room skylight. (NMM)

Charles Henry Evans Judkins, the legendary commodore of the Cunard Line until his retirement in 1871. Famously stern and gruff to all passengers, he made an exception for pretty young women, like Fanny Appleton of Boston. (CAL)

Captain, officers, and sailors on the *Teutonic,* in 1897. (NMG)

Daniel Dow, a popular Cunard commander of the early twentieth century, and a seafaring friend of Mark Twain's. (CAL)

Officers and sailors on the exposed bridge of the *Teutonic*. (NMG)

Cooks and bakers on the *Teutonic*. (NMG)

Chief engineer and engine room staff on the *Teutonic*. (NMG)

Stewardesses on the *Teutonic*. (NMG)

Chief steward and assistants on the *Teutonic*. (NMG)

Greasers, firemen, and trimmers on the *Teutonic*. (NMG)

Maritime Museum in Greenwich, records a young man's wide-eyed initiation into life on the ocean. He noted his first bouts of seasickness and, amid flying dishes, "the difficulties experienced by Landsmen in general while dining at sea in a gale." A few days later, he was feeling better and already drawing conclusions. "Here we have no restraint," he wrote, "and it is on the sea where you can judge your fellow men best for they soon show the white feather in danger—and a life at sea is most assuredly one of danger and anxiety. But I begin to experience its pleasures also, and feel very comfortable." Two weeks out, as the ship approached the Madeira Islands, he heard the lookouts' cries of "Land Ho"—"decidedly the pleasantest sound that has reached my ears since my departure from Liverpool." But they had a good wind and so could not stop long enough to go ashore. Ismay settled in for the long passage down South America, around the Horn, and up to Valparaiso; it took three months. Homeward bound a few months later, he savored whatever seagoing enjoyments he could find, and occasionally stood watches when sailors took ill. "It is very monotonous on board ship at times," he noticed, "—you are glad of anything to amuse you for you cannot at all times be reading and writing." Back in Liverpool after almost a year away, and permanently cured of sea fever, he pursued a shipping career on dry land.

Ismay went into the shipbroking business, at first with an established partner, soon by himself. He bought his first ships, of a few hundred tons, and sent them to Mexico and the West Indies. His initiative and success were duly noticed by the ubiquitous Gustav Schwabe. In 1867, Schwabe helped Ismay acquire the name, flag, and goodwill of the White Star Line, a bankrupt company that had been started to run fast sailing clippers out to the Australian gold rush. (The price was just £1,000.) Two years later, after that legendary game of billiards at Schwabe's home, White Star was morphed into a transatlantic steamship line.

The new enterprise began with a single ship, the *Oceanic*. Indeed, Harland's plans for this vessel were drawn up and agreed upon by all parties even before the Oceanic Steam Navigation Company, Limited—the official name—was registered for stock offerings. The *Oceanic* was joined within just two years by five other ships of essentially the same design. They formed an instant transatlantic line, capable of providing weekly service between Liverpool and New York. Building six expensive steamers so quickly depended on Schwabe's skill at finding investors. In the first stock offering, Ismay took fifty shares at £1,750 apiece, Schwabe and Harland twelve shares each. Like Robert Napier and David Tod before him, Harland bought into the steamship company that would purchase his wares—and thus was investing in himself. Bound together by Schwabe as the heartbeat and common denominator, White Star and Harland & Wolff did business without a traditional shipbuilding contract. Harland was simply assigned to build large ships that could run at least as fast as the latest Cunard and Inman liners. Instead of a set price, the builders would be paid their costs plus a 4 percent commission. The absence of any contracted specifications gave the builders more engineering freedom; Harland could let his imagination run. This unusual deal depended on a rare level of trust, clinched by strategic family ties, among Harland, Ismay, and Schwabe.

The *Oceanic,* launched in August 1870, was the first modern ocean liner, the most influential steamship of her time or (perhaps) of any time. The hull resembled those of Harland's recent Bibby ships: a box girder, straight stem, and, at 420 feet long by 41 feet wide, a ratio of length to beam of 10.2 to 1, making her the thinnest steamer crossing the Atlantic. Seen from the dock, she rose a deck higher than her typical contemporaries; instead of the usual clumps of deckhouses, a single house stretched along most of the upper deck, with a promenade on its roof for passengers who were better protected, up there, from dousing waves and spray. Looming overhead were four masts and a single funnel, raked slightly backward. Open iron railings replaced the usual high, solid bulwarks, allowing shipped water to be shed more quickly. In the wheelhouse abaft the bridge, the arduous duty of helming the ship through a contrary sea was substantially eased by John MacFarlane Gray's recently patented steam-powered steering gear.

The most striking innovations were inside. The main saloon and most of the cabins were shifted from their traditional places at the stern to midship, sparing passengers the exaggerated pitching motions and the constant noise and vibration of the screw machinery at the rear of the vessel. From the stern, the first range of cabins began 133 feet forward, followed by the saloon just past the funnel, then a second range of cabins which ended 104 feet from the bow. Cabin passengers thus spent most of their days and nights in the lightest, airiest, most stable section of the ship. From the funnel, placed slightly aft of the midpoint, the dirty coal smoke and smuts blew up and back, away from the most heavily used rooms. The saloon, at the widest point of the hull, was 82 feet long and 41 feet across, and brightened by larger than ordinary windows, fourteen inches in diameter, to port and starboard. The saloon's extra space left room for four long tables instead of two. In place of the usual padded benches, diners sat in individual revolving armchairs that left them freer to move about as appetites and digestive conditions required. Details included a piano and two coal-burning fireplaces with marble mantles. The separate smoking room was larger and more inviting than on competing ships. Cabins were almost doubled in size, with electric bells to summon stewards, more headroom, better ventilation, and steady, adjustable mineral-oil lamps instead of flickering candles. Water taps replaced jugs that rattled in their brackets (especially—it seemed—when people were trying to sleep). The entire ship reflected Harland's essential insight that comfort mattered, in every sense and to all senses.

By their technical improvements and lavish attention to comfort and luxury, the *Oceanic* and her sister ships recalled the Collins Line of two decades earlier. The Collins liners had pioneered a straight stem on the Atlantic, but since then Cunard and Inman had generally stuck with clipper bows. Two of the five Collins ships had, after all, come to deadly ends, in one case for unknown reasons that could have included the unorthodox stem. But now White Star was implicitly invoking the Collins tradition of plush and fearless innovations. Digby Murray, the White Star commodore, had served as first officer on a Collins ship—and thus embodied the links between the two lines. "The management has been essentially one of spirit and of 'go,'" said the *Nautical Magazine* of London, "reminding us somewhat of the bold and original commercial spirit of the Americans. . . . Each vessel possesses what every American considers to be a positive necessity, namely, a splendidly appointed barber's shop." After the *Oceanic,* the next White Star ships were to be called the *Atlantic, Pacific, Arctic, Baltic,* and *Adriatic*—exactly the same names as the five Collins liners. (On second thought, considering the grim fates of the earlier *Arctic* and *Pacific,* those White Star vessels were quietly renamed the *Celtic* and *Republic.)*

Digby Murray and Thomas Ismay took the *Oceanic* on her maiden voyage to New York in March 1871. After turning back to Liverpool when her main bearings overheated, she arrived late—which apparently sharpened the curiosity of New Yorkers. She was the largest steamship to reach Manhattan since the *Great Eastern,* yet looked light and graceful despite her bulk. "Her lines are as fine and as delicate as those of any first-class yacht," the *New York Times* reported. The straight stem, that "great Yankee improvement," was noted with satisfaction. Some 50,000 people came to see the *Oceanic* at the White Star wharf in Jersey City, across the river from Manhattan. Her rearranged accommodations seemed amazing, almost unnerving. Even the steerage quarters looked more convenient, clean, and agreeable. "The visitor's ideas are apt to get hopelessly muddled," warned the *Evening Mail*, "with such an entire reversal of accepted traditions in marine architecture." Ship people were by nature conservative, as though too many changes too quickly would taunt and provoke the timeless, unchanging mysteries of the deep. Why else had the saloon and cabins remained at the stern for so long, despite the obvious drawbacks of that location? Freed of any such constraining precedents, the *Oceanic* was literally revolutionary. "A masterpiece of naval architecture," said the *Herald.* "On every hand there are so many improvements on the old style of ocean steamers that are suggestive of comfort that the beholder is lost in admiration and astonishment."

Edward Harland watched his creation with the inevitable pride of a father. Given the advantage of having a decade to work out his ideas in the Bibby ships, he had established new standards and necessities with his very first Atlantic liner. The *Oceanic* reached New York two months before his fortieth birthday. He would go on to design other notable White Star ships, but he had already stamped his name on transatlantic history. "For the first time perhaps," he later wrote, indulging an excusable flourish of parental exaggeration, "ocean-voyaging, even in the North Atlantic, was made not only less tedious and dreadful to all, but was rendered enjoyable and even delightful to many."

"So far," said the *New York Times* in September 1871, on the first arrival of the *Baltic*, "the White Star line is all and more than all the public expected." The *Baltic* was the third White Star liner, to be quickly followed by three others. The booming transatlantic traffic was suddenly more competitive. The individual ocean races between Cunard and Inman, always officially denied by company spokesmen, now had another horse. In April 1871, the *Oceanic* and Inman's *City of Paris* had left New York on the same day. At Sandy Hook, the mouth of the harbor, the *Oceanic* was thirty minutes ahead, but the Inman ship beat her to Queenstown. (Both lines denied taking the shorter, riskier, more iceberged northern route.) A few months later, the *Baltic* and the *City of Paris* departed New York together, and this time the White Star ship reached home five hours sooner.

The three transatlantic lines were closely matched. In public their leaders all said they cared more about safety than speed. Formal instructions to ship captains conveyed the same prudent emphasis. But paying customers wanted to arrive as soon as possible, to cut short the trials of sea travel and to be associated with the newest and best on the ocean.

Steamship companies felt constrained to play this game, of acting one way and talking another. In this atmosphere of pretense, and high stakes of profits and human lives, relations among the lines became testy. In November 1871, as the *Atlantic* and *City of Paris* were together approaching the British coast, the Inman ship overtook and narrowly passed the White Star liner, coming within fifty yards of her in a dangerous area of partly submerged offshore islands. (Thomas Ismay sent an indignant protest to William Inman.)

The following May, the *Adriatic* set a new westbound elapsed record of seven days, twenty-three hours, and seventeen minutes, an average speed of fourteen and a half knots. (The old record, by Cunard's *Scotia,* had stood for nine years.) The *London Times* printed the news. William Inman sent a protesting letter to the *Times:* the *Adriatic*'s time was improperly computed and was actually a bit more than eight days. Thomas Ismay sent a rebuttal to the *Times:* hogwash, and here is a copy of the *Adriatic*'s log to prove it. Inman sent a surrebuttal to the *Times:* for determining such records, a ship officially arrives in New York not at Sandy Hook, as White Star was claiming, but later, at the quarantine station. Heading off another sally from Ismay, the *Times* added a wearied, capping conclusion: "We can insert no more letters on this subject."

For the time being, the upstart line had the fastest ships on the ocean. During 1872, the White Star liners averaged nine days, nineteen hours to New York and eight days, twenty-one hours home. Those averages beat the Cunarders by thirteen and five hours, and the Inmans by thirty and thirty-one hours. White Star did have the advantage of a brand-new fleet of ships, with clean hulls, immaculate boilers, and the most advanced engines and machinery. Cunard and Inman were competing with mixed fleets of new and old vessels, and their more venerable ships pulled down their overall averages. Still, the new transatlantic player had managed an impressing, emphatic feat during only its second year of operations. In October 1872 the U.S. Post Office, in recognition, gave White Star the contract to carry its Saturday mail from New York, bypassing the older, slower lines.

Digby Murray served White Star as its commodore for the first two years. In the management structure, he stood between Edward Harland and Thomas Ismay, telling both the builder and the owner how their ships were doing. His stature and experience gave Murray the ballast to speak the truth as he saw it. A rarity among transatlantic leaders, Murray was a titled British nobleman, the eldest son of the tenth baronet of Blackbarony, from

Peeblesshire, south of Edinburgh. (The family baronetcy dated back to 1628.) Educated for the naval service, as a young man he served in the Royal Navy for five years, then retired to a varied career in merchant shipping. He met his American wife while working for the Collins Line. Involved with the White Star Line from its first planning stages, but never an investor in the company, he was independent and outspoken. "I always expressed my feelings pretty clearly," he said later. Forty-three years old in 1872, described by a Liverpool newspaper as a "bluff, hearty, but most courteous sailor," he spoke in a deep bass voice that left no doubt of his meaning.

As commodore, Murray commanded each White Star ship on her maiden voyage. In February 1872, amid the worst stretch of North Atlantic winter weather in years, he took the *Republic,* his fourth White Star maiden, out to New York. His formal report to management, written at intervals during the voyage, was angry and exasperated, as though his previous recommendations had been ignored. The *Republic* fought contrary winds and foul weather the whole way. Water flooded into the saloon and staterooms through the newly redesigned ventilators. Sails were very difficult to handle, Murray reported, because the pulley blocks were too small. The ship was too heavily loaded for a winter passage, weighing down the fore end, making her impossible to drive hard till the coal bunkers lightened. The scuppers by the seamen's forecastle were, again, easily clogged by cargo ("I have represented this every voyage"). Seven days out, a ferocious gale destroyed all but two lifeboats and nearly killed James Agnew Williams, the second officer. A wave slammed him against the funnel, breaking his left thigh bone, dislocating an ankle, crushing ribs, and leaving him bleeding from the nose, mouth, and ears. Water poured through broken skylights, dousing and terrifying passengers. "The ships, with their present doors, are not seaworthy," Murray continued, "and if they are not altered we shall yet lose a ship by them; had we been running instead of hove to, I doubt if we could have prevented the fires from being put out. . . . I feel rather discouraged in again applying for stronger doors; and if I have expressed myself strongly, I feel very strongly on the subject." Murray wrote as a sailor, trying to instruct landsmen and pleading for improvements. "I fear Mr. Harland will say this is an exceptional gale," he anticipated; "these ships, during their career, will run long enough, I hope, to encounter many exceptional gales."

On this occasion, it appears, Murray got the full attention of his colleagues. A year later, when White Star policies were being questioned—

and after Murray had left the company for an important shore job with the Board of Trade—he was asked about his report on the maiden voyage of the *Republic*. "Everything recommended in that letter was done to the ship," he said. "If ever I wanted anything, or thought there was not coal enough on board, I went straight to the office, and always got it. . . . All my suggestions have been carried out; every one of them."

Across the ocean, the White Star Line did have a more persistent critic: Bradley S. Osbon, editor of the *Nautical Gazette* in New York. After a picaresque career at sea and in naval journalism, he was briefly employed by White Star in 1870, apparently as its first New York agent. Something disrupted this arrangement; after Osbon and the line for obscure reasons parted company, he started his weekly newspaper of maritime news, and Joseph Hyde Sparks, an original investor in the White Star Line, came over from Liverpool to run the New York office. Against this murky background, it is difficult to know just what to make of Osbon's relentless criticisms of his former employer in the *Nautical Gazette*. He railed at White Star's newspaper ads for exaggerating the size and power of its ships, indicted the service and food on them, asserted that Murray had faked a record-breaking log entry, and predicted disaster in December 1872 when the *Adriatic* departed New York with a partly damaged screw. (The *Adriatic* came home safely, in good time.) Much of this barrage was directed at Hyde Sparks, the man who had taken Osbon's job.

All this might be understood as the flailings of a jilted suitor. Osbon was on more substantial ground, though, when he doubted the long, thin structure of the White Star ships. Many well-respected British steamship experts, including William Fairbairn, John Scott Russell, and William Froude, shared those doubts—and their opinions were not complicated by tangled employment histories with White Star. Edward Harland's liners, still so new, had not been running on the fearsome North Atlantic long enough for any proven conclusions about their safety. "We do claim," Osbon warned in January 1873, "that these *narrow* ships are dangerous, and our prophecies will, one of these days, be verified, when some awful disaster overtakes one or more of them."

Two months later, the *Atlantic* left Liverpool for New York on March 20, 1873. For this late-winter passage, she carried light cargo and only 33 cabin passengers. After stopping at Queenstown, the steerage quarters—fore and aft of the more expensive midship staterooms—were 80 percent filled with about 800 emigrants. When the ship headed out to sea, 12 stowaways emerged from their hiding places. With the crew of 146, that brought the total on board to around 1,000 people.

Down in the engine room, John Foxley was making just his second voyage as the *Atlantic's* chief engineer. He was not happy with the coal in his bunkers. White Star ships normally used the best steam coal from South Wales: it provided a quick, hot flame, with less smoke, and left the boilers cleaner. But a recent strike in the Welsh mines had cut the supply and raised the price of South Wales coal. So for this trip, as for her previous one, the *Atlantic* was burning a mixture of three-fourths Welsh and one-fourth Lancashire coal, which cost less. Some engineers preferred such a combination because the highly friable Welsh coal tended to crumble in handling, and the more durable Lancashire coal would bind with it and keep the unburned pieces from falling between the fire bars. But the Lancashire also burned faster and less efficiently, requiring more coal and stoking to produce a given volume of steam. The *Atlantic* usually consumed about 670 tons of Welsh coal on a westward passage. On her last voyage to New York, Foxley's first as chief engineer, she had been coaled with 1,155 tons of mixed Welsh and Lancashire, of which she had used 810 tons—140 over her average. Now she went out with about 927 tons of mixed coal, not a generous margin of safety for a westbound winter run after the previous trip's unusually heavy consumption.

The captain was James Agnew Williams, recovered from his injuries on the harrowing maiden voyage of the *Republic,* and promoted on the recommendation of Digby Murray to command the *Atlantic.* About thirty-five years of age, he looked older. In his previous job, commanding ships for another steamship line, he was fired for reckless drinking. But the White Star Line, expanding so quickly, needed experienced captains, and he came with solid recommendations. On this voyage, for the first three days out from Cape Clear he had good weather and favoring winds. With all sails set and all boilers blazing, the ship ran 938 miles in those three days. Then she ran into a prolonged midocean gale, blowing hard, contrary winds from the west and southwest, that cut the next four days to a total run of only 747 miles. On Friday, the twenty-eighth, as the ship was approaching the dreaded Grand Banks of Newfoundland, Foxley advised the captain they might be running short of fuel.

The voyage turned on two questions: How much coal had they burned to that point? How much was left? In the adverse weather of those four stormy days, the screw would have slipped more in the water, consuming extra fuel in proportion to the distance run. Foxley's inexperience as chief engineer may have allowed wasteful stoking, with fires burning too thickly. Cornelius Brady, the third officer, noticed persistent clouds of excess steam being vented out the pipe behind the funnel. The men below, shifting responsibility from themselves, faulted the fuel supply as too full of sand, clinkers, and Lancashire coal. Edward Egan, a fireman, later said the stokers were driven hard to maintain the normal fifty-two pounds of steam pressure. "We had to keep up full fires and prick them," said Egan. "You have to keep throwing on more firing with the mixed. . . . A fire of Welsh coal would last half an hour." As part of his regular duties, Foxley each day told the captain how much coal was left—a subjective estimate, derived from simply looking at the bunkers, that was further compromised by the ongoing shipboard clash of cultures and priorities between engine and deck men. To protect himself and his machinery against emergencies, an engineer would typically underestimate his remaining supply by about 10 percent. ("In plain terms," Edward Harland later explained, "he likes to sing out before he is hurt.")

At noon on Friday, then, Foxley told Williams they had only 319 tons left. They probably had more, and Williams probably knew that, but he had to act on the figure he was given. He never went near the bunkers; that was engineer territory. If the weather continued as it had for the last four days, they might not have enough coal to reach New York. They thought about running up to Halifax for refueling: not an uncommon detour for a westbound ship in bad winter weather, but a prodigious waste of time and money that White Star steamers had as yet avoided. Williams decided to try for New York. The ship's speed was cut from twelve to eight knots. To save coal, the captain turned off the steam heat in the main saloon and cabins and ordered the chief steward to limit his fires in the kitchen. The *Atlantic* picked her way across the Grand Banks, running as fast as she could on reduced steam. On Sunday, Williams told a cabin passenger, Charles W. Allen of London, that a late arrival could mean a hefty penalty for White Star's mail contract. "We must be in New York in time to leave Saturday next," he said, as Allen remembered it, "or we shall have to pay the United States government 5,000 dollars for breach of contract."

At noon on Monday, Foxley said they were down to 127 tons. The ship was still 460 miles from Sandy Hook—at current speeds, about two days away. Halifax was only 170 miles to the north. The barometer was falling, the wind still from the west, and the ship was making poor headway. Williams, Foxley, and Chief Officer John Firth all agreed: they should change course and head for Halifax. The ship made a sharp right turn. With the revised destination now so close, Foxley lit all his furnaces and the *Atlantic* steamed up to twelve knots, her top speed without sails. Running flat out, she could make Halifax by morning, coal up, and race on to New York, perhaps in time to be turned around and depart with that Saturday's mail.

The evening started cloudy, but the cover broke up at eleven o'clock, leaving just light passing clouds and occasional stars peeping through. Williams navigated by dead reckoning, a combination of compass headings and throwing the log at regular intervals to estimate his speed. In a fatal error, he did not take soundings; that involved slowing the ship and dropping a lead weight overboard to measure the water's depth. Such soundings, when placed against detailed charts of the waters off Nova Scotia, would have told Williams he was approaching shore. He later explained that he had not thought soundings necessary because the night was so clear. (It also would have meant repeatedly slowing down.) The current, running hard from the north and west, was stronger than Williams appreciated, warping his sailor's sense of where he was going. He had never before come onto the dangerous Nova Scotian coast. The *Atlantic*'s lookouts peered into the darkness, trying to make out the bright Sambro lighthouse at the entrance to Halifax Harbor. They never saw the light because the ship was actually charging toward an unlit rocky coast some eight miles southwest of Sambro.

At twenty minutes past midnight, Williams, wrongly estimating that he was still forty-eight miles south of Sambro, left the bridge and went to nap in the nearby chart room, leaving orders to be awakened at three o'clock. (Apparently he was not drunk.) At two o'clock the quartermaster at the helm, guessing that they should have seen the Sambro light by then, questioned the course but was overruled by the second officer, in command on the bridge. The captain was still asleep shortly after three o'clock when a man in the wheelhouse, astonished, saw breakers to starboard. The lookouts shouted "Ice ahead!" and "Breakers or ice ahead!" The helmsman put the wheel hard to starboard. The second officer reversed the engines, full speed hard astern, signaling the engine room three times. His hand was still on the engine telegraph lever when, at practically top speed, the *Atlantic* struck heavily on a rock and ledge off Meagher's Island, at the tip of Cape Prospect. The bow was jammed into a large crevice in the rock, which held the snout like a vise. Succeeding waves lifted the hind part of the hull and drove and dropped it again on the ledge, as many as four more times, ripping mortal gashes in the starboard bilge. It was 3:15 in the morning of April Fools' Day.

The grinding impacts woke and startled the sleeping people on board. Many things were happening at once, shocking and puzzling, piling together so suddenly. In the cold darkness, people still groggy from sleep groped and stumbled around. One steerage passenger—hearing and feeling the jagged crunch of metal on rock, and supposing the ship had reached Halifax—said, "There goes the anchor." A strong gust of air shot through steerage cabins, blowing out the lights and upsetting some of the berths. A cabin passenger heard a steward calling everyone out. "What are we to do?" he asked the purser. "Do the best you can," said the purser. Some thought they heard an explosion: perhaps the boilers bursting, because of unvented steam or when cold water hit them, or maybe the distress rockets that were being fired. Waves at once swept up all the port lifeboats. Chief Officer Firth cleared two starboard lifeboats for launching, but another high sea washed them away too. Only one lifeboat was safely lowered. Down below, frigid water cascaded into the ship, blocking cabins, stairways, and exits.

After a very short time, only six to eight minutes from impact, the ship went down at the stern; still held at the bow, she slowly canted over onto her starboard side, throwing people from the deck into the ocean and swamping the only lifeboat in the water. The forward part of the upper deck and most of the masts remained barely above the surface. Many people, especially those in starboard cabins, were trapped below. A terrible mingled moan came up: the death cries of the drowning. About twenty steerage passengers pushed out of portholes on the exposed side. The ship had keeled over to seaward and windward, baring the deck and rigging to oncoming waves. (If keeled in the other direction, the hull would have offered some slight protection.) Captain Williams climbed into the main rigging, urging others to join him. It was very cold. Bare hands were shredded on the freezing lines. Several hundred people simply clung to the slanting deck, washed by waves, paralyzed by fear and numbing resignation. Williams told them to keep moving, not to give up or lie down to sleep, but many just dropped off and died of exposure.

The nearest safety was an outlying rock about forty yards ahead and, some hundred yards beyond that, Meagher's Island. A sailor took a rope and swam for the rock. He couldn't get a footing and was hauled back. Another man tried, made it, and climbed up on the rock. From there, a second rope was taken ashore. One by one, the hardier men left the railing and rigging and hauled themselves along the lifeline to the rock. Many were drowned or swept away in the attempt. Around two hundred people crowded onto the rock. With no space left for new arrivals, some proceeded to the island. About fifty reached shore, battered and freezing. The rest stayed on the rock, swept by wind and waves, and dying by slow degrees.

The ebbing tide left the *Atlantic*'s stern unsupported. At five o'clock, with her bow still wedged into the rock, the ship broke in two just behind the foremast. The rear piece, about four-fifths of the hull, sank deeper. At daylight, fishermen came out from the island in small boats and brought survivors from the rock to the shore. A few boats reached the ship and took people from the rigging, including the captain. (His hands were frozen and his legs were stiff, he later explained, and he was exhausted, still feeling the effects of his injuries from the *Republic*.) John Firth counted thirty-two people still in the mizzenmast rigging with him. For long hours, the rough sea and high wind discouraged any more boats from coming out to the wreck. Island residents gathered on the shore, could see the dwindling band of survivors holding on to the rigging, but were afraid to do anything. Finally, at two in the afternoon, when Firth had been clinging to the rigging for ten frigid hours, a local Anglican minister named William J. Ancient persuaded a few fishermen to row him out. A Royal Navy seaman in his youth, the Reverend Ancient pulled in Firth and John Hindley, twelve years old, the last two survivors from the remains of the *Atlantic*.

The cold, wet, windswept aftermath proved nearly as lethal as the sinking itself. Most of the survivors were young, strong men. All the women died, and all the children except John Hindley. Ninety-four of the 146 crew members lived; they had the seagoing experience and toughness to survive hours in the rigging and on that outlying rock. The final death toll was only approximate because the exact number that had left Queenstown could never be established. (The purser and his records were lost in the wreckage.) The most probable estimate counted 585 dead: the worst transatlantic steamship disaster of the nineteenth century.

" *O*ur prophecies are being fulfilled, and if some of our warnings are not heeded in time, even worse disaster will overtake some of these long, narrow ships," Bradley Osbon wrote in his *Nautical*

Gazette. "The White Star Line has been taught a fearful lesson, and Heaven help them to profit by it." In the immediate aftermath, some of the established critics of Edward Harland's slender vessels pounced on the dreadful news with unseemly satisfaction. Here was the apparent proof, they asserted, of a murderously faulty design—and nearly six hundred innocent bystanders had paid for it. "This vessel of snake-like form is most dangerous," William Fairbairn wrote to the *London Times.* "She was extremely weak for her length. . . . That unsound principles are at work in the building of iron ships I have not the shadow of a doubt."

A catastrophe of such proportions had not befallen the transatlantic trade for almost two decades, not since the disappearance of the *City of Glasgow* in 1854. The *Atlantic* disaster made no sense. She had not hit an iceberg in the graveyard of the North Atlantic, or collided with another ship in a fog, or disappeared into a raging tempest in midocean. On a clear night, with stars showing and moderate wind and seas, she had blundered at full steam onto a well-charted coast, miles from where her captain and officers believed her to be. What? How? On both sides of the ocean, ship men and the traveling public needed and demanded some sort of explanation. The daringly new shape of the White Star ships became a plausible culprit.

Formal inquiries in Halifax and Liverpool gathered testimony from all the surviving principal players, including Harland, who made a quick trip to the wreck site to pick through the remains. The problem, so the accumulated evidence seemed to show, did not lurk in the ship or her design. "She was one of the best built and strongest ships I have ever seen," a Board of Trade surveyor testified at the hearings in Liverpool. "She perfectly kept her shape, and I have not seen a leaky rivet or leaky bolt in her hull, or in her upper deck." The sinking of the *Atlantic* was a great tragedy without a single villain or cause. Instead it was a fateful string of unrelated events and situations: the price of Welsh coal, the stinginess of the White Star Line in supplying the ship with an insufficient mixture of Welsh and Lancashire coal, John Foxley's inexperience as her chief engineer, the four days of adverse weather, the lack of trust and candor between engine and deck men, the deranging effects of commercial emphases on speed and money, the wish to avoid a penalty on the mail contract, and—in particular—James Williams's ignorance of the Nova Scotian coast, his poor navigation, and his failure to take soundings. Each of these factors by itself was probably harmless. But put together, in that sequence, they killed 585 people.

White Star's two main competitors tipped their own conclusions by quickly imitating most of Harland's improvements, including his slender hulls. Cunard had already built a trio of ships in 1870—the *Abyssinia, Algeria,* and *Parthia*—which, on the model of Harland's Bibby liners, had straight stems instead of clipper bows. Intended mainly for cargo and steerage, and not especially fast or luxurious, these three ships still had conventional hulls with length-to-beam ratios of under 9 to 1. In 1871, when Cunard first planned two major new, fast mail steamers, the *Bothnia* and *Scythia,* the initial designs projected ships just a bit longer and thinner at 384 feet by 42 feet. During the following year, as White Star ships were outrunning Cunarders across the Atlantic, the new ships were extended 41 feet longer, at the same width, yielding a revised ratio of 10 to 1. When finally launched in 1874, the *Bothnia* and *Scythia* looked like White Star liners masquerading under a Cunard flag and paint job. Aside from the fashionably slender hulls, they had straight stems, cabins and the main saloons at midship, four tables in the saloons, and larger smoking rooms than Cunard tradition had previously allowed. A few years later, the two ships got revolving armchairs in the saloons, another borrowing from Belfast.

The Inman Line retained its traditional clipper bow but otherwise followed the new pattern. In 1872 it withdrew from service the *City of Brussels,* only three years old, for extensive alterations. She emerged with her saloon moved to midship, an extra deck that gave her a higher profile, electric bells in the main cabins, and other Harlandesque touches of comfort and luxury. Three new Inman ships that same year—the *City of Montreal, Chester,* and *Richmond*—also flattered Harland by imitation, though none was as fast or profitable as the White Star liners. These new ships, all delivered before the *Atlantic* sinking, could not have been affected by any chastened rethinking in the wake of that catastrophe. But the contract for another new liner, the *City of Berlin,* was signed six months after the disaster, and it called for a hull even longer and thinner than the *Atlantic*'s, of 499 feet by 44 feet, a daring ratio of 11.3 to 1. Starting her service in the spring of 1875, within six months the *City of Berlin* had beaten White Star's transatlantic records in both directions.

The slimming trend, pushed along by intense triangular competition, had reached its natural limits. Carried to an extreme, it violated Harland's insistence on passenger comfort; such vessels rolled excessively and shipped too much water in heavy seas. Stories were passed around, perhaps unreliably, of a grisly voyage of an unnamed, extremely thin

Liverpool liner during which eighteen sailors were washed overboard, and of another trip from which the entire crew deserted in New York to protest such dangers and discomforts. At the same time, new resistance studies by the freelance engineer William Froude showed that a beamier ship with a sharp bow actually moved through the water more easily. Returned to a ratio of around 8 to 1, a steamer gained interior space, stability, and longitudinal strength, with no loss (and perhaps even an increase) in speed and economy. As the design pendulum swung back, that became the revised standard model.

In other respects, the legacy of the *Oceanic* was more enduring. "It is to this line," the journal *Engineering* said of White Star in 1878, "we are in a great measure indebted for the rapid advance in marine architecture and engineering during the last few years." For centuries, oceangoing ships had been high at the ends and low in the middle. Harland made his White Star liners low at the ends, with turtlebacks to shed water, and high in the middle, to put passengers and their accommodations away from waves and spray. From this time on, ship superstructures grew steadily loftier, decks added on decks. Over the next few decades, in an era of sharply expanded competition and added players, an unequal share of the improvements came from Harland & Wolff in Belfast and White Star in Liverpool. "Other builders have adopted the type," the *Marine Engineer* of London noted in 1885, "though some of them have not had the grace to acknowledge it." Year after year, as judged by speed, comfort, economy, and space, the White Star liners ranked among the finest on the ocean. Even Bradley Osbon finally tempered his criticisms.

11.

Competition and Invention

*T*he transatlantic race was heating up. During the quarter century that followed the advent of the *Oceanic,* the North Atlantic was crossed by more fast steamship lines than ever before or since. Within this sharpened competition, the form and power of a typical transatlantic liner evolved with unprecedented speed. By 1895 the ocean steamship had essentially matured into the design it would retain through all future improvements. These swift changes were driven by a tangled, circular interplay between engineering and enterprise. On one hand, the naval architects and engineers, now organized into professional associations that enforced higher standards of education and rigor, aimed to raise their work from craft to science. On the other hand, ocean entrepreneurs of more commercial intentions hoped to draw added customers with the best, newest, fastest ships—that were also safe, cheap to run, and durable. These requirements and purposes often clashed with each other; speed, as always, raised issues of comfort, safety, and economy. But competition, as always, came down to speed. Modernity never slowed down. Everything had to go faster.

*I*n 1860, on the eve of its Civil War, the United States ranked fourth in the world in industrial production, behind Britain, France, and Germany. The long war demanded sudden increases, especially in factories and railroad building, that set off a remarkable explosion of enterprise in the decades after 1865. The uncommonly rich natural resources within the country—of soil, climate, forests, waterpower, coal, iron, and petroleum—were joined by human waves of immigration from Europe. Powered by steam, stitched together by tracks and trains, enabled by the prevailing fashions in politics and economic theory, the United States became a baby colossus. It soon boasted more railroad mileage than all of Europe together. By 1890 it led the world in industrial production, with a booming annual output of manufactured goods almost equal to the combined totals of Britain, France, and Germany. No nation had ever come so far so quickly, powered by such geometric leaps in capital investment, population, and productivity.

The American industrial and agricultural heartland was served by (and to) the Pennsylvania Railroad, the largest corporation in the world. Its lines ran west from Philadelphia to the coal mines and steel mills around Pittsburgh, the factories and farms of the Upper Midwest, the slaughterhouses and grain exchanges of Chicago—and out to the Mississippi River nexus of St. Louis, down to the national capital in Washington, and up to Jersey City, across the Hudson River from New York. This sprawling empire of rails and rolling stock was assembled by J. Edgar Thomson, the relentless president of the Pennsylvania Railroad from 1852 until his death in harness in 1874. Less flamboyant and well publicized than many other American business leaders of his time, but notably more effective, he created the first major integrated railroad trunk line, fed by webs of subsidiary lines. The Pennsylvania was one of the first corporations to be run by experts—engineers and professional managers like Thomson—instead of by owners. They kept the line profitable but also plowed capital into technical innovations (Westinghouse air brakes, steel rails, train dispatching by telegraph) and interregional expansions.

As the most important enterprise in the nation, the Pennsylvania affected any undertaking within its vast sphere of influence. Andrew Carnegie got his start with the railroad, rising from telegraph boy to superintendent of its western division. Its leaders financed his later ventures, and Carnegie named his great steel plant near Pittsburgh the J. Edgar Thomson works, after his best customer. The American petroleum business began in the oil fields of northwestern Pennsylvania, from the New York border to the outskirts of Pittsburgh. John D. Rockefeller's Standard Oil Company at first competed with the railroad, then secretly conspired with it to adjust transportation costs and drive out other oil refiners. (Today Carnegie and Rockefeller remain much more famous than Thomson, but it was his railroad and mutual favors that really launched the imperial fortunes of both men.)

From almost the moment in 1857 that the Inman Line switched its American terminus from Philadelphia to New York, Thomson wanted his own North Atlantic steamship company to connect his railroad with European markets. Geography did not favor his plans. Philadelphia lay far inland, some 103 miles up the Delaware River from the ocean capes at the entrance to Delaware Bay, and about fifteen hours longer by steamship from British ports than New York City. The Delaware River, twisted into sharp bends and powerful, shifting currents, was not easily navigated. Yet Philadelphia was also closer than New York to the cornucopia of goods flowing eastward along the Pennsylvania Railroad: in particular, to the oil, coal, and iron products that were being pulled out of the ground and processed in the western part of the state. If loaded directly from the railroad's wharves along the Delaware onto its own steamers, and then sent across the North Atlantic, this traffic would profitably bypass New York and complete Edgar Thomson's grand design for a self-contained system of transportation stretching over 4,000 miles, from Chicago and St. Louis all the way to Liverpool.

Thomson first lent his name to a scheme by a Philadelphian named Henry Randall to build wooden steamers with two paddle wheels hung on each side of the hull. It was a grandiose, Brunelian vision: the ships were to be 500 feet long and 8,000 tons, with iron bracings and bulkheads, and room for 3,000 passengers. They remained just a vision; Thomson lost interest. The Civil War impeded but did not halt his steamship plottings. In 1863 he had his board of directors promise the railroad's support for any feasible plan—simply as an act of good citizenship, he said, not just for the material advantage of the railroad. "If the traffic of the country passing over our Highways to the markets of Europe cannot be provided with complete facilities in our own Port," he pointed out, "it will of necessity seek other commercial centres." To encourage local enterprise, Thomson offered lower rates to oceanbound traffic on the Pennsylvania's lines west of Philadelphia than on the eastward run to Jersey City and the Hudson. "With this advantage," he asserted, "I can see no good reason why a Line of Ocean Steamers cannot be well sustained." The railroad's board formed an internal steamship committee under one of the directors, Edward C. Knight, a Philadelphia sugar refiner and shipowner. (Earlier Knight had endorsed Henry Randall's steamship projections.)

The Knight committee stuttered along for years, bringing proposals to the board that were discussed but dropped. Among the deterrents was the unresolved question of whether adequate steamships could be built in the United States. The American shipbuilding industry, formerly so rich and accomplished, had fallen into a headlong decline that even the general postwar boom could not reverse. The war, and fast Confederate raiders, had crippled American shipping interests; and ocean steamships could be built 20 to 35 percent more cheaply in Britain, given the lower prices of both materials and labor. But the most serious wounds to American shipbuilding were self-inflicted by its stubborn, baffling, outdated allegiance to wooden hulls, sailing ships, and wooden paddle wheels for steam propulsion.

Only one major American shipyard, the Cramps of Philadelphia, managed a successful transition from wood and sail to iron and screw propellers. While New York, the center of American shipbuilding, clung to the old technologies, the yards along the Delaware—closer to the mines and mills of western Pennsylvania—were quicker to adopt iron hulls and screws, but only for smaller vessels designed for river and coastal routes. "All New York ship-builders and marine engineers spoke of propeller engines with the most profound contempt," Charles H. Cramp said later. "The screw-propeller was sneered at by them as a low-down Philadelphia idea." As Cramp, from his perspective on the banks of the Delaware, came to understand the situation, American shipbuilding was ruined by the provincial obtuseness of New Yorkers. "Anything of Philadelphia to these engineers was the limit of low-down ability in a technical way," he maintained. "The influence of Philadelphia, as we had no large ships or large steamship companies, was not listened to."

The Cramps—William and his sons, Charles in particular—had been building ships in Philadelphia since 1830. For three decades they turned out wooden packets, clippers, and canal boats, and occasional screw vessels and steamers. Government contracts during the Civil War pushed them into iron construction. "Our yard became a sort of kindergarten, as most of the workmen had to be trained to the work and working appliances had to be designed," Charles Cramp recalled. "We soon found that practically new men would have to be utilized, as the boiler makers, mostly foreigners, who worked at the business were few in number and did not grasp the mechanical necessities of the new departure." Only the younger carpenters and joiners, not yet set in their ways, could be retrained to bend iron frames and plates, and to pound rivets home instead of nails. Cramp became one of the few American advocates of the new technologies so long established in Britain, while leading New York ship men like John W. Griffiths continued to proclaim the advantages of wooden hulls.

In 1870, thirteen years after the Inman Line had forsaken Philadelphia, the Pennsylvania Railroad finally took definite steps to replace it with two Atlantic steamship companies of different purposes: a line of high-speed steamers built in the United States to attract cargo and first-class passengers, and a second line of slower ships, built in Britain, to transport cargo and emigrants in steerage. After so many years of deliberation, the railroad intended to attract all aspects of the North Atlantic traffic. The central figure in both lines would be Clement A. Griscom of Philadelphia. In the railroad's tradition of professional management, he was not an owner but an expert employee, well experienced at running ships and goods across the ocean. For the next three decades, Griscom was the most important American leader in transatlantic steam. Forceful and creative, and quite persistent, he made his most telling moves toward the end of his career. Though ultimately defeated, he was far more successful—over a longer period—than Edward Collins of the Collins Line, his only real predecessor in the United States.

Of Quaker background, Clement Griscom was descended from several old Philadelphia families. (He was a great-great-grandnephew of Betsy Ross, born Elizabeth Griscom, the Quaker housewife who probably did *not* sew the first American flag.) Clement graduated from the Friends' Academy in Philadelphia when he was sixteen. His father, a prominent physician, expected his son to follow him into medicine. He arranged for medical school in Vienna, but Clement refused to go. "Very well, then," said the father, "if you won't be a doctor, then the sooner you go to work the better." Instead of studies abroad, he was dispatched to a clerkship at Peter Wright & Sons of 318 Walnut Street, importers of crockery and luxury goods. Within a few years, Griscom had shifted his employer toward a general shipping business, including buying and chartering transatlantic steamers. At twenty-two he became a full partner. James A. Wright, a son of the founder, remained titular head of the firm, but Griscom was already the main force.

Another Friend involved with ocean ships, Griscom was joining an extensive heritage, in Philadelphia and elsewhere. In general, Quakers had engaged in such enterprises at some risk to the purest tenets of their faith and practice. Samuel Cunard's father, Abraham, came from a century of Philadelphia Friends; but, once exiled to Halifax, he married a Roman Catholic and converted to Anglicanism before engaging in shipping. Sam, a nonbeliever, then showed no trace whatever of his father's ancestral faith during his career with ships. The first New York sailing packet line was started by Quakers in 1817. The main founder, Jeremiah

Thompson, worked to abolish slavery—but also felt no ethical obstacle to making a fortune from cotton picked by slaves. In 1850, John Grubb Richardson of Belfast and Liverpool cofounded the Inman Line, and as a good Quaker he made it available to the Irish emigrant trade. But then he felt obliged to sell his interest in 1854 because William Inman wanted to lend their ships to the Crimean War. A worldly business always called for hard compromises with such an unworldly faith.

Clement Griscom followed in this ambiguous tradition. A birthright member of the Race Street Meeting, raised a strict Hicksite Quaker, he later became a quite "Nominal Friend." He served liquor on his ships, drank a pint of champagne each day, and enjoyed a cigar after dinner. He looked like a riverboat gambler, with his high, florid complexion and a lush handlebar mustache. At work, plotting moves and fretting over his ships, Griscom was indistinguishable from any other driven American businessmen of his time, showing no hint of the temperamental caution often associated with Friends in trade. "The dominant personality is there," one observer noticed. "He is genial, yet you take no advantage of it; he is kindly, but his eyes can grow hard upon necessity." Griscom sent his sons to a Quaker grammar school but then on to the secular University of Pennsylvania instead of to the local Quaker colleges. At home, within his own family circle, some of the traditional ways persisted, such as the Friends' form of address. Once one of his boys came by his office unexpectedly. "Well," snapped Griscom, mixing his idioms, "what the devil does thee want?"

In February 1870, the firm of Peter Wright & Sons submitted a new steamship proposal to the Pennsylvania Railroad board. It was presumably Griscom's idea. The board had not discussed any steamship scheme since the fall of 1868; but, with the Wright proposal in hand, it now took a series of decisive initiatives. The Red Star Line, or International Navigation Company, combined the interests of the railroad, the Wright firm, and oil men from around Pittsburgh. Petroleum was then used mainly for lighting, as kerosene. The world's supply of whale oil was dwindling, and kerosene gave a steadier, less smelly flame than other kinds of oil. Sensing a burgeoning market, Griscom had moved the Wright firm into oil exporting in 1861. They sent ships to Europe loaded with barrels of oil, built petroleum tanks and wharves at Point Breeze on the Schuylkill River in Philadelphia, and spun off two subsidiaries for oil storage and refining. The Red Star Line was intended to carry Pennsylvania oil and other bulk products by steamship to the continent, and to bring back emigrants to Philadelphia and, by the Pennsylvania

Railroad, out to the Midwest and beyond. The line was financed by a few major, interested investors (mainly railroad managers) and was cushioned by a million-dollar bond guarantee, and other considerations, from the railroad.

To save initial costs, they decided to build their steamers in Britain. The American navigation acts forbade the registration of any foreign-built ships in the United States. The Red Star Line needed a European flag of convenience. According to the story later passed down in Griscom's family, he hung a large map of Europe on the wall of his office. With a red pencil he marked the key manufacturing towns of England, France, Germany, and Belgium. From the other side of the room, he aimed a pointer at the center of the red marks, walked toward the map, and landed squarely on Antwerp. By this simple process, supposedly, the Red Star Line became a Belgian company. But Antwerp was already the principal European port for Pennsylvania's oil exports (a trade that especially interested Griscom), with an annual volume twice that of Bremen, the second-ranked oil port on the continent. The Red Star Line probably landed in Belgium because of petroleum, not that map and pointer. In any case, the Red Star Line included only the necessary 1 percent of Belgian citizens among its investors. "We expatriated ourselves unwillingly and from commercial necessity," Griscom later explained. "The enterprise is American and always has been."

In the fall of 1871, Griscom went to England to arrange for the construction of his first ships. He signed a contract with the Palmer shipbuilding company of Jarrow-on-Tyne, near Newcastle, for three steamers of moderate size and speed. The *Vaderland*, 320 feet long and 2,700 tons, was the first Atlantic steamship designed to carry bulk oil in built-in tanks. (Previously oil was shipped in barrels which took a week or more to load and unload.) American authorities, however, refused to allow oil to be carried in passenger ships, so the next two Red Star vessels—the *Nederland* and *Switzerland*—were given conventional cargo holds. All three steamers had their first class-saloons and staterooms at midship, on the White Star model, and room for eight hundred passengers in steerage and over 2,000 tons of cargo. At the end of the *Vaderland*'s maiden voyage in February 1873, most of the eighty-three emigrants on board—from Germany, Switzerland, and Italy—landed at the new Red Star wharf at the foot of Washington Street in Philadelphia, crossed the street to the strategically placed Pennsylvania Railroad depot, and at once departed for points west.

The railroad invested more heavily in its other new transatlantic line, the American Steamship Company, buying most of the initial $700,000 worth of stock and guaranteeing a bond issue of $1.5 million. Of the ASC's nine directors, five also served on the railroad's board. Despite the immense resources and territorial reach of the Pennsylvania Railroad, it was taking a substantial risk. It was going to build four modern, iron, screw-propelled ocean steamships. As a true American company, it had to give the contract to an American shipyard. No such vessels had ever been built in the United States; the Collins liners of two decades earlier were wooden-hulled paddle wheelers.

Bids were invited from four Delaware River shipbuilders. The Cramps submitted the lowest offer, of $520,000 per ship, and got the contract. The ASC, in a revealing bow to the superiority of British ship-building, sent Charles Cramp and his top engineer, J. Shields Wilson, across the ocean to investigate "everything new in steamship construction" by touring the major British shipyards. They started at Birkenhead, dipped down to the Thames, and wound up in Glasgow. "Mr. Wilson and I were received with uniform kindness and politeness at every shipyard in Great Britain," Cramp said later, "and every opportunity was given us to investigate everything that had been done; if our trip had been a royal one, we could not have been better received." (Perhaps the British ship men did not take the American competitors seriously enough to hide anything from them.) Cramp and Wilson went home laden with the latest techniques, especially about marine engines, and with one imported expert: James Younger, whom they found working at the Clydeside shipbuilding firm of John Elder & Company. Twenty-eight years old, a graduate of the University of St. Andrews, and an accomplished engineer and draftsman, Younger emigrated to Philadelphia in 1872 to lend his Scottish expertise to the Cramps. "No problem in mechanics was too intricate or too subtle for him," the grateful ASC board later recorded. "His methods were entirely practical and though his purely scientific attainments were of a high order he never cared to exploit them except as his work required their use."

Work on the new ships went slowly, impeded by the scarcity of skilled workmen and delays in obtaining iron plates and large engine castings. To build four such large vessels, the Cramps laid out a new shipyard between Norris and York streets in Philadelphia, with a frontage on the Delaware over two blocks long. The dins of tools and riveters could be heard from some distance along and across the river. Work was broken down into a tight cluster of specialized buildings: of blacksmiths; boring tools, the boat shop—the largest structure at 200 feet long and 58 feet wide—for bending, punching, and cutting plates and frames; the machine shop for building engines; and the boiler shop, with shears able to lift loads of thirty and fifty tons. A patriotic visitor in early 1873 admired the first two ASC vessels rising in their slips. "Nothing could be more graceful than their pretty elliptic sterns," he wrote. "These ships are noble objects, and inspire one with confidence in the capability of Americans of turning out a finer model and finer workmanship than the much-vaunted Clyde steamers."

The ships were named the *Pennsylvania, Ohio, Indiana,* and *Illinois,* after the four states principally served by the Pennsylvania Railroad. They were essentially identical: 355 feet long and 3,000 tons, with accommodations for 76 first-class passengers, 800 in steerage, and 3,800 tons of cargo. Except for the obsolete placement of the main saloons at the stern, they were as up-to-date as any British steamships, with straight stems and modern engines, but a knot or two slower than the swiftest transatlantic mail ships. A proud American who took the *Indiana* to Liverpool in the summer of 1874 praised the food—"superior to that of all of its competitors"—and the roomy bathtubs, with hot and cold salt water. Through heavy seas and high winds, the ship "behaved like a living creature," parting head waves smoothly and obeying her helm like an agile pilot boat. "As an American, I could not other than be jubilant over the extension of the flag." A fellow passenger who had crossed the Atlantic twenty-eight times declared the *Indiana* the easiest, most comfortable ship he had ever experienced.

However, both the Red Star Line and the American Steamship Company steamed through oceans of red ink, with annual losses of up to $147,000. The Panic of 1873 shoved the American economy into a prolonged depression that hurt ocean business. The railroad, cutting costs and raising efficiencies, put both lines under Griscom's management. To make up their annual deficits, the two companies could usually extract "loans" from the railroad that would never be repaid. But that rich support also carried a loss of independence. In the fall of 1874, the ASC asked to withdraw two of its ships for the unprofitable winter season, as the Liverpool lines were then doing. The railroad, wanting to maintain full facilities for its eastbound cargoes and westbound emigrants, refused permission, so the ASC had to keep running all four ships, losing about $10,000 on every winter round-trip.

As a technically Belgian company, the Red Star Line could hire European crews and pay them less than the Americans earned on the ASC ships. Red Star also won a mail contract from the Belgian government worth $100,000 a year. As the business cycle crept upward, Griscom could gradually replace his prohibited oil shipments with other American imports that sold well in Europe, such as grain, meat, coal, and pig iron. Eventually the Red Star Line started turning small profits. And Griscom would be heard from again.

A dozen major steamship lines now carried the swelling traffic—of European emigrants in steerage and American tourists in first class—across the North Atlantic. The French Line, out of Le Havre, and two German lines, running from Hamburg and Bremen, all received certain favors from their national governments. Their steamers, for the time being, still came from British shipbuilders and did not yet compete with the speed and service of the crack Liverpool mail lines. The French and German companies would become more important later on, after they started building their own ships.

Two private lines operated outside the main course between Liverpool and New York. The Allan Line of Montreal was started by a Scottish seafaring family, the sons of a sea captain who took sailing ships from the Clyde to the St. Lawrence River. Two brothers emigrated to Montreal, two others to Liverpool. In 1853 they transformed the family shipping business into a transatlantic line, with their first four iron screws built by the Clydeside firm of William Denny of Dumbarton. The Allan liners carried the Canadian mail and provided the newly independent nation a vital steamship link with its mother country. Sir Hugh Allen, knighted in 1871, became a ubiquitous commercial presence in Montreal. His ships ran an especially dangerous route: the northern great circle passage (always thicker with fogs and icebergs than a more southernly course), then between Nova Scotia and Newfoundland into St. Lawrence Bay, and carefully up the difficult river, so strewn with natural obstacles, unpredictable winds, and yet more fog. During the long, icebound winters, the Allan ships could not even enter the St. Lawrence; instead they avoided Canadian ports and came south to Portland, Maine.

The Anchor Line of Glasgow emerged from the shipping firm of Handyside & Henderson, consisting of four Scotsmen. After early lives at sea, they ran sailing ships in the Mediterranean fruit trade, then small steamers to New York. Previous attempts to launch a transatlantic line from Glasgow—a logical step, given the nearby Clyde shipbuilders—had run aground on unrealistically ambitious plans. Glasgow lacked the vast railroad networks and manufacturing and commercial resources behind Liverpool. The Anchor Line began economically in 1856, with modest steamships of under 2,000 tons. In 1863, with the larger *Caledonia* and *Britannia,* it started competing against the established transatlantic lines. By the early 1870s, the Henderson brothers owned one of the most extensive shipbuilding and engineering works on the Clyde, and the only dry dock in Glasgow. As an integrated, self-contained enterprise, the Hendersons could build, engine, rig, fit out, and manage their own ships, all from Glaswegian facilities within easy walking distance of each other. (By contrast, Samuel Cunard, Robert Napier, and Charles Mac Iver had built and run their line's steamers from three distant cities.)

Both the Allan and Anchor lines survived dreadful strings of disasters. The Allan Line, committed to speed by its mail contract, lost eight ships from 1857 to 1865, causing hundreds of deaths. The murderous rigors of the northern route exacted their tolls; in April 1863 the *Anglo-Saxon* foundered on Cape Race, at the mouth of St. Lawrence Bay, killing 237 people. Eight Anchor Line steamers went down, typically during the winter season, between 1864 and 1873. Five of these disasters involved loss of life, notably the *Cambria* in October 1870, sunk off the Irish coast with 196 people. Both steamship companies managed to survive these streaks of hard luck and dead customers, in part because they enjoyed essential monopolies in markets not served by the Liverpool–New York steamers.

Two new private Liverpool lines competed directly with Cunard, Inman, and White Star. The National Line was founded in 1863, at a time when it appeared the Confederacy was going to win the American Civil War. British politics and public opinion were profoundly ambivalent about the conflict. The government in London acknowledged the South as a belligerent but withheld full diplomatic recognition. Before 1861, 80 percent of the South's cotton crop had gone to British textile manufacturers. The Union's naval blockade of the South was causing misery and unrest among British workers, especially in the Lancashire mills near Liverpool. Hoping to break this stranglehold, the Confederacy bought fast blockade runners from British builders—especially the Lairds of Birkenhead—which then were supplied and run out of British ports. During 1862 the Confederates inflicted an unbroken series of humiliating battlefield defeats on the Union. "They have made a nation," Chancellor of the Exchequer William Gladstone declared in a notorious speech that fall. "We may anticipate with certainty the success of the Southern states so far as regards their separation from the North."

In this climate, a group of Liverpool merchants led by William Rome and Charles Edward Dixon started the National Line to bring cotton from a newly independent Confederacy. Two of their first ships were named the *Virginia* and *Louisiana*. But then the Union started winning battles, and Lincoln's Emancipation Proclamation gave the war a moral dimension it had lacked. The *Virginia* and *Louisiana* were prudently renamed the *Greece* and *Holland*. After the war, the National Line found its niche with large, slow steamers, emphasizing cargoes and steerage passengers, and at cheaper fares than the older lines. "Really a marvel of accommodation and comfort at very low rates," Henry Morford remarked in 1867. Without competing for speed, first-class passengers, and mail contracts, the line made money—an operating profit of £120,000 in 1869—and paid plump dividends to its stockholders. By 1871 it had a fleet of twelve ships, topped by two new sisters, the *Spain* and *Egypt*, at around 440 feet long and 4,600 tons. The new ships each had room for 120 passengers in first class and 1,400 in steerage. Both vessels were built nearby on the Mersey, by the Lairds and the Liverpool Shipbuilding Company.

Stephen Barker Guion briefly worked for the National Line, managing its lucrative steerage business, before leaving to cofound his own Guion Line in 1866. He was a naturalized British citizen, born and raised in New York City, where in his early twenties he and his brother went into business with John S. Williams in the shipping firm of Williams & Guion. About 1850, as the only bachelor among the company's three partners, he emigrated to Liverpool to run its office there. In the waning days of American sail, they established the Old Black Star Line of sailing packets to specialize in the emigrant trade. (The "old" was an invention to impart a comforting patina of longevity, an admired quality in so risky an endeavor.) In 1859 the Brown Brothers office in Liverpool offered to sell the three remaining Collins Line ships to Williams & Guion, but nothing came of it.

Unlike most Americans involved with ocean ships, Williams & Guion understood where the future lay—and it wasn't with wooden paddle wheelers. "What we are seeing now was as apparent to me fifteen years ago as it is today," Williams said in 1869. "I saw that we were going to lose our commerce simply from the effect of the advantage which Great Britain had over us in the building of iron steamers. They had seen the advantage of those steamers before we did, and had facilities for building them, and pressed the business ahead." After the Civil War, as ocean commerce resumed, Williams & Guion shifted into iron screws. Most of the line's investors were British, but most of the stock was owned by Americans. As with the Red Star Line, its actual nationality was not made apparent. "Practically speaking, Mr. S. B. Guion is the line," the *Nautical Magazine* of London explained. "The American connections of the line have no doubt helped greatly towards its success."

The Guion steamers all carried American names. The first, the *Manhattan,* made her maiden to New York in August 1866. She was 343 feet long and 2,900 tons, with a clipper bow and two masts, built by the Palmers of Jarrow-on-Tyne. (Charles Mark Palmer owned forty shares in the Guion Line.) The *Manhattan's* hull and engines were designed by John Jordan, a noted engineer and shipbuilder in Liverpool. (He also owned forty shares.) As Guion's superintendent engineer for its first decade, he gave the line a reputation for technical innovations. By 1870, Guion had eight ships built by Palmer, all of around 3,000 tons, that were known for their easy seagoing qualities and—in particular—for providing adequate service to first class and steerage and cargo: the three components of the North Atlantic trade. (Clement Griscom gave his Red Star contracts to the Palmers because he wanted clones of their versatile Guion ships.)

The shipping lanes between Liverpool and New York were almost getting crowded. In the early 1870s, a traveler departing Liverpool on a Tuesday could choose between the Inman Line and Cunard's second service, to Boston; on Wednesday, between Guion and National ships; on Thursday, among Inman and the Cunard second line again, and the fancy new White Star liners; on Friday, National again; and on Saturday, the old standby, the top Cunard steamers to New York. (Inman and White Star ships, intense competitors leaving on the same day, would stage unofficial ocean races.) In 1873 these five Liverpool lines carried more than 236,000 passengers across the ocean, led by Cunard's 72,600 and Inman's 63,700—ample increases over prewar levels. Two decades later, those figures would look modest.

*T*he surging growths in commerce and competition were matched by sharper work from the shipmakers. The two spheres were quite independent, yet quite connected by lines that both goaded and retarded the activities of the other. The naval architects who designed the ships and the marine engineers who improved the machinery were all—at their best—practical artists. As such they worked for their own internal purposes of inspiration and creative satisfaction and would have invoked their iron muses almost regardless of external circumstances. But to find success, as the world understood it, they had to sell their wares and see them in profitable commerce across the unsparing North Atlantic. The artists bent over their drawing boards and workbenches, conducted experiments and ran trials, and then shared or hoarded findings with their peers—all within a small, enclosed world, boundaried by esoteric knowledge and language. The ultimate judge of the final product, meanwhile, lay out there, in the wider and less controllable milieus of competing transatlantic enterprises.

British shipmakers were now organized into professional associations, a sign of their growing specialization and self-regard. The Institution of Engineers in Scotland was founded in Glasgow in 1856; emphasizing ships from the start, it soon added "and Shipbuilders" to its name. The Institution of Naval Architects was started in London in 1860. Both groups established strict criteria and levels for membership, building walls around what they had come to regard as a profession, and held meetings once or twice a year to hear scholarly papers. The papers at these sessions provoked oral discussions, often vigorously contentious, that were then published with the papers in annual volumes of *Transactions*. Marine designers and engineers, formerly housed within sections of more general professional groups, now had their own forums—and invaluable printed records—for their particular feuds and issues, speeding and broadening the exchange of ideas.

Both institutions were initially organized around a single large, commanding personality. The moving spirit and first president of the Scottish group was William John Macquorn Rankine, the Regius Professor of Civil Engineering and Mechanics at the University of Glasgow. A man of stupefying intellectual range and industry, Rankine all by himself constituted a clearinghouse of engineering knowledge. His massive textbooks— on shipbuilding, applied mechanics, civil engineering, and the steam engine—remained standard authorities for engineering students well into the next century. Rankine also did original research and thinking in many unrelated fields, was one of the principal founders of thermodynamics, wrote an almost weekly article for the *Engineer* journal, and taught his classes at the university. Yet he carried this staggering workload quite lightly, "always ready to entertain or be entertained." He laughed easily, told stories, wrote songs, and sang them with gusto at convivial dinners with his friends. Perhaps his secret was that he so enjoyed his work that it did not seem like work. "His outflow of humour and music," a colleague noted, "was one of the striking aspects of Rankine's character or personal identity—striking because not often associated with such profoundness as his."

Largely self-taught, Rankine had the uncategorized mind and free-ranging intellectual play of an autodidact. He was born in Edinburgh in 1820, into a Scottish engineering family. At sixteen he won a gold medal for an essay on the undulatory theory of light. As an apprentice to the distinguished civil engineer John MacNeill, he worked on railroads, waterworks, and harborworks. Until his midthirties he made his living as a working engineer, spending nights on more theoretical studies and writings, notably in thermodynamics. His appointment to the Regius chair at Glasgow turned his attention toward hull design and shipbuilding, the city's main industry. Rankine's prolific writings, his application of the most abstruse reaches of higher mathematics to practical problems, and his many other activities—such as starting the Institution of Engineers in Scotland—helped keep Glasgow at the head of British naval engineering and shipbuilding.

Rankine's intellect took in realms and questions beyond the reach of most engineers. He read deeply in the contemporary fiction and poetry of his day and liked to hold forth about his favorite authors. He even ruminated over the ultimate meaning of nineteenth-century progress—the inexorable engine that he himself so richly stoked. Two of his own songs, published together in *Blackwood's Magazine* in 1862, suggest an unresolved internal dialogue. "The Coachman of the 'Skylark'" recalled the fastest horse-drawn stagecoach of its time, "Before your pestilent railways / Had spoiled all sorts of fun." Instead of a railway's dark tunnels and ugly embankments, the Skylark thundered across an unspoiled landscape, stopping at country inns for good food and foaming beer, and a smiling barmaid of sparkling eyes. Travelers breathed clean air, instead of choking on smoke and steam, and had time for friendly conversation and a good cigar. But "The Engine-Driver to His Engine" sang an ardent love song to speed, the century's most addictive drug. "My blessing on old George Stephenson! let his fame for ever last; / For he was the man that found the plan to make you run so fast." The train squealed along at sixty miles an hour, flashing past telegraph posts and milestones, soon bringing the singer home from London to his own true love, waiting by the cottage door in the North Country. Progress still allowed unspoiled rural pleasures. So which to choose?

The Institution of Naval Architects was founded at the home, and largely at the behest, of John Scott Russell. In January 1860 he stood at the peak of his career; the *Great Eastern* was recently finished, after eight grueling years, and its technical and commercial failings were not yet apparent. The INA grew out of Russell's long-standing distress over the lack of schools and institutions of naval architecture in Britain. (He had of necessity sent his own son to Paris for an education he could not obtain at home.) Earlier in the century, the Admiralty had twice established schools of naval architecture at Portsmouth, but they had not attracted enough students or support to continue. "France is snatching from us some of the valuable trades which we had fancied were exclusively our own," Russell warned. "I regret to say that in technical education we are utterly behind other civilized nations." A third British school was started in South Kensington in 1864, partly because of pressure from Russell and the INA. A few years later it was moved to Greenwich and took more permanent form as the Royal Naval College.

The INA became the leading organization of its kind in Britain because of its national reach. Though based in London, and considerably sustained by Admiralty officials and officers of the Royal Navy, its influence and activities stretched across the whole country. Among the new members admitted in 1867 were David Tod of Glasgow and Edward Harland of Belfast. The annual sessions, however, continued to reflect the verbose presence of Russell. As a founder and the first financial supporter of the institution, he got the floor whenever he wished during the oral discussions. He expounded at will, repeating his theories—now three decades old—of wave-line hull resistance, digressing freely into memories of his own career, and recalling episodes in the history of British steam navigation. ("Nothing can be more beautiful than watching the waves on the Lake of Geneva," he told his colleagues. "I have done so repeatedly.") After the *Great Eastern* finally ruined his shipbuilding business, he had plenty of time. Occasionally slight rumbles could be heard from younger members, impatient with the older man's meanderings. But in general he was treated with the affection and respect he had earned. "Whenever Mr. Scott Russell takes the trouble to tell us anything," an admiral noted at one session, "he tells it in such charming language that I almost feel myself in the position of Eve and the old serpent. I cannot find it in myself to resist those eloquent periods, that magnificent declamation . . . although sometimes I cannot help thinking it is my imagination, and not my reason, which gives way to his arguments." ("Your duty is to keep order," Russell advised a presiding officer, "and as I am the most disorderly man in the Institution, keep your eye on me.")

In the spring of 1870, Russell told an INA meeting that the Admiralty would not, as they had hoped, pay for resistance studies on full-sized ships. Instead it would underwrite experiments by William Froude, who was present, on ship models in a tank. That reminded Russell of his own tests on models, years earlier: "It was very interesting to me, and the most agreeable period of my life was that romantic period of about two years in which I was mainly occupied with the amusement of making pretty little experiments on a small scale." But the findings, he said, were worthless, contradicted by actual ships in actual running. So, within the restraints of collegial courtesy, he ridiculed the tests Froude had in mind. "You will have on the small scale," said Russell, "a series of beautiful, interesting little experiments, which I am sure will afford Mr. Froude infinite pleasure in the making of them, as they did to me, and will afford you infinite pleasure in the hearing of them; but which are quite remote from any practical results upon the large scale."

Poor Froude! A man of mild address and scholarly disposition, he found his project mocked, before it had even begun, by the formidable Russell and others at the session. "The feeling of the meeting is very much against experiments with models," he allowed. But he defended his intentions and methods, and the careful patience and persistence he promised to bring to the tasks. "I do not at all despair," he said. "We have still a great deal to learn, both practically and theoretically, and though I cannot set my reputation and credit against that of the various persons who have addressed you to-day, all I can say is that I believe, myself, I shall get a great deal of useful information from the experiments which I propose to make."

Froude was being too modest, as usual; he had already made important contributions to several fields, and his model tests would cap a peculiar career. He came from an English family of preachers and scholars. At Oriel College, Oxford, he studied under the theologian John Henry Newman, with whom he later exchanged intricate discourses about the nature of religious proof. He took first honors in mathematics but found his formal education of little value once he started working as a civil engineer under Brunel and others. Theory and practice seemed utterly unrelated. "When I became an engineer," he later reflected, "I had to unlearn much of what I had learned at college, and to shake off a lot of cobwebs which had accumulated in the old fashioned books . . . in which 'Theory' meant a something which nature <u>ought to have</u> conformed to, instead of solutions based on a sound apprehension of real natural laws as they in fact exist, and I had to pull down a lot of work and ideas built up on these imaginary lines of thought, and relay the old materials for myself by the light of working experience in a laborious though no doubt often a bungling form." He worked through the intellectual revolution of the nineteenth century, from received wisdom to the skeptical experimental method. "Our doubts in fact appear to me as sacred," he decided. "I believe that in no subject whatever is my mind capable of arriving at an absolutely certain conclusion."

At age thirty-six Froude retired from active, full-time work and went home to Devon, to care for his ailing father—an Anglican rector and archdeacon—and to be a country gentleman. The family income freed him to work or not, as he pleased, and to pick up any puzzle that struck his fancy. Emerging occasionally from this bucolic seclusion, he assisted Brunel with the *Great Eastern,* offering advice about ship rolling and wave action. Froude's later experiments in these fields, reported at early meetings of the Institution of Naval Architects, brought him his first taste of public recognition. Fame did not seem to interest him much. After his father's death, he moved down to an isolated home in Torquay, on the Channel coast.

Here he started playing around with ship models. He towed models from a launch on a river, measuring the resistance of various hull forms and lengths. To control his environment, he built a large masonry rainwater tank near his house, but it was too short; a cushion at one end stopped the models too soon to get useful data. He therefore obtained his Admiralty grant of £1,000, plus £500 a year for two years, in order to construct a truly ambitious tank. It was 270 feet long, 10 feet deep, and 38 feet wide, roofed over, with a railway running down the center line of the tank, suspended just above the water. An engine at one end pulled a "dynamometric" device along the railway at any speed, up to about 1,000 feet per minute; the device, attached to various models in the water, measured their relative resistance as they were dragged along. The models, 10 feet long, were molded of hard paraffin. Once measured, they were broken up and remolded in another configuration. "All this works very nicely," Froude wrote to a colleague in Glasgow, "and gives instructive results: which I am glad to show, and discuss with all true men, with whom discussion is instructive—so if you come this way please give me a look." His object, he insisted, was "simply to <u>know</u> all about the matter" by a disinterested search for scientific truth: "Why should we not all join in making these essential investigations."

Froude's experiments, meticulously reported and widely discussed, revised previous notions of ship resistance and hull design. Naval architects had believed that resistance was proportional to the midship sectional area: that is, to the width and depth of the thickest part of the hull, and not to its overall length. Froude instead emphasized the retarding effect of skin friction between the water and hull surface, over the entire submerged extent of the hull, and especially as related to its length. The skin friction increased with velocity, at a knowable ratio. By his "law of comparison," Froude showed how a full-sized ship's resistance could be deduced from its model. Shipmakers could thereby predict, much more precisely, the hull and engine power they needed in order to reach a desired running speed. Froude's studies also changed hull design. The wave-lines of Russell gave way to the stream lines of Froude and of Macquorn Rankine (who was also working this field): a sharp entry, then full, straight lines instead of hollow curves, broadening out to the maximum width of the ship.

Russell, skipping past his earlier scornful remarks, came to embrace Froude as his true intellectual disciple. "I can confirm many of the statements made by Mr. Froude from experiments conducted by myself long ago," he insisted at an INA meeting in 1874. "All the knowledge I have of the forms of ships, and their virtues and vices, was obtained in this manner. . . . I was often laughed at because I believed in experiments on my small models." One hopes Froude allowed himself a modest smile of vindication.

*In the machinery of propulsion, this period saw the advent of the compound engine, the most important advance in steam power since James Watt had his epiphany while crossing Glasgow green. The concept was already well known; its practical application was delayed for almost a century, until various mechanical improvements made it feasible. When steam was introduced into an engine's cylinder, and cut off before the stroke was completed, the steam kept expanding and exerting force. Instead of then being vented away, it could be injected into a second, larger cylinder, at lower pressure, and continue to expand and create steam power. It thus did more useful work from a given volume of steam, with less waste. The temperatures inside the cylinders of a compound engine cooled and heated within a more narrow temperature range than in a single cylinder, which also made it more efficient. Cutting such wastes meant a ship could carry less coal and fewer firemen, reducing expenses and leaving more space for cargo and passengers.

A compound engine was easily used on land, with endless supplies of fresh water nearby. To fit its more complex machinery into the tight confines of an ocean steamship was more difficult. The multiple cylinders needed higher boiler pressures to drive them; the finer tolerances of high pressure required clean freshwater for the boilers. Surface condensers, first introduced in the 1830s, were supposed to provide pure water. But the lubricants of the time, based on tallow and other animal fats, clogged the condenser tubes and decomposed into fatty acids and soluble copper salts that attacked the iron plates and tubes. Petroleum-based lubricants, first drawn from the Pennsylvania oil fields, solved this problem because they did not deteriorate under high temperatures. At about the same time, the old box-shaped tubular boiler yielded to James Howden's cylindrical or "Scotch" boiler, smaller and lighter (always desirable qualities on a ship), yet stronger and safer at the higher pressures demanded by compounds. New machine tools, especially hydraulic riveters, allowed boilermakers to turn out tighter, more consistent products.

More than anybody else, John Elder of Glasgow made the compound engine usable on ocean steamships. As the third son of David Elder, Robert Napier's long-established works manager, he grew up in the smoke and iron din of Clyde shipyards. At high school he excelled in mathematics and drawing, then briefly partook of a college engineering course. Most of his education happened during a five-year apprenticeship in the Napier yard. In 1852, at twenty-eight, he left his father's supervision to join Charles Randolph in a firm that built steam engines. A few years later they started building ships as well, pushing constantly for cautious shipowners to try compounds. Elder was always in a hurry; in a little more than a decade he took out eleven patents on compound engineering. "He was handsome beyond the average of men, with fine features and dark clustering hair," according to one contemporary account. "His skill and quickness in overcoming difficulties of all kinds were marvellous. Few could move so fast as he." Elder sped through a brief life before dying of liver disease in 1869, at the age of forty-five. His firm, eventually renamed the Fairfield Shipbuilding and Engineering Company, went on to new feats under the dynamic management of the naval architect William Pearce.

Ocean steamers with compound engines were first applied to longer routes, out to China and the west coast of South America, for which a compound's reduced coal use—by a factor of 30 to 40 percent—made compelling arguments. In the late 1860s, the Anchor Line's *India* and the National Line's *Holland* became the first transatlantic ships powered by compounds. Once they demonstrated their durable economies, compounds became the Atlantic standard almost overnight. The White Star Line ran, from its inception, on fast but frugal compound engines designed by Edward Harland. Those engines were crucial to White Star's sudden leap into transatlantic prominence. When Charles Cramp toured British shipyards in 1871 for the new American Steamship Company, his visit to the Elder company—and the particular snap and bustle of work there—convinced him to adopt compounds for the ships he was about to build in Philadelphia.

John Elder had bridged the two major phases of transatlantic Scottish shipbuilding in the nineteenth century, from the Napier yard that engined the first two decades of Cunard liners (and built the *Persia, Scotia,* and *China*) to the Fairfield company that produced most of the famous Clyde steamers from the late 1870s till the end of the century. Alexander Carnegie Kirk, another Fairfield engine man, first introduced the triple expansion engine that became standard in the 1880s. His successors at Fairfield, Archibald Bryce-Douglas and Andrew Laing, were each in turn the most prominent British marine engine specialist of his time. These men formed a remarkable, unbroken line of succession, from David Napier through Andrew Laing: eighty years of Clydeside engineering leadership down a single steam lineage.

*I*n this time of accelerating enterprise and engineering, the Cunard Line was not keeping pace. The newer Liverpool companies were quicker to try technical improvements, and in general they offered customers faster passages with better food and service than Cunard. At earlier times, before the transatlantic trade started brimming with competition, Cunard could coast along on its unique, remarkable reputation for safety. That no longer sufficed.

The American writer Samuel Clemens (Mark Twain), a constant traveler with an old Mississippi River rat's fond interest in any steam vessel, took the Cunarder *Batavia* home from Liverpool in November 1872. At midocean, after two days of a brutal storm, she came upon a ruined, dismasted sailing ship with nine men, still barely alive, lashed to her rigging. Seven brave members of the *Batavia*'s crew volunteered to attempt a rescue. As Clemens watched, "keeping an eye on things and seeing that they were done right," he later explained, "and yelling whenever a cheer seemed to be the important thing," the volunteers rowed three hundred yards through a still-stormy sea, plucked the nine men from the rigging, and brought them back safely to the *Batavia,* frozen and exhausted after more than a day and night of wet exposure to the wintry North Atlantic.

The episode so impressed Clemens that he wrote an homage to the Cunard Line, published in the *New York Tribune.* "It is rather safer to be in their vessels than on shore," he declared. "Those practical, hard-headed, unromantic Cunard people would not take Noah himself as first mate till they had worked him up through all the lower grades and tried him ten years." After passing through such a perilous storm, and watching that dangerous rescue, Clemens wondered how the line maintained its safety record, year after year. The answer, apparently, was that Cunard so clung to its own traditions and resisted most changes. "It is a curious, self-possessed, old-fashioned Company," Clemens decided. "When a thing is established by the Cunarders, it is there for good and all, almost. . . . Forty years ago they always had stewed prunes and rice for dinner on 'duff' days [Sundays]; well, to this present time, whenever duff day comes around, you will always have your regular stewed prunes and rice in a Cunarder."

A year later, however, Clemens crossed to Liverpool on the Inman Line's new *City of Chester.* His wife, Olivia, who had accompanied him on the *Batavia,* stayed home in pregnancy. "Livy darling," he wrote her when three days out from New York, "you really don't know what a steamship is. The Batavia is 316 feet long. This ship is nearly 500." He recounted all the dazzling Inman luxuries: the main saloon stretched across midship, brightly lit by windows and skylights, with a piano and bookcases; the smoking room, light and upholstered, and its marble-topped card tables; his cabin's "delicious" spring mattress and generous window, which gave him a sense of being outdoors; the steam heat and steady ship's motion. "At night I can read with perfect ease (and all night long), for a swinging lamp hangs above my head. I can lie there and pull a knob and a flood of clean water gushes into my wash-bowl. At any hour of the whole night I can turn over and touch an electrical bell and a steward comes in a moment." The hallways were wide and comfortable, the stairs elegant and easily navigated, the library ample and well chosen. "She is a much more comfortable ship than any Cunarder," Clemens told his wife. "I do so regret taking you in the Batavia. . . . Your journey would have been a hundred times pleasanter."

Many other transatlantic customers echoed Clemens in these years. "The Cunard Company seem, in my humble judgment, to be making a somewhat unfair use of their *prestige* for safety," said an Englishman in 1875. "They overcrowd their ships, give inferior accommodation to first-class passengers, and supply second-rate provisions (to speak mildly)." The Inman, White Star, Guion, and National lines were establishing new standards for how an ocean passage should <u>feel</u>. Inman and White Star had suffered dreadful disasters, that was true, but a realistic traveler could estimate the slight chance of dying on the newer lines against the virtual certainty of enjoying the experience more if the Cunarders were avoided. "Every year more people are finding out that the Cunard ships are the least comfortable of those which cross the Atlantic," wrote George W. Smalley, the influential London correspondent of the *New York Tribune*. Perhaps, Smalley suggested, Cunard should make amends "by some slight attention to the wants of passengers, or even by occasional civility."

Cunard tradition, so long a source of safe stability, had grown rigid and ossified. The company was still controlled by the families of the three original founders. When Samuel Cunard died in 1865, his shares passed on to his two sons. But Ned died young, in 1869, and William remained distant and uninvolved, without the interest or personality to lead. George Burns had retired in 1860 and left his shares to his children. Of all these offspring, John Burns—forty-six years old in 1875—seemed the most likely future leader, a man of force and utter self-confidence.

Charles Mac Iver was still very much on hand, patrolling the Liverpool docks every day, watching the details, and maintaining his own standards—as he had done, almost without pause or rest, since inheriting his brother's interest in 1845. Now entering his fourth decade of authority, Mac Iver believed he remained the absolute boss of the Cunard Line. "I have the personal direction of the company," he said in 1873, "in all matters connected with it, both outdoor and indoor." His vigilance, more than any other single factor, had sustained the line's safety record. It had worked, had always worked, and Mac Iver saw no compelling reason to change it. Or, indeed, to change anything. "There is, I think, often more loss by pushing than waiting," he advised one of his sons. "That was my elder brother's experience as it is mine." In the 1850s, the Cunard management had waited for the Collins Line to fail, and so it had. But the Cunarders were now sailing on a different ocean, and the line's traditional caution and conservatism had become dangerous. A bold Cunard employee came to Mac Iver with a good idea: replace the padded benches in the saloons with separate armchairs, as White Star and its imitators had done. Mac Iver ordered the man out of his office. "The next thing that would be wanted," he said, "would be lace for the sailors' jumpers." All the best and fastest transatlantic liners of the 1870s were built for other companies. The *Scotia* of 1862—a paddle wheeler!—was the last Cunard ship to hold the speed record in either direction. The famous Cunard fleet now seemed old and slow.

Finally, in 1880, John Burns took full charge. The Cunard Line went public, offering stocks worth over half a million pounds. "The growing wants of the company's transatlantic trade," the prospectus explained, "demand the acquisition of additional steamships of great size and power, involving a cost for construction that might best be met by a large public company." Cunard still retained a measure of its old reputation; the shares were quickly bought up by investors. "A splendid stroke for the Cunarders," a Guion official admitted, in private. Mac Iver accepted a reduced role, at last, and soon retired completely. Burns, the chairman of the new board, announced that the line would not be second on the Atlantic. Cunard was rejoining the game.

Ships as Buildings:
Two Cycles to Cunard

*S*een from the outside, at a dock or on the ocean, a transatlantic liner looked severely functional. Every external piece—of funnel, steam pipe, deckhouse, bridge, ventilator, mast, winch, cargo crane—served a particular utilitarian purpose and was designed simply with that function in mind. It had to work right and be strong enough to bear the North Atlantic in winter. Its appearance did not matter much. Inside the liner, though, the ship turned into a building, a temporary residence for—in this era—1,000 people or more. In the cabins and public rooms of first and second class, design and decoration took on commercial significance for this ever more competitive ocean traffic, luring customers by making the dangerous ordeal of a sea passage just a bit more agreeable.

As the liners grew larger and more elaborate, they became showcases of fine architecture. Ever since the *Great Western* of 1838, enthusiasts had called these ships "floating palaces." Now that label seemed strictly accurate. The Atlantic steamers of the late nineteenth century at last provided what designers most needed to show off their fanciest wares: money and space. The *Campania* and *Lucania,* new Cunard sisters in 1893, had ten times the interior room of the first Cunarders in 1840. Such generous dimensions allowed naval architects to imagine dining saloons two or three decks high, brightened by vaulting skylights and large stained-glass windows; suites of staterooms with sitting rooms (and even private

baths); grand staircases and hallways of indulgent width and ornamentation; and myriads of smaller rooms, each displaying its own look and
function.

So the crack steamships of this period flaunted expensive prodigies of
high Victorian style, lavish but fleeting. The special requirements of shipboard design encouraged architects to strive for a light, airy openness
instead of a dark, heavy, overstuffed look—the best of Victorian aesthet-
ics, that is, instead of the worst. In our time, seen from the distance of
more than a century later, the styles of the late 1800s have emerged from
the minimalist scorn of the modernists. Many fine Victorian buildings on
land have survived, to be rescued and reverently restored to their old glories, sparkling with details and bright colors. But old steamships were just
too valuable as scrap metal. After careers on the North Atlantic of about
twenty-five years, on average, and occasional service in less demanding
ocean routes, they were inevitably sentenced to the ship breakers. Of all
the great transatlantic steamers built between the Civil War and the turn
of the century, not one has survived. They endure today, at their best, as
lost masterpieces of Victorian architecture.

*T*he design history of these ships-as-buildings may be followed
through two decades of the fastest ships on the North Atlantic,
starting with the *Britannic* of 1874 and her virtually identical sister, the
Germanic, of 1875. Named for the two nations of their owners and
builders, they comprised the second generation of White Star liners built
by Harland & Wolff, substantially bigger and more powerful than the
founding cluster of six steamers. Conceived and built only four years after
the *Oceanic* and her sisters, the *Britannic* and *Germanic* were already
bursting with improvements; they expressed the fearless engineering
imagination of Edward Harland at the top of his game. The largest ships
on the North Atlantic (for a while), 470 feet long and 5,000 tons, in profile these liners nonetheless looked like fast, racy, overgrown steam yachts:
four high masts and two low funnels, all raked sharply backward and suggesting an image of speed even at rest, and turtle backs at the bow and
stern to shed water and protect the crew and steerage passengers as they
moved around their exposed quarters at the ends of the vessel.

The social and structural heart of the two liners was the grand saloon,
placed just forward of midship, on the middle of the three principal
decks. This room was fifty-two feet, nine inches long and forty-two feet,
six inches wide, across nearly the entire width of the hull. Three long

tables ran down the middle of the saloon, with four smaller tables against each of the exterior walls; eleven separate tables helped break mealtime conversations, among up to two hundred diners at a time, into more manageable clusters. The individual armchairs, a quickly imitated White Star hallmark, were safely anchored to the floor but could be turned 180 degrees toward the aisles for easier movements and minglings between meals. A marble fireplace with an open coal fire warmed and glowed at the back of the saloon; from there, a line of three fluted columns bisected the room lengthwise, leading forward to a four-sided cabinet full of recent books, and on to an upright piano against the front wall.

In good weather and friendly seas, the saloon was packed from breakfast time until late at night. The designer's challenge was to make the room attractive despite its constant, noisy, crowded patronage. The ceiling was a little more than eight feet above the floor—agreeably high, by shipboard standards of the time—and painted white to encourage an impression of light airiness. Across the room, tones of deep red and light brown stood out, from the crimson velvet upholstery of the chairs and of the sofas against the exterior walls, and the polished teakwood of the tables and chair arms. The parquet floor presented oak, ebony, and black walnut in geometric patterns. Along the exterior walls, the portholes— round windows well fortified by iron foul-weather shutters that were hinged at the top and secured up and away—alternated with panels of embossed papier-mâché, an admired parlor decoration of the period. Various salmon-colored embossed designs were arrayed against delicate green backgrounds. These wainscoting panels were edged by teak-and-gold moldings and a base of bird's-eye maple that ran all around the saloon, just above the dado formed by the backs of the crimson sofas.

The steward's pantry, serving food and storing tableware and utensils, was directly aft of the saloon, on the larboard side, with a lift running up to the galleys on the next deck. Cooking odors were thus largely insulated from the main-deck staterooms fore and aft of the saloon: a characteristic Harland step toward improved sensory comfort, keenly appreciated by passengers fighting seasickness. (In this case the ship-as-building flipped the arrangement of many substantial houses of this time, in which the kitchen was downstairs and the dining room above it.)

The two other main public rooms were pitched at the incompatible tastes and needs of Victorian men and women. Near the grand saloon, the smoking room was a sturdy masculine retreat of dark woods and comfortable seating upholstered in black haircloth. The marble-topped tables, with small, circular indentations for securing glasses and ashtrays against the North Atlantic, could be moved around to accommodate gambling and card games. Under each table, four thin legs tapered down toward the center and then outward at the base to form an hourglass shape, allowing extra legroom to the card players. From a vestibule adjacent to the saloon, a carpeted staircase led up the center line of the ship to a paneled landing, then split into two wing stairways to the promenade deck. Daylight streamed down from skylights and portholes into this heady space, two decks high, the tallest opening on board. Women had their own retiring room on the promenade deck, nineteen feet long and twelve feet wide, softened by a plush Brussels carpet and more sofas of crimson velvet. The wall decorations were brighter but more subtle than in the manly smoking room, with large mirrors in elaborate frames. Footstools encouraged reclining, and the location at the top of the ship (above the men and their tobacco smoke, and two decks removed from the steerage) allowed healthy ventilation and relative quiet for genteel women sailing at low ebb. This daily variety of shipboard environments—stateroom, saloon, and the two gender havens, along with as much walking and sitting on the promenade deck as weather and dispositions allowed—helped stave off creeping boredom, at least through the early stages of the voyage.

Steam propelled the ship, heated the rooms, and performed many other tasks. It kept food warm in the steward's pantry. It ran small donkey engines for weighing the five-ton anchors, and for powering the winches and cranes when the ship was loading and unloading, speeding these tedious operations and relieving the crew. It assisted the helmsman at his wheel beneath the captain's bridge, by the second mast. (If this patent mechanism, as yet fairly novel, should fail, a backup wheel, sailor-powered, could still control the rudder from its traditional place at the stern.) A fan turned by its own steam engine blew fresh air, hot or cold as desired, into the staterooms on the main deck—even when foul weather closed most of the ports and hatches. A storm at sea now at least need not stifle all ventilation. In the barbershop adjacent to the saloon, a tiny steam engine turned the tonsorial brushes, lathering cheeks and jowls at the barber's command: a whimsical touch, not really necessary, but another bow to the universal force and symbol of the century.

Down at the bottom of the ship, the biggest engine of all pulsed and chuffed in a massive, unrelenting heartbeat, lightly noticeable until the ringing silence when it finally stopped on arrival. The engines of the *Britannic* and *Germanic* were built by Maudslay, Sons and Field. (Harland & Wolff didn't make its own engines until a few years later.)

Steam from eight boilers, at up to sixty-five pounds per square inch, entered the two high-pressure cylinders, each four feet in diameter, and then expanded into the two low-pressure cylinders, each almost seven feet across. Engineers now calculated the "indicated" horsepower as the energy generated when an engine was running hard, under maximum steam pressure. These two ships were each rated at 760 nominal horse-power (by the old Watt method, derived from the dimensions of the engine) and at 5,000 indicated horsepower (a more telling measure of the actual driving force of modern compounds and high-pressure boilers). The two independent sets of cylinders allowed an extra measure of safety; if one broke down, the other could safely bring the ship into port, at reduced speed but still under power.

The *Britannic* and *Germanic* were the fastest, most popular transat-lantic liners of the late 1870s. They snatched the North Atlantic speed records back from Inman's *City of Berlin,* cutting them to about seven and one-half days in both directions. "Taking them as a fleet," said a disinter-ested expert at a meeting of the Institution of Naval Architects in 1877, "there is nothing that will compare with the Oceanic, or White Star Line." After the American tycoon J. P. Morgan crossed on the *Germanic* in 1880, he and his influential, bicontinental friends decided henceforth to embark only on White Star ships. "Of the eleven at our table I don't think anyone except Mrs. Egleston missed a single meal altho it was very rough," Morgan wrote of this passage. "Our table was the envy of all the passen-gers on account of the apparent joviality always at hand to say nothing of the good things which served to fill up the gaps of our five meals."

Edward Harland's two best liners even got faster with age—an extraordinary reversal of the usual declines of older ships and worn machinery. In the summer of 1891, the *Britannic* came home from New York in under seven days, seven hours, and the *Germanic* took only forty-five minutes longer: the best times yet for both ships, and still with their original engines and boilers. Their long, profitable careers were especially remarkable at a time of such ruthless, multiplying competition on the North Atlantic. They steamed on and on, each clocking over 2 million miles on the roughest ocean passage in the world, before being sent to the breakers.

William Pearce of John Elder & Company lusted to build a new Atlantic flyer in 1878 that would bring the speed records back to the River Clyde. Since 1840, Clyde-built ships had usually held those records, except for America's Collins Line interlude in the early 1850s and the current White Star supremacy. The former triangular rivalry for the leadership of British steam merchant shipbuilding, so vociferously contested among Glasgow, London, and Bristol in the 1840s, had long since been conceded to Glasgow. The recent triumphs by the Belfast shipyard of Harland & Wolff had therefore surprised and alarmed Clydeside builders.

Pearce was a Scotsman by adoption. As a naturalized citizen of the Clyde, he perhaps felt especially patriotic about returning the transatlantic records to his chosen home. ("We have always in this part of the country been partial to self-help," he said later, "which is the best help of all.") Pearce grew up in Brompton, Kent, on the southeastern edge of England—quite the opposite corner of Great Britain from Glasgow. Apprenticed at the nearby Chatham Dockyard of the Royal Navy, trained by the noted shipwright Oliver Lang, he worked on the first iron steamers built by the Admiralty. Too impatient to abide the slow promotions of government service, he became a surveyor for Lloyd's shipping registry and was posted to Glasgow in 1863. Within a year he was hired away by Robert Napier and Sons, no less, to be its shipyard manager.

Pearce was then thirty years old, bristling with a young man's energy and ambition. The great Napier firm, under the faltering leadership of Robert's son John, needed regeneration. The Napiers and their best client, the Cunard Line, had stuck for too long with their traditional side-lever engines and moderate boiler pressures. Pearce recognized that the real future of marine engineering lay downriver, at John Elder & Company. "In spite of the splendid success of the compound engine and surface condenser, the great steamship companies and ship-owners generally were very slow to adopt it," Pearce said later. "So wedded were the leading engineers of the day and the marine superintendents of the great companies to the old type of low-pressure engines that for many years they continued to fit them on board steamships, to the serious loss of their owners." Pearce spent his career as a restive artist seeking an amenable patron, prodding steamship companies to let him realize the engineering visions in his head. When he started building ships, and the Scotia's transatlantic records stood at eight days and a few hours, he recklessly predicted five-day boats within his own lifetime.

After the premature death of John Elder in 1869, Pearce took control of the most progressive engineering and shipbuilding firm on the Clyde. Officially he was only one of three copartners in the reorganized Elder company, but he was always the most creative, dynamic leader. He wanted to build crack transatlantic liners. As the most celebrated ships of their time, they would bring fame and contracts to the Elder firm—and provide Pearce and his associates a kind of artistic satisfaction. For prospective clients, however, Pearce had to fashion arguments that emphasized commercial returns. Shipowners worried about the extravagant tons of coal burned in the many boilers of the fastest ships; coal consumption increased more sharply than cruising speed. In rebuttal, Pearce maintained that if a crack ship carried only passengers and light cargo, it could spend less time in port and more on the ocean; loading and unloading heavy cargoes kept a ship unprofitably moored for days. A fast ship with little cargo would make twice as many voyages, and a reputation for speed would ensure full complements of paying customers. A fast ship also arrived much sooner, reducing by days the fixed costs of supplies and crews. The food for two crossings of a slow ship would stretch—said Pearce—for three passages of a fast liner. In adding up these numbers, he insisted, "the compensating advantages more than obliterate the extra cost."

Pearce first approached the Cunard Line with his plan for a record-setting transatlantic steamer. Charles Mac Iver, resistant as ever to major changes and to spending money, still swayed Cunard's management in the late 1870s, so Pearce was turned down. He took his scheme on to the Guion Line. Under John Jordan, its well-known superintendent engineer, Guion had boldly embraced many technical innovations. But the line's first attempts to set transatlantic records, with the *Montana* and *Dakota* of 1875, were embarrassing failures. Jordan gave them excessively hollow lines at the bow, engines of advanced design but poor performance, and daring new water-tube boilers that failed in trials and cost Guion some £60,000 for conventional replacements. In addition, three Guion ships had sunk between 1872 and 1878, wrecked on the Welsh and Irish coasts and in a collision on the Thames. After this string of debacles, Stephen Guion and Jordan's successor as superintendent engineer, John G. Hughes, lacked the cash and confidence to risk the costs—and novel machinery—of building a new champion.

Pearce therefore proposed that the client and builder share their initial expenses. "Find me X," he said to Guion, "and I will find the rest, and build a steamship which will beat the fastest on the Atlantic." So the joint deal was struck, cheaply, and the new challenger—the *Arizona*—was built for a contracted price of only £140,000. (The *Britannic* and *Germanic* had cost £200,000 each.) Following in the traditions of Robert Napier, David Tod, and Edward Harland, William Pearce got to build the transatlantic ships he wanted by becoming partners with the client, binding himself to the commercial success of the enterprise.

The *Arizona* was designed mainly by Pearce, with the help of his top engine man at Elder, Archibald Bryce-Douglas, and the advice of John Hughes of Guion. Essentially they built a duplicate of the *Britannic* and *Germanic,* but faster. Even before the contract with Pearce, Hughes had been musing about an ambitious new ship, perhaps 450 feet long, with turtle backs at the ends "similar to the White Star steamers," he noted. The concave bow of the unfortunate *Montana,* he added, "has brought hollow lines into disfavor here, the style of White Star, and other fast steamers being preferred." (William Froude's model experiments and Macquorn Rankine's studies were also pushing naval architects away from Russell's old concave wave-lines toward straighter lines at the bow.)

In profile the *Arizona* looked like the Clydeside twin of her Belfast rivals: four masts, two funnels, and a similar bow and stern. The *Arizona* was twenty feet shorter and a bit more narrow, but deeper, and therefore a few hundred tons roomier. The biggest difference was Bryce-Douglas's engine. A single high-pressure cylinder, sixty-two inches in diameter, expanded steam into two low-pressure cylinders of ninety-inch diameters. The boilers drew clean water from the largest surface condenser yet fitted on a steamship. At trials on the Clyde in May 1879, the engine generated almost 6,400 indicated horsepower—400 beyond the contract's stipulation, and 1,400 more than the *Britannic* or *Germanic.* The *Arizona* was a souped-up transatlantic hotrod, the first of a new breed.

The ship-as-building offered accommodations for three classes of travelers, then an exclusive feature of Guion's Liverpool service. On the main deck, forward of the machinery, staterooms for 140 saloon passengers had the biggest portholes, a foot across, and the fanciest (though shared) waterclosets, with dark mahogany seats and hardwood gratings. Just aft of the machinery, up to 70 second-class passengers looked out through ten-inch portholes, as did 140 steerage denizens at the stern. As needed, 1,000 additional steerage patrons could be housed on the deck below, more dimly lit by nine-inch portholes. The grand saloon, smaller than on the White Star rivals, was lighter overhead, the ceiling painted white and gold and opened up by a generous domed skylight in the middle of the room. Beams at the middle and ends of the dome held potted plants and flowers, like the interior greenhouses then becoming popular within fine homes on shore. On the upper deck, about a third of the ship's length back from the bow, a "Social Hall" featured a bank of large, heavy plate-glass windows, beneath the bridge and facing forward.

Instead of just peering out through single portholes to the side, passengers could actually look ahead through a whole row of windows that gave a protected view of the entire sweep of the approaching ocean. "This arrangement will be a great novelty," the *Nautical Gazette* predicted, "and this saloon will be the most attractive place on the ship. It will serve as the grand pilot-house for all the saloon passengers."

The *Arizona's* combination of speed and improved customer comforts was positively Harlandesque. In June 1879, she made the fastest westbound maiden voyage ever, and the quickest round-trip time for any ship. The fastidious Hughes reported some minor problems to Bryce-Douglas—overheated bulkheads, a broken air pump, recalcitrant bells, a faulty scupper pipe on the steerage deck—but no defects in the major machinery. The heavy coal consumption, averaging 135 tons a day, was "not very alarming," Hughes conceded. "We are very satisfied." Easy in her motions, steady even in rough seas, the *Arizona* was an immediate hit, crowded on each voyage with saloon passengers. Later that summer she dropped the eastbound record to seven days, nine hours, and twenty-three minutes. She soon broke her own mark by seventy-five minutes.

In November, crossing the Newfoundland Banks, the *Arizona* at nearly full speed ran headfirst into a massive iceberg. It was ten past nine at night, in calm seas and clear visibility under a light cloud cover. In the main saloon, an English lady was playing the piano to accompany a gentleman who was singing "See Our Oars with Feathered Spray." Up in the Social Hall, the nightly auction of bets on the next day's run had just finished. The pooled bets, piles of gold and silver coins, lay on one of the marble-topped tables. The bright lights inside the Social Hall kept anyone from seeing, through the plate-glass windows, the iceberg looming ahead. About five hundred feet wide at its base, it had three conical masses rising sixty or seventy feet high. The lookouts thought it was just a cloud low in the sky. (Iceberg season had ended months earlier.) The *Arizona* hit in a sudden explosion of noise, recoiled, hit again as the propeller kept turning, stopped dead, and trembled from stem to stern. Her snout was stuck in the berg. Heavy chunks of ice continued to fall onto the forward deck. In the Social Hall, the tableful of coins was thrown to the floor, along with many startled passengers. A woman in the saloon fainted, and others sobbed hysterically. When the reversed engine pulled the bow from the iceberg, the ship started listing to starboard and down at the front. A panicked sailor announced that she was sinking. "Steady, gentlemen," somebody said. "Keep cool."

The captain rushed to the bridge. "My God, men, where were your eyes?" They lowered a boat so the first officer could survey the damage: substantial but perhaps not mortal. Stephen Guion, who was aboard with his sister and two nieces, agreed with the captain that the ship should make for St. John's, Newfoundland, the closest landfall, about three hundred miles to the west. She proceeded carefully through the next day, in calm, lucky seas, and arrived late at night. The system of watertight compartments—long standard on most transatlantic liners—kept the leak from spreading back through the hull.

The next morning, seen from the dock, the crumpled bow looked like the inside of a railroad tunnel. The turtleback and upper deck had absorbed most of the impact; vital sections below the water line were less damaged. The stem was broken off down to about seventeen feet above the keel. Five sailors in the forecastle were injured, but nobody was killed. It seemed improbable that the *Arizona*—presenting such a blunt, open tangle of twisted iron plates and beams to the ocean—could have reached St. John's at all. Fitted with a temporary wooden bow, she limped home to the Clyde for repairs. After two months in the Elder shipyard, she was back racing across the North Atlantic. As when Cunard's *Persia* slammed into an iceberg in 1856, the *Arizona* by surviving such a direct, high-speed impact only enhanced her reputation for unsinkable strength. The next spring, she made yet more Atlantic records in both directions.

Once entered, the race to claim the fastest of all the fast transatlantic liners left its participants no relief. The process defined itself. Among the competing lines and builders, rumors were always floating around of another new ship, or a better kind of engine, or a more frugal boiler sure to relieve the eternal problem of coal consumption. In July 1879—one month after the record-breaking maiden voyage of the *Arizona*—John Hughes already started thinking about his next Guion ship. He sent a letter (emphatically marked "<u>Private</u>") to Archibald Bryce-Douglas at the Elder yard, asking what size engine and boilers he would recommend for a steamer just a bit larger than the *Arizona*. "In case we decide to go on," he explained, "I would like you and I to start upon the same basis. Of course a new ship would have to be made to beat the Arizona."

Over the next year, as Hughes kept track of what the competition was planning, the size and power of the new Guion liner—the *Alaska*—kept expanding like steam in a low-pressure cylinder. "There is nothing under weigh yet that will be much ahead of the Arizona," he wrote Bryce-Douglas in March 1880. "Alaska would take the lead." Then four days

later, pushing and redundant: "If we do go on with her we must be prepared to keep the lead in speed." Bryce-Douglas, a devoted engineer rising to a technical challenge, needed no convincing. But the contract
details, to be nailed down between William Pearce and Stephen Guion,
delayed progress. Hughes wanted his new flyer, and soon! Reaching for
leverage, in June he warned Bryce-Douglas that another shipbuilder, the
Barrows, also wanted the job of building the next Guion record-breaker.
"If your firm mean business," he scolded, "somebody should see Mr.
Guion at once." The deal was done a few weeks later.

The *Alaska,* ready for sea in late 1881, shared most of her design and
interior features with the *Arizona.* But she was seventy-six feet longer,
3,000 tons roomier, and her engine put out 10,000 indicated horsepower,
a huge increase of some 3,600 horses—all these leaps in less than three
years. Pearce and Guion, taking no chances, put the bar much higher
than any existing ship could approach. In 1882 the *Alaska* made the first
passages, east and west, in under seven days, cutting the eastbound mark
down to six days, eighteen hours, and thirty-seven minutes. An impressed
journalist, Thomas Dykes, called her "the greyhound of the Atlantic," a
vivid honorific—suggesting both inbred speed and intentional pitched
races—that was later routinely applied to the ocean record-breakers that
followed her.

One final improvement, not begun until the last stages of her construction or completed until after her maiden voyage, helped keep the
Alaska up-to-date in the speeding company of the North Atlantic. While
she was being built, early types of electric lighting suddenly became
required equipment for the crack Atlantic liners because of their many
advantages over candles and oil lamps. The "glow lamps" (incandescent
bulbs) recently invented concurrently by Joseph Swan in Britain and
Thomas Edison in the United States gave a steady, pure, white light
instead of a sputtering, yellow, flickering flame, and with no odor or
residue, an improvement especially welcomed by queasy passengers. The
glow lamps left no smoke or carbon to tarnish paint and gilt work, and
did not require the constant, individual attention of the overworked
stewards. Entire decks of lights could be controlled by the flip of a single
switch. Needing no flame for ignition or illumination, they also presented less risk of starting a dreaded shipboard fire.

In 1879, within a year of Swan's breakthrough patent, William Inman
placed the first transatlantic electric lights—just six bulbs in the main
saloon and engine room—on his *City of Berlin.* After Swan improved his
bulb in 1881, extending its average life, Cunard and White Star electrified

a few of their ships too. Yet another steam engine had to be squeezed into precious shipboard space, and a dynamo's delicate machinery was vulnerable to the motion and moisture of the ocean. For marine engineers like John Hughes, wedded to iron and steam, electricity was an exotic new form of power, not easily understood or trusted. Competitive pressures forced Hughes to accept, belatedly, electric lights on his *Alaska.* The bulbs were still quite fragile; given recent advances, the manufacturer claimed, only 10 percent of the lights would have to be replaced after a round-trip on the Atlantic. "Let it be clearly understood," Hughes wrote to his lighting contractor, the Siemens firm of Manchester, "that we want the oil lamps left to be ready in case of defects with Electricity." (In this aspect of the ship-as-building, the sea preceded the shore. Most passengers encountered those mysteriously glowing bulbs on ships well before they were much used in homes or businesses on land. The Cunard office in Liverpool did not acquire electric lights until 1888.)

The *Arizona* and *Alaska,* so astonishing when they first appeared and set their records, were soon passed by newer ships. Their careers on the North Atlantic were relatively short, only fifteen and eighteen years. In that time, they developed their own loyal followings: passengers who crossed on them year after year when they could have taken faster, more fashionable ships, but preferred the Guion liners because of some sentimental affection, difficult to comprehend or to explain without courting derision. A steamship on repeated voyages across the intensely variable North Atlantic assumed a certain knowable personality, more than the sum of its inert metal and wooden components. It would call up a particular cluster of tangy memories, sensations, encounters, made especially full and sharp by the unique circumstances of life on a transatlantic steamer. When the ship reached the end of her useful life, it could feel like—not a death in the family, quite, but perhaps the demise of a distant friend or of a special dog. When the *Alaska* was sent to the breakers in 1899, the *Marine Engineering* journal of New York keened over the loss of "the best known vessel afloat," one so especially handsome, not bulky-looking despite her great size. "A feeling of regret," the journal acknowledged, "strange perhaps in connection with an inanimate object, but none the less real. There is, however, something almost animal about such a ship."

"So great is the demand for berths on the fast steamers that they sometimes have to be engaged months beforehand," said the *New York Times* in the summer of 1883. "The success of the swift steamships which have recently been placed in the transatlantic trade has led the managers of other lines to order vessels of the same pattern." In the past, admirers might praise a new steamer by exclaiming "Isn't she a beauty?" or "What a graceful model!" But now, connoisseurs of the North Atlantic race course were more liable to enthuse "Isn't she a grand old snorter?" or "Did you ever see such a huge old buster?" According to the *Times,* the rich Americans who dominated first-class accommodations during the tourist season brought along their snapping Yankee impatience and national will to win. "Speed is the only thing which they talk, think, or dream of anywhere between Sandy Hook and Roche's Point. Whenever their vessel distances some other steamship which is bound in the same direction, they are thrown into ecstasies. The males conceive a brotherly affection for the officers, who are regarded with almost sisterly tenderness by the female devotees of rapid transit. But they all love the huge, swift steamship."

The *Servia* was the first major Cunard steamer to appear since the company went public in 1880. As the symbol and flagship of the renewed Cunard Line, she carried a heavy load of hope and expectation. "The speed will be greater than that of any ocean steamer afloat," predicted John Burns, chairman of the Cunard board. "We have adopted in every detail of the ship and engines the most advanced scientific improvements compatible with the safe working of so great a vessel." That balanced statement—of progress hedged by safety—expressed a Cunard tradition that went back to the origins of the enterprise. Cunard waited for other steamship lines to try new inventions; if they worked and endured, under all the conditions the North Atlantic could throw up, then Cunard might copy them. Since 1840, this cautious tortoise had outlasted many reckless hares. "Science is, if anything, tending to tread too hastily upon the heels of commerce," Burns said at the launch of the *Servia* in 1881. "There is no portion of a ship, from stem to stern, from truck to keel, which is not subject to suggested improvements by scientific men; and I can assure you that it requires all the proverbial blandness of a chairman to decline the overtures of hosts of suitors who come to ask his favour in adopting their varied wares." (The Cunard-friendly audience laughed and applauded.)

The *Servia* was big and fancy, but not (it turned out) fast enough to beat the *Alaska.* She was designed and built by the Clydeside firm of James and George Thomson, Cunard's main shipbuilder since 1867. The construction of two other new Cunarders from the Thomsons, the *Aurania* and *Pavonia,* took too long—as Cunard management saw it—and those ships also failed to perform as hoped. The fabled Cunard luck no longer favored the company. Its decisions, furthermore, could not still be immured within the private deliberations of the three founding families. Cunard leaders now had to face an annual meeting of disgruntled stockholders, unhappy with their dividends and curious as to why competing ships kept leaving even the newest Cunarders in their wakes. "It seemed to me unfortunate that the <u>later built</u> vessels of the Company should not surpass in speed the <u>older ones</u> of the Guion Line," a London investor wrote to Cunard headquarters. "If they are not superior, it would seem the money spent on them has been literally thrown away."

All these circumstances pushed Cunard away from its entrenched loyalty to the Thomsons and toward the hottest shipbuilder on the Clyde. To beat the Guion ships, go to their source. William Pearce of the Elder company, rejected by Cunard's former management a few years earlier, now held all the cards. In the spring of 1883, Burns asked him to build the fastest ship on the Atlantic. Make that an order for two such ships, said Pearce. All right, Burns replied, on condition that you don't ever build another greyhound for the Guion Line. I won't, said Pearce—until at least eighteen months after the delivery of the second Cunarder. (As negotiations came down to agreement, the *Aurania*'s engine blew up at sea, providing another reason for Burns to switch from the Thomsons.) To seal the deal, Burns went to Glasgow instead of Pearce coming to Liverpool: a revealing sign of who was courting whom. "Directors give you full power to close," the Cunard board telegraphed to Burns, "though holding strong views that P is driving a hard bargain. Get the G matter placed on as satisfactory a footing as possible." The contract, signed at £616,000 for the two ships, was the richest British merchant shipbuilding pact yet. Cunard had to scramble to raise the money; it tried to sell some pricey Manhattan real estate in New York, then pursued a loan from an insurance company, and finally settled for a debenture loan of £250,000 for three years at 5 percent interest. The rest would come mostly out of future profits.

The *Umbria* and *Etruria* did bring the transatlantic speed records back to Cunard. They made Pearce "probably the best-known shipbuilder in the kingdom," the journal *Engineering* noted a few years later, and the Elder firm "the most widely renowned establishment in the world." John Elder, not long before his death, had moved his shipyard three miles down the Clyde from Glasgow, to the country town of Govan. On the south side of the river, he converted seventy-four acres of arable farmland into the soon-famous Fairfield works, laid out with wide spaces and room for future expansions. The name made nostalgic, ironic reference to when that land was green, quiet, and unpolluted: Fairfield. "Now the verdure has been trampled down," a visitor noticed in the 1880s, "and the face of the earth is hidden by paving-stones and iron rails. The river is inky, and the smoke lying in a brown fog overhead is ever being replenished from the high chimneys of the neighborhood. The scene within the high brick walls which keep out idlers is exhilarating but scarcely picturesque."

Running at top speed, the Fairfield yard employed 7,000 men. The general office and brass foundry were placed outside the main entrance at the southeast corner. The major complex of shops and sheds was surrounded and crisscrossed by networks of railroad tracks for moving heavy materials around the yard. Underground pipes delivered water, gas, and blasts of air as needed. Beyond the main complex, running down to the river, were building slips for up to fifteen vessels under construction at once. Supplies of metal and lumber flowed from the inland side, through the noisy shops of the machinists, boilermakers, blacksmiths, carpenters, and cabinetmakers, to the ships, and finally into the river. The marine engineering department filled a block of buildings about 300 feet square and 50 feet high. Inside, four traveling cranes lifted loads as heavy as forty tons. "It is doubtful," said another visitor, "if a similar collection of ponderous tools is to be found anywhere else in Great Britain." Screw-cutting lathes, slotting machines, drills, turning lathes, punches, shears, planing, bending, and trimming machines remade rough chunks of metal into precise pieces of a steamship, amid a constant cacophony of shrieking, creaking, pounding, grinding, and groaning, a spray of sparks, and the frictioned odor of hot iron and steel. One particularly dangerous and mysterious-looking machine, delicately called "the Devil" by its awed attendants, flourished a metallic disk, eighteen feet in diameter, ringed with sharp steel cutters. It could shave through over two inches of solid iron plate in four minutes. Invented by John Elder's shipwright father, David Elder, for the Napier yard several decades earlier, it stood and whirred away as a working monument to the stable, ingrown continuities of Clyde shipbuilding.

In a transition nearly as significant as the shift from wood to iron, steel was now the essential material for building the finest steamships. Just as an iron ship was both lighter and stronger than a wooden vessel of the same size, steel offered the same advantages over iron; but, like iron, for years it was resisted by ship men as too expensive, of uncertain quality, and unfamiliar and difficult to handle. In general manufacturing, it was at first used just for cutlery, tools, springs, and small machinery parts, not for major structural purposes. Refined from molten iron by an erratic, expensive process of hand puddling, steel would become soft and weak if overheated and cooled too slowly, or hard and brittle if cooled too quickly. It cost up to £60 a ton, too much for any shipbuilder. Henry Bessemer's invention of 1855—blowing a blast of air through the fluid pig iron—industrialized the process, producing much greater quantities of steel in a given lot, and with lower-paid, less skilled workers, thus bringing the final costs down. Later variations on Bessemer's method, especially the Siemens-Martin technique, further improved the quality and dropped the price.

By the late 1870s, steel was still twice as expensive as iron, but the difference was narrowing. Shipbuilders could reasonably weigh steel's many virtues against its extra cost. More ductile than iron, it could be flanged, joggled, bent, and drawn out without breaking—even when cold, a great practical advantage for shipyard operations. A piece of steel could be quickly refashioned on the spot, at the hull, with no pause for heating and protected handling. Strong yet elastic, steel would recover its original form and dimensions whenever a disturbing force stopped acting on it. A steel ship was about 14 percent lighter than an iron counterpart, a weighty difference in such an enormous structure. That saved fuel and left more interior space, and extra strength meant more safety in rough seas, collision, or grounding. Steel boilers allowed for higher steam pressures, safely enclosed by tougher shells. Though more susceptible to corrosion and rust than iron (a real shortcoming at sea), steel parts could be adequately protected by galvanizing and then a labor-intensive maintenance program of frequent cleaning and painting.

The *Umbria* and *Etruria* were built of steel, including even the three masts and their spars. The engines—the components that made them record-breakers—drove crankshafts of Vickers compressed steel, and steel propellers. The cylindrical steel boilers generated steam at 110 pounds of pressure to drive three mammoth compound cylinders to generate more than 14,000 indicated horsepower—almost three times the mechanical force of White Star's *Britannic* and *Germanic,* a mere ten years later. Andrew Laing, the latest in Elder's line of engine wizards, designed and built the machinery for the *Umbria* and *Etruria.* Only in his late twenties in 1885, he went on to produce more record-setting greyhound engines, over a longer period of time, than anybody else. After apprenticing as a millwright in his native Edinburgh, he had come to Glasgow and was hired by William Pearce as a draftsman. While taking college engineering classes at night, after just four years Laing was made head draftsman for the Fairfield yard—"an astonishing measure of promotion," said a contemporary account, "but a measure which he won through sheer, unaided ability, and force and determination of nature." That focus and forcefulness made him a brilliant engineer but a bristly, difficult employee, never inclined to accept much direction. Pearce sensibly left him alone, "one of the most eager and untiring of men," to tinker at will with his engines.

As buildings, the *Umbria* and *Etruria* balanced their darker wood tones with more skylights and electric lighting. Each ship was brightened by about 850 Swan glow lamps, the most extensive shipboard installation yet. ("The lamp is of great simplicity," it was explained, "and very suitable for those not well acquainted with electrical matters.") For the music room on the upper deck, the general style of Renaissance revival included dark Spanish mahogany and satinwood, Venetian mirrors at the ends, carved pillars, and seventeen elliptic arches along the semicircular walls, rising from fluted and paneled pilasters into carved capitals, cornices, and beams. A vaulted skylight flooded daylight down into the room and, through a wide, circular opening in the middle of the floor, on to the main saloon on the deck below. People in the music room could lean on the wide railing around this opening and watch the saloon scene underneath them; from the saloon, the vista lifted more than twenty feet through the deck overhead, up to the shared skylight. The smoking room and engine room had their own skylights.

The new Cunarders were about the same length as the *Alaska*, but seven feet wider. That allowed them roomier corridors and staircases, and more space to work around the machinery down in the engine room. The main saloon was fifty-two feet wide and seventy-six feet long, with seventeen tables seating from ten to fifty-two people each. Here the style was simpler and more subdued than in the music room, just slightly carved oak wainscoting, a dark walnut sideboard, and armchairs of dark wood legs and arms. The ceiling was painted a creamy, lighter shade of the wainscot oak. At night, these somber tones were lightened by 103 glow lamps, most of them in three-bulb, silver-plated clusters suspended over the tables or in tight pendants hanging about thirty inches from the ceiling. The rest were placed in wall brackets above the side seats. The exotic quality of the unflickering new light, so bright yet soft, gave the polished woods a steady, restful gleam.

These two ships were the last of their breed, the final single-screw flyers on the North Atlantic. In the Cunard manner, they summarized recent progress in marine engineering: compound engines, high-pressure boilers, domed skylights, steel construction, and electric lighting. One of their few real innovations was aesthetically unfortunate. The two smoke-stacks were double-jacketed: an inner funnel for the smutty exhaust of the boilers surrounded by an outer stack that released foul air from the ship's steam-powered ventilation system. However functional this may have proven, it made the smokestacks—each about eighteen feet in diam-eter—too stumpy and portly, and awkwardly out of proportion with the rest of the vessel. Seen in distant profile, the ship looked like a gigantic deck cleat, just two stubby knobs plunked in the middle of a long, thin piece of metal.

But the *Umbria* and *Etruria* were fast and fashionable. In the sum-mer of 1885, an American writer well-named John Gilmer Speed dined at the Reform Club in London on a Saturday, caught a fast train to Liverpool, embarked on the *Etruria,* and on the following Saturday sat down to dinner at the Manhattan Club in New York: the first transat-lantic traveler known to have spent only a week away from his clubs. A few years later, the two Cunard greyhounds cut the Atlantic records in both directions to six days and a few hours, averaging almost twenty knots all across the ocean. In just a decade, the fastest passages had been sliced by almost a day and a half.

*T*he pace of progress kept speeding up, quicker and costlier, exhausting and exhilarating. A shipowner would boldly contract for the biggest, newest, swiftest ship on the ocean; in two or three years it would be obsolete. "When," asked the owners, "is this progress and devel-opment to cease?" The builders, competing with each other, kept improv-ing their ships. They assumed infinite progress. At a session of the Institution of Naval Architects in the summer of 1886, the shipbuilder William John reported on recent milestones in transatlantic steamships. French and German lines, generously subsidized—he said—by their national governments, were challenging British supremacy on the Atlantic. And another threat, looming still more dangerously, lay across the ocean. "We must march forward or else we shall bring even more for-midable rivals into the field," John warned, "—I mean: from the other side of the Atlantic."

A few months later, the Inman Line was bought up by an American group dominated by the Pennsylvania Railroad and John D. Rockefeller's Standard Oil Company. Two of the most powerful corporations in the

United States swallowed one of the three major British transatlantic lines. Flourishing American money, the reorganized Inman Line then built the next two Atlantic greyhounds, the legendary *City of New York* and *City of Paris*. (Both were named for earlier, well-remembered Inman liners.) They were soon recognized as the most beautiful steamships of their time, fast and plush, uncommonly graceful and well proportioned; though built on the Clyde, they were financed and controlled from Philadelphia and New York.

These startling developments were masterminded by Clement Griscom. As the manager of the Pennsylvania Railroad's two Atlantic ventures, the American Steamship Company and the International Navigation Company (or Red Star Line), he had been running second-class ships in second-class markets. To step up to the crack steamers racing between Liverpool and New York, he needed the continued, generous support of his railroad friends. The Pennsylvania could not reasonably embrace another steamship fling—at least if it were measured in strict business terms. The ASC kept losing money ($107,000 in 1881, $115,000 a year later), and the Red Star Line didn't do much better. But the railroad drew less countable benefits from its ships. They linked its trainloads of passengers and cargoes to European ports in long, profitable loops that stretched from the American heartland on to Antwerp and the continent. The keystone logo of the Pennsylvania, familiar on thousands of American boxcars and locomotives, adorned the funnels of its ocean steamers, spreading its name and reach across 4,000 miles of international markets and travelers.

Griscom also brought along his own connections with Standard Oil. For two decades, he had taken particular interest in the gushing flow of petroleum exports from western Pennsylvania to Europe. Oil money helped launch and sustain the Red Star Line. In 1881, Griscom and Standard Oil put together the National Transit Company, a safely vague name for a trust that combined all of Standard's pipeline affiliates. The largest single unit in the Standard empire, National Transit united 3,000 miles of pipelines, cutting costs and eliminating competition. The pipelines took major customers away from the railroads by moving crude oil more cheaply and efficiently. But, in the spirit of the times, National Transit and the Pennsylvania Railroad worked out their differences to mutual benefit. Bridging them, Griscom served as president of the pipeline trust while he sat on the board of directors of the railroad. He could tap these oddly joined allies for another transatlantic venture.

The Inman Line was available. It took a series of blows in the early

1880s: the death of William Inman, the founder; an expensive new steamship that performed so poorly she was returned to the builder; the sinking of the *City of Brussels* at the mouth of the Mersey, causing ten deaths; and a fire that destroyed the Inman pier in New York. Inman reorganized its engine department and hired an experienced marine engineer to run it. But in the tightened competition on the North Atlantic, the wounded company could not recover, despite its good name and three decades of steady profits. ("The Inman Line always paid the best salaries to its employees," a newspaper reported, "and it had the reputation of paying well for any service rendered to it.") When a saving issue of debenture bonds didn't attract enough investors, the Inman management gave up and went looking for a buyer.

"Our Company," Griscom recalled in 1891, "thinking it desirable to get a conspicuous footing in the passenger traffic between Great Britain and the United States, resolved to extend its service about 1884." The Inman Line, "with important goodwill and agencies all over the world, was tottering and anxious to sell out. After reflection, we concluded it would be more economical to buy out this old Company, get its place on the Atlantic and its agencies, and equip it with modern steamships, than to fight for places in the trade." To start, Griscom's International Navigation Company loaned Inman £70,000 in exchange for weekly service between New York and Liverpool. As the Pennsylvania Railroad directors noted, in private deliberation, this secured Inman's business and cooperation without the "hostility" bound to follow "the establishment of a new and competing line." From that beachhead, in the fall of 1886 Griscom went on to purchase the Inman Line for £600,000. The money came in nearly equal chunks from the railroad and from Standard Oil pipeline interests represented by Benjamin Brewster (one of Rockefeller's closest cohorts and his near neighbor on Fifth Avenue in New York). These financial sources were kept secret; the *Liverpool Daily Post* assured any of its readers who were fretting over Yankee incursions that the Pennsylvania Railroad had "no interest whatever" in the deal. The two Atlantic lines, the overt players, blended their names into the new Inman and International Steamship Company.

Griscom had his own ideas for the first steamer of the merged line. From his years in the business, he nurtured an amateur's enthusiastic curiosity about marine design and engineering—not the arcane details, but the broad tendencies in this era of exploding progress. (He later helped found the Society of Naval Architects and Marine Engineers, the American counterpart to Britain's older Institution of Naval Architects, and served as its first president.) In February 1887, Griscom crossed the ocean and spent the next several months sifting tenders from British ship-builders. The liner he had in mind would launch the next generation of Atlantic greyhounds: powered by triple expansion engines (which extended the compound principle by expanding steam into three succes-sive cylinders), strengthened by watertight compartments running both lengthwise and across the hull, and driven by twin screw propellers.

The power of the engines of the *Umbria* and *Etruria* had made twin screws inevitable. At more than 14,000 indicated horsepower, those ships imposed nearly intolerable strains and torques on their single shafts and propellers. Any additional power for the next greyhound would have to be distributed between twin screws. Double propellers also seemed safer—one might break down at sea but probably not both—and made the ship more maneuverable, able to turn on her heel by running the screws in opposite directions, and still steerable even with the rudder knocked out. These advantages had inspired many smaller twin-screw vessels since the beginning of steam navigation. Over recent decades, double propellers had shown up in harbor tugboats, fast Confederate blockade runners during the Civil War, and British warships. But their added weight, expense, and coal use, the doubled space needed for their engines, shafts, and machinery, and the larger crews required to feed and tend them had delayed twin screws for the major Atlantic steamers.

Griscom wanted them, and quickly. In March 1887 he came to terms with the Lairds of Birkenhead for a twin-screw greyhound, 500 feet long, of 8,500 tons and 17,000 horsepower. Griscom knew the Lairds; four years earlier, they had built two much smaller steamers for his Red Star Line. However, it belatedly became clear that the Lairds would have to expand their yard facilities to make room for the new behemoth, delaying its completion. Griscom couldn't wait. The White Star Line was rumored to be planning its own twin-screw greyhound. So Griscom broke his deal with the Lairds, went up to Scotland, and switched to the Clydebank firm of James and George Thomson.

This was not the same Thomson yard that produced those disap-pointing Cunarders of the early 1880s. James Rodger Thomson, son and nephew of the two founders, had hired a new engineering manager, John Gibb Dunlop. Apprenticed and employed for fifteen years at the John Elder shipyard, Dunlop brought its state-of-the-art methods and stan-dards to the Thomsons. Their chief of staff was John Harvard Biles, one of the brightest young naval architects in Britain and a frequent contributor to the proceedings of the Institution of Naval Architects. At an INA meeting in the previous July, he had ticked off the persuasive advantages of twin screws for merchant steamers. (Perhaps that helped lead Griscom toward the Thomsons.) Biles was thirty-three years old in 1887, just reaching maturity and prominence in his field of mingled science, craft, and art. "Naval architects are exceedingly sensitive on the question of the form of the ships they design," he said later. "It is in the form of the lines of ships that the artistic sense is displayed and . . . artists are exceedingly sensitive of criticism of their own work." The Griscom contract would allow Biles, still young, to reach the artistic peak of his career.

The money sources behind Griscom had authorized him to strike a deal for one new greyhound. At some point in the protracted dickerings of that spring, he decided he needed a matched pair, like the recent Cunarders. He got assents for two ships from Benjamin Brewster, on behalf of the Standard Oil interests, and from James Wright, the titular president of the International Navigation Company. Wright then took their arguments to the final arbiter, George Roberts, president of the Pennsylvania Railroad. "We have very urgent letters from Mr. C. A. Griscom, pressing us to agree that he shall contract for a second new ship," Wright told Roberts. "We agree entirely with his opinion." The railroad, already so deeply entwined with the merged steamship line, surely could not fail it now. "We invited new capital into our concern [from Standard Oil] at your instance," Wright pointed out. "We made our contract for the [Inman] Liverpool Line at your instance, and to give you a connection that you thought was necessary. We believe that a first class ocean connection such as this will be can render you as great service, and is as worthy of your favorable aid as an inland connection." The railroad's board, weary from years of steamship deficits, was not delighted. "It is believed to be the policy of this Company," the board grumbled, "to withhold further investments in Ocean Steamship Lines." But given the railroad shares already sunk in the expanded enterprise, it too fell into line and came up with another half-million dollars.

The *City of New York,* flagship of this new Anglo-American company, was launched in March 1888. (She was appropriately christened by Lady Randolph Churchill, an American woman who had married an English lord and produced their Anglo-American son, Winston Churchill.) The *City of Paris* followed seven months later. They were designed mainly by John Biles, with suggestions from Griscom. The two ships retained the traditional Inman clipper bows, figureheads, and bowsprits. Debatable in sailing qualities, and long since spurned for straight stems by other Atlantic lines, the elaborate forefoots did impart a dramatic, soaring look to the bows, romantic and nostalgic. In all other respects, the ships leaned hard into the future: the biggest (560 feet and 10,500 tons) and most powerful (20,000 horsepower) Atlantic liners crossing the ocean, propelled by twin screws and two triple expansion engines. For extra safety, they had double bottoms extended up to the first keelsons, both transverse and longitudinal bulkheads, and sealed, doorless barriers between the watertight compartments, so no sudden emergency would find the bulkhead doors jammed open. The reckless term "unsinkable," heard in marine discourse since the 1860s, was routinely attached to these ships.

They were so pretty, inside and out. Biles gave them three masts, and three funnels placed between the first and second masts. (Even with double engines and screws, it still seemed imprudent not to carry sails.) The sixty-foot funnels were longer and thinner than those on the recent Cunarders, and in more pleasing visual proportion with the high, steel masts. The ships had five decks: the orlop, just above the hold, for cargo; the lower and main, for cabins and public rooms; the upper, for sheltered promenades and the plushest staterooms; and a final deck on top of that, giving access to the steamer chairs, masts, funnels, and lifeboats (which, breaking from the usual practice, had space for everybody on board). About twenty special stateroom suites included couches, armchairs, extra beds, toilet stands that became card tables, and even—in some cases— private bathrooms. A discreet or bashful party could hide away in these rooms through the whole voyage, even taking meals there. The private suites, costing up to $600 during tourist season, were booked for months in advance.

In a striking departure, not prudently respectful of the North Atlantic's blunt battering power, the main saloon was placed under a long dome toward the front of the ship, between the first mast and first funnel. There the saloon's windows and skylights were cleanly ahead of the spewing funnels—but at times too vulnerable to waves crashing over the bow. The saloon itself was the most gorgeous room yet seen on a transatlantic liner, 55 wide and 100 feet long. The walls, to a height of about 9 feet, presented dark, polished wood lightened by windows, mirrors, and vertical niello panels. Above this dark lower band, the white ceiling soared up into a long, rectangular, vaulting arch that ran most of the length of the saloon, peaking at 22 feet above the floor. Stained-glass windows in intricate Renaissance styles curled up and down across the arch, protected by an outer barrier of thick external glass. At night, electric bulbs between these inner and outer windows made the stained glass glow with rich colors. At the bases of the arch, just above the dark walls, robust mermaids held brackets on their shoulders, between bas-relief panels of nymphs, tritons, and dolphins. The electric lights in the side alcoves had burnished-brass reflectors in the shape of scallop shells. At the ends of the saloon, in the capacious half-moon of gabled space between the lower walls and the top of the arch, were an organ loft at the front and, at the rear, an ornamented oriel window of twelve panes, bowed slightly outward.

On the other side of this window was the ladies' drawing room—placed above and just aft of the saloon, and well removed from the smoking room, for the sake of better ventilation. The portholes here featured two slides: stained glass for daytime and mirrors for night. The couches, settees, stools, columns, rugs, walls, drapes, and ceiling panels were all crowded with silken, light-colored French floral and geometric designs—a Victorian notion of what would appeal to genteel women, but a too-busy excess of decoration that helped make the adjacent saloon look especially restrained and tasteful. The smoking room off the saloon was just straight lines and right angles, the rows of leather armchairs and their classical scrolled arms in stiff, manly alignment, everything squared off in somber wainscot oak and dark panels of American walnut above a tiled floor. The library, intended for both genders, mingled these two wildly varied feminine and masculine styles under a central skylight. Of the 800 volumes on the shelves, richly bound and displayed, some 250 were by American authors—a major British cultural concession and probably a sign of Inman's Yankee ownership.

The *City of Paris,* the slightly swifter sister, repeatedly set new transatlantic records in both directions. In the summer of 1892 she steamed to New York in just under five days and sixteen hours. "All of the 265 cabin passengers felt that it was in some way to their credit," the American journalist Richard Harding Davis wrote in *Harper's Weekly,* "that the boat on which they chanced to cross beat the record, as individuals seem to feel they are deserving of your admiration because they have not slept for two nights. . . . As the writer of this article was one of the 265, he feels this pride, and though possibly had he been left at Liverpool the *City of Paris* might still have done as well, he wants the fact that she did so well, and that he was on board when she did it, generally circulated through the medium of this paper."

*T*he White Star Line had not built an important new ship for a dozen years. The escalating rounds of furious transatlantic competition began with the line's *Oceanic* and her sisters, then the *Britannic* and *Germanic* that soon followed. In their roiling wakes, the Guion, Cunard, and Inman and International lines had all built their own champions, often in imitation of Edward Harland's designs. His White Star ships, so advanced when they first arrived, kept running routinely fast voyages across the Atlantic, relatively stingy with coal, well patronized, and quite profitable. White Star had no compelling financial reason to enter new horses in the race. As a naval engineer and shipbuilder, Harland always guarded the purses of his shipowners. When he calculated the rising fuel consumption of modern engines and boilers, the boggling cost of building the next record-breaker, and her too-brief life at the top, the game made no practical sense to him. "From a commercial point of view," he predicted in 1885, the Atlantic greyhounds "would be proved to be an egregious mistake."

By then, however, Harland was no longer running his own shipyard. In his mid-fifties, having passed the leadership of Harland & Wolff to William James Pirrie and Walter H. Wilson, he went on to a second career as the mayor of Belfast, high sheriff for County Down, and member of Parliament for North Belfast. He was knighted after entertaining the Prince of Wales on his royal visit to northern Ireland. Back at the shipyard, Pirrie courted clients and managed the business while Wilson took charge of design and engineering; both had come to Harland & Wolff as young men and risen through the ranks. Breaking with their mentor, and coming to quick agreement with Thomas Ismay (still the head of the White Star Line), they decided to risk the costs and uncertainties of building new greyhounds.

In their general form and proportions, the *Teutonic* and *Majestic* looked like Harland designs—constructed on his box-girder system, turtle-backed at the bow and stern, and showing higher ratios of length to beam than other recent Atlantic liners. They were longer but more narrow than the new Inman greyhounds, with a bit less interior room and 3,000 fewer horsepower: joining the game but not intent on bigger numbers in all categories. (White Star was, as ever, going its own way.) The *Teutonic* and *Majestic,* driven by triple expansion engines and twin screws, still had three masts. But, for the first time on a major Atlantic steamer, the masts carried no spars or yardarms, and thus essentially no sails. The chief designer, Alexander Carlisle, trusted the doubled machinery to avoid a complete breakdown at sea. Fifty years after the first transatlantic steamships, these White Star greyhounds finally relied on steam power alone. Their foremast stood cleanly unencumbered, its tapering length broken only by the girdling crow's-nest for the lookouts—another novel touch by Harland & Wolff in the crack Atlantic service.

The public rooms were assigned to the old London firm of George Trollope and Sons, interior decorators. In the spirit of Victorian revival, they drew on distant architectural traditions for the White Star ships. Staircases, passages, and doors all bristled with wood carvings inspired by the Italian Renaissance. The main saloon, more elaborate than its counterpart on the Inman greyhounds, resembled the palace of a fifteenth-century Venetian prince. "Certainly the most magnificent apartment that has yet been seen on the water," wrote one journalist. "From the construction and decoration it appears to be twice as large as it really is." The walls and ceiling gleamed in tones of ivory and gold. The windows were lined with Renaissance-like brasswork and fitted with stained-glass blinds bearing the arms of the principal cities and states of Europe, Canada, and the United States. Columns rose from the floor, passed through a neo-Egyptian style of ornamentation, and into flutings topped by lush Corinthian capitals. Above them, the dome opened upward, over a circling band of large, three-dimensional panels—marked off by more Corinthian columns—that depicted, through ivory models on gold backgrounds, the history of sailing ships from the ancient Romans onward. Instead of hanging lamps, electric lights were integrated into the ceiling ornamentation. The dome soared overhead in stained glass, the colors denser and more detailed than on the recent Inman rivals.

On the upper deck, in a masculine cluster toward the stern, were the smoking room, barbershop, and main gentlemen's lavatory. "A somewhat special feature in these ships," the journal *Engineering* said of the smoking room; "an effort having been made to do away, as far as practicable, with the customary somewhat stiff seats of steamers' smoke-rooms, and in their place to substitute comfortably upholstered couches." In another bow to the priorities of the absent Harland, the room was more civilized and luxurious than previous seagoing tobacco havens. The leather couches were curved, sleek, supple, rather feminine in fact, not hard and angular. Two skylights rose above bands of classical, partly human sea creatures. The walls were covered with paintings of additional ancient sailing ships and with panels of richly gilded embossed leather in reproductions of Flemish *cuir repoussé*. Shallow niches in the walls held small statues carved in high relief in pear wood, after the style of Donatello. The ceiling followed an old English plaster motif of eccentric patterns punctuated by modeled rosettes. And then two utterly modern touches: on the outside wall, an electric fan for drawing off fogs of tobacco smoke, and, on the floor, gleaming brass cuspidors, another concession to American ways.

The best staterooms were designed to resemble bedrooms on shore, not cabins on an ocean steamship. Floral-printed curtains hung at the rectangular windows; no bare, circular portholes here. Instead of upper and lower berths, real beds with real mattresses stretched out in a single tier, two pillows wide, and framed by real headboards and footboards of brass piping or carved wood. Homelike details included easy chairs, chests of drawers, and hanging cupboards. An electric light in the ceiling, positioned to allow reading in bed, was controlled by a bedside switch. The ivory bell button, also close at hand, summoned the steward when unruly ocean conditions intruded.

Designed in obvious response to the recent Inman greyhounds, the *Teutonic* and *Majestic* started service in 1889 and 1890. They at once became the main rivals of the *City of New York* and *City of Paris.* The two lines scheduled their finest ships against each other, sending them from Liverpool and New York on the same days of the week. Arraying Glasgow builders against Belfast, and American owners against British, the four ships ran many brisk ocean races, unofficial but zealously pursued, boiling flat out all across the ocean. They were very closely matched. The White Star greyhounds took the westbound record away from the *City of Paris* in the summer of 1891, but she soon took it back. The struggle for supremacy went unresolved—and therefore stayed interesting, profitably drawing ocean-crossing partisans back again to their favorite ships.

*A*nd they weren't Cunard ships. The advent of the Inman and White Star twin screws eclipsed the *Umbria* and *Etruria,* the Cunard single screws of only a few years earlier. The Cunard management in Liverpool and New York fretfully tracked the next rivals, hoping they would fail. Vernon Brown, Cunard's agent in New York, sent the cabin plans and advertisements of the new Inman greyhounds to his colleagues in Liverpool—an unusual display of careful, worried attention. The Cunard men leaped on early rumors of problems on the *City of New York:* of poor food and appointments, and disorganization in the engine room, and a story that an engineer had quit the Inman greyhound in disgust and applied for a Cunard job. "So far," a relieved Cunard employee informed his boss, John Burns, "she has certainly not proved herself an unqualified success—except as an advertisement for the 'Umbria.'" But the second Inman sister, the *City of Paris,* immediately zoomed past the Cunarders. Vernon Brown soon reported that she, alas, was carrying more passengers than the *Umbria*—proving, said Brown, that the fastest ship always got the most customers, and in effect asking the home office in Liverpool what it was going to do about the matter.

At the same time, agents of Cunard and White Star fell into petty squabbles, accusing each other of promotional exaggerations and unethical behaviors. John Burns asked his friend Thomas Ismay about it; Ismay assured him that the White Star agents "have instructions to keep perfectly straight on all points." The top leaders maintained gentlemanly pretenses, sending out formal congratulations on their rivals' new ships and records, saying nothing offensive in public. In this spirit, White Star asked Cunard for a copy of its traditional Sunday religious service, intending to start a similar custom on White Star ships. "They do not hesitate to flatter us by imitation," Cunard's secretary, Albert Moorhouse, confided to Burns, "whenever they think it advantageous to do so."

An ironic comment, given that Cunard had built its line on slow, selective technological imitation. In 1891, after typically deliberate consideration, Cunard decided to add its own twin-screw greyhounds, the *Campania* and *Lucania.* The contracts went to the builders of the *Umbria* and *Etruria,* the John Elder firm (now called the Fairfield Shipbuilding and Engineering Company). The planning process for the new ships reflected meticulous attention to what the main rivals had done. James Bain, Cunard's superintendent engineer, was told to find out what kinds of safety valves and electric lights the White Star and Inman greyhounds were using. The prism lights in the dining saloon were to be about twenty-eight inches long, "equal in size to those on 'Majestic,'" and the ventilators in the ceiling "at least equal in size to those" on the White Star ship. The India rubber cubes along the principal companion ways should be the "same as used in 'Majestic.'" In the older Inman greyhounds, the twin propeller shafts were extended beyond the hull and held in place by struts that proved troublesome at sea; but the shafts of the newer *Teutonic* and *Majestic* were supported within the hull structure, another White Star idea that Cunard deemed worth borrowing.

In the fall of 1892, the *London Times* printed a long article on the *Campania* as she neared completion. Edward Harland, retired from his shipyard but still a most interested observer, read the article and sent the

Times a rebutting letter that fairly cackled with faint praise. The new Cunarder, he wrote, "will no doubt prove to be a great stride in advance of the Umbria and Etruria of that line." (Well, yes.) "She seems to be almost a facsimile of the Majestic and Teutonic, not only in the model and general arrangements of her superstructure, but also in the mode of construction, especially in the shell plating and the method of supporting the twin screw shafts in the 'run' without the aid of the usual brackets. . . . The larger hull, on more modern proportions and lines, and propelled by more economical machinery than in the Umbria," (thrust and parry) "will no doubt be a great 'pull' for the owners, whilst giving greater speed and comfort to the travelling public." And to the jugular: "Indeed, I think, too, that the owners and builders of the White Star Line may feel quite flattered by the spirit of imitation evinced in this the crowning effort of the Cunard Company."

The engines of the *Campania* and *Lucania* were also imitations, but in their case Andrew Laing was only copying himself. Recently promoted to a seat on the board of the Fairfield company, in charge of all engineering matters, Laing was still prickly—but tolerated, even favored, because of his inventive talents. He earned a bonus of one shilling for every horsepower of the engines built by his department. Some years earlier, he had devised a more compact type of triple expansion engine. He placed two high-pressure cylinders above and in tandem with two of low pressure, and a single intermediate-pressure cylinder between them. This design eliminated one cylinder and reduced the size of the working parts, but still kept the machinery in proper balance. Laing first applied this engine to a yacht owned by William Pearce, then to an Atlantic steamer for a German line. With the bugs thus worked out and his invention well proven, for the two new Cunarders Laing built the most powerful merchant-ship engines yet: cylinders 37, 79, and 98 inches in diameter, pushing through strokes of 69 inches, pumping out 15,000 horsepower per screw and 30,000 per ship—more than twice the power on the *Umbria* or *Etruria*. (Under a tight construction schedule, buried in his babies, Laing neglected to treat a hovering Cunard representative at the shipyard with all expected politeness; he was afterward urged to "avoid friction.")

The passenger accommodations, the most plush yet seen on a Cunard ship, generally followed the styles of the recent Inman and White Star greyhounds. The public rooms on the promenade deck were decorated by the Glasgow firm of Wylie & Lochhead, makers of fine furniture, cabinetry, and upholstery. The library, toward the bow, revived the French Renaissance. Against the curved back wall, the books were behind glass doors and mahogany astragals, above a lower cabinet of carved amboyna panels and pilasters. The armchairs and two large ottomans in the middle of the room were blue velvet, on a Turkish carpet over oak parquetry. A ring of writing tables hugged the forward wall, in relative seclusion. The ornate ceiling was painted in two shades of ivory. At the middle of this deck, the drawing room—also in a Renaissance style—had a coal-burning open fireplace, with a brass grate and a hearth of Persian tiles topped by three arched mirrors. The walls, organ, and grand piano were all made of satinwood, the chairs and ottomans upholstered in rich velvets and brocades of various colors. The smoking room, aft of the second of the two funnels, suggested the ambience of an old Scottish baronial hall, dark and subdued. That required another fireplace, with a bronze dog grate and blue tiles surmounted by an overmantel of carved oak. The bar beckoned from the other side of the room. Alcoves and recesses broke the space into small, convivial groupings of card tables and chairs bottomed in natural pigskin.

The first-class dining saloon on the main deck was designed and decorated in-house, by Fairfield's own craftsmen. One hundred feet long and sixty-two feet wide, it only served meals; the expanded space of a modern liner allowed such specialization, with general socializing reserved for the public rooms two decks above. Described as generally Italian in style, the saloon was actually dominated by Spanish mahogany woodwork in the walls, columns, and sideboard. The chairs had dark red velvet seats and mahogany backs and arms. The ten-foot ceiling—a foot loftier than the norm—was divided into ornate, ivory-painted rectangles. Corinthian columns sprouted among the four main tables. A central well, about sixteen by twenty-four feet, broke through the ceiling, poked into the upper and promenade decks, proceeded through the middle of the drawing room (screened off by beveled-glass lattice windows), and ended up in a distant curved dome of stained glass, thirty-three feet above the floor of the dining saloon: the tallest open space yet seen on a liner. The grand staircase, paneled in solid teak and gold Japanese leather paper, descended in wings and landings from the rear doors of the drawing room, past the staterooms and facilities on the upper deck, and down to the aft entrances of the dining saloon. Wide enough for six people to pass abreast, the staircase included a center handrail for those at sea in the middle of the stairs, too far from the side railings. Together, the central well and grand staircase—each under its own skylight—allowed the dining saloon a comforting, quieting sense of lofty headroom, and natural light and air, despite its position deep within the ship.

The *Campania* and *Lucania* capped a quarter century of furious racing and rivalry that was dominated, from 1879 on, by a single Clyde shipbuilder. As the final, ultimate British greyhounds of the nineteenth century (600 feet long and 12,500 tons, and just two vestigial masts for signaling, lifting the lookouts, and handling cargo), these Cunarders brought the transatlantic records down to under five and a half days in both directions. According to a list compiled by the *Marine Engineer* journal, the fastest Atlantic passages were reduced on forty separate occasions between April 1884 and September 1894. Ships built by Harland & Wolff did it five times; Thomson ships, nine times; and Fairfield ships, twenty-six times. (And this accounting did not even include the many earlier marks set by Fairfield's *Arizona* and *Alaska*.) As a young shipbuilder, at the start of his career, William Pearce—the man who made Fairfield the launcher of champions—had predicted five-day boats within his own lifetime. Pearce died suddenly, of heart disease, in 1888, at the age of fifty-five. He missed his reckless prediction by just a few years.

*T*he ships-as-buildings steamed into historical paradox. In their structure and machinery, they became the largest, most complicated artifacts of the Industrial Revolution. In their furnishings and decorations, they were seagoing galleries of the art and architecture of the Renaissance. They mingled nineteenth-century Glasgow with sixteenth-century Florence—William Pearce and Andrew Laing in collaboration with Michelangelo and Leonardo—a steam hammer and a painter's brush. One was driven by coal and water, the other by Christian inspiration and a patron's wishes.

This grand fusion, reaching across three hundred years, was essentially a historical coincidence. The great Atlantic liners of this era were built at a time of Renaissance revival in late Victorian architecture. "We continually turn our eyes to Italy . . . the beginning of modern architecture," said the *Architectural Review* of London. "In the Italian Renaissance, for the first time, the architect steps into the light of history, and his buildings bear a stamp of personality." The leaders of transatlantic steamship companies inevitably lived among such historic references in their own homes and offices. Thomas Ismay's ornate mansion, "Dawpool," was designed by the architect Richard Norman Shaw as a Renaissance palace on the outskirts of Liverpool. When Ismay and his peers chose interior designs for their steamers, they naturally settled on the lushest, most fashionable styles then available on land. They were practical men, not oversteeped in architectural history, who only wished to make their ships more attractive to discriminating customers.

Yet they chose better than they knew. The major artists of the Renaissance were also engineers, builders, architects, inventors. The explosion of unfettered inquiry and creation that began in fifteenth-century Florence led implicitly to the Industrial Revolution and hence to the *Britannic, Arizona, Umbria, City of New York, Teutonic,* and *Campania.* A literal "Renaissance man" had recognized no distinctions between art and machinery; such walled categories belonged to a later time. The crack transatlantic steamships of the late nineteenth century restored that distant union and harmony.

The main dining saloon may be perceived as a cathedral sanctuary. On the *City of New York,* it was 55 feet wide and 100 feet long, with smaller alcoves off to the sides. Vaulting overhead was a dome of Renaissance stained glass; in the gables at the ends, an organ loft and the projecting oriel window of the ladies' drawing room. "Here then," said the *Engineering* journal, soaring into an atypical flight of religious musing, "are all the requisite structures for a church service—the oriel window as the pulpit, the organ loft for the choir, and the dining saloon as the nave and transepts." Even a nineteenth-century engineer, in this instance, could still recognize the timeless worship encased within the hubris of modern progress.

Ships as Towns:
Officers, Crew, Steerage

*B*oard the ship and look around. Avoid the architectural showcases in the grand public rooms and cabins of first class. The promenades on the upper decks stretch along the port and starboard sides for over four hundred feet on the biggest liners of this era. These avenues offer access to many rooms and deckhouses of workaday functions: the chart room, purser's office, quarters for the top mates and engineers, the captain's cabin, pastry chef, main galley, other galleys, bathrooms, lavatories, hospital, butcher, barber, apothecary, carpenter, printer (for the ship's newspaper and the daily menus), and the new electrician (who has replaced the lamplighter). On the lower decks, where first-class passengers seldom go, are dank quarters for sailors in the forecastle, stewards in the "glory hole" toward the stern, and steerage passengers in any available space. At the bottom of the ship, making everything run, are the engineers and engines, firemen and boilers, and bunkers on bunkers of coal.

Toured in this way, the ship doesn't seem like a floating palace or any kind of single structure. Instead she has streets and buildings, shops and tradesmen, electric lights and a bridge telegraph, dynamos and horsepower (without horses), water supplies, garbage, and sewage. Her population embraces separate neighborhoods of quite assorted people engaged in unrelated tasks and pastimes: rich and poor, workers and drones. Taken all together, the ship and her inhabitants become a crowded, busy town—which just happens to be steaming across 3,000 miles of the North Atlantic, days into nights, on through storms, fogs, and ice fields at a steady twenty knots.

*T*he captain may be considered the mayor, though his dictatorial whims at sea far surpassed the powers of any urban tyrant. A successful captain held office for a long time; the range of human and technical skills demanded by the job were not easily found or replaced. A typical transatlantic commander would work for a particular steamship company for decades, switching from ship to ship as needed, but always bringing his own methods to each new posting. A liner took on the personality of her captain because his moods and decisions reached down into the smallest details of shipboard culture. Loyal customers kept returning to their favorite steamer as much for the captain as for the ship. (The newspaper notice of a sailing might advertise the names of both.)

The captains of the late nineteenth century—even edging well into the twentieth—were still marked by their early careers on sailing ships. They usually came from seafaring families, grew up inhaling the ocean in coastal towns, and scarcely had to choose their livelihoods. The lore and lure of salt water flavored their earliest memories and ran through the rest of their lives; they went to sea as a matter of course. To ship out under sail was not just another job. The unique skills needed to handle sails and move around the rigging, long absences from home, risks and adventures, and the faith to trust an extended ocean passage to nothing but the wind all made sail its own peculiar calling. Wind-sailors knew they were special: not many men could or would do it. Samuel G. S. McNeil of the Cunard Line started out on sailing ships when he was thirteen. After about ten years, he thought about switching to steamers—but recoiled at the prospect. "There were too many people on board who were not sailors," he later snorted, ". . . who, apparently, only went to sea for money."

Sandy McNeil had other reasons. He grew up in Liverpool, the son of a chief engineer on Cunarders who later settled into a shore job as superintendent engineer for another steamship line. After grammar school, McNeil apprenticed himself to a firm that traded between Scotland and Calcutta. He always retained sharp memories of the odors of those old sailing ships—of Stockholm tar, manila, and canvas—which seemed redolent of all the world's oceans and of the daring hardships of sailing them. The experience fired his youthful imagination and hooked him for life. But the food on those ships was terrible: monotonous and coarse, just salt beef and pork, bully and pea soup, rancid butter, and hard, wormy biscuits that would get up and run away if not held down. Bad as it was, even this miserable fare usually ran out before the end of the voyage.

So steam had its appeal, despite all those nonsailors on board. The crack transatlantic liners no longer carried much sail, but they did offer better food, warmer berths, more regular watches, and shorter passages. "I always liked speed and smartness," McNeil recalled. "I loved the lines of the fastest mail ships belonging to the Cunard and White Star." Succumbing to the "dirty and hateful steamers," he served as second officer on the *Etruria* and *Lucania* at the start of a distinguished Cunard career. After those long, slow passages to India, the pace and wetness of the North Atlantic racecourse astonished him. The ships "pitched and lurched and rolled and snorted, and sometimes they seemed almost to stand up on end." Driven at top speed through nearly any conditions, absorbing severe structural damage as a routine extra cost of the trip, they buried their noses off Fastnet and scarcely came up again until Sandy Hook.

Junior officers like McNeil learned to maneuver around the outsized personalities of their captains. One commander arrived on the bridge every morning at 7:45, brimming with complaints. He always grumbled, just on general principles, even with no adequate cause: the boatswain and his mates were worthless, the first officers could not tell a good man from a bad one, somebody wasn't wearing a cap, the bridge-boy's hair was too long, and the whole ship was intolerably slack. And so on, for hours, trying the patience of anybody within earshot. McNeil decided to give him a legitimate irritant. He had the steward collect some cigar and cigarette butts, and he dropped them outside the captain's cabin at 7:30. The captain stormed up to the bridge on schedule, face flushed and eyes aflame in vindication. For ten minutes he damned at will the unknown culprits responsible for this outrage and then, well spent, complained no more for the rest of the day. McNeil thereafter made sure to leave something awry in the captain's path, keeping him happily brief in his complaint. "There is no place in the world like a ship," McNeil declared after a lifetime at sea, "for providing opportunities of seeing men exactly as they are and not merely as they would like to appear."

Personality emerged on the ocean. The steamship company gave the captain his sailing orders. Once clear of land, though, in these years before wireless communication, the ship was alone on the vast North

Atlantic, and the captain ran his seagoing town just as he wished, revealing himself in his ship. William H. P. Hains came from the seaside town of Devonport, near Plymouth. The son and grandson of sailors, after training in sail he joined the Cunard Line in 1857, at the age of thirty-four. Over the next forty years he commanded many Cunarders, finishing on the *Campania,* and retired as commodore of the line. Hains became a particular favorite of women passengers and was known as "the calico captain" and "a sort of Bunthorne of the sea," after the aesthetic poet in Gilbert and Sullivan's *Patience* who was adored by all lovesick maidens. He was tall and distinguished-looking, with curly hair, a lush mustache and muttonchop whiskers, and a properly sensitive look to the eye. Smiling and laughing easily, he would favor the traditional ship's concert with his famous, full-throated rendition of "I'm afloat, I'm afloat, and the rover is free." (When not roving, he also had a wife and six children back in Liverpool.)

Hains's popularity among the smitten women crowding the captain's table demanded diplomatic finesse. As he later told the story, he nearly ran aground on feminine jealousy when the Swedish operatic soprano Christine Nilsson crossed on his ship. Hains placed her, as the most distinguished woman on board, at his right hand at table and otherwise paid her meticulous attention. Hains sensed rumbles of discontent from other women in his temporary orbit. A delivering storm blew in; Hains had to spend seven straight hours on the bridge; in his cabin beneath the bridge, Nilsson sang inspiring songs that floated up to the bridge through an open porthole. (A troublesome detail to the story: all portholes would have been clamped shut in such heavy weather.) In any case, the storm over, Hains told the other women they had all been saved by Nilsson's heroic singing, which sustained him in the crisis, so they must not resent her place at his right hand. And they did not. So he said.

Samuel Brooks made nearly seven hundred passages across the Atlantic, and a record overall total, as of 1894, of almost 2.5 million miles by sea. He started as a cabin boy on a sailing ship, passed through the grades, and took command of his first Inman liner at twenty-seven. Aboard the *City of Chester* in 1878, a passenger described him as "urbane, genial, and devoted to his ship—a well-built, handsome man, with a rosy face and a full, yellow beard frosted with gray, a disciplinarian and yet a gentleman, whether he is on the bridge or below." When time and weather allowed, at the captain's table he combined lobster meat, lettuce, and olive oil into his celebrated lobster salad. On Queen Victoria's birthday, he stood up after dinner, raised his wineglass, and ringingly proposed a toast to the queen: "Whom all Englishmen revere, not only because she is a good queen, but also because she is a good mother and was a good wife." Three cheers followed, shaking the ship, and a heartfelt chorus of the British national anthem.

After twenty-two years with Inman, Brooks took command of the Guion Line's *Arizona,* then the fastest ship on the ocean. He brought his reputation for safety—of never losing a passenger at sea—to the breakthrough Guion flyer. The coloratura soprano Adelina Patti, a great transatlantic star in her time, always preferred to cross with Captain Brooks. "He is a favorite with every passenger," said a constant traveler, "who blends suavity with a dignity that never repels, but prevents intrusive familiarity." When the actress Lillie Langtry, notorious for her affair with the Prince of Wales, made her first trip to America in 1882, she sat at Brooks's right hand. (For her thespian gifts, surely, not her sexy reputation.) One night a fearful storm brought cabin passengers up on deck in fright; Brooks spent most of the night, he told Langtry, "arming" women in various stages of dress from the deck back to their cabins.

During the first three decades of transatlantic steam, Cunard captains in the mold of Charles Judkins had earned, and richly deserved, their renown for grumpy unfriendliness to passengers. In the late nineteenth century, that frosty, dubious mantle was assumed by White Star captains. Of Hamilton Perry, the White Star commodore in the 1880s, it was said, "It depends upon how you take him." Indeed. Captain Charles William Kennedy of the *Germanic* was "best liked by those who know him best." No doubt. "I won't say he's a darned disagreeable cuss," a passenger allowed about Benjamin Gleadell of the *Celtic,* "but he's too dignified to suit me." Peter Irving of the *Republic* was "still another of the White Star's galaxy of quiet men." The food and service on White Star liners were generally the best on the ocean; but their captains would seldom unbend from duty and try to charm.

Cunard captains, by contrast, now frequently resembled William Hains, talkative and approachable. (Perhaps this reflected the belated shift in Cunard shore leadership from flinty Charles Mac Iver to smooth John Burns.) William McMickan grew up in southern Scotland, the son of a shipbuilder. He went to sea at sixteen, on the inevitable sailing ship, and joined the Cunard Line as first officer on the *Africa* in 1859. Over the next quarter century he worked his way up to the command of the *Umbria.* He looked exactly like an old sea dog, rotund and weatherbeaten, standing with his legs spread wide and toes pointed out for stability. At ease in his cabin, eyes sparkling, he would spin seafaring yarns by the hour. "He likes a glass of wine, an iced American beverage on a warm summer's night, or the hot spiced grog of his native land on a winter's night; he is fond of a genial companion, has an ear ever open for the last good story, and an eye of admiration for the gentle sex."

Horatio McKay, another Scotsman, came from the east-coast town of Stonehaven, south of Aberdeen. He grew up on the water, first shipping out at age nine on a sailing ship skippered by his father. After sixteen years in sail he switched to steam as fourth officer on Cunard's *Asia*. He toiled on Cunarders for four decades, encountering his full share of dangers and deliverances. In quiet intervals he taught himself French and German, learned to paint and play the violin, read extensively, and wrote many magazine articles that were published on both sides of the ocean. He paid informed attention to the dangers of ice and fog and the need for prescribed shipping lanes across the North Atlantic. "The much-dreaded ghostly-looking iceberg is as silent as a tombstone, and like that emblem, death reigns in its vicinity," he wrote in 1881. "There is a general but erroneous impression among landsmen that seafaring men do not fully realise the dangers to which their calling exposes them." He never ended a voyage, he said, without a prayer of thanks. Yet he also accepted the universal lust for speed: in October 1894 he brought the *Lucania* to New York in a new record time of five days, seven hours, and twenty-three minutes.

Daniel Dow was born of Scottish parents in County Cork, Ireland. A rare Irishman among Cunard captains, he was nicknamed Paddy Dow. On one of his earlier postings, his ship was passed at sea by the Cunarder *Aurania;* as he watched her stern receding in the distance, he felt an ambition to command her. Dow took his initial Cunard commission at twenty-eight, after a dozen years on the ocean. In due time he indeed became master of the *Aurania* and one of the prominent Atlantic captains of his era. Dow stood tall on the bridge, showing expressive features and a pointed gray beard. "He is a genial gentleman, full of lively anecdote, which he can tell in a racy Hibernian style," one Liverpool newspaper commented. "You can't imagine his getting flurried. . . . Modest, unassuming, he nevertheless possesses a strong will and an inexhaustible fund of shrewdness." Exceptionally popular among passengers, he claimed close friendship with Mark Twain, still a fan of ocean steamers, and displayed photographs in his shipboard cabin of their yarning encounters over the years.

Something kept pulling these commanders back to their ships, voyage after voyage, usually until they were too old to keep doing it. "I like the post and I like the life," explained Angus Macnicol of the Allan Line.

"We get a good deal of nasty weather; but in a good ship there is nothing to be afraid of. It is a healthy life, you know, with sufficient excitement to make it interesting." The transatlantic captains were paid well by the standards of the time. The top Cunard salaries increased from £750 in 1880 to £1,000 ten years later. But, as Sandy McNeil insisted, they didn't go to sea just for money. "You remain away awhile," Paddy Dow reflected after four decades of ships, "and go pottering round your garden, when suddenly the call comes back to you. You want to throw off the land; there doesn't seem enough air. You feel that you must brush off at any cost the land cobwebs, and get back into your nostrils the smell of the ship, like which there is nothing else."

But the life must not be sentimentalized or shorn of its rough edges. The situation can scarcely be overemphasized: These captains were driving ships hard across the North Atlantic, in all seasons and weathers. The bridge was still an exposed platform with just a waist-high railing for protection. At the front of the vessel, about forty feet above the sea, it was routinely deluged by oncoming waves. On White Star ships of the 1890s, officers might spend the whole voyage in oilskins and sea boots. Just to reach the bridge from their quarters down on the foredeck, they would wait until the ship briefly steadied herself, then slosh through swirling water to the ladder. Under duress, the powered steering gear often broke down. Men would lie prone on a deck covered with black oil from the steering engine to reconnect the chains around the quadrant. The guy wires that secured the funnels and held them upright needed careful watching; if they ever looked slack, they had to be tightened at once, under any conditions.

Standing on the bridge, the captain generally faced forward, watching the slice of ocean toward which he was heading. But he always had to check the weather side as well. On an eastbound ship, that usually meant turning around. The Red Star Line's *Rheinland*, on a voyage to Antwerp in the early 1880s, was chased all across the ocean by a very powerful gale from the west. One morning, Captain William G. Randle, after spending the whole night on deck, spotted a gigantic wave rolling up astern. He shouted a warning through the speaking tube to the quartermaster at the wheel and leaped from the bridge up into the adjacent mainmast. The sea hit the stern and rolled over the entire ship. Looking down, Randle could see nothing but the foremast and smokestack. "My God," the captain thought, "they are all gone!" After long moments, the ship rose through the roiling foam and spray, coming up like a giant bird. A cloud of steam spewed from the engine room through its broken skylight. The lifeboats were all gone, and the railing around the promenade deck. Seven crewmen were severely injured, with broken bones, and one was swept overboard and lost. The passengers, safely barricaded below, only noticed that it went dark for a while.

In wintertime, a four-hour watch on an open bridge became a torturing ordeal. "Having ploughed across all oceans and on all seas," wrote Charles Terry Delaney, a transatlantic officer, "I have no hesitation in declaring that the North Atlantic during the winter months is the worst place in the world for continuous bad weather. Cape Horn is completely outclassed." As a liner approached the American shore, the prevailing westerly winds bit into the naked faces of the men on the bridge, combining with the forward speed of the ship to make a cutting, penetrating cold unlike anything ever felt on land. The driving bow burrowed into the sea, sending purling waves up and backward, drenching the bridge again. Ice covered the bridge deck, preventing even the warming exercise of walking back and forth. A man had to stand in one place, beating his arms for warmth, peering into a hard northwest squall and longing for the end of his watch. "The men on the bridge," Delaney concluded, "belong to that class of men who have the curse of the gypsy blood in their veins: the blood of wanderers, practically untamed men who cannot brook a quiet life."

Captain R. W. Grace went to sea at the usual young age. He served as on officer on Cunarders and then commanded ships of the National Line for the next twenty-three years. In October 1886, as commodore of the line, he took the *America* on a westward run. He was known as "the sleepless captain" because he stayed on deck for so long in foul weather. Soon after clearing Fastnet, the *America* ran into persistent, heavy westerly gales. Grace remained on the bridge from six o'clock Thursday night until noon on Saturday, forty-two straight hours without sleeping. When the ship seemed out of danger, the captain finally went inside and rested. He felt a congestive chill, he admitted, but he still ate a hearty dinner that night, followed by a long conversation with passengers. Again he said he felt chilly. He repaired to his cabin and died a few hours later, sixty-two years old. A sailor's death at sea, in the stern line of duty.

*T*he working population of the seagoing town was having labor troubles. The workforce was still divided into the three broad groups that had staffed vessels since the start of transatlantic steam, but their relative proportions were shifting. The *Campania* of 1893 carried a crew of 415 people. Of these, 195 men tended the engines and boilers, 159 men and women did the stewarding and catering work, and just 61 men—including only 40 sailors—comprised the entire deck department, the traditional backbone of ships at sea. For over fifty years, the spread of canvas had shrunk while the engines and boilers got bigger. Each generation of ships raised fewer masts and more smokestacks. The advent of twin screws and double engines, and their redundant, canvas-free safety, had capped a process of long history and inexorable momentum. The ship was now purely a machine.

"Men in steamers do not sing at their work," said the transatlantic captain James Bisset. "Only the thudding of the engines, driving the ship on, regardless of wind and weather: and men the servants of the machines." Passengers no longer woke to the distant music of sea chanteys as sailors hauled on lines or pulled in sails. There were no sails. Seamen had little to do but stand their watches, sweep and paint, scrape and scour, and handle cargo at the start and end of the voyage. They no longer got to deploy their old skills on deck, of mending and carpentry, or the athletic daring of dancing around the rigging in any weather or emergency. Like the home weaver on shore reduced to tending a noisy power loom, or the handy craftsman who now stood at a machine in a factory, doing only one mindless, numbing task through long, time-clock hours, the sailors had lost their varied, challenging, satisfying work to the march of nineteenth-century progress. Stripped of the unique life they had chosen despite all its dangers and hardships, they were just janitors and stevedores, for four or five pounds a month.

No surprise, then, that steamship companies complained about an acute scarcity of sailors, and good sailors, to fill even the reduced crews of modern liners. What true seaman would take such a job? The forecastles, traditionally British, of the transatlantic steamers were obliged to admit more foreigners, especially Scandinavians. The officers, also British, still preferred their own tars whenever they could be found. "We are all British sailors, and we do not want anything else," said Captain Thomas Thompson of the White Star Line. "I want the best sailor *for all weathers,* and in that respect the Englishman has not his equal anywhere. Of course, I do not exclude the Scotchman. There is nothing to choose between him and the Englishman." (Thompson did apparently exclude the Irishman.) "But let there be bad weather or danger of any kind, and the Englishman is all there. Your foreigner, on the contrary, is likely enough to have to be sought for. In bad weather you have never any trouble with the English sailor." Perhaps because a storm at least gave him more interesting tasks.

Though the crew now tended no sails, it had to tend more people. The ship's passenger capacity in both cabin class and steerage kept increasing. The *Campania* needed 114 stewards and stewardesses, and 45 cooks, bakers, and other galley workers. The bakers started work at four in the morning to produce breads, cakes, and hot rolls for breakfast. (The cooks did not have to appear until 5:30.) Coffee was brought to first-class passengers who wanted it, in their staterooms or on deck, at six. Four meals were served between breakfast at eight and the final end of supper, fourteen hours later. During the brief times between meals, passengers nursed their digestions with ginger nuts, prunes, oranges, and cakes. The cooks wound up at ten o'clock, a scant seven and one half hours before they were due back at work. A staggering quantity and variety of food passed through the galley and down the throats of those on board. The *Etruria* went to New York in August 1886, on a run that lasted eight days, with 547 cabin passengers, a crew of 287, and 12,550 pounds of fresh beef, 760 of corned beef, 5,320 of mutton, 850 of lamb, 1,750 of ham, 500 of bacon, 350 of veal and of pork, and 2,000 of fresh fish; and 600 fowls, 300 chickens, 100 ducks, 50 geese, 80 turkeys, and 200 brace of grouse; and 15 tons of potatoes, 30 hampers of assorted vegetables, 11,500 eggs, 1,000 pounds of butter, 750 pounds of cheese, 1,000 quarts of milk, and 220 quarts of ice cream. Not much was left on arrival.

The stewards and stewardesses began at six in the morning, cleaning the main saloon and public rooms after the amusements of the previous evening. They served meals, tidied the cabins and made beds, drew baths and kept the lavatories in good order, fetched and carried, all day long: the busiest people on the ship. They worked the longest hours, without breaks except brief mealtimes, and had the most frequent contacts with passengers—who were unpredictable and not to be refused. According to Violet Jessop, a stewardess on the White Star Line, "the type of passenger who patronized it expected all the service the company could give, and got it." In storms and high seas, fretful passengers—sleepless and seasick, and fearing the worst—demanded special attention and comfort. On the *Arizona* in 1882, during a heavy blow one night, Lillie Langtry splashed through ankle-deep water in a corridor and found a steward cleaning shoes at the foot of a gangway. "Is the ship going down?" she asked in a scared whisper. "D'ye think if the ship was in danger," the steward replied, "I'd be here brushing boots?" ("This seemed such a sensible process of reasoning," Langtry decided, "that I returned, with complete confidence, to my berth.")

To the officers and sailors, the steward department seemed an alien intrusion, just a passel of household servants and food servers that properly belonged on shore, not on ship with their hands out. A typical steward's wage was about four pounds a month, a stewardess's only two pounds, but their incomes were considerably fattened by tips at the end of the voyage. Tips! Seagoing men didn't take tips! "Pampered and spoiled by passengers," Charles Terry Delaney said of stewards, and therefore immune to ship discipline; "the liberal tipping gives to servants aboard a ship too much money." The presence of women among the ship's workers also seemed incongruous to traditionalists. They were so outnumbered—eight stewardesses among the 407 crewmen on the *Campania*—and so overpowered by the dominant rough, masculine style on a ship. Violet Jessop was appalled by the bugs and bad behavior she saw in the steward quarters: "No place can be so utterly devoid of 'glory,' of comfort and privacy and so wretched a human habitation as the usual ship's glory hole." In 1886 the Cunard Line suspended Mr. Ball, chief steward on the *Gallia,* and Mrs. Clegg, one of his stewardesses, for their "misconduct" together. Again, in 1899, Mr. Reid, the fourth officer on the *Pavonia,* and Mrs. Cowan, a stewardess, were dismissed for "impropriety." The women might have indulged these flings willingly; or they might not have felt allowed, or secure enough in their jobs, to refuse their superior officers. The scant historical record, which by its nature underreports the dynamics and frequency of such incidents, does not say.

After an especially swift and smooth passage, and devoted service, an excited or grateful passenger might bestow a tip of up to a sovereign, a gold coin then worth twenty shillings. But a steward could never count on any reasonable exchange of reward for effort. "There's some as does and some as doesn't," a steward explained in 1892. "And them as does is them you wouldn't think would. Last voyage there was a gentleman at my table, and prop'ly rich he was, too, so I've 'eard—him, and his wife and six children. No end of trouble they was, and he very harsh like in speaking, and nothink quite good enough for him, and the children reg'lar nuisances, coming from their table to carry off the fruit and the sweets from the saloon. But he did keep me busy! At the end of the voyage says to me, 'You've been very kind and attentive, William, and I want to make you a small present.' 'Thank you, sir,' says I. And with that he gives me a shillin'! You could have knocked me down with a feather!"

Down below, double screws and engines meant more engineers, stokers, and trimmers. In recent years, steam or hydraulic power had taken over a ship's steering, winches, cargo cranes, anchors, refrigeration, ventilation, lighting, and even brushes in the barbershop. Yet the most constant, exhausting physical labor on a steamer—moving coal around—was still done by hand: a sweaty, resistant holdout of muscle power on this prime symbol of modern progress. Coaling a ship before a voyage went on for days and nights of constant noise, amid sifting clouds of fine coal dust that covered every surface in the living quarters. Each shovelful was lifted and thrown repeatedly. In the harbor, a small barge was loaded with coal and came alongside the steamship. Four men with shovels filled a bucket lowered by a crane up on deck. The bucket was dumped in the ship, either into a side port or on the deck. Men shoveled the coal into the hold, where other men picked it up again and distributed tons into various bunkers, balancing the weights for stability. At sea, trimmers shoveled the coal into wheelbarrows or baskets and dumped it on the floor by the boilers. Firemen lifted it, one last time, and finally threw it onto the fires. On each voyage, about 2,000 tons of coal were passed along this time-frozen, unimproved, laborious, inefficient process.

The bottom of a ship resembled the subterranean parts of a town: dark, dirty, and dangerous, given over to functions that were necessary but best not seen or smelled by those of delicate sensibilities, populated by fearless rats and hard men of indoor pallor. Descending from the passenger decks, the rumble of machinery got louder, level by level, the temperature warmer, the air more acrid and smoky, stinging eyes and nostrils. It was like approaching a live volcano from above. The iron gratings and ladders were smeared with oil, greasy and slippery, and sticking to hands, shoes, and clothing—all to prevent corrosion in such a moist environment. In this shrine to steam power, the force of the century, the machined needs of iron and steel trumped any human concerns.

The huge leaps since 1870 in horsepower and boiler capacity required more fires in more furnaces. They brought the boiler-room air temperature up to 130 and even 140 degrees—for men engaged in wrenching exertions. The room was fitfully illuminated; spaced cones of light from overhead electric glow lamps alternated with dark shadows. It felt murky and claustrophobic. A stoker called for more coal from the trimmer by rattling his shovel against the furnace door. The sharp ring of metal on metal cut through the background roar of the fires. When a furnace door was opened, horizontal flames might lick out, reaching and nickering, from the bright red and orange glare within. Flinching from the stab of heat and sparks, the stoker threw in his load, or poked the coals with his slice bar, and slammed the door shut; all done in a snap, to keep the fire working within the furnace, raising steam, instead of making the room even hotter. Men took breaks whenever they could, as permitted by the relentless steam gauge on their boilers. If the needle dipped below a certain prescribed point, the break was over. Steam pressure meant horsepower meant speed, and speed measured everything else. During a westward voyage of the new greyhound *Majestic* in 1890, it was noted with surprise that not a single fireman or trimmer had to be carried out of the stokehole, overcome by the work and heat.

At the end of their four-hour watch, craving cool, fresh air, the firemen and trimmers came up on deck "like demons out of hell," James Bisset recalled. Among the other people on board, most of them so clean and rested, the coal handlers indeed looked like inhuman underground creatures, zombies barely alive, dressed in filthy dungarees, flannel shirts, hobnailed boots, and drenched sweat rags around their necks. Their boots squeaked with pooling perspiration. Clumped among themselves, silent and exhausted, they did not invite easy association. "They were tough-looking characters," as Bisset remembered them. "Their faces, blackened with coal dust, and streaked with sweat, had a dulled, animal-like look, and they seldom smiled. It was killing work." Eight hours later, they were back at it; and so on, every twelve hours, through the voyage.

They had to be tough even to consider such employment, and the work only made them tougher. Like the stewards, they lacked true seagoing tradition and had not grown up within ship discipline and hierarchy. Like the sailors, with whom they uneasily shared the bottom of the ship class structure, they were hardened, muscular men, not happy about their five pounds a month, and bristling with inarticulate resentments that could easily boil over into random violence and outlashings. On the Guion liner *Montana* in 1877, firemen and sailors started brawling after a series of incidents. The leaders of the two groups, John Quinn and John Kelly, squared off on deck; Quinn stabbed Kelly over the eye. Later another fight broke out in the forecastle. A seaman, John McSorley, was badly hurt, and three stokers were arrested by New York waterfront police. "The average engine crews of the present day," said the *Nautical Gazette* in 1882, "are the veriest trash, both as men and skilled laborers." On the *Etruria* in 1889, eight firemen refused to muster for a lifeboat drill. Placed in irons, they served a day in jail when the ship reached Liverpool. In 1898 a group of firemen on the American liner *St. Louis* attacked a stoker named Thomas Cummings. He shot and wounded five of them, but they declined to prosecute. Just another day in the stokehole; such steam-venting brawls, usually only involving shovels and fire bars, often marked the start or end of a voyage. A *Lucania* fireman staged a one-man mutiny in 1900. The Liverpool police court gave him a month at hard labor for disobeying the third engineer and another month for assaulting him. Thirty-three firemen on a White Star liner in 1905, unhappy about their pay, stopped working during an eastbound run; they were jailed in Liverpool for a week at hard labor. On the Cunarder *Ultonia,* firemen broke into the wine cellar on New Year's Eve, then defended their quarters with fire axes. They were briefly locked up at sea, but the ship still needed them, so they were released to their toil. "Drink and women," one Atlantic captain said later, looking down from the airy perspective of the deck, "were all those firemen lived for, their work being merely means to those ends."

As with the sailor shortage, steamship companies periodically wondered why they had such trouble hiring enough stokers, of sound habits and tractable dispositions. "I can account for the falling off in good men," said Thomas Wilson Sewell, chief engineer on the *Majestic,* in 1890. "I do not believe the pay has anything to do with it. Men will not go to sea for any sum when they learn that their work is to be confined to air-tight fire-rooms and inclosures stifling with heat." A White Star veteran, Sewell was one of the best-respected transatlantic engineers. He insisted that "not until more attention was paid to the bodily comforts of the firemen and less to getting high speed could good men be expected to enter the service." The comfort-sacrificing zeal for speed, driven onward by competitive pressures beyond any stokehole rebellion, would not change. After a brief strike by its firemen, the Cunard Line at times considered various wage incentives—a bonus of one pound for each of four consecutive round-trips, in good conduct; or, less generously, a bonus of six or seven shillings after a man had made three straight voyages; or ten shillings per stoker, and five per trimmer, for the watch that generated the highest average power in a voyage—but none of these measures addressed the real problem: the everyday, twice-a-day working conditions in the stifling stokehole.

The engineers nursed their own grievances, again more about work than money. As steam eclipsed sail, and engine displaced deck, the chief engineer and his assistants made fifteen to thirty pounds a month, twice what the officers below the captain were getting. Yet officers remained the ultimate authorities on ship. They enjoyed better quarters and food than engineers and ranked higher in social prestige, mingling at will with first-class passengers while engineers stayed below, covered in grease and grime. "Captains should remember that engineers, with few exceptions, are their equals by education," one engineer wrote in 1885, "and that they are entitled to be treated with decency. They obtain their certificates by an examination on what is certainly a far more intricate and difficult subject than navigation. If anything occurs in a steamship, the captain generally sends for the chief engineer, and still it appears to be the opinion of captains that engineers are a necessary evil"—even though captains, "if they were placed in an engine room, could not, were it to save their lives, start or stop the engines." When a greyhound set a new speed record, newspapers would praise and quote the officers and neglect the engineers, who noticed the omission.

A transatlantic steamer was, once again, a tight little microcosm of the larger social transformations of the nineteenth century. In this case it arrayed the traditional, nearly medieval ship hierarchy against the newer, knowledge-based, more egalitarian aristocracy of the machine. Engineers believed that they knew more than the officers, in more crucial areas, and had to pass more exacting tests; therefore, their status should at least equal the deck's. "The older established the line and the more influential the company is," groused a former engineer for an Atlantic mail ship in 1890, "the more conservative and unfair its administration as far as it affects the engineer's interests and status. . . . In every conceivable way the engineer on board ship is placed at a disadvantage." The Institute of Marine Engineers, established in 1889, aimed to raise standards and give its members true professional rank. It at least allowed them a forum and voice.

Trained to manage boilers and engines, an engineer faced his roughest challenge when supervising his stokers and trimmers. "Leaving port, he is not surprised to find them half drunk and altogether unruly," an engineer recalled. "The fires are not properly worked; there is confusion and trouble and ugly looks and words in the fire room. . . . No trouble is more annoying and serious than the friction and heat developed between the engineer and his unruly subordinates. You really cannot do much with a fireman who cannot or will not fire. Generally it is impossible to watch them all, and you get no satisfaction until they get down to a seagoing basis." The engineers were typically Scottish Protestants; the firemen and trimmers, Irish Catholics. Divided by religion and national allegiance, the engineers stood over their hard-pressed men, giving them orders, always pushing. The coal handlers were far more numerous, often harried and angry—and armed with shovels and slice bars. "A fact which is well known in the Mercantile Marine," the Nautical Magazine commented, "but which very few engineers will acknowledge, is that they are afraid of their men." That fear, understood by all parties, kept them stale-mated. Accidents could always happen, a shovel or slice bar dropped in a certain way; just a sad misfortune in the roaring, twilit boiler room, and no one to blame or punish.

No matter the provocation, an engineer could not directly discipline his men. He could only later present a case to the captain, who alone claimed the authority to levy fines or put the man in irons. In the moment, in the stokehole, an engineer might rail and threaten a fireman, but he could not follow through. "His word should be sufficient for entry into the log," one engineer urged, "without having to take an offender before the commander and chief officer, causing loss of time and tedious explanations. . . . Being on the spot and knowing the harm done, he is in a better position than the captain to inflict the fines. How many black-guards get off without any punishment through the deck and engine-room departments not pulling together?" Shipboard culture, always resistant to changes, had been coerced into many adjustments under steam. But captains, beset and mystified by steam technology, did not welcome any further dilution of their commands. Engineers were still encouraged to stay below, just taking care of their machines.

*T*he steerage was the tenement district of the seagoing town. Here a patchwork quilt of nationalities and religions lived in dirty, over-crowded rooms, with questionable air and light, eating crude, monoto-nous food. Lacking power and money, and the hopeful self-possession to object, the people squabbled impotently among themselves and scurried meekly before authorities. Steerage comprised the largest distinct neigh-borhood aboard ship, bursting with up to 1,000 temporary residents. They shared a working-class social status with the sailors and firemen, but strict walls—literal and occupational—were supposed to keep them apart and protected. Steerage people were paying customers, like the swells up in first class, but the goods and services they received bore no resemblance whatever to those dispensed in stateroom and grand saloon. They were all in the same boat but in starkly different parts of town.

The Inman Line started a new steerage era in 1852 by offering cheap fares to Irish famine emigrants fleeing to America. For the next three decades or so, the steady improvement in Atlantic liners—bigger, faster, more luxurious—also meant better steerage accommodations, as the else-where-focused priorities of shipbuilders and designers spilled over, quite incidentally, into the steerage. At some point in the 1880s, these two historical trends stopped traveling together, hit a sharp fork, and diverged. The ships continued to grow and speed up while general steerage conditions actually got worse: a surprising, puzzling twist in the transatlantic story. The Inman liners of the 1850s had, by their breakthrough machinery, made steerage more tolerable. The steerage of the late nineteenth century, flipped and reversed, was made less bearable mainly by people, not machinery.

Pick up the story in 1872, on one of the new White Star ships that were revolutionizing liner design. A middle-class London clerk, down on his luck, booked steerage on a White Star steamer to America. Afterward he wrote a long account of the voyage that appeared anonymously in *Catholic World,* a monthly magazine published in New York. The place of publication matters: *Catholic World,* edited by the influential journalist-priest Isaac Hecker, articulated the interests of American Catholic emigrants; it therefore would have welcomed a critical report on steerage conditions on steamships run by British Protestants, like the White Star Line's. The clerk's account began at the landing stage in Liverpool. For ten shillings he had bought his steerage kit of a plate, basin, drinking and water can, and a knife and fork; a mattress and thick blanket, quite necessary, cost extra. The clerk entered the single men's quarters at the fore end of the ship (single women and married couples were placed aft). He descended stairs to the main deck, then more stairs down to the steerage 'tween-decks. When his eyes adjusted to the dim light, he saw an open area filled with three narrow wooden tables, each seating about twenty people. Around this space were shelves of bunks: the lower two feet above the floor, the upper three feet from the ceiling, and three feet between them. Each shelf held four parallel berths, six feet by two feet, of unpainted boards, recently hammered together (and therefore free from vermin). Men assigned to the inside berths entered by the foot or head. The lavatory for the forward steerage, of adequate size, had a tiled floor, eight closet pans, four hand basins, and four sinks. "Putting aside the absence of any privacy," the clerk wrote, "the arrangements were suitable, and the fittings generally clean"—though the crew's inattention made the water supply erratic.

Meals were served from big pails and eaten at the tables or in one's berth. Breakfast consisted of watery coffee, doubtful potatoes and biscuits, and fresh, warm bread, the best part of the meal. Dinner at noon began with soup—"a hot compound with a faint reminiscence of gravy and mutton bones, some grains of barley, and fragments of celery and cabbage; sometimes, instead, a thick mixture of ground peas"—and proceeded to heavy chunks of meat (edible half the time) or dreadful salt fish, steamed potatoes of the taste and consistency of tallow candles, and boiled rice sweetened by coarse sugar or treacle. At five, tea was served with biscuits and butter. A barrel of biscuits was always available at the head of the staircase on the main deck. The biscuits were defiantly hard; but once broken and crushed, they tasted sweet and nourishing. Sundays brought a treat of plum duff pudding with raisins. "The victuals seemed generally to be of good quality," the clerk decided, "and, except in the case of the fresh bread and sugar, were provided with lavish if not wasteful abundance"—though the food was often carelessly prepared and roughly served.

The worst aspects of steerage were certain shared human failings. Most of the five hundred emigrants "treated their abode on shipboard as a time when the ordinary rules of civilized life were temporarily suspended, and eschewed washing, shaving, and all the vanities of dress until they again felt themselves on terra firma." (The reluctance to wash became pungent.) In the evening, after tea, some of the cooks, engineers, and firemen lingered on the upper deck and bantered, not delicately, with passengers. The sailors behaved worse, flirting roughly, at times obscenely, with women from steerage and threatening any protective men. Many women just remained below after dinner to avoid such encounters. "Most of our troubles," the clerk concluded, "arose from the crew and attendants rather than the arrangements of the ship itself."

In 1873, after a series of alarming newspaper stories, the U.S. Congress authorized a formal investigation of steerage conditions. At the time, the major steamship lines—except for Clement Griscom's brand-new American Steamship Company and Red Star Line—were owned by British and German interests. The U.S. government's inquiry therefore proceeded under no homegrown commercial pressures for positive findings; indeed, international rivalries might have even encouraged sharp criticisms. The report, by John M. Woodworth of the U.S. Marine Hospital Service, drew on findings by investigators who had taken steerage trips. "Much of the cruelty, ill-usage, and general discomfort of the steerage passage belong to the history of the past," the report concluded. The newer ships, as well ventilated as a typical hotel or hospital on shore, offered improved space, light, and food. Stewards kept the floors clean, sweeping them out every day and sprinkling sawdust and sand. The food was more ample than required by law. The air did get much worse at night and during storms. Water closets were adequate, with national differences: in British ships, "a number of seats are frequently placed together without division, and in the closets for men a urinal is commonly placed alongside of them. The Germans to a large extent, though not universally, have the more decent arrangement of single-seat closets, an arrangement preferable not merely for the sake of decorum but for sanitary reasons."

What this report lacked was testimony by actual emigrants, speaking in their own voices and from their own experiences. The investigators were middle-class professionals, slumming for a while, taking notes, and testing the air. The general tone of distant detachment rang especially harsh in the findings of Helen M. Barnard, who inspected vessels at eight ports in four countries and crossed in the steerage of an Inman liner. "The treatment of steerage-passengers on the responsible and respectable English and German lines is generally good," she declared, "judged by the popular standard of what is due the poor and ignorant classes in return for value given by them." (Emigrant witnesses might have illuminated that point, of how the experience squared with what they had expected or thought they deserved.) Barnard blamed many problems in the steerage on the emigrants: "Another fact, that the sentimental humanitarian, in his efforts at reformation, often loses sight of, or does not know, is that the large majority of this people are of the lowest order of humanity. They are filthy in their habits, coarse in manner, and often low in their instincts. . . . A great amount of discomfort among these emigrants arises from their own ignorance and life-long habits. Many of them have lived in hovels to which the steerage of a steamship, in comparison, is a palace." Though her language was obnoxious in its condescension and too-sure conclusions, Barnard was making an argument not so different from the clerk's in *Catholic World*. When assaying steerage conditions, consider the people as well as the ship.

In this era of accelerating transatlantic competition, all the major lines courted the profitable steerage traffic. Their agents and advertisements were spread across the emigrating nations of Europe. When a line's steerage was criticized in print, the company's self-interest could force quick improvements. In 1881, an Irish writer named Charlotte O'Brien published racy, sensational allegations in the *Pall Mall Gazette* of London: that single women and married couples were berthed in the same compartment of a White Star ship's steerage, causing "sin, full of ravening wickedness, and all manner of uncleanness." The story provoked a Board of Trade investigation and sudden reforms. "The ships are vastly improved," O'Brien reported a year later. "The single women are always quite separate from every other class. . . . Single men are almost always berthed, as the

English law originally intended, in compartments bulkheaded away from the other sections." The major lines had also started providing stewardesses and more convenient sanitary facilities for steerage women.

In the summer of 1884, Thomas I. Wharton—a Philadelphia gentleman, lawyer, and writer—contemplated a vacation trip to Europe. The steerage fare was only fifteen dollars from New York and four guineas (twenty-one dollars) from Liverpool. "It seems as if it must be more profitable to travel than to stay at home," Wharton mused. Twenty-five years old, in good health, and a previous voyager to Europe in first class, as a lark he booked steerage on the Cunarder *Oregon* to Liverpool, returning in the steerage of the Guion Line's *Alaska*. In general, he was surprisingly pleased. Men and women were protected from each other; "the police system was very good." The food was simple and hurriedly served, but ample and, at times, tasty. "Really it is good fare. I was treated better by the Cunard company, I think; yet I do not wish to be unjust to the Guion victuals." The bread and meat were excellent—unlike the salt cod on Friday. The sleeping quarters were cleaned every day after breakfast. "This was always fairly well done, as far as I could tell. The floor was well swept and scrubbed, and the steerage generally smelt sweet enough by the time we returned to it. In certain particulars the sanitary arrangements were not the best possible. It is, of course, in great part the fault of the emigrants." The bakers, stewards, and sailors were always more or less drunk on porter, but only one crew member—the bartender—was reliably unpleasant. Wharton achieved real friendship with one sailor, Ned Kennedy, after sharing his flask one morning.

Crossing in summertime, Wharton enjoyed good weather and no tedious, prolonged sieges of being held below. His berth—of canvas on an iron frame, with a straw mattress—was all right. "None of us were really uncomfortable at all, except the grumblers." On a calm, starlit night, shipboard life became more open and vivid. Wharton liked to stand near the bow, looking forward. With all the lights and motion behind him, the ship seemed nearly at rest, mustering just an occasional slow rise at the head and a gentle ripple from the water. Strolling aft, Wharton peered into the lit windows of the various rooms and hallways, catching snatches of talk and domestic scenes. The night lights and shadows seemed romantic, more friendly. Men and women gathered by the single men's companionway to sing, mostly seafaring songs, and some current music-hall hits. The women led until stewards sent them below. The men kept singing, fifty or so on deck, as the women sang below for another hour before falling asleep.

"The steerage passenger, in spite of ropes and placards, is really free of the ship," Wharton insisted. "What cabin grandee is invited into the forecastle, or walks welcome into the quartermaster's room? has the warm friendship of the baker, or is even on speaking acquaintance with the firemen? knows when the watches relieve each other, or what his dinner is going to be before he gets it? sees the ice dragged up from the ice-house, and the beef hauled into the galley? hears what is going on up on the bridge, in the engine-room, in the second cabin, down in the stokehole? . . . For my own part, I am ashamed of the airs I have given myself in days when I was a first-class passenger, and the next time I travel first-class I shall be less proud."

From their different angles, the clerk in *Catholic World,* the U.S. government inquiry of 1873, Charlotte O'Brien's revised impressions, and Thomas Wharton's steerage crossings converged on a single point. During the 1870s, and at least until 1884, steerage across the Atlantic was generally quite bearable—especially in fair weather—on the major steamship lines and on the finest new ships, like the White Star liners in 1872 and the *Oregon* and *Alaska* in 1884. And even without these ideal circumstances, steerage was still not bad. Over the next three decades, though, it got a lot worse. The ghastly stories of terrible steerage ordeals, later to be told and retold for generations, mostly came from this later period. Why such a marked and unexpected deterioration in the steerage?

What happened was a fundamental change in the demographics of the emigrants themselves, and the effect that had on their treatment aboard ship. Before 1880, European emigrants came overwhelmingly from northern and western Europe, mainly Britain, Germany, and Scandinavia. Thereafter, the emigrants increasingly came from southern and eastern Europe, especially Italy, Russia, Poland, and Austria-Hungary. Year by year, the proportions from these areas grew steadily, from 8 percent of all emigrants to America in 1880, to 35 percent in 1890, and 72 percent in 1900. In the earlier period, the British and German men who ran and crewed the major steamship lines could still regard most emigrants as people of their own stock, or race, or blood, as the terminology of the time put it. These same men considered the newer emigrants— Italians, Jews, Slavs—as of different stocks, probably inferior, and surely not as worthy of respect. General steerage conditions began to decline not because of any overt, declared shifts in policy but simply as the everyday expression of routine, unexamined ethnic and religious prejudices. In 1888 a Cunard official assured prospective travelers that his line, at least, mainly carried British and Scandinavian emigrants, and hardly any Italians, and therefore was obviously more desirable.

Of all the new emigrants, Russian Jews met the least friendly welcome in steerage. Mostly of peasant background, of different "stock" and "blood," they were—after all—not even Christians. "The naturally dirty habits of these people," said Vernon Brown, Cunard's New York agent, "make them always undesirable passengers." For a time, the German lines refused to carry Russian Jews on their finest ships. White Star, citing the danger of cholera infection, briefly refused their business entirely. At other times, other lines shunted Jewish emigrants into their own separate compartments, offering kosher food and creating transient ghettoes within the seagoing towns. ("Never travel by cheap steamers," Thomas Wharton had warned, "if you wish to avoid Jews, and especially Polish Jews, *en bloc.*")

*T*he Anchor Line steamer *Utopia* departed Naples for New York on March 12, 1891, fully loaded with 880 passengers and crew. She was an old tub of just 2,700 tons, seventeen years in service, and demoted in her dotage to the Mediterranean emigrant traffic. On this trip the *Utopia* was carrying 813 Italian men, women, and children, mostly from the southern provinces of Calabria and the Abruzzi, bound for America. Though the Board of Trade had recently mandated enough life belts for everyone on board any British transatlantic vessel, the ship's equipment included only 160 belts. (It seems unlikely that one of the crack Liverpool ships would have sailed with so many passengers and so few life belts.)

Laboring slowly across the Mediterranean, after five days the *Utopia* approached Gibraltar Bay early in the evening. She was going to stop there for the night and then go on to the Atlantic. It was raining hard, with a rising gale, but not yet dark. The situation demanded delicate navigation, entering from the east against a southwesterly wind. After rounding Europa Point, at the southern edge of the Rock of Gibraltar, Captain John McKeague reduced speed and noticed that the anchorage was crowded with British naval vessels. To clear the warship *Anson,* McKeague spun his helm sharply to starboard, then to port. But the warship's submerged ram, invisible and unconsidered, caught the *Utopia* at midship, ripping a hole twenty-six feet long and fifteen feet wide. The *Utopia* pulled free, drifted briefly, and in five minutes began to sink.

Many passengers were on deck, straining for their first glimpse of land. The ramming caused a sudden rush to the hatchways, as people hoped to retrieve their belongings and warn family members below. At the same time, other passengers in the 'tween-decks steerage, panicked by the water pouring in, jammed into the hatchways from the opposite direction. Nobody could move in or out. Hundreds were trapped below; some tried to push through portholes, sticking their heads through, but got no farther. Something exploded in the forecastle. Only twenty minutes after impact, the ship settled at the stern in shallow water, a quarter of a mile from shore. The bridge remained awash, with the funnel and masts still protruding above the sea.

It was so sudden and unexpected, in such chopping waves, that no lifeboats could be lowered. Other naval vessels in the bay came up and shone their searchlights. Given the conditions, it seemed too dangerous for their boats to come alongside the sinking ship; so they lay to leeward, picking up survivors. On the deck of the *Utopia,* people were falling or jumping into the water, or climbing into the rigging—all in a ragged crescendo of screaming and crying. The insufficient life belts were fought over and seized, mostly by men. "The Italians were thrown into a state of complete and cowardly panic," the *New York Times* reported. "They yelled frantically and fought madly to reach the forecastle. A few of the married men dragged their wives with them, but the bulk of the single men were heedless of the piteous appeals of the women and children. . . . The only instances of manliness occurred among the people in the rigging. Many men and nearly every woman clasped children to their breasts, but they were gradually overcome by sheer exhaustion and cold. . . . The majority of the Italians, however, behaved more like beasts than like reasoning men." (They behaved, that is, like the American and British crewmen on the Collins liner *Arctic* in 1854, pushing women and children aside in a howling frenzy to save themselves.)

The toll came to 562 dead—the worst transatlantic disaster since the *Atlantic* ran aground off Nova Scotia in 1873. All but a few of the eighty-five women and sixty-seven children died. About three hundred emigrants survived; half returned to Naples, and half took the next boat to New York. They had lost all their money and belongings except the clothes they were wearing. A recent American law, intended to filter the swelling floods of Europeans entering the country, barred emigrants without baggage. The law was waived, this time, for the bereft survivors of the *Utopia.*

*T*he annual totals of emigrants to America increased markedly at the start of the new century, reaching more than 1 million a year after averaging less than 370,000 during the 1890s. This sudden leap in volume exceeded the capacity of the major steamship lines; in 1904 the White Star liner *Celtic* arrived in New York overloaded with 283 more passengers than her available berths, amid charges of overcrowding and extortions by the crew. Two years later, so many emigrants came aboard the *Celtic* at Queenstown that some of the women were consigned to the men's quarters. In 1905, the Red Star Line's *Vaderland* left Antwerp jammed with 615 in steerage, far beyond safety. Conditions were so packed and unclean that ten emigrants died at sea and three more soon after landing in New York. This unprecedented emigrant flood—peaking at nearly 1.3 million in 1907—forced much of the traffic into minor lines and independent ships. These smaller, older, less regulated vessels, cashing in while the market boomed, pushed along the general freefall in steerage standards and facilities.

A careful investigation for the U.S. Immigration Commission in 1908 found the typical steerage much dirtier than those described by Thomas Wharton in 1884. "Sweeping is the only form of cleaning done," said this report. "No sick cans are furnished, and not even large receptacles for waste. The vomitings of the sick are often permitted to remain a long time before being removed. The floors, when iron, are continually damp, and when of wood they reek with foul odor because they are not washed." Washrooms and lavatories were, by law, supposed to be kept clean and serviceable. "The indifferent obedience to this provision is responsible for further uncomfortable and unhygienic conditions," the 1908 report went on. "The number of wash-basins is invariably by far too few, and the rooms in which they are placed are so small as to admit only by crowding as many persons as there are basins. . . . Floors of both wash rooms and water-closets are damp and often filthy until the last day of the voyage, when they are cleaned in preparation for the inspection at the port of entry." The same sinks were used for personal hygiene, laundry, dishwashing, and disposing of vomit. As for the sleeping quarters of a typical steerage, the dirt, air, and crowding had again become intolerable: "a congestion so intense, so injurious to health and morals that there is nothing on land to equal it. That people live in it only temporarily is no justification of its existence."

Even on the best new ships of the main Liverpool lines, the prejudices of the British crew might overcome the improved steerage facilities of the ship herself. As Helen Barnard and the *Catholic World* clerk had stressed, people sometimes affected conditions more than the actual vessel did. "The stewardess very evidently had no sympathy for suffering Hebrews," wrote one of the American investigators after a steerage trip on one of the finest liners in 1908. "The repeated complaints of a Russian friend did secure some attention for the sick Hebrew women from the stewardess. This same Russian who constituted himself a friend of the slighted Jewish women and had entered several complaints on their behalf insisted that there was a strong anti-semitic spirit among the crew." On many ships, stewards would freely touch and fondle women in the steerage, entering their quarters on pretexts, peeking and leering as the women changed clothes. "The atmosphere was one of general lawlessness and total disrespect for women." Steerage stewards probably took such liberties in the past—but not to the same degree, it seems, as when the female victims were merely Italians, Jews, and Slavs.

Once again, the voices of the emigrants are largely missing. Coming from grim circumstances, in some cases fleeing religious or political persecution, the emigrants traveled with low expectations. The steerage often brought their first exposures to other cultures, even to such modern amenities as electric lights and indoor plumbing. Just what went through their minds *at the time, on the ship,* was essentially lost to history. "The average steerage passenger is not envious," wrote Edward Steiner, a Slovakian emigrant to America, in 1906. "His position is part of his lot in life; the ship is just like Russia, Austria, Poland or Italy. The cabin passengers are the lords and ladies, the sailors and officers are the police and the army, while the captain is the king or czar." But once they were released from steerage to America, the new land, the world of the emigrants started to open and beckon. When they later looked back at the experience, and told their stories, the ocean voyage seemed a last, lingering, unlamented trace of the old country.

The various parts of the seagoing town went about their daily routines in contained ignorance of each other. The cabin passengers knew nothing of the simmering workplace discontents among the sailors, firemen, and engineers. The steerage was not acquainted with first class. In the evenings, the wealthy travelers might toss coins or small treats of fruit or candy down to the emigrants on the lower decks, to feel generous and watch the amusing scramble: just a distant, muffled contact. (Let them eat candy.) The deck officers, up here, did their best not to know much about steerage conditions, down there. The sailors and engineers spat mutual resentments back and forth. The ship steamed on, a grid of unrelating urban neighborhoods. Only a disaster like the *Utopia* sinking reminded all hands, again, of their common fate at sea.

· 14 ·

Anglo-Americans

$\mathscr{A}$ transatlantic steamship of the late nineteenth century carried, among the passengers, two distinct worlds heading in opposite directions. Down in steerage, emigrants from an ever-wider variety of European countries were coming to make the United States steadily more diverse, adding new rivals to the country's old dominating group of British-descended white Protestants. On the return passage, the same ships were taking old-stock Americans, up in first class, back to find their British roots. A few decades earlier, Americans would tour Britain—the nation that invented the Industrial Revolution—in order to glimpse their own present and future. Now every summer brought to Britain platoons of American tourists who were seeking their past: the cultural landmarks and traditions that defined their shared Anglo-American histories. Ever since the first voyage of the *Great Western* in 1838, Atlantic steamers had been hauling a dubious cargo of utopian hopes for reconciling enemies; the faith that simply by making the voyage swifter, surer, and easier, transatlantic steam would bring feuding nations together in recognition and peace. Over the final decades of the century, something like that actually happened for many Americans and Britons.

This tectonic shift may be refracted through the complex mind and life of Henry Adams. No American brought more extended history (and resistance) to the possibility of Anglo-American reconciliation. His great-grandfather and grandfather, John and John Quincy Adams, had both served as president of the United States, fighting wars and endless diplomatic crises against the British for two generations. His father, Charles Francis Adams, was Lincoln's minister to England during the Civil War, working desperately to limit British support for the Confederacy. Henry spent the war in England as his father's secretary. "It was the hostility of the middle-class which broke our hearts," he said later, "and turned me into a life-long enemy of everything English." After the war, Adams crafted a unique career as a writer, traveler, omnivorous reader, and—not least—as perhaps the greatest American historian ever. For a dozen years he toiled on his masterwork, a nine-volume history of the United States during the Jefferson and Madison presidencies. Upon completion, it included so many unforgiving judgments on the mother country that Adams decided he should not send it to his old friends in England. His enthusiasm for Anglo-American rapprochement was restrained by deep, unhealed wounds—personal, familial, and national.

One of the layered paradoxes of Henry Adams was that he might loathe something in general, in the abstract, yet embrace it in the particular and actual. "He is queer to the last degree," a bemused English companion concluded; "cynical, vindictive, but with a constant interest in people, faithful to his friends and passionately fond of his mother and of all little children ever born." So Adams vigorously disapproved of all things English, but he often steamed over—amid squawks of complaint—for extended visits, and he remained devoted to a small circle of particular Englishmen he had met during the Civil War, sending them frequent long, updating letters for the rest of their lives. He hated England but still loved a few of its people.

Easily displeased under the best circumstances, Adams could not even begin to abide boredom, discomfort, or danger. He therefore dreaded crossing the North Atlantic. In the summer of 1872, Adams and his wife, Clover, joined the herds of migrating Americans to Britain. They embarked from Boston on the Cunarder *Siberia,* a single-screw vessel, 320 feet long and 2,500 tons. The voyage called up his queasiest memories of previous crossings. After two days of placid seas, the deep-ocean swells got stronger, higher, sliding Adams into seasickness. "Cursed the sea and life in general," he wrote in his shipboard journal. The next day was worse. "Wretchedness aggravated by idea of a week more of it. Can't read. Can't talk." A day later, "Worse than ever. . . . Wish I were dead! Wish I'd never been born!" A pious family sang psalms on deck. ("Damn 'em.") He felt better the next morning, the sixth day at sea, breakfasting on herring and chops and hoping eventually to get ashore again. But the next day was cold, gray, and windy. "Ugh! We passed the evening together in our stateroom, in the dark. Very cheerful!" Two days later, "How good to contemplate even at such a distance the possibility of freedom. . . . Everyone has arrived at the inevitable grumbling period, the sign of nearing the end." A day later, "Rain and more wind. Everything uncomfortable. . . . Our whole souls are occupied by idea of getting ashore. We reach Queenstown tomorrow night. Today has been long and disgusting; wet, cold and gloomy. But the old tub goes a little faster with the wind." One more day, and then released at last from an average run of eleven days—"disgusting while it lasted, but not worse than usual."

Twenty years later, Adams came home on the *Teutonic*. In the interim, he had crossed the Atlantic repeatedly, expecting the worst and reliably finding it. ("Nothing short of life or health will ever make me endure that misery again," he announced after an easy winter passage in 1880. "Henceforth I mean to stay ashore.") He boarded the *Teutonic* in February 1892 with the usual morbid forebodings. But the new White Star ship barely resembled the *Siberia* of 1872: twin-screwed, twice as fast, nearly twice as long, and with almost four times the interior space. Adams was surprised and smitten. "The voyage was less trying than I expected," he admitted. "The ship was so big and so fast, and relatively so comfortable, that as I lay in my stateroom and looked out of my windows on the storm, I felt a little wonder whether this world were the same that I lived in thirty years ago. In all my wanderings this is the first time I have had the sensation. All the rest of the world seems more or less what it was, and Europe is less changed than any of the rest; but the big Atlantic steamer is a whacker." He marveled that he could still relish dinner during a winter gale, and repose in his well-ventilated deck stateroom, reading all through the night by electric light. Nothing impressed him as new or good, he declared, "until I saw the Teutonic; which knocked me silly." Eight years before his famous worship of the dynamo at the Paris Exposition, Adams already adored the *Teutonic* as an undeniable, defining wonder of the age.

*A*dams was carried away, at last, by the quarter century of headlong steamship progress that followed the advent of the *Oceanic* in 1870. The quick, recurring vaults in size, power, speed, and luxury made a transatlantic passage perceptibly different from the experience of only a few decades earlier. The five typical phases of midcentury voyages—curiosity, misery, relief, boredom, and deliverance—might still be distinguished. But they happened faster, spread across as few as six days instead of two weeks or more, jumbling together into a more chaotic pattern. Everything went faster, careening up to the edge of control in the constant, reckless juggle of speed and safety, always accelerating toward the end of the steaming century.

Amid these human improvements, the timeless natural context did not change. The great circle route between Liverpool and New York remained the most challenging long ocean passage in the world—especially in winter, and then during the iceberg season that followed. Even summer passages promised few balmy days for walking the deck in bright sunshine. "The ocean, particularly the northern Atlantic, is never warm," the Inman Line warned passengers in 1888. "It is impossible to take an excess, on going to sea, of overcoats and wraps, and especially of the thickest under-clothing, and these will be needed, not only on deck but often in the staterooms, and usually throughout the voyage. There is a delusion that people do not take cold at sea." The prevailing westerly winds and currents still fought westbound ships all the way across, and blanketing fogs and storms on the dreaded Newfoundland Banks still buried ships in the graveyard of the North Atlantic.

Even in this era, then, passengers' accounts of transatlantic voyages do not often sound cheerful notes. "There are people who actually profess to enjoy a steamer passage to Liverpool," an American woman wrote in 1874; "I always think how unhappy they must have been before they left home." Nothing to do, and too many people holding fixed, determined smiles; only men seemed to enjoy the experience ("the taste for barbarism and old coats, latent in all of them, comes to the surface"). The American humorist known as Petroleum V. Nasby crossed on the *City of Richmond* in 1881. "Ocean travel is either monotonous or dangerous," he concluded. "A man is not a fish, and no man takes to water naturally." The only advantage of traveling by ship, said Nasby, was the absence of dust in the road.

In 1889 one ship beat through a storm for half the voyage, then followed in the storm's wake until arriving. Hundreds of passengers fighting seasickness lay miserably in their steamer chairs on deck, swaddled from head to foot like helpless, dying patients in a floating hospital. "A torpid, lethargic, pulseless congregation it was," one man noted, "without energy, interest, or pleasure." On a run in 1894, a suffering man sat up in his berth for the first time in three days. His head felt like molten lead, his legs like pieces of string, but he toiled up to the deck and gazed at the ocean, tumbling and breaking in white-topped green slabs and troughs: "Its very silence strikes me as ominous. It is like watching a man in a fit of dumb, inarticulate rage. It reminds me of seeing people dance, through a window, when you don't hear the music." After two impervious young men walked by, brazenly enjoying themselves, the sufferer wished he had a gun. "Nobody is at their ease," an English writer named Martin Morris concluded in 1896, after his first crossing. "Their expressions are as blank as mummies, and their manners as frigid and congealed as the frozen mutton below. Not a pore is open that can be shut. . . . For even the best ship can hardly be said to be a cosy or comfortable place. Like some huge lavatory, there is everywhere a cold, petrified aspect."

And yet. "There were times in this rather squalid life," Morris went on, "when the soul could emerge and rise to the purest exaltation." Fresh, beautiful, electric mornings, and calm days of "sea-dreaming," and fiery sunsets as the descending blaze spread a glow across the sea, and long, star-gazing nights of slowly drifting meditation between the infinite ocean above and the fathomless ocean below. "Rare moments when the true purposes of life become evident and conscious realities . . . bathed in the eternal mystery of life and death." The North Atlantic still blew its travelers through many extreme moods.

The ocean and its trials defied human improvement. But within the ship—that tight, churning capsule of constant invention and progress— the experience did change. Life on a steamer was most obviously affected, at every moment of the voyage, by the bold, ongoing structural and mechanical leaps in hull size and engine power. "The novice aboard a big steamship," wrote a man on one of the new twin-screw liners of the 1890s, "looks wonderingly around the broad sweep of the deck, where swarms of people wander about as comfortably as on spacious city streets. He sees wide doorways opening into great halls, and grand staircases descending into vast depths. . . . He is struck with the apparent disregard of those very narrow limitations of space which he has always associated with ships. There seems to be plenty of room, length and breadth, height and depth." As the ship superstructure added a fourth deck, and then a fifth, the perils of the sea included getting lost on board. On the *City of Paris* in 1891, a man hoping to reach his stateroom on the main deck from his steamer chair on the promenade deck had to proceed aft and down the main stairway to the saloon deck, then down a small flight from a nearby entry to the upper deck, and wind through a series of uncharted passages to another entry, down more stairs to the main deck, and then forward at last to his cabin—not a journey to be undertaken lightly or too many times a day.

Twin screws and more powerful engines meant added speed. Passengers made clear, by their choice of vessels, that they preferred the fastest: and for a more practical reason than modernity's instinctive addiction to speed, or fatuous bragging rights, or basking in purchased prestige. Taking the swiftest liner of the moment sliced hours or days from the usual ordeal of a North Atlantic passage. The fastest greyhound got you back on dry, solid land that much sooner. This lust for speed—the unboundaried quest that ship designers, builders, and owners risked such time and money to provoke and satisfy—was ultimately driven by its paired opposite, the eternally troublesome wind and weather of that most coveted ocean racecourse.

Speed mattered so much that travelers would usually overlook the extra discomforts that came with it. When a greyhound was pushed at maximum steam pressure into the rough seas and dirty weather of a typical North Atlantic day, it was not an easy ride. Eating and writing, changing clothes, climbing stairs, and traversing hallways all became messy, bruising adventures. The ship rocked and rolled in no apparent pattern; a passenger could not anticipate the next lurch because it might so vary in force and direction. The hard, pounding collisions between fast ships and high waves forced travelers from the open decks down to their cabins, keeping them confined and sick for longer periods, and closing down the ventilators that might have provided some fresh-air relief. The *Germanic* sped into a heavy gale in 1885, meeting brutal waves that knocked out the captain, killed several people, and swept away the forward steering gear, pilothouse, and all the lifeboats. At half speed the ship might—perhaps—have nursed a better chance of escaping such death and destruction.

On the westbound *Etruria* in the late winter of 1887, "We had one fairly fine day," a man wrote in his journal, "but all the rest of the time has been wet, cold, and rough." The ship buried her nose into the oncoming waves, wriggled through them, and buried again. The funnels were coated with salt up to their tops. "Driving a boat at over twenty miles an hour cannot be made comfortable, except in smooth water," the man concluded. "The stern of the boat comes out of the water every few minutes, and allows the screw to run round in such a way that you would fancy everything would break to pieces." One night, the *Etruria* charged against an especially high wind, amid frequent lightning. "As the vessel dashed into the waves at her enormous speed, the water flew in all directions. I stood on the deck under a shelter for some time and never saw anything grander. I should say this boat could not last very long, the strain on her must be too great; she shakes from stem to stern, and as you sit in her cabins, with everything creaking and twisting, you want good nerves."

Since 1850, when David Tod's *City of Glasgow* proved that an iron screw steamer could profitably handle the North Atlantic, passengers on screws had put up with the noise and vibration of the shaft and propeller. At first it was just a mild trembling, most felt in rooms toward the stern, and a muffled burr or buzz. Voyagers got used to the sound and motion, noticing them most when they stopped at ominous times, such as in midocean. After 1870, marine engine designers squeezed more power from engines of the same size by increasing their velocities. Over the next two decades, average revolutions per minute rose from 56 to 64, and average piston speeds from 376 to 529 feet per minute. For passengers, that imposed extra noise and vibration, both from the engine room and as the propellers spun and thrashed through the water. Twin screws and double engines compounded the annoyance; the first Inman and White Star ships so equipped were swifter but noticeably more shaky. The reciprocating masses of their triple-compound cylinders—larger, heavier, pounding faster through their abrupt, jarring, in-and-out cycles—defied much dampening.

During the initial trials of the *Campania*, the first Cunarder with twin screws, she vibrated so violently that deck plating and stanchions were broken. The distinguished Glasgow physicist and inventor William Thompson, later Lord Kelvin, was asked to investigate. He recommended loading the ship down with massive deadweights to steady her. Three thousand tons of coal went into her holds. For the public trial, a glittering party of invited guests and journalists came on board to savor the largest, fastest, most luxurious Cunarder yet built. The party sat down to lunch as the ship steamed up to full speed. The resulting shudders so alarmed the guests that some abandoned their tables and rushed up on deck to see what was wrong.

Everyone was baffled. James Bain, Cunard's superintendent engineer, and Andrew Laing, who designed the engines for the Fairfield company, went along on the *Campania*'s maiden voyage to New York. Afterward Bain reported no vibration until the engines reached 71 rpm, but "very marked" movement from 71 to 76, which gradually subsided to the top speed of 80. So the problem was isolated to just that narrow range. To fix it, Bain strengthened the ship's stern and changed the pitch of her propellers, giving them less water-gripping surface and more slip so the liner would run more smoothly at speeds above and below the most agitating level of engine revolutions. The vibrations sank to more tolerable shivers. Later on, marine engineers learned how to balance weights, forces, and engine cylinders to contain this annoyance. But the new greyhounds of the 1890s still palpably shook and shimmied: another tradeoff of speed and comfort that passengers generally accepted in return for shorter, earlier releases from the unimproved ocean.

Speed—the universal, deranging drug of the nineteenth century—enticed and endangered, seduced while it disrupted. Its lure was never simple, to be broken down and weighed by clear lists of gains and losses. It shortened an ocean passage (cutting time off one agony), while it also made a passage less agreeable (deepening another agony). Was it preferable to suffer more in the short run, or suffer less for longer? Which advantage exacted the lesser penalty? "It is better to be comfortable for seven days," declared a fan of the older White Star ships in 1890, "than to be miserable for six." Well, perhaps; but the cream of the paying traffic inevitably drifted toward the new twin-screw greyhounds. "They insist in crowding into the steamer that makes the fastest passage," said another skeptic, "innocent of the dreadful discomfort and knocking about they will get before they reach the other side."

Among its seductive appeals, speed gave the Atlantic racecourse a brisk sporting aspect. When two greyhounds departed at the same time on the same route, a transatlantic passage could turn into a thrilling, high-stakes race. Wagers were placed and odds quoted, both aboard the racing vessels themselves and among sporting observers in Britain and America. At a time of few international games—shortly before the modern Olympics were established—an ocean race became a kind of athletic contest, pitting two closely matched machines and their human attendants against each other. Not just a made-up sport, it was quite serious and real, with daring, practical risks and ramifications hanging on the outcome. To a unique degree, this transatlantic sport skirted sudden, deadly calamities unknown to shore athletes who just played golf or tennis or baseball.

More than a game, it carried a mean, dangerous edge; for a pitched race, every member of the crew from the captain to the firemen tried even harder, pushing up to the limits of boiler pressure, metal fatigue, and human exhaustion. If the sweating firemen came up on deck for a quick blow in cooler air, and "a rival steamer should be in sight and gaining on them, or even keeping up with them," a frequent traveler wrote in 1892, "down they go again without asking, to renew their toil, to rake the fires, to 'chuck' in more coal, to coax the giant to do a little better and leave the rival astern."

Ocean racing had marked transatlantic steam ever since the *Sirius* first puffed across with the *Great Western* in pursuit in 1838. The major matchups—especially Cunard-Inman in the late 1860s—were savored by all the players and bettors (and piously denied by the steamship companies). The sharp competition of the period after 1870 produced no fierce, persistent racing for about two decades. Then Inman and White Star trotted out their first twin-screw greyhounds at nearly the same time. The *City of New York* and *City of Paris* were owned by Americans and built by the Thomsons in Glasgow. The *Teutonic* and *Majestic* were owned by British interests and built by Harland & Wolff in Belfast. Straddling the Anglo-American reconciliation, pitting New York against Liverpool and Scottish builders against Irish, the four ships were identically new, fast, luxurious, and popular. Ocean races became more intense and frequent than ever before or since.

The games began with the *Teutonic*'s maiden voyage to New York in August 1889. The *Teutonic* and *City of New York* crossed the ocean neck and neck. Twice the White Star ship passed the Inman liner, but twice machinery problems forced her to slow down and pull back. The *City of New York* came in first, steaming up the harbor in front of a noisy, swarming morning crowd, and landing her passengers before her rival even reached the White Star pier. But the *Teutonic* did clock a time of six days, fourteen hours, and thirty-three minutes: the fastest westbound maiden passage yet by a British steamship. The rivalry started with this characteristically split decision. Later that season, the *Teutonic* passed Roche's Point, outside Southampton, forty minutes behind the *City of New York*. They boiled along in sight of each other, the distance between them narrowing, until the *Teutonic* overhauled the Inman liner near Fastnet Light. The *City of New York* veered to the south and then—in a burst of speed—crossed the *Teutonic*'s bow and settled into a parallel course to starboard. The lights of each ship remained visible to the other through that night. In the morning, the *Teutonic* started pulling ahead. By four o'clock that afternoon she disappeared into the western horizon. For the next three days the *Teutonic* steamed along unaccompanied, against persistent westerly gales, bucking heavy swells and cross seas. Locked in the race, ignoring the weather, the White Star liner clicked off heedless daily runs of 456, 431, and 471 miles and charged into New York as the winner, in eight hours less than her maiden time. In November, the *City of New York* started west with an hour's lead. After nearly 3,000 miles, they were essentially tied. Spectators on Fire Island in New York watched them racing together into the home stretch, both vessels flat out and pouring smoke, down to the wire. The *City of New York* arrived nine minutes sooner, but the *Teutonic* won on corrected time.

After repairs and improvements over the winter, the two ships sailed on different days until August. When they left Liverpool together on the sixth, the news zipped along the transatlantic telegraph cable to New York. "Great interest was aroused in shipping circles here," the *New York Times* understated, "and some money was wagered upon the result, the Teutonic being a slight favorite." (The Inman and White Star agents in both countries denied any knowledge of an ocean race.) The *City of New York* passed Roche's Point forty-five minutes ahead. Again the two ships spent the night watching each other's lights. "It was a stern chase in which every soul on board took a deep interest," the *Times* reported, "the greatest ocean race which has yet taken place." The *City of New York* held the lead through that day and night, the ships still within sight of each other. The *Teutonic* came abreast, stayed there, pulled in front, and blazed out of sight. She arrived in New York four hours ahead. Gliding in triumph up the bay, she received salutes of steam whistles and horns from other ships, all the way to her berth at the foot of West Tenth Street. But two weeks later, the *City of New York* beat her home by two hours and thirty minutes.

So it went, back and forth. During the entire spring-to-fall season of 1890, each ship crossed the Atlantic sixteen times. The *City of New York* averaged just under six days and five hours—seventy minutes faster than her rival's mean time. But the *Teutonic* claimed a new single-passage westward record of five days, nineteen hours, and five minutes. The Inman Line, not convinced, accused the White Star captain of cooking his log on that trip. "There is nothing the matter with our log," replied the chief officer of the *Teutonic,* in the offended tones of a sporting man accused of unsporting conduct. "The log was not doctored. We could not have done it" (because too many interested gentlemen were on board, checking their own watches against the announced times)—"and, besides, there was too much depending on it."

The sporting element on any liner, an expanding presence, took over the smoking room in the morning of the first day at sea and held it for the rest of the voyage. All day and well into the night, the drinks, cigars, and falling cards never stopped. Storms and passing vessels, even icebergs, held no interest for the players; they did not walk the decks, or peer out the windows, or even look up from their games. Gambling at sea was, like an ocean race, a blaring but formally unacknowledged fact of transatlantic life. Both depended on the same human instincts for tight competition, and the elusive play of chance and fortune. Gambling and racing prospered together on the ocean, each encouraging the thrills of risk and bravado, climaxed and defined by significant winning and losing. (At the very least, they promised a jolt of relief from ocean boredom.)

During the final two decades of the century, the games in the smoking room became richer and more dangerous—and more annoying to nonplayers among the passengers. The enormous new ships carried up to six hundred people in first class, a much larger and more anonymous group of wealthy travelers through whom professional gamblers could move about undetected, marking their pigeons and setting traps. The players—pros and amateurs both—were mostly Americans, intent on making the traditional American game of poker a patriotic ornament of ocean-crossing culture. The smoking room came to resemble a high-rolling casino at Saratoga Springs, New York, or a gentleman's game on a gilded Mississippi River steamboat. On the *City of Richmond* in 1881, the best poker player was a young man who claimed he was just the son of a Wisconsin farmer, making his first trip to England. "He was uneasy till a game was organized in the morning," a fellow traveler recalled, "and he growled ferociously when the lights were turned down at twelve at night. He was impatient with slow players, because, as he said, all the time they wasted was so much loss to him. He could drink more Scotch whiskey than any one on the ship, and he was the pet of the entire crew, for his hand was always in his pocket."

In many British and American hotels of the time, gambling might take place in private rooms. Even if technically illegal, it was tolerated so long as it remained discreet and out of sight. On the high seas, beyond any meddling legal jurisdiction except the captain's whims, the games in the smoking room became more open and constant. The *Servia* sailed from Liverpool in November 1883 with Hen Rice, a small-time New York City politician, and Robert Solomon, an English gambler and diamond merchant, among the passengers. They spent the voyage in the smoking room, relieving other players of stakes from $50 to $1,500, before gambling with each other. Rice took Solomon for about $3,000. When the ship reached New York, Solomon accused Rice of cheating with marked cards and demanded his money back. The case made the newspapers, embarrassing the Cunard Line. Its management asked Captain Theodore Cook of the *Servia* about the matter; "of an entirely exceptional character," he said, just happened toward the end of the voyage, and surely wasn't a problem in general.

Cook was telling his bosses what they wanted to hear. The gamblers drank oceans of alcohol (very profitable for the steamship company), and they tipped stewards generously for their service and cooperation (which made them popular with the crew). Gambling also attracted passengers and—within limits—brightened the shipboard tedium. Only nongambling, nondrinking travelers seemed to mind. Thirteen passengers on the *Umbria* in 1887 objected to "the noisy scenes" and "the gambling and intemperance now so general on ocean steamers" for dangling those linked dangers before innocent young men: "If the smoking room is to be made a gambling hell, it should be labelled as such." After this protest—not the first in recent years—a stern but ambiguous sign was hung in Cunard ships. "This Smoking Room is intended for the pleasure and convenience of all," it warned, "and gambling or offensive conduct of any kind cannot be permitted. Passengers are requested to aid in carrying out this rule by promptly reporting to the Captain any serious infraction of it." But just what amounted to a "serious" infraction? That was left politely undefined, and the games could roar on unabated.

Gambling took occasional casualties. On the *Westernland* in 1887, one young man lost heavily, got drunk, and jumped overboard. More broadly, the intrusion of this rough American presence on mainly British ships irritated the Anglo-American reconciliation of the time. "The Atlantic steamers are the resorts of professional gamblers from all parts of the United States," wrote Charles William Kennedy, former commander of the *Germanic,* in 1890. When a refined gentleman, perhaps a clergyman, entered a smoking room and sat down with his book and cigar, he soon felt assaulted by the lurching, chaffing sounds of the gamblers. "Voices rise higher and higher," Kennedy reported. "Coarse, vulgar, abusive language, mingled with oaths, is shouted across the room. Rising from his seat, the thoroughly disgusted traveller thrusts into his pocket the book he has been attempting to read, throws away his cigar, and rushes indignantly on deck, wondering why it is that the smoking-room is allowed to be monopolized by the lowest class of men that cross the Atlantic."

The riskiest gamble of all, shared by all, was simply setting forth on the ocean. The astonishing size and power of the new twin screws obscured the immutable dangers of crossing the North Atlantic. The ships *seemed* safer and stronger; Henry Adams, serenely reading in his *Teutonic* stateroom by electric light, felt less vulnerable on the ocean than ever before. But size and power remained deceptive, ambiguous advantages. Even the grandest steamer still felt tiny when the ocean rose in full cry. Many more ships and lines now plied the transatlantic steamer routes, crowding the lanes and making collisions more likely. Bigger, faster liners inevitably collided harder and more lethally, against icebergs or other vessels. And their increased carrying capacity also raised the potential death toll in a disaster.

The Cunard Line brought its unique safety record into this era, proudly proclaiming itself "the Ocean Service that has never lost a man." In 1880 an article in *London Society* magazine repeated the familiar boast that "from 1840 to the present time, not one of their passengers has lost his life by accident on any of the thousands of voyages that have been made across the Atlantic in their ships. They have not lost a single vessel." Such puffery slid past the sinking of the *Columbia* in 1843, and it ignored the quiet, occasional loss of a passenger to an unexpected rogue wave. The first such death on record occurred in May 1867: while passengers were sitting on the deck of the *China,* a high sea rushed over the ship and killed James C. Cogswell, a prominent Halifax banker. (Other, unrecorded such incidents may have happened earlier.) Later on, at least six steerage passengers were killed by seagoing accidents from 1888 to 1902. In October 1905, at around midocean on a westward run, the *Campania* rolled heavily on her lee side just as a huge wave swamped the fore deck and washed five steerage passengers overboard to their deaths. The Cunard tradition of safety, though unarguably remarkable, was not quite as advertised.

The Cunarder *Oregon* had almost reached New York from Liverpool in March 1886 when she collided with a sailing schooner, probably the *Charles H. Morse.* The *Oregon,* 500 feet long and 7,300 tons, was another Fairfield greyhound: one of the biggest, newest, swiftest ships racing on the North Atlantic. The *Morse* was a three-masted vessel, only 150 feet and 500 tons. Just before daylight, five miles off Fire Island, in safely clear weather and moderate wind, the *Morse* suddenly appeared to the south of the *Oregon.* The small wooden schooner inexplicably bore down without changing course and slammed into the mammoth steel liner on her port side, just forward of the bridge—like a mouse assaulting an elephant. The steamship trembled, stopped, and then began to drift. While the *Morse* disappeared and presumably sank, the sea poured into the *Oregon,* gradually overcoming her watertight compartments. She was carrying about nine hundred passengers and crew.

After some initial panic, the officers and crewmen steadied themselves and started loading women and children into lifeboats. Three hours passed. A New York pilot boat happened by, looking for a customer; she took the women and children, freeing the lifeboats to transfer everyone else from the *Oregon* into a passing coastal schooner. About nine hours after the collision, the Cunarder sank by the head into eighty feet of water. Only the second Cunard ship lost in the North Atlantic traffic, the *Oregon* carried down her cargo and most of the mail—but everyone on board escaped alive.

In December 1892 the *Umbria* struggled through a hard winter passage westward, fighting sharp gales and heaving seas. Six days out from Liverpool, about five hundred miles from New York, the single propeller shaft broke and the ship was disabled. Captain Horatio McKay lowered three sea anchors and brought her head to sea, so she rode the swells in no immediate danger. Chief Engineer Lawrence Tomlinson thought he could fix the shaft. The fracture had occurred in the thrust collar section, where a series of metal flanges attached to the shaft spun against horseshoe-shaped blocks to brace the forward-acting force of the propeller, and thereby push the ship ahead through the water. The break zigzagged diagonally between two thrust collars without fully severing the shaft. Tomlinson hoped to attach these surrounding collars to each other, thus splicing the fracture into an interlocking bond that might prove strong enough, for long enough, for the *Umbria* to limp down to New York.

The shaft tunnel was a tight, confined workspace, dark and echoing, and dripping oily water from condensation on the steel hull. The thrust collars were made of thick, hard-tempered steel, designed to contain massive forces at high speed and friction—and not to be worked with simple hand tools. As the ship bucked and rolled, Tomlinson and the other engineers used ratchet drills, then hammers and chisels, to cut slots into the collars. Organized into two watches of six hours each, they drilled and hammered for seventy-two hours. It was hot, exhausting, exasperating work; a man could chip away for only fifteen minutes before needing to rest. Meantime two special clamps were forged from steel plates. Finally the two collars were clamped and bolted. The engine started running slowly; after two hours, a bolt broke. The repairs took another sixteen hours. Again the engine was started, and this time the bolts all held. The *Umbria* headed for New York at four knots. As Tomlinson, hovering and hoping, kept watch over his mended shaft, he brought the ship gradually up to ten knots—still only half speed—and she arrived safely on the last day of the year.

In some measure, it was Cunard luck, once again, that nobody died when the *Oregon* sank and the *Umbria* was disabled. The *Oregon* foundered slowly, in daylight and calm seas, close to land, and in a heavily traveled area, so that her lifeboats could be safely lowered and retrieved by passing vessels. If she had sunk in midocean, at night, in typical North Atlantic swells, the outcome might have turned catastrophic. The *Umbria* had beaten across the ocean through rough weather and seas, conditions so demanding they may have caused the shaft to break. But once Tomlinson had finished his repairs, the sky brightened and the swells subsided, allowing the wounded ship a much easier course to safety in New York.

Lucky, yes, but the old Mac Iver traditions of strict discipline and ruthless standards also lasted into the late nineteenth century. "In the Cunard service we were never encouraged to make any needless effort," Captain James Anderson said after retiring from the line. "I never remember any commander being blamed for a long voyage nor praised for a short one. There was everything in those fine ships which a man could desire, to insure safety." Before each voyage, every member of the crew was assigned to a particular lifeboat, with identifying badges, and all hands were drilled in procedures for boats, fires, pumps, and watertight doors. When steamship experts traveled on Cunard vessels, they noticed reassuring differences from other lines. "I found prevailing on board a very superior state of things to what I have noticed in too many steam-vessels in other trades," wrote the British shipowner William S. Lindsay in his classic four-volume history of merchant shipping, published in 1876. "Everywhere the most perfect order and quietness prevailed." "Never was discipline more perfect, or order more complete," recalled Henry Fry, a Canadian shipping official, in 1896, after more than fifty years of watchful, appraising travel on transatlantic steamers. "At sea, in all weathers, everything went like a well-regulated machine."

Cunard service and food remained notoriously inconsistent. "Very sorry, sir," said the Cunard steward in a widely circulated joke, "can't have another towel, but—we never lost a life on this line." Not quite; yet it was still true that the Cunard Atlantic service had never lost a life *by ship-wreck*. When the usual boast was thus hedged, Cunard luck backed by Cunard discipline maintained a special, life-protecting history on the dangerous North Atlantic. "'Cunard luck' has become proverbial," the *Nautical Magazine* commented in 1891, "but it is well known that this good fortune is largely due to the systematic and careful management."

"*Y*our flag," said an Englishman to an American woman aboard the National Line's *Canada* in 1878, "why, it is made up of minus marks; nothing but subtraction, while the English contains the two signs of plus, which mean increase and multiply."

"But, you must remember," the American replied, more than a century after the Declaration of Independence, "that we did one very grand example in subtraction when we subtracted our nation from Great Britain, so our flag bears minus marks in honor of our victory. What became of your plus marks then?"

*H*enry Adams was uniquely steeped in the history, politics, and culture of Anglo-America. From his rarefied perch, he might—perhaps—forgive the rancid historical quarrels with England, so vigorously pursued by three generations of his ancestors ever since the American Revolution. After an international tribunal settled, in 1872, the U.S. government's claims against the British for their collusions with the Confederacy during the Civil War, the two nations entered a long period of relative calm in their diplomatic relations. As history and politics receded, the cultural irritant persisted: the vast, surprising gulfs between two nations so similar, yet so different. "London never was so low, so stupid, so clean daft, so bankrupt of wit, humor, sense and decency," Adams informed an English friend after a visit in 1879. "Your people are dull, sleepy, stupid, heavy, wearysome, and at the same time more self-satisfied and opinionated than ever." Like many of his countrymen, Adams felt especially patriotic when traveling abroad. As a historian, though, and a skeptical but passionate observer of his times, he sensed that England and America remained natural, inevitable allies in a dangerous world. They were stuck with each other. Adams therefore hoped—wearily, in spasms, always with doubts—for a cultural reconciliation. "We ought not to be harsh towards the poor little island," he allowed. "It would like to improve if it knew how."

Steamships shuttled the two cultures back and forth. Americans were much keener to visit Britain than Britons were to see the United States. Most of the cabin passengers, up to 80 percent on a typical voyage, were Americans; so the steamship lines catered to Yankee tastes. "The impetus of comfort in ocean travelling," noted the *Nautical Magazine* in 1875, "comes from America, and . . . the wealthy Americans." From March well into October, the liners filled up with American tourists seeking rest and culture in Europe, and coming home laden. The young American woman abroad became a regular staple of contemporary Anglo-American fiction. In 1879 Henry James had his first international hit with *Daisy Miller,* the prototype of many such guilefully innocent heroines to come. The ever-improving steamers made transatlantic exchanges faster, easier, and more routine; still not reliably comfortable but at least briefer. The advance of steamship technology played its own nudging part in the Anglo-American cultural reconciliation. "Each successive season seems to draw closer the bonds that connect the English-speaking nations throughout the world," said the *Newcastle Daily Chronicle*. "Steam has bridged the Atlantic, and the passage to and fro of the citizens of America and England is so common that it has almost ceased to be noteworthy."

In London, American tourists favored the Langham Hotel, in the West End near Regent's Park. A Yankee island in the British sea, it was managed by Colonel James M. Sanderson of Philadelphia, late of the Union army. The Langham was an imposing structure, in a palatial Italian style, offering a main hall five stories high and even thirty-four suites with private bathrooms. Its supply of drinking water—a perennially sore matter for American tourists anywhere in the world—was pumped by steam engines from its own artesian well. American men were drawn to the familiar smoking room and billiard hall. Their wives especially liked the hotel's location near the best London shops. "Gradually the Langham is becoming the headquarters of our countrymen," wrote an American visitor, "and these attract many of the natives, especially those who are admirers of American institutions, a class that has rapidly and astonishingly increased since the overthrow of the rebellion."

The scale and pace of British life surprised Americans. The entire nation could have fit into just a corner of Texas; "how so small a country," one tourist wondered, "could have made such a prodigious noise in the world." After the brisk, restless tempo of America, then accelerating into a prolonged industrial boom, Britain seemed slow, staid, and too orderly. Waiting for their meals in restaurants, hoping for service at inns and shops, Americans drummed their fingers and muttered darkly about old-world sluggishness. They disliked the cold hotel rooms, the lighting by candle instead of gas, the shocking scarcity of ice water, and—in particular—the unavoidable British custom of "feeing," the constant expectation of tips for the slightest favors. The food seemed bland and monotonous, the coffee hardly drinkable. "I am so tired of joints, and boiled vegetables, and milky puddings," said a Yankee woman after staying in Kensington, "that I would give my immortal soul for a good American dinner."

The tourists believed they looked, dressed, and behaved just like the natives. They puzzled over being so immediately recognized as Americans—before revealing themselves when speaking, even by children who paused and stared in the streets. The British welcomed their money and lavish spending habits but doubted their manners; Americans seemed pushy, self-assertive, and braggy. Their most refreshing qualities might reach into distressing extremes. Americans' frankness could become vulgar and impudent; their friendliness, garrulous and tiresome; their energy and enthusiasm, impulsive and fanatical; their humor, flippant and irreverent; their eagerness and freshness, superficial and raw; their embrace of novelty, foolish and reckless; their knack for making money, crassly materialist. Thus in general. On a personal, more individual level, many Britons welcomed Americans into their homes and clubs. "There is not a peculiarly offensive ass or snob in this hemisphere," Henry Adams sneered in 1878, "who has not been entertained by the Duke of Something or the Marquis of Something Else." Almost anybody who seemed authentically American, from Buffalo Bill to the most blue-blooded of Boston Brahmins, might become an exotic overnight success in London society. Regardless of their relative stations or reputations at home, the visitors only had to look and act like true Yankees, within limits. "If they didn't eat peas with their knives," an American recalled, "and spit on the carpet, they were welcome."

Beyond the small differences and irritations stretched a broad common ground. Americans found themselves in another country, palpably foreign, yet the people spoke English—in many peculiar variants, to be sure, but it was still their own language. The cultural landmarks, historic sites, noted authors, and famous buildings seemed as well known and comforting as anything in Boston or New York. "Nothing in London strikes the American quite so strangely as the absence of strangeness," wrote a tourist. "One is amazed each instant at the familiar aspect of names and places." Any American of even middling education arrived well steeped in knowledge of British government and customs. The British, by contrast, often seemed remarkably uninformed and incurious about American life. (At a boardinghouse, a young English woman told some tourists that she had supposed Americans spoke an Indian language. But what about the American books printed in English? "I thought they were translated," she said.)

As the mutual bitterness of the Civil War gradually healed, the shared ground looked wider and firmer. By the 1880s, certain American products and personalities had become quite fashionable in England. "In London the word American," a tourist noted, "applied to a thousand commodities, from beef to baby-jumpers, is so common as to cause wonder that the bulk of the English people are yet so ignorant of our geographical boundaries and social status." James Russell Lowell, the U.S. ambassador to England, had long nursed his patriotic grudges from the war; but now he was embraced by English aristocrats: "the pet of countesses," as Henry James noted, sounding jealous, "the habitue of palaces, the intimate of dukes." Bret Harte, the American writer serving as his nation's consul in Glasgow, disliked Britain when he arrived in 1878. Two years later, he admitted that he couldn't help approving the place. "Just now, there is quite an American 'boom' in England," he wrote in 1880. "American artists, writers, actors, and even American ladies and gentlemen they have discerned to be as good as themselves, with an added originality, which although unprecedented is not altogether dangerous or subversive!"

In the fall of 1886, an Englishman named Cecil Spring Rice who was sojourning in North America wished to meet Theodore Roosevelt, a patrician young New York politician of some promise. Learning that Roosevelt had booked passage on the *Etruria* to Liverpool, Spring Rice took the same sailing and, in his best Eton and Oxford manner, introduced himself. At sea, with ample, oceanic time on their hands, the two men sized each other up and decided to become friends. When they reached London, Spring Rice served as best man at Roosevelt's wedding. That spring, the Englishman joined the staff of the British legation in Washington and was duly admitted to Henry Adams's tight inner circle: "An intelligent and agreeable fellow," Adams judged. "He has creditable wits. Mad, of course, but not more mad than an Englishman should be." (Decades later, the Anglo-American friendship plotted and launched on the *Etruria* climaxed with Roosevelt in the White House and then Spring Rice as the wartime British ambassador in Washington.)

The most influential American interpreter of Britain to Americans was George W. Smalley, the London correspondent of the *New York Tribune*. For almost thirty years, his urbane, well-informed reporting—published in the most important American newspaper of the time—shaped and shaded public opinion in the United States. Smalley's writing career traced his own Anglo-American reconciliation. He arrived in London in 1867 as a Radical Republican, still angry about the war, distrustful of Englishmen, and critical of most British institutions and traditions. As he settled into his job, Smalley began to pick up the tones and politics of his surroundings: "well-bred men and women," as he put it, "so many delightful houses in town and country, such hospitality." Tall and erect, with a square-jawed, deeply seamed face, he cruised through British society in utter self-confidence, cultivating his contacts and informants. Lillie Langtry noticed that he loved gossip and could command attention even without raising his low, steady voice. He dined out three times a day and knew everyone he thought might be useful. His own dinner parties, small and distinguished, were coveted Anglo-American events. "A singular specimen of a successfully anglicised Yankee," noted Henry James, using a phrase that described himself as well. "He is more British than the British," snapped Theodore Roosevelt, who did not at all approve. Over the years, the U.S. ambassadors and consular staffs in Britain came and went with the shifting political tides at home. Smalley remained on hand, preaching Anglo-American amity from his conspicuous pulpit: the steadiest, best-connected, most astute American presence in London.

Cultural reconciliation in Smalley's favorite milieu of high society was pushed and pulled by new financial circumstances. Old, landed wealth was having money problems; cheaper American grain imports were cutting the price of wheat, reducing rental incomes for titled landowners, and forcing the sale and mortgaging of ancient family estates. As the aristocrats declined, British society had to admit more bankers, builders, and manufacturers. It thus became more "American," defined by wealth and class mobility as well as inherited land and titles. The government now elevated peers from industrial, commercial, and financial circles. (John Burns, head of the Cunard Line, became the first Lord Inverclyde.) Even plain Americans, if polite-mannered and rich enough, might be formally presented at court—a traditional privilege previously limited to the well-born British aristocracy.

The Prince of Wales, whiling away the long years before ascending to the throne, came to prefer the luxury and opulence of the new rich (both British and American) to the shabby, fading country houses of old, spent unwealth. "He's more like an American than an Englishman," Bret Harte decided. Rebelling against his severe mother, the queen, the prince loved the unbound parvenu tastes for gambling, extravagance, and sensual self-indulgence. He was known to savor smart American women like Jennie Jerome Churchill, the mother of Winston. A bold woman from New York once spoke to him without deference, and the prince found it quite charming. "He began to covet American society," George Smalley reported—so much that princely taste thereafter would include the unavoidable presence of "one American woman, or more than one, who, in the current phrase of London, had to be asked if you wanted the Prince."

As British society became more American, its colonial counterpart looked and sounded like a distant outpost of London. The upper reaches of social life in both countries were merging toward each other, meeting somewhere out in midocean. Henry Adams noticed strivingly elegant New York women rolling their *r*'s, holding them, trying to imitate the Prince of Wales. ("They affect the British aristocracy without understanding it.") Cecil Spring Rice dropped by the very particular Century Club in New York, "where all the learned men go. They are exactly like English literary men, and the club is like a glorified Savile." In the show-off summer "cottages" of Newport, Rhode Island, the dress, manners, and speech were all slavishly copied from the prince's scandalous Marlborough House set. Spring Rice, visiting Newport and pursuing his bemused studies of rich Americans at play, heard many savory morsels of gossip about London society, "which they talk about much as Scotch ministers talk about heaven, half familiarity and half awe."

In these years, the center of world finance was shifting from London to New York. The banking house of Morgan wielded muted, telling power in both cities. J. P. Morgan, of Yankee background, was perhaps the most significant Anglo-American citizen. Educated in Europe and the United States, fluent in three languages, equally at home in New York and London, he exhibited all the most highborn attitudes and behaviors of the British aristocracy. He crossed the ocean at least four times a year—usually on White Star ships—for business and pleasure, buying art treasures and taking the healing waters. Morgan and his great rival William H. Vanderbilt coveted the same staterooms on the *Britannic;* at one point Vanderbilt reserved those favorite rooms for five years in advance. Later on, Morgan would pay for several sailing dates on several ships, holding them all till he knew, in the final rush of business and acquisition, which one he would grace with his inexorable, imperial presence.

The most intimate form of cultural reconciliation was an Anglo-American marriage. By the late 1880s, the young American woman—unafraid, outspoken, available for marriage, and rich—had steamed over and stamped her own style onto the London social season. "They are so bold and forward," clucked an English matron, watching the Prince of Wales dance a flirty quadrille with an American beauty, "and they do—such American things." George Smalley wrote knowingly of their "invasion" and "conquest" of British society. "The American girl who marries in England has begun life earlier than her English cousin," according to Smalley. "She has met men, and even talked to them while yet unmarried; a thing which few English girls venture to do." All her life, she has enjoyed more natural and genuine relations with the opposite sex. "Her ideas are not bounded by the horizon of Mayfair. She is fresh, original, independent." Not born into a walled social system that was still so defined by inherited rank and deference, "she strikes therefore at once that note of equality."

What she brought to an English marriage, though, was not so much her democratic charm as her comforting American dowry. The classic exchange wedded inherited Yankee wealth to a threadbare British title; each spouse gained something, and sometimes they even grew to love each other. In one notorious transaction, Consuelo Vanderbilt of New York, in love with another man, was forced by her mother into marriage with the duke of Marlborough, who needed bushels of dollars to maintain his Blenheim Palace in proper baronial style. This cynically coerced union ended after eleven years and two children. By the turn of the century, about seventy American heiresses had married titled British gentlemen. "For the most part they turn out very happily," an informed observer said of these marriages. "In many cases they are, no doubt, love-matches; but where they are not matches of the heart, love of title and the power and position that it brings is often a counterfeit which is mistaken for the genuine article."

After the mysterious suicide of his wife in 1885, Henry Adams lived alone in a house at 1603 H Street in Washington. He never went out on the thriving capital social circuit, or left a calling card, or invited anyone to dinner. Instead, various Anglo-American friends and relatives would simply appear, uninvited, for late breakfast or dinner. To brighten the meals, he encouraged a steady flow of pretty young women. One of these frequent guests, his niece Abigail Adams of Boston, thought he resembled the White Rabbit of *Alice in Wonderland,* with his tiny hands and small kid gloves. "No one could have been a more delightful host than Uncle Henry—gay and amusing with a wonderful faculty for keeping the ball rolling and stimulating light conversation," Abigail Adams recalled. "His rule that serious things must be discussed lightly and light things seriously made on the whole a pretty satisfactory formula."

"My breakfast-girls," Adams called them. Among his favorites were two Marys who wound up in quite conspicuous Anglo-American marriages. Mary Endicott was descended, in the ninth generation, from the first governor of the Massachusetts colony; her mother was a Peabody, one of the ruling families in the ocean-trading city of Salem; her father was a justice on the Massachusetts Supreme Court, then secretary of war under President Grover Cleveland. Mary Endicott impressed Spring Rice, who met her at the Adams table, as "very English—no accent, quiet, pretty, very dignified and staid for 21 years." In the fall of 1887, the English politician Joseph Chamberlain, bound for higher office, arrived in Washington to negotiate another squabble about fishing rights in Canadian waters. Spring Rice brought him along to 1603 H Street. "Chamberlain amused and interested me," Adams reported to one of his old English friends. "He was a success in society, and . . . flirted desperately and openly with the young girls, and especially with one of my own little flock who come to breakfast with me."

Endicott was twenty-three, Chamberlain fifty-one (and twice widowed). It was not a typical Anglo-American match: she had a distinguished family but not much money; he had earned his own fortune in business but had no title. He nonetheless pursued her. She first turned him down but then reconsidered, and they were married in Washington in November 1888. Three years later, on one of his periodic complaining visits to England, Adams called on them in Birmingham. "Our Mary was extremely pretty, sweet, decidedly improved, and apparently very happy," Adams observed. "The Chamberlain house is very big, very costly, and" (noted with approval) "rather American than English."

Mary Victoria Leiter of Chicago was one of the reigning, ringing belles of Washington society. Her friendship with the young wife of Grover Cleveland had brought her inside the doting Adams orbit. The daughter of an unvarnished department-store tycoon, she glided into any room in a floating, undulating gait—so poised and confident, tall and slim, with a beautiful oval face. Charles Dana Gibson used her as one of the first models for his Gibson Girls. In 1890, twenty years old, she steamed over to Europe for the fourth time. "My young women here are now all married and settled," Adams mourned. "I have no longer one left to my breakfast-table, except Mary Victoria Leiter, and she has gone to Europe, like all other vertebrates."

In England, at a country house party given by Lady Brownlow at her Ashridge estate in the Chilterns, Leiter sank into rapt conversation with George Nathaniel Curzon, one of the brightest, most ambitious young Englishmen of his generation. She took him out into the rose garden. He felt like kissing her (so he later wrote) but managed to restrain himself. She invited him to the opera and to tea. For the next ten days, they met as often as they could—exchanging gifts, photographs, and daily letters—and fell in love. In November she boarded the fashionable *Teutonic* for New York. "We have had a merciless voyage from Queenstown," she wrote him from mid-Atlantic. "We sailed into a furious storm, and for four days floated on a turbulent sea of misery." Two days later the storm was over, but the roiling currents loosed within her had not subsided. "I am still sorrowing for dear England," she continued, "and it is the first time that my patriotism, no not patriotism, but my delight at seeing 'Liberty' has been changed by a regret and fondness for any other shore. . . . What I would not give to have a glacier . . . to act as a magnet to draw you three thousand miles! Oh! dear oh! dear!"

It was a long, delayed courtship, interrupted by distance and Curzon's implacable dedication to his work and career. They were secretly engaged for two years and finally wed in 1895, at the same Episcopal church in Washington where Mary Endicott and Joseph Chamberlain were married. "Another of my breakfast-table is taken over to you by this event," Adams informed an English friend. "Surely the British empire is bent on bankrupting me by one means if not by another. . . . It goes long on silver and short on girls." Mary Curzon adapted slowly to her new life; she missed her father and her American friends and felt beset by English servants. In 1898 George was appointed viceroy of India, the most important position in British life after the monarch and prime minister. As vicereine, presiding over two grand residences and many public ceremonies, and assisting her husband in quiet but crucial ways, Mary held the highest rank in the British Empire that any American, man or woman, ever attained. Her prominence symbolized the Anglo-American reconciliation at the loftiest levels.

If his two Marys could find happiness in English political marriages, then perhaps Henry Adams might manage his own private peace with the former enemy. During his frequent visits in the 1890s, he started to soften. "Suits me fairly well for the moment," he conceded, "and I breathe rather more peacefully here than elsewhere." He noticed a waning of the old grand style; the landed nobles, scrimping within tightened budgets, did less entertaining at their country houses. Their footmen wore no powder, and the railroads were now jammed with third-class passengers. Democracy was quite visibly advancing. "England is much less smart; much more American, and decidedly less queer and old-fashioned than it used to be." The ancient Adams family loyalties still flickered on occasion. "England always presents itself as the stupidest, most ignorant, and most indolent if not the laziest society I have ever seen or read of." (An aging steam engine at high pressure, slowing down and blowing off steam.) "Still I have passed most of my life in it, and now I am back again, and there is still a certain novelty and charm about the life." For Adams, that amounted to an Anglophile love letter.

As the century ended, the two great English-speaking democracies were more united than ever before. After one last diplomatic crisis, over a boundary between Venezuela and British Guiana in 1895, the governments in London and Washington came together in mutual interests. "The most astonishing feature of the present time," wrote Sir Julian Pauncefote, the British ambassador to the United States, "is the sudden transition in this country from Anglophobia to the most exuberant affection for England and 'Britishers' in general." Andrew Carnegie, the ruthless little Scotsman who had emigrated to the United States and grand wealth, urged that Britain and America federate into a single nation. At a large, representative banquet in Washington, the guests toasted the healths of the queen and president, sang "God Save the Queen," and tried to sing "The Star-Spangled Banner." From the merged folkways of high society down to the sentimental American popular songs heard so unavoidably in the streets of London, the transatlantic culture cruising back and forth on steamships was mixing and marrying. A few years later, this new Anglo-American union would be tested—and then assume its most serious form—in the general catastrophe that followed the guns of August 1914.

15.

Germans

*T*he German naval squadron steamed through the English Channel and approached Portsmouth, on the southern coast of England. The invading squadron consisted of six iron-armored battleships, two cruisers, and three smaller vessels: the newest and finest units of the German navy. At their head was the royal steam yacht *Hohenzollern,* the particular pride and toy of Kaiser Wilhelm II. The first oceangoing yacht ever owned by a Prussian monarch, the *Hohenzollern*—at 4,500 tons and driven by twin screws, with engines of 9,000 horsepower—was larger and faster than many transatlantic steamships of the day. Toward the bow, the kaiser commanded his own bridge, reached up a mahogany stairway. His private study amidship included a telephone for contacting any section of the yacht; this nautical sovereign was very much in charge. To a greater degree than any of his predecessors on the Prussian throne, Wilhelm favored ships and sea power as instruments of the national will.

The German emperor and his squadron were coming to England in the summer of 1889 on a parading mission of overt peace: a grand Anglo-German naval review at Spithead, the channel between Portsmouth and the Isle of Wight. Only thirty years old, and just beginning the second year of his reign, Wilhelm was collecting initial impressions of his European rivals. The *Hohenzollern* was met near the Nab lightship by the Prince of Wales, aboard the British royal yacht *Osborne.* The kaiser and the prince steamed past the formidable ranks of British naval vessels lined up at Spithead, receiving salutes, and then came ashore at East Cowes.

There Wilhelm paid a formal visit to his grandmother Queen Victoria, who bestowed on him the rank of admiral in the British navy.

Two days later, with the mightiest British battleships assembled for his inspection, Wilhelm first boarded the merchant steamer *Teutonic*. The gleaming new star of the White Star fleet, she had just completed her trial cruise from Belfast; directly after the naval review, she would return to Liverpool for her maiden voyage to New York. Under the terms of an Admiralty subsidy, she was built for quick conversion to an armed cruiser brandishing twelve guns. For the review she was fitted with just four five-inch guns at the bow and stern. Thomas Ismay of the White Star Line greeted the kaiser, introducing him to Sir Edward Harland and Gustav Wilhelm Wolff, Harland's German-born partner in their Belfast shipyard. (The kaiser was presumably pleased by Wolff's middle name and by the German name of the steamship.) The size, power, decoration, and furnishings of the *Teutonic* were overwhelmingly impressive, but the kaiser paid special attention to the guns. Firing a twelve-pound cartridge and a steel shell of forty-five pounds, they could penetrate a five-inch plate of wrought iron at a distance of two hundred yards, or hit a target three feet by three feet from a mile away. Wilhelm examined the guns in detail and watched the firing drills carefully. According to the account next day in the *London Times,* he ordered an adjutant to obtain a similar gun for him, and quickly.

The kaiser's visit to the *Teutonic* in 1889 generated a persistent myth in transatlantic history. During the following decade, German ship-builders and steamship lines suddenly overtook their British rivals, building bigger, faster ships that set new speed records and held them for years. This unexpected vault to supremacy later acquired a simple explanation in a memoir published in 1925 by a White Star captain named Bertram Hayes. When he joined the *Teutonic* as fourth officer in January 1891, Hayes apparently heard a parochial story from his shipmates about Wilhelm's visit: that the kaiser, so dazzled by the White Star liner, had told one of his men, "We must have some of these." This imperial remark, Hayes claimed, provoked the startling German ascent of the 1890s in merchant shipping. A good story, crisp and personal, it entered the literature of transatlantic history and has been passed along from book to book ever since. But it wasn't true. The kaiser's remark probably referred to the guns, as reported by the *Times,* not to the ship herself. And in any case, the forces behind that fast ascent gathered long before the reign of Wilhelm II, even before the unification of the modern nation of Germany.

*A*t the southeastern corner of the North Sea, the German coast-line runs eastward about 130 miles from the border of the Netherlands, then turns sharply to the north and extends another 100 miles up the Danish peninsula to the border of Denmark. This short, jagged stretch of coast provides the only German access to the North Sea and thence to the Atlantic. Barricaded by the North and East Frisian islands, lacking deep, protected harbors, and often closed by winter ice, it offers few natural advantages for oceangoing ship traffic. But two major rivers here empty into the North Sea: the Elbe, which rises more than seven hundred miles to the southeast and flows through the German heartland, and the Weser, which begins five hundred miles to the south. The busy port cities of Hamburg (on the Elbe) and Bremen (on the Weser) grew up along these rivers, some forty or fifty miles inland from the inhospitable coast.

The two cities had lengthy commercial histories as both rivals and allies. From the twelfth through the sixteenth centuries, they were part of the Hanseatic League, the north-German confederation for mutual trade and defense. The Hanse towns prospered in part by deploying superior vessels—the cog and then the hulk—for their sea commerce. After the league disintegrated in the 1600s, Hamburg and Bremen held on to their alliance with the Baltic port of Lubeck. Though much tossed around by the European wars of the next two centuries, the people of these German cities still thought of themselves as *Hanseaten,* proudly invoking the medieval glories and independence of the Hanse. (They expressed what the German historian Michael Stürmer has called "this perennial desire of Germans to distance themselves from the center, from Bonn or Berlin, from being German.") Prussia, the dominant German state, was traditionally warlike, agrarian, and autocratic. In determined contrast, Hamburg and Bremen were peaceful, commercial, cosmopolitan, and relatively democratic. Focused on their ancient mercantile ends of *Soll und Haben* (debit and credit), they remained ferociously autonomous and self-sufficient. When the Zollverein, the German customs union, was stitched together by Prussia in 1834, Hamburg and Bremen refused to join, preferring to maintain their historical status as free ports.

Facing westward across the North Sea, instead of inward toward Prussia, both cities forged trading and investment ties with Britain and the United States. British capital and technology flowed in, along with raw American cotton and tobacco, while German textiles and emigrants flowed out. To carry these trades, in 1847 a group of Hamburg merchants started the Hamburg-American Line. The founders announced their intention to establish "a regular connection between Hamburg and North America by means of sailing ships under the Hamburg flag. [The nation of Germany did not yet exist.] The ships are in the first instance intended for the route from and to New York." Adolph Godeffroy, from one of the old, ruling shipping families of Hamburg, ran the line for its first thirty-three years. At his direction, four German-built wooden ships of about 700 tons, with room for twenty cabin and two hundred steerage passengers, started service to New York. Each liner boasted a small library, crockery engraved with the ship's name, an ample supply of linen, uniforms for the captain and mates, and even "a bed for each passenger," as Godeffroy explained. "Since a similar packet service does not yet exist here, it has been the aim of the directors to create something extraordinary, and for that reason the whole undertaking was planned from the outset, being internally solid and respectable through and through, whilst outwardly also providing it with the necessary glamour."

For six years the line's sailing packets did good business, buoyed by the surge in German emigration that followed the failed revolutions of 1848. But after the Inman Line started carrying emigrants by steam power, the Hamburg-American Line had to modernize its fleet. In 1854 Godeffroy urged the necessity, "if we do not wish to be overtaken by our active neighbouring city Bremen," of acquiring large screw steamers. "Furthermore, the paramount competition of England, although more distant, must not be overlooked, where currently steam shipping is expanding in a gigantic way." Since no German shipyard could yet build such steamships, the line went to Scotland and contracted for two iron ships from Caird & Company of Greenock, an established Clydeside firm. To supervise the construction and guard the line's interests, Godeffroy sent Jacob Diederichsen, an engineering veteran of the Prussian navy, to Greenock.

The *Borussia* and *Hammonia*, 280 feet long and 2,100 tons, cruised at twelve knots. In June 1856 the *Borussia* left Hamburg on her maiden voyage to New York with cargo and 402 passengers, about two-thirds of capacity. The officers and crew were German, and Diederichsen ran the engine room. After fighting strong westerly gales for most of the way, groping through constant fog over the final four days, and skirting icebergs on the Newfoundland Banks—in short, a typical westward passage across the North Atlantic—the *Borussia* came safely into New York after a run of sixteen days. A month later the *Hammonia* followed with 394 passengers, again in sixteen days. The line's sailing ships had averaged around forty days westbound. Godeffroy soon ordered two slightly larger steamships, the *Austria* and *Saxonia,* from the Cairds. Following in the venerable seafaring, enterprising traditions of the Hanse, keeping pace with progressing ship technology, the Hamburg-American Line was steaming steadily ahead, safe and profitable.

*T*he *Austria* departed Hamburg on September 4, 1858, for her second trip to New York. After stopping at Southampton, on the southern English coast, she steamed out to sea with 400 tons of freight, 68 passengers in the first cabin, 111 in the second, 241 in steerage, and 103 officers and crew. (The departure from Southampton included a grim omen: the anchor ran out uncontrollably as it was weighed, spinning the capstan hard and hurling sailors around. Two were badly injured, and one man was thrown overboard and drowned.) On this voyage of late summer, for the first week the ship beat against strong head winds and high, crashing head seas. Her daily runs fell to poky totals of 120 and 150 miles, much less than expected. On Sunday, the twelfth, the morning broke more agreeably, but the seas remained unsettled, keeping most passengers below. Finally, on the thirteenth, the sun rose warmly into a clear sky, and the winds subsided at last. Passengers crowded up on the deck, the first and second cabin on the poop at the stern, the steerage toward the bow, inhaling their first real chance at fresh air since leaving England.

Given eight days of closed-down hatches and virtually no ventilation, the steerage quarters 'tween-decks were intolerably fouled and fetid. The captain and ship's physician ordered a fumigation; at a time when many diseases were thought to be carried by unclean air—"mephitic humors"— a suffusing cloud of thick smoke was supposed to prevent sickness, even the shipboard plagues of cholera and typhus. The fourth officer and boatswain descended to steerage. The boatswain was going to heat the end of a chain and plunge it into a bucket of pine tar to produce the cleansing cloud. The chain got too hot. The boatswain dropped it, hastily and carelessly, overturning the tar bucket and setting the wooden floor on fire. The blazing tar oozed across the deck, the flames fanned by open hatches and ventilators. Though the *Austria's* hull was iron, her internal structures, gangways, stairways, deckhouses, masts, and spars were all made of wood—a series of dry, welcoming matchsticks for the fire exploding below.

It was around two o'clock in the afternoon. Passengers up on deck heard a cry of "Fire!" At the stern, a group of gentlemen playing a game of shuttlecock assumed a false alarm. Looking around from their game, they saw thick smoke and flames bursting through midship. The captain bolted from his cabin. He ordered the men of the fire brigade to their posts, but the fire had already melted the lead pipes of the steam pump, so their hoses were disabled. Two strings of fire buckets—under the top-gallant forecastle, and on the spar deck, near the funnel—were locked in place by a chain, secured and useless. Fire and smoke killed the engine room crew and kept the chief engineer's controls beyond reach. The ship steamed ahead, unstoppably, against a light westerly breeze, her own motion sucking in more oxygen to feed the rising fire. The foremast, burned loose from below, fell like a lone pine tree. The ship's magazine of guns and ammunition exploded. "We are all lost!" screamed the captain. "My God! We are all lost!"

A beautiful, relieving day at sea, in calm weather and bright sunlight, with no icebergs or other vessels in sight, had changed within fifteen minutes into a roaring, inescapable disaster. The fire burned higher along the starboard side, leaving the lifeboats there aflame or inaccessible. Passengers rushed toward the larboard boats, still on their davits. Overwhelmed crewmen ordered them out so the boats could be lowered; other passengers at once crowded into the vacated seats. The captain, apparently unhinged, suddenly barged forward to save himself. Boats were lowered at odd angles, spilling people into the water, overturning and foundering on the ocean. Only one boat got safely afloat. The screw propeller, still thrashing away, chopped into boats and swimmers. Flames spread toward the jammed throng of passengers cowering on the poop. Smoke and sparks blew backward and engulfed them. When the heat and pain became intolerable, people jumped from the deck, clothes ablaze, in effect choosing a less agonizing way to die. A Hungarian emigrant told his wife to jump, then his six older children, one after the other, and then leaped himself, holding his baby. The *Austria*'s untended boilers gradually lost steam pressure. The ship slowed down, stopped dead, and burned on.

At five o'clock, a French bark alerted by the smoke happened by and picked up the survivors. The toll came to 471 dead—a higher body count than in either of the recent Collins Line disasters, and almost as many as the 480 victims on the vanished Inman liner *City of Glasgow* in 1854. The captain was criticized for deserting his ship, and the crew for losing control of the passengers. The captain had at least died; the surviving crew members insisted they had done their best in impossible straits. "We leave

The main saloon of the *City of New York* (1888), the most beautiful room yet seen on an Atlantic liner. Its unusual placement, too far forward, left the dome's stained-glass windows vulnerable to heavy seas coming over the bow. (UGA)

The ladies' drawing room of the *City of New York,* behind the oriel window—visible in the previous photograph—that overlooked the aft portion of the main saloon. (UGA)

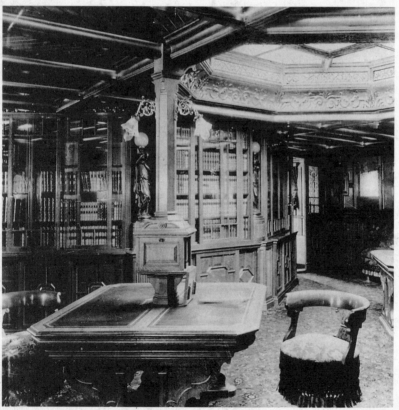

The library of the *City of New York.* (UGA)

The smoking room of the *City of New York*. (UGA)

One of the 10,000-horsepower triple-expansion engines of the *City of Paris*. (UGA)

The main saloon of the *Teutonic* (1889), decorated by the London firm of George Trollope and Sons. (NMG)

The library of the *Teutonic*. (NMG)

The smoking room of the *Teutonic*. (NMG)

RIGHT:
The grand staircase
of the *Teutonic*. (NMG)

BELOW:
A first-class
stateroom of the
Teutonic. (NMG)

The main saloon of the *Lucania*. (1893) (NMM)

The library of the *Lucania*. (NMM)

The bridge of the *Lucania*—at last, an enclosed pilot house for the helmsman,
though the other men on the bridge remain exposed to waves and weather. (NMM)

The turbine age begins: the revolutionary turbine-powered *Turbinia*, designed by Charles Parsons, crashing the parry at the Grand Naval Review at Spithead in June 1897. (NMG)

Cunard's *Lusitania* (1907), designed by William J. Luke and built by John Brown & Company, was the first British ship to hold the Atlantic speed records in a decade and, in May 1915, would mark the end of seventy-five years of Cunard luck and safety on the North Atlantic. (NMM)

Cunard's *Mauretania* (1907), the greatest steamship in transatlantic history, was designed by Edwin De Rusett and built by Swan, Hunter & Wigham Richardson; here photographed alongside the *Turbinia,* it was the first vessel to show the potential of marine turbine engines. (NMM)

Two of the *Lusitania's* four screws, during the liner's construction at the Clydebank shipyard of John Brown. (NMM)

Cargo-handling and anchor equipment toward the bow of the *Lusitania*. (NMM)

The verandah cafe of the *Lusitania,* decorated by James Miller, with wall lattices in the style of his fellow Glasgow architect Charles Rennie Mackintosh. (NMM)

A first-class stateroom of the *Lusitania.* (NMM)

A third-class public room of the *Mauretania*. (NMM)

A third-class foyer on the shelter deck of the *Mauretania;* the stairway leads down to the dining saloon on the upper deck below. (NMM)

The third-class dining saloon of the *Mauretania;* white tablecloths, revolving chairs instead of benches, and polished-ash panels were all relative luxuries for the steerage. (NMM)

The first-class
pantry of the
Mauretania. (NMM)

The second-class
dining room of the
Mauretania. (NMM)

The second-class
smoking room of the
Mauretania. (NMM)

The first-class smoking room of the *Mauretania*, decorated by Harold Peto. (NMM)

The first-class library of the *Mauretania*, also by Peto; the windows on the far wall are bowed outward to form a shallow bay. (NMM)

The first-class lounge of the *Mauretania,* decorated by the Melliers of London, in the 1920s, after it was converted to a ballroom. (NMM)

CREDITS

CAL   CUNARD ARCHIVES, SPECIAL COLLECTIONS, UNIVERSITY OF LIVERPOOL

MMA   MARITIME MUSEUM OF THE ATLANTIC, HALIFAX

NMG   NATIONAL MUSEUMS & GALLERIES ON MERSEYSIDE, LIVERPOOL

NMM   NATIONAL MARITIME MUSEUM, GREENWICH

PANS   PUBLIC ARCHIVES OF NOVA SCOTIA, HALIFAX

UGA   UNIVERSITY OF GLASGOW ARCHIVES AND BUSINESS RECORDS CENTRE it to every one," nine of them said in a statement from London, "to judge whether we have done our duty or not."

*B*remen had recently started its own transatlantic steamship line. The city, sited on a less important river than Hamburg's Elbe, was smaller and less progressive than its ancient rival, like a younger brother always striving to keep up. Bremen did enjoy older, stronger ties to America; since 1783, Bremen ships had been trading with Boston and Baltimore, and later with New York. In 1827 Bremen's mayor established the port of Bremerhaven at the mouth of the Weser to provide his city improved dock facilities on the North Sea. The river between there and Bremen was dredged and locked to allow access for bigger ships. Still, the city retained anachronistic traces of its medieval, Hanseatic past for well into the industrializing nineteenth century. The railroad and telegraph did not reach Bremen until 1847, a year before the governing constitution was modernized.

More than anyone else, Hermann H. Meier poked and prodded Bremen into the steam age. Sometimes called "the King of Bremen," he was a relentless presence in many different local enterprises. As a young man, in 1834, he started working for his family's Bremen-America Packet Line, shipping out to English and American ports and learning the business. Later he helped start the Bremen Bank. By 1850 he was considered the most influential merchant prince in town. Even before the Hamburg-American Line switched to steam power, Meier wanted his own steamship company. After years of plotting, in 1857 he raised capital from regional banks and cobbled together three small Weser steamboat lines and a local insurance business—all of which he already served as a director—into the North German Lloyd Line. (The name was simply a marketing ploy, borrowed from the famous British insurance company and shipping registry to bestow instant history and legitimacy on the new line.) "We cannot reckon on support from the Government," Meier declared, "as most other companies can to a large degree. We are obliged to look to our own energy, activity and carefulness and a reasonable economy to keep a hold on our capital."

The first ships, as well as the name, were British. Like Adolph Godeffroy, Meier had to cross the English Channel to acquire oceangoing steamers. The *Bremen* and *New York* were built by the Cairds at Greenock, the *Hudson* and *Weser* by the Palmer brothers at Jarrow-on-

Tyne; the balanced names invoked the cities and rivers of the proposed transatlantic traffic. The four ships were essentially identical iron screws of 318 feet and 2,700 tons—and significantly larger than the first steamships of the Hamburg-American Line. At every step, the histories of these two lines unfold with reference to what the rival line had recently done or was about to do. The old Hanseatic competition was merely updated and transferred to the high seas. In the details of that regional quarrel, and not in any national impetus or imperial decree, lay the real force behind the German ascent in merchant shipping.

All four Lloyd ships started crossing the Atlantic in 1858. It was not a promising beginning. The first ship in service, the *Bremen,* arrived in New York with just 115 passengers and 150 tons of freight—though she could carry up to 571 passengers and 1,000 tons. During the initial year of operations, the *Hudson* was burned at anchor in Bremerhaven and had to be sold, the *Weser* suffered catastrophic storm damage and also had to be sold, and the *Bremen* was laid up for six months after breaking her crankshaft. The promising new Lloyd fleet was down to only one ship. The line built more steamers, but misfortunes continued. Of the first ten Lloyd vessels, six were lost at sea or were quickly sold at a loss. Hermann Meier refused to seem discouraged; his regal stature in Bremen still retained the confidence of his board of directors, and he cut a saving deal with his principal bank. The Lloyd company kept buying ships from the Cairds, sending them out, and hoping for better luck.

Both German lines grew up in the context of comprehensive political and industrial changes in Germany. With the power of the Prussian throne and army behind him, Chancellor Otto von Bismarck pulled eighteen separate states and free cities into a new German empire, the Second Reich. Hamburg and Bremen, scorned by Bismarck as a "foreign bridgehead on German soil," were the final holdouts. In 1870 the recently united Germany fought a war against France that the French were expected to win. Instead the Prussian-led forces swept to a swift, complete, and astonishing victory. Germans began to rise on their hind legs, look around, and regard themselves differently. Instead of a fragmented, old-fashioned backwater, "the land of writers and thinkers," they now formed a new force in Europe. The hero of Friedrich Spielhagen's novel *Sturmflut* caught the rising spirit, recalling "what it means to belong to a country that is not a nation and, because it isn't, is not regarded as complete by the other nations with whom we trade, yes is really scorned by them." But after the Franco-Prussian War, said the hero, those nations knew "that they were no longer conducting commerce and concluding treaties with Hamburgers and Bremeners, with Oldenburgers and Mecklenburgers, or even with Prussians, but with Germans who sail under one and the same flag."

Visitors to the new Germany noticed a difference. In 1885 an American tourist, the Reverend William Leonard Gage of Hartford, Connecticut, spent three weeks in Berlin, marveling at the transformations since he had first seen the city thirty years earlier. Then it was backward, preindustrial, with unpaved streets and inadequate lighting, not even a city water system. Now it looked rich and modern. And Berliners had changed in similar ways: "Then they used to hang their heads, to look provincial, to act as if they had no right to be; now they hold their heads as high as the highest, and each one looks as if he had a special right to wear a crown. They have lost the deferential manner, the old-fashioned simplicities of dress; they have put on the smoothness, the sleek, gay manners of the nation they have conquered; and now Berlin seems little other than Paris."

Industrialization happened later in Germany than in Britain and the United States. For most of the nineteenth century, German industrialists had to import British technology and engineers to build and run their factories. The victory over France delivered to Germany the province of Lorraine, with its rich deposits of low-grade phosphoric iron ores. Combined with the coal fields of the Ruhr, Upper Silesia, and the Saar, this conquered iron supply quickly sent German iron and steel production whizzing past that of every other European nation, including Britain. Elementary schools with required attendance had begun in the 1700s in Germany, long before Britain and France; as industrialization took off, it exploited the key advantage of a highly literate workforce. In just a single generation, accelerating toward the end of the century, Germany became the foremost industrial and trading nation on the continent. It led the world into the new electrical and chemical industries; the international language of science was German; its universities and many fields of academic scholarship provided models that were copied everywhere. In their phenomenal speed, range, and inventiveness, German achievements of this era had few rivals.

Except, until the 1890s, in shipbuilding. During their first two decades, the Hamburg-American and Lloyd lines did not try to match the size and speed of the crack steamers racing between Liverpool and New York. The German lines kept running their Caird-built ships of moderate size and power, setting no speed records but pulling in sound profits. In 1880 Hermann Meier—catching the new German spirit of competing to win, not just to place—decided to build his first *Schnelldampfer* (fast steamer). Some of his Lloyd colleagues wanted to give the contract to a German shipyard, but Meier insisted on going to the hottest shipbuilder in the world: William Pearce and the Fairfield works, at Govan on the Clyde. Their *Arizona,* built for the Guion Line in 1879, was then the most celebrated new ship on the Atlantic. Aside from making the fastest trips yet across the ocean, she had also collided head-on, at full speed, with an iceberg—and survived. Meier wanted his own version of such a paragon of strength and swiftness for the Bremen line.

So, in order to compete with the fastest transatlantic ships, Lloyd still had to buy its goods in Scotland. Over seven years, Pearce built nine express liners for the Bremen company. The first of these, the *Elbe* of 1881, did indeed look like a slightly smaller copy of the *Arizona,* with four masts and two funnels. At 4,500 tons and sixteen knots, she was substantially bigger and faster than any previous Lloyd ship. On her third voyage out, she ran from Southampton to New York in eight days, a new record for that course, at a speed almost matching the best times from Liverpool. The British government, reaching for speed over national pride, switched its homeward mail from a too-slow Cunarder to the new Lloyd flyer. (Stung and embarrassed, Cunard soon went to Fairfield for its own Pearce greyhounds, the *Umbria* and *Etruria.*)

The second cluster of Fairfield ships for Lloyd, the *Aller* and her two sisters of 1886, had the largest triple-expansion engines yet on the Atlantic. By expanding steam into three successive cylinders, they drove the *Aller* over eighteen knots but still used coal more stingily than a simple compound engine. German craftsmen were sent to Glasgow to execute the final decorations and furnishings; the style was more opulent than previously applied on the Clyde. "The main saloon is a floating art-gallery, decked in the richest upholstery and emblazoned in gold and tints such as have never before been seen on any transatlantic steamer coming to this port," said the *Nautical Gazette* of New York. "She is more like an imperial yacht than a simple mail and passenger steamer. . . . Every new appliance worthy of adoption and tending to the efficiency of the ship has been introduced into her, and these happy combinations tend to make her the ship of the age. She is beautiful, strong, fast and economical."

The ship of the age was surpassed two years later by the *Lahn,* the final Fairfield ship for Lloyd in this sequence. Built of steel and teak, at 448 feet and 5,700 tons she could carry over 1,000 passengers. Andrew Laing installed his new triple-expansion engine, a single intermediate cylinder in a tandem arrangement with the four high and low cylinders. For a year the *Lahn* was the fastest ship on the Atlantic—except for her Fairfield cousins, the *Umbria* and *Etruria*. Johannes Georg Poppe, an architect and designer already well known in Bremen, finished her main dining saloon in grand baroque style. (During the next two decades, Poppe would continually surpass himself in decorating Lloyd ships, freely mingling architectural styles from the Italian Renaissance through Viennese Secession to Art Nouveau.) As both ships and buildings, these nine Fairfield liners lofted new standards for the German merchant marine.

In the mid-1880s, Bismarck's imperial designs forced a decisive new direction for the German presence on the North Atlantic. He finally bullied Hamburg and Bremen—the most resistant components of the Second Reich—into the Zollverein. With the empire secured at home, he shifted to an imperial policy abroad as well, coveting markets and colonies in Africa and the Far East in order to keep pace with the European rivals carving out their own faraway territories. For such distant ambitions he needed reliable steamship mail connections. This linked policy, of starting an empire with a steamship mail subsidy, met political opposition among liberal elements of the Reichstag. In March 1885 Bismarck addressed his fractious parliament. "As the Germans had withstood the foe in 1870 as a nation of brothers," he said, "a spirit of party strife and confessed dissension must not now be allowed to ruin the newly founded empire." By a close vote, not the thumping patriotic mandate Bismarck wanted, the Reichstag authorized subsidized lines to Australia, East Asia, and Egypt.

Hermann Meier, approaching the end of his career, had spent a long lifetime promoting Bremen by resisting Bismarck and Prussian encroachments. He had spurned earlier offers of a government subsidy because it would have meant losing his guarded autonomy. Bismarck now gave him no choice. On July 4, 1885, he was forced to sign the contract presented by the "Iron Chancellor." In the midst of the quite successful Lloyd string of Scottish-built greyhounds, Meier had to agree to run eastern mail steamships of a specified size and speed—and to build them in German shipyards. "Today, the day of the American Declaration of Independence," said Meier, "I am signing the contract for our slavery."

The new policy presented a barbed gift to German shipbuilders, part boon and part challenge. The principal yards—at Hamburg and Bremerhaven on the North Sea, and Kiel, Stettin, and Danzig on the

Baltic—had no experience with major oceangoing steamships. Though German mines, foundries, and steel mills were booming, they were placed and focused on inland industries. The Vulcan shipyard in Bredow, near Stettin, had to build locomotives in between orders to produce ships. Lloyd gave Vulcan a contract for six steamers of modest size and speed: three of 1,500 tons at twelve and a half knots, and three of 3,600 tons and fourteen knots. (On the long run to China, economy mattered more than speed or luxury.) The first of these German-built, subsidized ships, the *Preussen*, started service in the summer of 1886. She and her sisters were closely watched, in hope and skepticism, by ship men in Germany and Britain.

*M*eantime, back on the North Atlantic, Hamburg's response to the Lloyd *Schnelldampfer* fleet of Fairfield greyhounds arrived with the advent of Albert Ballin, the greatest figure in the entire history of German merchant shipping. His stature eventually would rise to the level of Samuel Cunard, William Inman, and Thomas Ismay, the major British leaders of transatlantic steamship companies. Ballin first surfaced when German steamship lines and shipbuilding still lagged behind their British counterparts. Within two decades, keyed by Ballin as the sleepless, ubiquitous catalyst and tactician, Germans matched and then even surpassed the British. "I am the only German," he later declared, "who can say with truth that he has been fighting the English for supremacy in the shipping world during the last thirty years. During this long period I have, if I am allowed to make use of so bold a comparison, conquered one British trench after the other, and I have renewed my attacks whenever I could find the means for doing so." Yet for most of his career Ballin was perceived as a conspicuous Anglophile, with many friends and contacts in Britain. In Hamburg, his favorite daily newspaper was the pro-British *Correspondent*, and he spoke and wrote English fluently. He often went to London and New York on business. Amid the increasingly bellicose politics of Wilhelmine Germany, Ballin characteristically spoke for peace and restraint. His personality unfolded in complicated layers.

Albert Ballin grew up near ships, on the northern shore of the Elbe in the western part of Hamburg, where most vessels were then docked. His Jewish ancestors, originally from Frankfurt, had migrated to Denmark, then—in his father's generation—to Hamburg. Albert was the youngest of nine children. At seventeen he went to work in the family emigration agency, booking passengers from Germany through British ports to

America. The business struggled along until the marked growth in German emigration that began around 1880. Ballin persuaded the Carr steamship line to add plain emigrant accommodations to its cargo steamers on the North Atlantic. By cutting out the lavish overhead costs of first-cabin passengers, Ballin and Carr could offer cheaper emigrant fares than the Hamburg-American or Lloyd lines. After a long, brutal fare war, Hamburg-American bought out the Carr company—and welcomed the enterprising Ballin as the head of its new passenger department. From that base, he quickly became, in effect, the leader of the entire Hamburg line.

Thirty years old in 1887, Ballin was a short, balding, homely man, his appearance redeemed only by the warm, intelligent eyes. He moved easily in society, well-mannered and elegantly dressed, a noted gourmet and raconteur. But he was absorbed, every day, in the shipping business; nobody among his contemporaries knew more about the details of ships, markets, and schedules. His board of directors had to insist that he not come to the office on Sundays. (Instead he worked at home.) An accomplished insomniac, his mind forever grinding and fussing, he took progressively larger doses of drugs to sleep. At sea, though, he slept easily and never felt seasick. Though physically better suited to a career as a steamship captain, he instead spent most of his driven life moored to a shorebound desk. "One needed to cross the ocean with him on one of his splendid liners," a friend recalled, "to appreciate to the full his geniality and charm and his brilliant conversation."

As a loyal Hamburger, Ballin undertook his first major initiative by challenging the *Schnelldampfer* fleet racing out of Bremerhaven. Lloyd's nine Fairfield greyhounds were fast—but they were single-screw vessels. Even before Inman and White Star had started running the first British twin screws across the Atlantic, Ballin decided on double engines and propellers for the next ships of the Hamburg-American Line. Late in 1887, contracts for two such liners were signed: one with a British builder, the Lairds of Birkenhead, and the other—at the urging of Bismarck, and buttressed by a supporting memorandum from young Prince Wilhelm— with the Vulcan shipyard of Stettin. (At the time, Wilhelm was only the presumed heir to the presumed heir of the reigning kaiser, so it was Bismarck's request for a German builder that probably leaned more heavily on the Hamburg-American leadership.) For the Vulcan company, the new contract meant another step up: a challenging promotion from the less ambitious ships it was building for Lloyd's subsidized line to China.

The Vulcan-built twin screw, the *Augusta Victoria,* was ready first.

She began a new German express service, stopping at Southampton in a direct challenge to the traffic that Lloyd had monopolized for two decades. In May 1889, the *Augusta Victoria* made the fastest maiden trip yet to New York. She brought 734 passengers in steerage, 25 in second cabin, and 13 (including Albert Ballin) in first. Her time of seven days, two hours, and thirty minutes from Southampton was the equivalent—for a Liverpool boat timed from Fastnet—of just six days, eight hours, and thirty minutes. With that single swift crossing, averaging almost twenty knots, German shipbuilding loudly came of age. At 460 feet and 7,700 tons, the *Augusta Victoria* was by far the largest ship yet built in Germany. She lay in shiny triumph at her dock in Hoboken, across the river from Manhattan, attracting praise and many visitors. In five hours one afternoon, some 30,000 admirers (mostly proud German-Americans) toured the ship, noting the lavender-tinted ladies' saloon and music room, and the first-class cabins with their twelve feet of headroom. She had three funnels, a straight stem, and three masts without spars; the double propulsion meant no need for emergency sails. "Her lines are unusually graceful," declared the *New York Times.* "Dividing bulkheads form air-tight compartments that make the vessel practically unsinkable."

The second Vulcan-built twin screw, the *Furst Bismarck,* appeared two years later. Another three-stacker, longer and more powerful, she lowered the maiden record to New York to six days, fourteen hours, and fifteen minutes. In tandem with the *Augusta Victoria,* she gave the Hamburg-American Line a brace of homegrown twin-screw greyhounds as swift and popular as Inman's *City of New York* and *City of Paris* or White Star's *Teutonic* and *Majestic.* German food and service on the ocean were generally praised and included a band of about a dozen musicians recruited from the stewards. The atmosphere struck Atlantic travelers as distinctly German: drenched in music, food, drink, and warm spirits. The *Furst Bismarck* offered a luxurious dinner in first class of seven to ten courses, with wines and ales available at 40 percent lower cost than in New York. "Everything about the ship has a military air," a British passenger noted. "The stewards enter in regular order, and when a change is ordered they all march out keeping time to the band, making, with their neat uniforms and snow-white gloves, a goodly sight to see." The band then played a formal concert at nine in the evening, suggesting a German beer garden at sea.

Instructed by its Hanseatic rival, Lloyd also now bought its crack Atlantic steamers from the Vulcan works. The first two, the *Spree* and *Havel,* started service in 1890 and 1891. Mark Twain, still roaming the

North Atlantic, took passage on the *Havel* in 1893. He compared her to all the previous ships of his experience and allowed that he was impressed. "I should not have supposed that the modern ship could be a surprise to me, but it is," he wrote. "The menu is choice and elaborate, and is changed daily." Everything seemed so costly and sumptuous, quite secure, and remarkably *quiet.* "The absence of hubbub, clatter, rush of feet, roaring of orders. That is all gone by. The elaborate manoeuvres necessary in working the vessel into her dock are conducted without sound; one sees nothing of the processes, hears no commands. A Sabbath stillness and solemnity reign, in place of the turmoil and racket of the earlier days." The ship, 460 feet long and 7,000 tons, cruised at eighteen knots. The decks rose so high and dry, and the *Havel* plowed ahead in such undeflectable bulk and power, that it even seemed the ocean and its old terrors were finally subdued. "The giant ship of our day does not climb over the waves, but crushes her way through them," Mark Twain concluded. "Her formidable weight and mass and impetus give her mastery over any but extraordinary storm-waves."

*T*he *Elbe* left Bremerhaven for Southampton and New York at three in the afternoon of January 29, 1895. For this cold, winter, westward passage she was carrying only 50 passengers in first class and 149 in steerage. The officers and crew brought the total on board to 352 people. The *Elbe* was almost fourteen years old. As the first of the North German Lloyd fleet of Fairfield greyhounds, she had started the recent ascent of the German merchant marine toward parity with the best British liners on the Atlantic. Aging well, she made her fastest passage on her eighty-third outward voyage, in 1889, by steaming to New York in just seven days and twenty hours. Though now demoted to secondary traffic by newer Lloyd ships, and well past her prime according to the unforgiving standards of the North Atlantic, she was still considered staunch and seaworthy.

At shortly after five o'clock in the morning of January 30, about thirty miles off the Dutch island of Terschelling in the North Sea, the *Elbe*'s bridge crew sighted the lights of an approaching ship off the port bow. She was the Scottish steamer *Crathie*, one-tenth the size of the *Elbe*, heading up to Aberdeen from Rotterdam. The two ships were closing toward each other at a right angle. It was dark; the waves high; the air was clear with just occasional clouds; the wind, strong from the east. The crew of the *Crathie* saw the lights of the *Elbe* from three or four miles away. By the rules of the sea, the *Crathie* was supposed to give way to a ship coming up to starboard. She should have slowed down and crossed behind the *Elbe*. But the mate in command of the *Crathie* did not change speed or course. The *Elbe,* within her rights, also kept coming fast and straight. She sent up warning rockets that had no effect. The final split second made all the difference: which ship would hit, and which would be hit. The *Crathie* buried her nose in the port side of the *Elbe,* just abaft the engine room bulkhead. Striking at a slant, the assaulting prow plowed toward the stern of the *Elbe,* tearing a long, lethal gash. The ships stuck together for an instant. When the *Crathie* pulled free, ice-cold seawater poured into the *Elbe.* The doors of her seven watertight compartments, in the sudden panic and confusion, were not closed down.

The Lloyd liner would sink in twenty minutes. The engine room was flooded at once, dousing the fires. As people below decks snapped awake, they found the floor already knee-deep in water, the corridor blocked by broken timbers and woodwork. A cabin passenger, Carl Hoffman, awoke to a sound like a gunshot, then heard scuffling feet and hoarse shouts overhead. He hurried into his clothes and went up on deck. "I never saw anything like it," he said later; "everybody seemed to have lost his head." The ship was listing sharply to port, obviously sinking. "Men, women, and children were running about madly, the women screaming with terror and every man getting in the other's way. The darkness increased the confusion and fright." People pushed and clawed over the precious, inadequate supply of life preservers. Jan Vevera, another cabin passenger, had to fight off men who wanted the preserver he was wearing.

On the port side, the lower deck was almost down to the water. The captain ordered women and children to the starboard boats. The ropes were cold and tangled, hard to handle. Sailors hacked at them with axes. The ship canted over to port, leaving the starboard boats marooned high on the side. Only three boats were launched, of which two foundered and sank. Jan Vevera climbed up on the rail. When the only lifeboat still afloat rose on a wave, he jumped in, landing on crewmen who did not welcome him. "One of the occupants tried to shove me out," as he told the story, "but I hung to him like death, thinking, 'If I go, you go, too, old man.' He seemed to understand this after he felt my grip a few times, and let me stay." People were bobbing in the water, crying for help—but the lifeboat was already full and stayed away. Up on the sinking ship, knots of desperate men and women hung over the rail, weeping and hoping.

The *Crathie,* with her bow stove in and prospects not clear, circled the *Elbe* slowly, her men watching and deciding what to do. She had only a few small lifeboats; the sea remained rough; it was still dark. So the *Crathie* did nothing. The *Elbe* settled down at the stern, her bow rose, and she sank. The *Crathie* steamed back to Rotterdam. At eleven that morning, about six hours after the collision, a fishing smack spotted the lone lifeboat and picked up the only twenty survivors. Cold, soaked, and benumbed, they barely had strength enough to climb aboard. Fifteen of the twenty were crew members, from the chief stoker to the third officer: an awkward disproportion that would further cloud the aftermath. All the rest—332 people—were lost, including every child and all but one woman.

"A more shocking and, apparently, a less excusable collision never occurred," said the American nautical journal *Seaboard.* "The grossest neglect of the rules of the road, and the most heartless disregard for the promptings of humanity, controlled the actions of the officers in charge of the *Crathie* before and after the disaster." A later inquiry by the British Board of Trade collected testimony from all the pertinent survivors from both vessels. (The entire bridge crew of the *Elbe* had gone down with the ship.) The case was so stark that the board disregarded national allegiances. It found the British mate of the *Crathie* in default and solely responsible, and therefore lifted his certificate.

*G*ermany kept rising. The ocean was never subdued; the progress of technology, the huge increases in size and power, still left any ship vulnerable. The German lines, like all the British lines except Cunard, suffered occasional disasters. That uncanny Cunard record aside, every steamship company on the North Atlantic expected to lose ships and hundreds of passengers to weather or human error or mechanical failure. It was inherent, inevitable, just another cost—if the least tolerable cost—of doing transatlantic business.

The aggregate tonnage of the German merchant fleet overtook that of the United States in 1884, and of France in 1889. Its only real competition lay across the channel. "The jealousy with which the English companies regard their German rivals is notorious," George Smalley wrote in a London letter to the *New York Tribune* in 1885. During the 1880s, Hamburg-American and Lloyd steamers brought almost 1.3 million passengers to New York—about the same total as the four main British lines: Cunard, White Star, Inman, and Guion. In the following decade, Lloyd became the largest steamship company in the world, with vessels running to every part of the globe. A few years later, it was replaced at the top—by its Hanseatic friend in Hamburg. The London-based Institution of Naval Architects, implicitly recognizing these changed circumstances, in 1896 held its first meetings in Germany. "The British people and the German people divide the honour of being the two great commercial races of the world," said the INA president, the earl of Hopetown, to open the meetings. "We earnestly pray that it may always remain a friendly and peaceful rivalry. We regard you as our cousins; the two nations are descended from the same stock; their languages are derived from the same root, and have a common origin. We have always been allies in the past. . . . Surely the world is large enough to hold us both." (The rest of the world could only watch and hope that was so.)

The British competitors sometimes took refuge in complaining that the German lines prospered on lush subsidies from their national government. Its merchant marine did receive certain indirect benefits. Imported shipbuilding materials came in free of duty expenses, and government railroads transported raw materials to shipyards for just the cost of handling. The central government also invested heavily in infrastructures of harbors, docks, and channels. But direct subsidies were largely limited to Lloyd's mail contract for its lines to Asia and Australia. As Albert Ballin pointed out, his company had never enjoyed a subsidy until 1900—and then it came to only £13,000 a year, for another Asian service. The German lines were still driven essentially by that ancient quarrel between Hamburg and Bremen (each in turn lusting to top the other), and the concurrent explosions of German industry, technology, and self-regard. By the 1890s, the old students were ready to start instructing their quondam mentors. In 1892 the Fairfield yard needed exceptionally wide steel plates—twenty-two feet by eleven feet, six inches by one and a quarter inch thick—for the rudders of the new Cunard twin screws *Campania* and *Lucania*. Every British steel mill declined the job as impossible, so Fairfield went to Germany and got its rudders from the Krupp works in Essen.

For almost six decades the center of transatlantic shipbuilding had rested on the banks of the Clyde. Now it was moving, perhaps irresistibly, toward the mouth of the River Oder on the Baltic Sea. Back in 1851, at Bredow, a river village a few miles north of Stettin, two local men started building iron steamships. Their first vessel was a 125-foot paddle wheeler of 102 tons. (At that moment, the Collins Line was sending ships of 2,800 tons across the Atlantic, and Isambard Brunel was beginning to think about his mammoth *Great Eastern.*) After six years, the Bredow firm was taken over and enlarged by a group of six Stettin and two Berlin businessmen. This Vulcan company built small men-of-war, then larger ironclads for the Prussian navy. The shipyard and the new Germany grew up together. In 1886 Vulcan built its first ship for the North German Lloyd: the loyally named *Stettin,* 260 feet long and 3,400 tons, mainly to carry cargo. Soon Vulcan's breakthrough fast twin screws for the Hamburg-American Line raised the yard to a much sharper level of ambition and expectation.

In its intellectual and material origins, the Vulcan works of the 1890s represented a melding of British and German experience. Robert Zimmermann, the director of shipbuilding, had first trained as an engineer at Berlin University. Moving from classroom to shipyard, he left Germany for training stints at three separate centers of British shipbuilding: with the veteran Scotts at Greenock on the Clyde, the newer Palmers at Jarrow-on-Tyne, and at Barrow-in-Furness on the northwest coast of England. In this last position he served as chief naval architect under William John, a distinguished builder and naval engineer. Zimmermann then went home and worked at the Germania yard in Danzig for ten years before coming to Vulcan. As he entered the major phase of his career, contemplating his most important ships, Zimmermann could draw on a comprehensive range of education pulled from both countries.

The Vulcan works ranged across sixty-two acres on the west side of the Oder. The engineering yard filled sixteen acres with the forge, foundry, and boilermaking and turning shops. Across a street, running down to the river, was the shipbuilding yard, Zimmermann's domain. Its 600-yard frontage on the river included seven building slips, roomy enough for ships up to 700 feet long—which nobody, as yet, had built in Germany. Four of the slips were aligned at acute angles to the river, to allow more space for construction and to send major launches down the Oder instead of across it, onto the opposite shore. A few hundred feet downriver, the lumberyard included a sawmill near the water, a joiners' shop, and large sheds for storing and curing cut wood. Basic materials arrived by water or by railroad, from the pine forests of Poland and steel mills inside Prussia. Many of the machine tools in the Vulcan yard, such as a hydraulic flanging device from a Glasgow firm, were of British design or manufacture. In contrast with British practice, which concentrated kindred operations in a single area, each ship under construction had some of its own tools, such as for machining plates, right at hand—gaining efficiency at the cost of duplicating functions. An official French visitor was impressed by the order and cleanliness of the works, the elegant, well-spaced arrangement of the different parts, and an amazing calm silence among the 6,000 workers: "One hears neither cry nor song, not even a loud conversation."

The American journalist Ray Stannard Baker watched a group of Vulcan workmen reshaping straight L-shaped beams into the arching curves of a hull's ribs. The man in charge pegged out the shape of the rib by inserting iron pins into a wide, perforated iron floor. A furnace sixty-five feet long, big enough to swallow the entire beam, heated the metal to red-hot malleability. At a signal, the furnace door opened, belching light and heat, and a man grappled the beam with oversized pincers and pulled it onto the floor. "With infinite deftness and fearlessness, with swiftness and yet without hurry, this flaming bar is crowded against the pegs of the curve, the workmen smiting it with hammers, driving other pegs, straining at levers, and smiting again. . . . In two minutes' time, a simple L of iron had become a ship's rib."

The German workers, with their baggy blue blouses billowing from the shoulders and hanging loose around the waist, and their wooden-soled shoes clacking on the cobblestones and iron floors, looked different to Baker than their American counterparts. Yet he recognized "the strong cousinship of sweat and grime and strength. But for a little more, perhaps, of stoop and stolidity, a little more of patience in their faces, these might be the men of an American shop." They punctuated their eleven-hour workdays with five meals. Rising very early, they bolted black coffee and rye bread before leaving for work. At eight o'clock they stopped toiling long enough for more coffee and bread, this time with a bit of sausage or cheese. At noon they broke for a long lunch brought to the yard by a wife, a child, or an old man. For ninety minutes they sat along the sunny sides of buildings, resting and munching their meals in silence and apparent contentment. At four they stopped for tea and a slice of bread covered with lard, in a fifteen-minute break called vespers. After work, at home they ate more bread and coffee, perhaps with smoked fish or cheese, and beer if they could afford it. Like life on a German steamship, a Vulcan shipyard worker's day seemed organized around eating.

In May 1897, with the kaiser in attendance, Vulcan launched the *Kaiser Wilhelm der Grosse*. Named for Wilhelm II's grandfather, during whose reign the modern nation of Germany was established, the ship was the North German Lloyd's response to Hamburg-American's twin-screw greyhounds. Her dual triple-expansion engines pumped out 28,000 horse-power. In September she set a new maiden record from Southampton to

New York of five days, twenty-two hours, and thirty-five minutes. Two months later, she raced home at a speed of 22.3 knots, beating the *Lucania's* two-year-old standing mark for the eastbound voyage. The following spring, she lowered the *Lucania's* westbound record as well. At 625 feet and 14,350 tons, the *Wilhelm* was twenty-five feet longer and 1,850 tons larger than the reigning Cunard twin screws. For the first time, a German liner had swept past all competition, even the crack Liverpool boats, as the biggest and fastest ship on the Atlantic. "In elegance as well as in speed," the British *Engineering* journal regretted, "the vessel is an object lesson."

Designed by Robert Zimmermann, with advice from Lloyd's leadership, the *Wilhelm* broke from many previous patterns in her external and internal arrangements: the most fearlessly innovative transatlantic liner since Edward Harland's breakthrough *Oceanic* for the White Star Line in 1870. It was apparent that German shipbuilders were no longer just imitating British models. The bow was a sharp, straight edge, sweeping back fifty feet, and up another fifty feet, into a more narrow hull than in the Cunard twin screws. The distinctive profile presented four funnels (for the first time on an Atlantic liner), arranged in two clusters of two instead of a single equally spaced series. The six decks were called the sun, promenade, upper, main, lower, and orlop. The walkway for first-class passengers, usually open to the sky on the highest full deck, ran along the promenade deck under a roof provided by the sun deck above. This protected structure may have felt confining on a rare warm, sunny day, but it suited the more typical gray realities of North Atlantic weather.

Inside the *Wilhelm,* the grouping of facilities recognized the logistical implications of great size. With so much interior space, both linear and vertical, related functions needed to be placed near each other, in distinct regions, so that passengers and crew did not expend too much time and effort merely walking from point to point. Following standard practice, first-class accommodations were amidship, second-class toward the stern, and steerage toward the bow. The upper deck amidship, usually available to the crew for moving between the ends of a vessel, was reserved exclusively for first class. In planning stages, the Lloyd nautical experts doubted this arrangement. But it succeeded because the crewmen's quarters were placed near their respective work areas, in particular neighborhoods of the seagoing town, so they had little need to traverse the upper deck from one end of the ship to the other. The deck crew bunked in its traditional forecastle, the engineers and firemen close to the engine and boiler rooms, the stewards near the first- and second-class cabins, the kitchen and bakery staff next to their own galleys, and the officers on the bridge deck. Officers and sailors still had the run of the promenade and sun decks, where most of their work was done. Quarters for the firemen were divided into three compartments, one for each watch, so the exhausted stokeholers could rise and collapse without disturbing their cohorts.

The long space between the two clusters of funnels left more room for an unobstructed main saloon, fifty-nine feet long and sixty-four feet wide, which was dropped from its usual place on the upper deck to the main deck below because the upper deck was devoted to first-class state-rooms. Up there, queasy passengers could keep their windows open in almost any weather without being doused. The special suites of state-rooms, one level higher on the promenade deck, were dryer still, and removed from even the general rich run of first-class passengers. With a sitting room of sofas, chairs, and tables, a tiled bathroom, a bedroom fea-turing a full-sized brass bed, and a private promenade and adjacent ser-vants' quarters, the suites allowed all the garrisoned seclusion that a bash-ful or discreet voyager might desire. The drawing room filled up the middle of the promenade deck, capped by a dome overhead, and opening down through a well to the main saloon two decks below. This vast area, extending almost four decks through the core of the *Wilhelm,* resembled the inner courtyard of an especially lavish Venetian Renaissance palace. From one end of the floor of the saloon, the grand staircase ascended to a landing, then curled left and right to the upper deck, where arched gal-leries ran around the four sides, overlooking the scene below. (Concealed in one of these galleries, the ship's orchestra played classical and popular music during meals.) The well continued through a riot of gilded orna-ments and flourishes up through the drawing room to the dome: Johannes Poppe, Lloyd's favorite decorator, in full, uninhibited cry.

The ship could carry 602 passengers in first class, 324 in second, 800 in steerage, and 458 officers and crew—a maximum population of 2,184. Novel devices and cues helped keep all these people fed and comfortable. In the main galley, an automatic boiler lifted eggs from the water after any timed interval the cook selected, while electric hoists brought food from the refrigerated rooms below. "A mechanical dish cleaner," *Engineering* reported, "consisting of two propellers in separate compart-ments, . . . so effectively set the water in motion as to insure that dishes will be thoroughly cleansed." Engine vibration, such an affliction to pas-sengers on the Cunard twin screws, was dampened by a system invented by the German engineer Otto Schlick; the *Wilhelm* marked its first use on a high-speed Atlantic liner. Bathrooms and lavatories for women passengers were placed on the port side, marked by a green light, while those for men beckoned from a starboard red light; so even after a late night in the convivial smoking room, a gentleman in befogged and confused condition, not quite sure of his bearings, might still steer by these running lights to his proper refuge.

Among other purposes, the *Kaiser Wilhelm der Grosse* was a floating, steaming monument to the rise of the Second Reich. The German imperial eagles, old and new, looked down from the stained-glass dome. At the four corners of the main saloon, behind sliding doors that disappeared into walls, were four smaller dining rooms for parties of about two dozen. Each was named after a prominent figure from the epochal years of German unification: the grand Bismarck; Field Marshal Helmuth von Moltke, who planned the enabling victory in the Franco-Prussian War as the Iron Chancellor's chief of staff; and the wife and mother of Wilhelm I. Each of these rooms was decorated in a different style, Italian Renaissance or rococo or Queen Anne, including a life-size portrait of the honored personage. Up in the drawing room, beneath a frieze picturing the poets and composers of many nations, the visual focal point was a full-length portrait of the first Wilhelm, muttonchopped and imperial, and arrayed in his coronation robe, crown, sword, and scepter.

The new pride of the Lloyd fleet held the Atlantic speed records until 1900, when Hamburg-American replied with its *Deutschland*. Also designed by Zimmermann and built by Vulcan, she was the last notable transatlantic liner of the nineteenth century. The *Deutschland* was (of course) longer, bigger, and faster than the *Kaiser Wilhelm der Grosse*. From Southampton, she knocked the maiden westward record down to less than five days and twelve hours, and she soon flew across in both directions at over 23 knots. Despite her size of 16,500 tons, she had hardly any cargo space. Instead she was pitched at the high-end passenger traffic, with—remarkably—more than twice as many beds in first class as in steerage. The 266 first-class staterooms, spread across three decks, ranged from suites at midship on the upper deck (for £247 per person) down to cabins on the lower deck (available for £22 11s). Toward the stern on the promenade deck, for the first time on the Atlantic a "grill room" offered an American-style bar and quick, à la carte meals of chops and steaks—a shrewd beacon for the Yankees who filled most passages. In the stained-glass dome above the grill, the symbolic arms of both Germany and the United States were encircled by the anchor of the Hamburg-American Line.

Appearing so quickly in the wake of the *Wilhelm*, from the same designer and shipyard, in most respects the Hamburg liner looked like an overgrown big sister of her Lloyd rival—including the signature profile of four stacks in two clusters. Two major differences distinguished these German champions. Instead of engines of four cylinders and triple expansion, the *Deutschland* was powered by newer six-cylinder, quadruple-expansion engines. (At 33,000 horsepower, they drove the ship to her speed records, but—unproven and resistant to much tinkering after the fact—they also made the *Deutschland* vibrate uncomfortably.) Instead of Johannes Poppe excess, in her interior furnishings and decorations the *Deutschland* was relatively modern and restrained. Designed by the Hamburg architect George Thiele and executed by the Berlin firm of J. C. Pfaff, the public rooms also reflected the taste of the omnipresent Albert Ballin, who was known for his informed appreciation of beautiful forms and colors. The Hamburg liner replaced Poppe's overstuffed architectural revivals with lighter, cleaner styles—which were still quite plush and brimming, in the manner of the day, with fine-art paintings and sculptures.

"Instead of a profusion of mirrors on the walls and heavy detail decorations on the ceilings and gilt on the furniture," as a German engineering journal explained what it called a "radical" innovation, "the aim has been to eliminate the oppressiveness caused by such profusion in the comparatively cramped spaces on shipboard and to produce an effect such that the capacity of the different spaces is magnified. This has a soothing effect in comparison with the never ending motion outside. . . . The rooms, with their modern architectural forms and artistic decoration, produce a very imposing impression, and at the same time have a comfortable appearance." The main saloon (carved white ceiling, dark mahogany furniture, and Japanese red silk on the walls) showed paintings of river scenery from the Elbe, Spree, and Rhine in Germany, and the Hudson in the United States. The drawing room on the promenade deck, draped in silks of light colors, was dominated by a full-length portrait of the reigning kaiser. The smoking room on the upper deck had a painting, over the entrance, of the home port of Hamburg and broad, unadorned ashwood panels stained in a mild gray-blue. The overall impact, simpler and more restful than Poppe's style, helped soften the furious, pounding rush of the *Deutschland*'s fast runs.

Thus the transatlantic nineteenth century, so dominated by the British, ended with a shock. The two new German liners carried off the cream of the traffic. "A source of regret, I might almost say of humiliation," said the veteran Scottish shipbuilder Robert Caird. "I do not think it is generally realized that the crack German mail steamers are so enormously ahead of our latest and best." (Caird felt this turnabout especially hard because his family firm had built most of the early steamers for both German companies, sustaining them for decades.) Worse than a national embarrassment, the German triumphs harbored grave possible military risks for Britain. The best merchant vessels of both nations were also designed for potential service as fast armed cruisers in a war. "The swiftest Cunarders can neither catch the Germans nor run away from them, still less can the White Star ships," a *London Times* article warned in the fall of 1900. "If, then, the Germans know how to build record-breaking Atlantic steamships and run them at a profit, have the English lost that art? . . . Do the English lines mean to contend with the German, or do they leave to them their present supremacy on the Atlantic unchallenged?"

16

The Two Finest Cunarders

$\mathcal{A}$s the twentieth century began, major strands of transatlantic history were converging toward an unexpected climax. The British domination of North Atlantic steam, virtually unchallenged since the *Great Western* of 1838, was broken. The reputation of Cunard as the major, oldest, most stable steamship line—the steady bedrock, since its *Britannia* of 1840, in an inherently unstable business—was shaken. The traditional superiority of British shipbuilding, best executed along the Clyde and in Belfast, needed renewed demonstration. The eternal quest for speed on the North Atlantic—without undue sacrifices of safety, comfort, or economy—came down once again to the search for an improved means of propulsion: a more powerful, more efficient way of harnessing steam on water. All these imperative circumstances would be stitched together to produce the two finest Cunard steamships ever built. In the most exacting traditions of Samuel Cunard, Robert Napier, and Charles Mac Iver, these great liners presented the ultimate melding of the ship as enterprise and engineering and building and town.

$\mathcal{C}$lement Griscom provided the final push. It was not his intention; he only meant to assert the American interests of his own company, as he had been doing for four decades. Instead he wound up illustrating the inscrutable indirection of history: meaning to do X, but instead doing Y, with the unforeseen consequence of Z. His brazenly ambitious International Mercantile Marine, an amalgamation of major American and British steamship lines, barged into the unsettled transatlantic situation at the turn of the century. Approaching the unsteady, newly piled-up circumstances, Griscom tipped them over, creating a fruitful chaos that subsided into unforeseen measures—and ultimately led to those two new Cunarders.

This closing irony finished the most notable American career in transatlantic steam. He had started, in the last days of sail, as a Philadelphia importer and ship broker. In 1870 he induced the Pennsylvania Railroad to finance two steamship lines across the Atlantic; in 1886, bankrolled by the railroad again, and his friends in the Standard Oil Company, he bought the Inman Line; soon he built the *City of New York* and *City of Paris,* the first twin-screw liners on the ocean, and the first to offer suites with private bathrooms. (He understood the American demand for luxury.) At about this time he conceived his grand design of an Anglo-American steamship trust. That came later; in the immediate future he needed two additional Atlantic greyhounds, as big and fast as his breakthrough twin screws, to provide full express service on his merged Inman and International Line.

Griscom turned fifty in 1891. A rich merchant prince of the Gilded Age, he belonged to the toniest clubs of Philadelphia and New York and served on the boards of two dozen major banks and corporations. He owned a fine steam yacht, shot quail in Florida, and collected art—especially Corot and others of the Barbizon school—on his frequent trips to Europe. Griscom and his wife (one of the eminent Philadelphia Biddles) had five children. Their favorite home was a country estate, out on the Main Line in Haverford, called "Dolobran" after an ancient family seat in Wales: another link in the Anglo-American reconciliation of the time. Here Griscom watched over his prize herds of blooded Guernsey and Ayrshire cattle. He liked to insist that the care of his cows, and the marketing of their milk to neighbors at ten cents a quart, took more of his attention than his ships did. But the uncertainties of doing business on the North Atlantic left him little respite. "Our household peace was always being disrupted," his son Lloyd recalled, "by the arrival of messages and telegrams—engines broken, boilers burst, rudders or propellers off, fires in cotton cargoes, ammonia leaks in cold-storage plants, ships aground." When a vessel was overdue, Griscom would pace up and down, stewing and waiting for news; "the loss of a ship meant weeks of gloom."

To get his two additional greyhounds, Griscom had to pick his way through unfriendly U.S. navigation laws and politics. He needed the subsidy of a mail contract. Only ships registered in the United States were eligible; but foreign-built vessels, like the *City of New York* and *City of Paris,* were denied American registry. For years Griscom lobbied for congressional relief from this bind, offering in return the military services of his ships in a future war. "We would take chances greater than the public would ever appreciate if we swung our ships under the American flag," he wrote a key Republican operative in 1891. "If the Government attaches any importance to having our fine fleet under the flag, available for cruisers and transports, they must take hold of the question with courage and vigor, and with a determination to give us the flag." The three Griscom lines were by far the most experienced American companies engaged in transatlantic steam. "The importance of a trained organization such as ours, is not appreciated in considering this question," he went on. "Running steamships is a profession, new people can no more take it up and make a success of it in competition with the bright minds of all nations now trained for generations in the business, than a boy can play the violin because he wants to. . . . Building ships will not necessarily produce the organization to run them successfully."

Griscom's proven expertise, and his well-nurtured political contacts in both parties, got him the compromised deal he needed. In May 1892 Congress allowed the *City of New York* and *City of Paris,* considering their American ownership, to be registered in the United States—provided that Griscom's line would also build two comparable vessels in an American shipyard. If these and other conditions were met, the deal included a mail contract worth $750,000 a year. Hedged and limited, it was still the boldest gesture by Congress toward transatlantic steam in almost half a century. Recognizing the symbolic moment, Griscom dropped the old Inman and International name in favor of "the American Line," with a rampant house flag of the American eagle in blue on a white background. He also departed from Inman tradition by shortening the names of his two best ships. Early in 1893, President Benjamin Harrison came up from Washington on a special train with Griscom and personally raised the American flag over the rechristened *New York.* Two weeks later, Griscom's thirteen-year-old daughter, Frances—"whom her friends persisted in calling 'Pansy' in spite of the dignified function to which she had been called," a reporter noticed—performed the same ceremony on the *Paris.* In her red frock and blue jacket, dotted with snowflakes, she suggested an American flag. Pansy and the president made the American Line official.

Most of the money to build the two new ships came from private sources. The Pennsylvania Railroad, after decades of absorbing annual steamship losses, was not interested. (By 1892 the railroad had literally sunk about $3 million in its poor American Steamship Company.) So Griscom took a $6-million mortgage from the Fidelity Trust and Safe Deposit Company, on whose board he sat, and issued bonds for that amount. About half these bonds were bought by representatives of Standard Oil. Ever since the start of his career, Griscom had tied his fortunes to the blooming American petroleum industry, especially the Rockefeller interests. Their substantial investments in his home-built greyhounds renewed the connection.

The contract went, all but inevitably, to the Cramp shipyard in Philadelphia. (Griscom and Benjamin Brewster, his associate from Standard Oil, served on the Cramp board of directors.) No less than Griscom, Charles Cramp had pursued his own campaigns for the future of American steam on the Atlantic. His misfortune was to build ocean steamers at a time when national commerce was turned inward. "The generation now active on the stage of affairs is purely and solely a railroad generation," he lamented, ". . . trained to regard the development of railways as the one great absorbing field of enterprise in this country." When the United States started to rebuild its navy in the 1880s, new American foundries, forges, and steel mills had to be built as well. Now the vast country was densely crosshatched with railroad tracks, apparently filled up to capacity. As the United States played more seriously on international stages, Americans were again looking outward. To Cramp this demanded a revival of American steamship building. "In no other structure appears such a combination of science and skill, such a conspiracy of brain and brawn," he enthused. "She is a thing of life, an autonomy within herself, and, once off the land, is for the time being a planet. . . . No other thing made by human hands can appeal to the sentiment of men like a great steamship."

The new *St. Louis* and *St. Paul* were slightly larger than the *New York* and *Paris*. Named for heartland cities on the Mississippi River, they emphasized the ardent Americanness of the venture. In profile they resembled the new Cunard twin screws, the *Campania* and *Lucania,* with straight stems, two masts, and two thick funnels. Their main saloons seemed copied from the Inman twin screws: dark lower woodwork surmounted by white, mermaids and gorgons supporting high brackets soaring into a long dome, and an organ in a loft at one end. British influences also showed up in their designers. John Harvard Biles, who had planned the *New York* and *Paris* for the Thomsons, was hired as a consulting naval architect on the new liners. Their principal architect, Lewis Nixon, an

American graduate of the Naval Academy at Annapolis, had finished his education with three years in England at the Royal Naval College in Greenwich. Still, Cramp could plausibly drape his new liners in proud red, white, and blue. "These ships are American from truck to keelson," he declared in 1893. "No foreign materials enter into their construction. They are of American model and design, of American material, and they are being built by American skill and muscle. . . . After many years of practical expulsion from the ocean, the Yankees are coming again, and coming to stay."

Starting service in 1895, the new ships were the first American transatlantic fast mail steamers since the Collins liners of the 1850s. "They have the graceful, sweeping sheer and lively rise of bows which characterize American ships everywhere," said the marine journal *Seaboard* of New York. The main saloon was moved back from its vulnerable forward location on the *New York* and *Paris* to a more protected, conventional place between the stacks. The oak-paneled library, adjacent to the saloon, was called the largest reading room afloat. The Philadelphia architectural firm of Furness, Evans installed nearly 2,400 decorative panels, mostly in delicate tints of ivory and salmon, in the public rooms and cabins, and used ground and leaded glass liberally. The sculptor Karl Bitter, who had just made his name with the Astor memorial gates for Trinity Church in New York, and notable work at the Chicago World's Fair in 1893, executed the Neptune and other mythic figures in the main saloon. The cabins in first and second class were served by "an unusually large number of baths and water-closets" nearby—more concessions to American travelers. All the steerage quarters except one were divided into smaller rooms, allowing more privacy and security. The thirty-two lifeboats had space for everybody, up to five hundred passengers in first and second class, nine hundred in steerage, and three hundred officers and men.

On the ocean, the *St. Louis* and *St. Paul* could not satisfy all the patriotic hopes they were carrying. Designed only to meet the contracted stipulations to match the size and speed of the *New York* and *Paris,* they were given quadruple-expansion engines of 20,000 horsepower. The *Campania* and *Lucania,* with 30,000 horsepower, cruised at more than a knot faster. The American liners also steamed into a peculiar string of initial problems. On her second voyage, the *St. Louis* developed a deep crack in her rudder, from the post just above the upper pintle extending downward in a lateral direction. "Instead of breaking a record, the ship broke her rudder," the *New York Times* jibed. For the final four days of the passage, she was kept on course by constant manipulations of the twin screws. Late in 1895, as the *St. Paul* lay at her dock in New York, the main steam pipe burst in the engine room, killing nine men. A month later, while racing the *Campania* into New York, the *St. Paul* got lost in a dense fog and ran hard aground on the New Jersey shore near Long Branch, far off course. (The American Line of course denied any racing.) All hands were saved, and tugboats eventually pulled the ship—heavily laden and planted deeply in the sand—out to sea with no major damage.

From these dubious starts, the American liners went on to long, respectable careers. Their speeds improved after refittings, and both ships repeatedly set new records for the course between New York and Southampton, down to six days, ten hours, and fourteen minutes eastward and six days, thirty-one minutes westward. In the summer of 1897, the *St. Louis* fell in with the *Campania,* running out of Queenstown, on a passage to New York. The American ship overtook the crack Cunarder and beat her to the Fire Island lightship by ten minutes. ("Heave over a hawser," shouted a passenger on the *St. Louis,* "and give the Englishman a tow.") Henry Adams, switching from the aging *Teutonic,* settled on the *St. Louis* as his latest favorite way to cross: a patriot despite himself. Bob Hope emigrated from England on the *St. Louis* as a five-year-old boy in 1908. He remembered standing up on deck—dressed in knickers and a cap, with his nose running—as she entered the harbor in the cold early morning. He beheld the Statue of Liberty and the lights of the city. Hope's signature theme song, "Thanks for the Memory," sometimes would remind him of that vivid moment.

With the American Line as his prominent base, Griscom drove toward his grand design. The audacity of the scheme depended, first, on a friendly political situation, then on financial resources of a scale far beyond the usual run of Atlantic steamship transactions. For politics, Griscom's best resource was Senator Mark Hanna of Ohio, the brains and money behind President William McKinley. For capital, Griscom would connect with J. Pierpont Morgan himself, the most powerful banker in the world. Griscom could not have found more potent allies.

Griscom and Hanna shared practical interests. Hanna had made his fortune in Cleveland by building ships that carried coal and iron ore across the Great Lakes to the Pennsylvania Railroad. (Many of these shipments wound up consigned to Griscom's transatlantic steamers in Philadelphia.) Though Hanna retired from business when he became a political boss, shipping remained a preoccupation for him. On a European vacation in 1899, he was appalled by the foreign domination of ocean steam. "It's just a shame!" Hanna exclaimed. "Where the hell are our ships?" The trip made him a relentless advocate of congressional subsidies for American merchant shipping. "My presentation of the Merchant Marine question," he urged McKinley, again, after a political tour in the fall of 1899, "in connection with trade expansion in the east took with the farmers as well as the workingmen everywhere—Don't forget a favorable notice in your next message."

Griscom now had a friend at the highest level of national politics. The general climate in Washington also favored his plans; the McKinley administration did not object to a late-1890s surge of mergers and combinations, as many industries were consolidated into fewer, bigger companies to rationalize markets, cut costs, and eliminate competition. In prosperity and approval, the Republican convention gathered at Philadelphia in June 1900 to nominate McKinley for a second term. For the week beforehand, Hanna stayed with the Griscoms out at Dolobran, holding meetings and pulling strings. The party platform did not neglect the merchant marine. "I know you can not fail to be gratified with the harmony and good-will that prevailed," Griscom wrote the president afterward, "and no one contributed more to that than our mutual friend, Senator Hanna, whose masterly skill was a marvel to me, notwithstanding I thought I knew him intimately well, but I never before had been alongside of him under such circumstances." McKinley swept to reelection in November, apparently presenting Griscom four more years of an agreeable administration.

Pierpont Morgan, in his sixties, was entering the final episodes of a dazzling career. His power to manipulate the American banking system generally exceeded that of any president in his lifetime. Most celebrated for his control of railroad lines, mixing and matching systems as he wished, he switched to steel when it replaced railroads as the most important industry in the United States. In 1901 he bought out Andrew Carnegie's holdings and created United States Steel, the newly anointed largest corporation in the world. Less wealthy than many of his more destructive business contemporaries, Morgan succeeded on the hard work of his bright young partners and his own reputation for smart, quick decisions, made without advice. "A man of whom it is said," as Albert Ballin noted, "that he combines the possession of an enormous fortune with an intelligence which is simply astounding."

In any situation, grand or trivial, Morgan got his way. "He must, by nature, be absolute dictator or nothing," Ray Stannard Baker concluded his magazine profile. In conference at the White House, or considering the purchase of another rare painting or manuscript, pursuing his next mistress, or just booking a cabin for one of his frequent Atlantic passages, Morgan expected simple obedience. "I took it for granted that I had the Captain's room on the 'Britannic' for the 16th April," he once wrote the White Star Line's New York agent. "As you are well aware, myself and party have had the same room on the same trip for 4 years; and under the circumstances, I think that independent of any application from us, it should not have been given to anyone else without consultation. . . . If I cannot have that room, I prefer not to go in the ship." Morgan got what he wanted and sailed as scheduled.

After mulling it for years, Griscom started putting together his ambitious steamship trust in 1899. To build six large new ships for the American Line, he arranged a mortgage bond issue with Morgan's Philadelphia subsidiary. That apparently marked the first major business contact between the two men. Later that year, Griscom and Morgan conferred at Aix-les-Bains in southeastern France, the banker's favorite European spa. The substance of their talks is unknown; but by the fall of 1900, with the help of Morgan's firm, Griscom was merging his steamship operations with the Atlantic Transport Line of Baltimore (the second most important American steamship company on the Atlantic)—along with two other lines, as yet unidentified. One of those lines, it soon transpired, was the large British Leyland Company. For the Leyland purchase, the house of Morgan advanced $11 million in cash from its own money, thereby linking itself more seriously to the enterprise.

The glittering prize of the White Star Line came next. Thomas Ismay, founder of the line, had died in 1899, and he was succeeded by—alas—his doofus son Bruce. William James Pirrie, chairman of Harland & Wolff (the line's shipbuilder) and the holder of the second largest block of White Star stock, could not bear to contemplate a future of working with Bruce Ismay. So Pirrie maneuvered the line into a deal with Griscom and Morgan. By June 1901 the Americans had agreed to buy the White Star Line—the most profitable steamship company on the Atlantic and second only to Cunard in historical importance—for the generous price of $32 million. For the time being, the transaction, so laden with political and commercial significance in both countries, was kept as quiet as possible. The White Star leadership came along with the package; Bruce Ismay took his seat at the table.

Still connected to his railroads, Morgan tried to arrange their cooperation with the steamship trust. A "through bill of lading" could offer shippers special rates, from inland points by rail to the ocean and then across it. With Leyland and White Star in hand, Morgan steamed home on the *Deutschland* and met with the leaders of the Pennsylvania Railroad, the biggest line in the United States. He explained that he wished to form a syndicate of $20 million drawn in three equal parts from the steamship lines, the house of Morgan, and those railroads especially interested in transatlantic traffic: the Pennsylvania, Vanderbilt's New York Central, and the Erie roads controlled by Morgan. Alexander Cassatt, president of the Pennsylvania, rejected the overture ("We had had a rather bitter experience in the steamship business," noted one of his vice presidents). At a second meeting two days later, Morgan said, "Why, Mr. Cassatt didn't understand me. In the first place, I spoke of twenty millions, but I think ten millions is quite sufficient. But I do not know that I want any money, that is not what I am after, what I want is the assurances that the Vanderbilt, Pennsylvania and Erie will join me in an effort to promote such a scheme. . . . What I want to do is to assure myself that a few important people controlling important interests will work with me to bring about the establishment of this enterprise."

In a revealing bow to Morgan's ego and power, everyone else in the room—Griscom, S. M. Prevost of the Pennsylvania, and several Morgan partners—acted as though that was what Mr. Morgan had actually said two days earlier, though they all knew that in fact he had quite reduced his proposal after Cassatt's rejection of the first overture. Prevost asked Morgan, in order to avoid any future "misunderstanding," to summarize the conversation in a memorandum. Morgan refused. Perhaps, then, Morgan's partners could do it. "No," said Morgan, "I will tell you what I will do, you make a memorandum showing your understanding and send it over to me and I will correct it." Prevost did as told. Cassatt still resisted any further steamship ventures. "It is not so much a question of financial responsibility," he wrote Griscom, "as one of policy, and I still feel it would be a mistake for railroads to take any part in promoting the undertaking."

Griscom later confided to Prevost that he realized Morgan had changed his proposal, but did not speak up at the meeting "because he thought it was more prudent to avoid any further discussion of our misunderstanding." The episode reflected a subtle shift in the top leadership of the steamship trust: from Griscom to the imperial Morgan, who so insisted on dominating any circumstance. When the International Mercantile Marine was formally announced to the public in the spring of

1902, the American press generally called it "the Morgan combine" or "the Morgan trust." Control passed from the ship men to money men. As president of the IMM, Griscom would still run the merged firm and explain its goals. "Our object is to try to give a better transatlantic service at a decreased cost," he said. "Through the magnitude and diffusion of its business, such a company can guarantee a reasonable stability of rates." But Griscom was inevitably swallowed into Morgan's blazing sun. (Later historical accounts of the IMM would, in general, slight the founding ship magnate in favor of the legendary banker.)

The IMM's own stability was threatened, in private, by Bruce Ismay's usual erratic behavior. Pirrie had pushed him to accept a merger he did not want, and he continued in small ways to sabotage the venture. Clinton Dawkins, the head of Morgan's London office, warned a New York partner in the fall of 1902 that Ismay was "torn by two opposing forces, desire to get the 'consideration', and reluctance to give personal control of a business built up by his father. As time went on the latter force very nearly mastered him. His conduct resembled that of a lunatic, and if you wish to be charitable you had better assume that the internal struggle affected his brain. He really wished, I believe, to find a pretext for breaking off." At one meeting he declared the deal dead, left the room, and slammed the door. His attorney, looking scared, was sent to retrieve him. He returned in twenty minutes to insist on some new details. Dawkins mused that two of his exasperated associates might wish to shoot Ismay. "He is a calamity; possibly not quite a sane one." Later they went to see a banker about extending an overdraft. Dawkins had warned the man that Ismay might do something strange. Ismay, sure enough, said that he doubted any bank had the right to loan on such security. That was his business, the banker replied, and granted the request. "My arms are nearly worn off," Dawkins complained, "with whipping my extraordinary friend Ismay round the post."

None of these problems reached the public. To outsiders, the IMM looked like a massive, irresistible force. It engaged in extended merger talks with the two German lines, then finally settled on agreements to cooperate in setting rates and schedules. It had almost managed to buy the Cunard Line as well. When combined with the recent rise of the American Line, the record-setting flyers built and run in Germany, the enormous presence on the Atlantic of the Hamburg-American and North German Lloyd companies, and the White Star and Leyland lines snatched away into American ownership, the IMM left British transatlantic interests feeling besieged and reeling. As William Forwood, a director of the Cunard Line, later recalled, "It appeared as if the whole Atlantic trade was destined to pass into the hands of the Germans and Americans."

Circumstances forced the British government to bold, desperate action. In the fall of 1902, the Admiralty gave the Cunard Line a huge subsidy to build two new liners to beat the world. For a loan of up to £2.6 million, at only 2.75 percent interest, Cunard would provide two greyhounds of 24 to 25 knots. The Admiralty would then pay Cunard £150,000 a year to maintain the ships. The question of how to design and construct flyers of such speed was left up to the line and its builders. At that level, each additional fraction of a knot required grand vaults in power and boiler capacity. The current champion, the North German Lloyd's new *Kronprinz Wilhelm,* needed 36,000 horsepower to reach 23.5 knots. For the subsidized fast Cunarders, perhaps three sets of quadruple-expansion engines, pumping out a total of 59,000 horsepower, would be required. Or maybe a new kind of steam engine.

*C*harles Parsons grew up at his family's ancestral Birr Castle in Ireland, the youngest of six brothers. Their father, the third earl of Rosse, invented an improved reflecting telescope that bore his name. He also sat in Parliament and presided over the Royal Society, so the family spent two months of the year in London. Charlie was a shy, strange boy, very quiet, overawed and overpowered by his five older brothers. To support Lord Rosse's tinkerings, Birr Castle had elaborate workshops with wood-turning lathes and other fine tools. Charlie and his brothers built their own steam engine to run a lens grinder. In season, the family kept yachts on the English Channel, at Cowes or Southampton. From an early age, Parsons inhabited a world balanced between machines and boats, inventions and water. As a country gentleman, raised on landed wealth and timeless certainties, he didn't have to stray far beyond this fixed background to produce his astounding new marine engine.

Parsons went to Dublin to attend Trinity College, then on to St. John's College at Cambridge University for mathematical courses. Studious and silent, modest and retiring, he left no strong impression on his classmates. In the 1870s the university, still emphasizing a classical curriculum, had no engineering school. Parsons attended lectures on mechanism and applied mechanics and, on his own time, filled his room with models of inventions. After graduation he apprenticed for three years, under the eminent mechanical engineer William Armstrong, at the Elswick Works in Newcastle. (Parsons invented an epicycloidal compound engine with four opposed cylinders.) Next he briefly tried an experimental shop in Leeds, working on a balky system of rocket propulsion for torpedoes, before landing at a firm in Gateshead, across the Tyne from Newcastle. There a problem of electricity (the next big thing) led him back to steam power (the last big thing): he was put to the challenge of generating electricity on ships with a steam turbine.

The reciprocating steam engine—the power behind the industrial nineteenth century—remained fundamentally inefficient because, to run almost any device, it had to convert lateral motion into rotary action. The great mass of the piston and shaft went in, stopped, and reversed, came out, stopped, and reversed—a constant, punctuated, wasteful interchange of weight and inertia. The shaft then, by a noisy, complicated system of cams, crankshafts, rods, and linkages, propelled a circle which ran the machine. These clanking, contrary forces wore out bearings, required constant lubrication and watchfulness, and were thought to cause the deep, penetrating vibrations at sea—more noticeable as ships and engines kept getting bigger—that so disturbed the sleep and equanimity of transatlantic travelers.

A turbine engine was another ancient gadget whose practical application to complex modern machinery had, as yet, eluded inventors. Related in concept to the water wheel and paddle wheel, a turbine directed some non-solid force—air or water or steam—against a contained but moving object to create direct rotary motion. This simple, continuous linking of energy and spin eliminated the inherent inefficiencies of a reciprocating engine. By 1850 over thirty patents for steam turbines had been granted. Yet the device was still just a tantalizing toy, an enticing theory constrained by stubborn difficulties in practice. Its motion seemed too unusably fast; it needed too much steam and fuel; its machinery required precisions and tolerances beyond the reach of mass, economical manufacture. And, perhaps most seriously, its striking differentness from familiar kinds of steam engines collided against vast, entrenched industrial structures of material and belief.

In 1884, only months after first turning his attention to turbines, Parsons came up with the essential improvement. Within a tubular enclosure, he ran the steam through a series of rotating blades arranged perpendicular to the vapor. As the steam pressure moved along and diminished, it turned progressively larger blades, producing an aggregate motion that was both slower and more powerful than from other turbines. Like a compound reciprocating engine, it used a given quantity of steam repeatedly and efficiently; so Parsons called it a compound turbine engine. He started his own company in 1889 at Heaton, about two miles from the center of Newcastle, to manufacture dynamos and turbines for use on land. The inventor's personality, so at ease with machinery, was never well suited to succeeding at business. His handshake was "curiously vertical, limp, and undemonstrative." In conversation, his attention would wander as he gazed up at the sky, apparently tuned out. His judgment was often impulsive and unreliable. Tall and lean, gentle and abstracted, he looked like a poet or scholar, not a man who spent his life in the din and grime of a machine shop. Yet he worked all the time, happily absorbed at his desk or workbench. "He has not looked upon commercial success as the only end," explained his friend and fellow engineer John Harvard Biles.

When Parsons built the first condensing turbine engine, in 1892, it moved his invention on to other purposes. After the steam performed its useful work, it was converted to water and returned to the boiler: the closed system that had made transatlantic steamships possible. The new Parsons engine was smaller and lighter than a triple-expansion reciprocating engine of the same power. Because ship designers always wanted to save space and weight, Parsons sighted a rich new market—which needed much larger engines than the municipal dynamos he had been building. Nearing his fortieth birthday, Parsons launched another firm, the Marine Steam Turbine Company, with his oldest brother and a half-dozen associates. To produce a prototype vessel, he first made two-foot and six-foot models. The goal was to attract attention with the fastest boat anyone had ever seen. His unfettered engineering curiosity considered a different type of boiler ("avoid as far as we can," he advised himself, "all experiment excepting the main issue of increased speed") and an improved screw ("we are following the ordinary lines, only making it a wee bit finer pitch").

The first turbine-powered vessel, the *Turbinia,* was 100 feet long and 9 feet wide, with a displacement of 44.5 tons. On a preliminary trial in the fall of 1894, she ran down to Tynemouth and back at a top speed of nearly 20 knots, generating 300 horsepower with—according to Parsons—"an almost complete absence of vibration." But the propeller spun so quickly that it created a vacuum, called cavitation, and lost purchase on the water. To study the problem, Parsons put a model screw and shaft in a tank of water. He heated the water to just below the boiling point, so that increased vapor pressure would simulate actual conditions even at lower speeds, and lit the screw by an arc lamp reflected from a revolving mirror on the screw shaft, so the cavity could be seen as though stationary: a rudimentary strobe light. Over the next two years, Parsons tried a larger screw, different kinds of screws, and multiple screws; nine different sets of propellers in all. He also tore out the machinery and installed a new engine, then several engines. Finally he settled on three engines, three shafts, and three propellers on each shaft. In the spring of 1897, turning 2,000 total horsepower, the *Turbinia* ran a measured mile at a scorching 32.6 knots.

The engines represented elaborations of Parsons's basic improvement of 1884. Within the containing structure, a series of inwardly projecting guide blades was attached to the inner surface, surrounding and nearly touching a light steel drum mounted on the shaft, which ran on bearings at both ends. Alternating with the guide blades were outwardly projecting rotor blades attached to the drum. Steam entered at the forward end and, in series, was aimed inward onto the rotors, then outward onto the guides. As it passed from ring to ring, the steam fell in pressure and expanded into steadily larger rotors. The turbine shaft, from the exhaust end, was coupled directly to the screw shaft. At high speed, the engine emitted only a steady hum or roar—"like a very strong gale of wind," Parsons suggested. Instead of clanking, it thrummed. An observer standing next to it, sensing no vibration, had to touch the exterior surface to feel the engine working.

"If you believe in a principle, never damage it with a poor expression," Parsons said later. "You must go the whole way. I had to startle people." So he contrived a bit of theater on an exceptionally visible stage, a Grand Naval Review—part of Queen Victoria's Diamond Jubilee—at Spithead in June 1897. Between two long lines of cruisers and battleships, drawn up in panoplied display, the inventor and his *Turbinia* crashed the party without permission. "At the cost of a deliberate disregard of authority, she contrived to give herself an effective advertisement," the *London Times* reported. "Perhaps her lawlessness may be excused by the novelty and importance of the invention she embodies." The bow of the thin, low, brick-red hull rose high from the water as she came to speed. The stubby funnel threw up a twenty-foot plume of flame and smoke. On the assembled warships, the officers and crews, royal observers, and Admiralty brass watched a marine spectacle unlike anything they had ever witnessed. A small picket boat, sent to intercept the raider, could not catch her or stop her. Looking "like a whale with a man on his back to steer," she dashed around and through the lines of ships. "Here, there and everywhere the little craft flew through the water," wrote a witness, "demonstrating to all the possibilities of the future."

All but overnight, the obscure Parsons was now courted, and customers came calling. In the next two years, the Admiralty ordered two turbine-powered torpedo boat destroyers, the *Cobra* and *Viper,* vessels of 10,000 and 13,000 horsepower; Parsons predicted they would reach thirty-five knots. A Glasgow steamboat owner ordered the first turbine-powered passenger vessel, the *King Edward,* for service on the Clyde. "It would be difficult to recall any instance in which like results have been attained so rapidly, and with so little of trial and error in the process," said the journal *Engineering.* Parsons had strikingly joined pure scientific knowledge to sharp, practical skill. "He broke most of the accepted rules of the marine engineer, and traversed the lessons of half a century, and yet he produced a boat which was, and still is, the fastest vessel in the world."

Then a double blow. In August 1901, muffled in a Channel fog, the *Viper* ran aground on a rock near Alderney and was wrecked. Everybody was saved. Just six weeks later, the *Cobra* encountered heavy seas off the coast of Lincolnshire, broke in half, and sank. Only twelve of the seventy-nine men on board survived. Neither disaster could fairly be blamed on the Parsons turbine engine; the *Viper* simply went astray, as any vessel might, and the long, thin hull of the *Cobra,* not strong enough for violent weather, was apparently lifted by high waves at the bow and stern so that she sagged and broke her back. But because of their novel machinery, both ships had been closely watched. In the aftermaths, the gross, blaring fact of their losses, one so soon after the other, registered more heavily than the detailed, excusing causes.

Not discouraged, Parsons rather quixotically hoped to sell his engine to the Cunard Line. In compiling its unique record of safety and stability, Cunard had always adopted new technologies quite slowly, preferring to watch competing companies try them first. After a business pitch from Parsons, the line's veteran superintendent engineer, James Bain, visited the Marine Steam Turbine works on the Tyne. He also took passage on the Clyde steamer *King Edward* from Greenock to Campbelltown and back. Bain's report induced the Cunard board of directors to invite Parsons to its regular meeting in December 1901. He explained the potential advantages of turbines for large cargo and passenger ships and then submitted his terms for a liner of 725 feet and 48,000 horsepower: bigger and more powerful than anything on the ocean. The board decided against it. When the Admiralty agreed to underwrite the two new Cunard flyers in the fall of 1902, Parsons tried again. He was informed "that the Board does not see its way to adopt Turbine Machinery."

*C*unard asked for tenders from four shipbuilders. Two were Scottish (the Fairfield Shipbuilding and Engineering Company and John Brown & Company, formerly the firm of James and George Thomson), and two were English (Vickers, Sons & Maxim of Barrow and Swan, Hunter & Wigham Richardson of Newcastle). All but Vickers had previously built major steamships for the Cunard Line. The initial specifications called for two vessels, 750 feet long and 76 feet wide, with three screws driven by reciprocating engines of 59,000 total horsepower. The builders generally agreed, at the outset, on the difficulty of making such a vessel strong enough. Yet as the planning process continued over the next eighteen months, all the first specifications were knocked upward, making the structural challenges even more acute. Both national pride and the future of the Cunard Line were riding on design and building puzzles of unprecedented ambition and complexity.

George Arbuthnot Burns, the second Lord Inverclyde, had been chairman of the Cunard Line for only a year. Forty-one years old in the fall of 1902, he had taken office ten months after the death of his father; the late John Burns, the first Lord Inverclyde, was no easy act to follow. A large, dominating personality, awarded the title for his achievements, he led the company through three decades of ferocious competition on the North Atlantic. The previous record of sons succeeding major fathers at the head of transatlantic steamship companies—after the death of William Inman, for example, or when Bruce Ismay bumbled into power at the White Star Line—was not encouraging. But according to the memoirs of Cunard director William Forwood (written years after the death of the second lord), he soon proved himself "a man of conspicuous ability, with a big grasp of affairs," handling the delicate negotiations for the crucial Admiralty subsidies with "much energy, tact, and knowledge of shipping." At a difficult time, Cunard had the leadership it needed, in direct descent from the founding triumvirate of 1839.

Lord Inverclyde and the Cunard board settled on Brown and Swan to build the two ships. A decision complicated by politics, it split the contracts between Scotland and England. Various hull forms were tested at the Admiralty tank in Haslar, the tank of William Denny and Brothers at Dumbarton, and at Brown's own recently built Clydebank tank. These extended procedures yielded a single set of external dimensions for the two ships. Each builder was allowed more freedom with the insides: the arrangements of rooms, furnishings, and decorations. For the most vital element of all, the engines, the discussion wound back to turbines.

James Bain, recently promoted to general superintendent of the Cunard Line, started the rethinking process. The new turbine steamer *Queen,* built by Denny for the Channel traffic between Dover and Calais, had just started running. At 310 feet and 8,500 horsepower, she had reached 21.7 knots in her trials, well beyond the contracted speed. In July 1903 Bain tried her to France and back. His enthusiastic report, along with a letter from Charles Parsons and a supporting statement by Edwin W. De Rusett of the Swan shipyard, all endorsed the new technology; the engineers were nudging the civilians toward turbines. The Cunard leadership appointed a special commission to study the matter. Its makeup implied that the decision was already made: of the seven members, five— Bain, William H. White of Swan, Thomas Bell of Brown, Henry W. Brock of Denny, and Rear Admiral H. J. Oram of the Admiralty—were known turbine advocates. And Parsons hovered nearby, offering suggestions and assistance.

The commission compared the performance of the new Channel turbine boat *Brighton* against her conventionally powered sister. It examined dynamo turbines at various city power stations and at the Neptune works on the Tyne. "Every point was calculated and tested, and Mr. Bell did not take anything for granted," Parsons said later. "He constructed a set of marine turbine engines at Clydebank, coupled with electrical generators as dynamometers, and from these he worked out the efficiencies." The commission's final report, closely held, was leaked and published by the *Glasgow Herald* in March 1904. The summary said that, despite expectations, turbines would not save much space or weight in the new Cunarders. They would consume vast quantities of coal, over 1,000 tons a day, and run wastefully at low speeds. But at top speed—which the new flyers would obviously maintain all across the ocean—they would need less steam and coal than reciprocating engines and require fewer men in the engine room.

Speed, as ever, carried the day. The final specifications called for ships 790 feet long and 88 feet wide, of 32,500 tons, driven by four screws and four turbine engines humming out 68,000 horsepower. All these dimensions leaped far past any other vessel, either running or planned. Swan would build the *Mauretania,* named for the ancient Roman province in northwest Africa, across from Gibraltar. Brown would build the *Lusitania,* so called for the Roman name of Portugal. (Such names intently connected the ancient empire with its modern British counterpart.) In a final revision, aimed at keeping up with the Germans, they would be the first British ships topped off by four smokestacks: an outward, coveted, unmistakable sign of transatlantic status and power.

The Sheffield steel fabricating firm of John Brown had bought the Thomson shipyard in 1899, for just under £1 million. The transaction merged Brown's large Atlas works in Sheffield (makers especially of engine shafts, guns, and armor plate) and Brown's 7,000 subterranean acres of coal mines in Rotherham with the famous Clydebank shipbuilding facilities: steel, coal, and ships, together in one symbiotic enterprise. John Gibb Dunlop, who had built the *City of New York* and *City of Paris* for the Thomsons, now served as managing director of the Brown shipyard. Thomas Bell was director of engineering, and William J. Luke was the naval architect. This team designed and built the *Lusitania*. To produce such a behemoth, the yard had to remake itself by spending £6,500 on a new gas plant, £6,500 on a new electrical plant, £8,000 to dredge the Clyde, £18,000 to extend its dock, £19,000 for a crane to lift 150 tons, and almost £20,000 more on other machinery and equipment.

Construction began in the fall of 1904. No previous ship had so resembled a building in scale and space; so Cunard broke another of its traditions. Over the years, the line had occasionally subcontracted some ship decorations to landbound specialists such as Wylie & Lochhead of Glasgow, who did most of the public rooms for the *Campania* and *Lucania* while Fairfield still executed the main saloons. For the grand new liners, the Cunard directors wanted to cede major roles to building architects with little previous ship experience. Lord Inverclyde, as the son and grandson of Cunard leaders, wasn't sure. "My inclination is to leave matters very much in the hands of the Builders, as we have done in the past," he wrote, "but I know that the Board are inclined to try something else. . . . I do feel that it is somewhat of a risky experiment." After winnowing a list of eight architects, amid much deliberation and fretting, Cunard gave the *Lusitania* to James Miller, known for his designs as the architect of the Glasgow Exhibition in 1901. He had never worked on a ship before. Miller was urged to stress simplicity and lightness by avoiding extravagance—and to tour the recent German ships for both inspiration and warning.

The point of seagoing interior design, always, was to make the ship not look like a ship. For this broad goal, the dry-land background of Miller may actually have helped his *Lusitania* work. She was given the prettiest interiors of any Cunard liner yet. The veranda café at the rear of the boat deck, light and airy, had rattan furniture, latticed wall patterns in the style of the Glasgow architect Charles Rennie Mackintosh, hanging plants against the high, skylit ceiling, large rectangular windows placed low enough to allow ocean vistas to seated travelers, and an open wall at the back, bringing the outside inside (though vagrant seas did not always cooperate). Forward from the café, past the smoking and music rooms, the lounge lightened its late Georgian style with satinwood furniture and subtle greens in the carpets, cushions, and draperies. Here Miller used enameled panels by Alexander Fisher of London over the two marble fireplaces, and twelve monthly stained-glass panes by Oscar Patterson of Glasgow along the extended dome that rose twenty feet above the floor. The two regal suites, on the promenade deck, each included a dining room, pantry, drawing room, lavatory, and two bedrooms. The regal suite on the port side invoked Louis XVI, with details borrowed from the Petit Trianon at Versailles. The furniture was Italian walnut and gold, the walls and ceiling white and gold. The chimney piece of the dining-room fireplace was sculpted from Fleur-de-Pecher marble. One bedroom featured rose-colored silk panels; the other, Wedgwood cameos. Even in second class, a few worlds away, the Georgian dining saloon and smoking room and the Louis XVI drawing room paid lavish, intricate homage to classic designs.

Electricity—the coming force that had first led Charles Parsons to turbines—ran all through the *Lusitania*. Four turbine-driven generators lit 6,000 lamps spread around the ship, connected by some two hundred miles of cables. In the kitchens, electricity boiled eggs, cut bread, peeled potatoes, sliced meat and sandwiches, and washed dishes. Electric power ran the ventilation and refrigeration, detected fires and steered the ship, turned winches, and closed the watertight compartment doors. ("It may be claimed for her," said the Cunard Line, "that she is as unsinkable as a ship can be.") Two electrically powered elevators, the first on a British liner, shuttled passengers among the five decks of major public rooms and cabins. The black-and-gold metalwork of the elevator gates and frames, and the polished mahogany woodwork, stood out smartly against the ornamented whiteness of the passing halls, floors, and ceilings.

The *Lusitania*'s trials in the summer of 1907 revealed the most unexpected of problems. She ran fast, averaging up to 25.3 knots, and all the machinery operated well. But at high speed, vibrations could be felt, especially in second-class cabins above the shelter deck and in some first-class staterooms. Turbine engines were supposed to eliminate such tremors; it turned out they could also be caused by propeller resistance and turmoil in the water, factors unrelated to the whirlings in the engine room. The ship returned to Liverpool for thirty tons of strengthening in certain pillars and beams. While the work was being done, no visitors were allowed onto the *Lusitania*. Superintendent Bain took her out, found less vibration, but had to report that "the result is not at all that might be desired." Passengers later would complain of vibrations in the lounge and upper level of the main dining room, and in other areas toward the rear of the uppermost two decks. Replacing the three-bladed outside propellers with four-bladed screws finally subdued the irritation, but only up to a point.

The ship was bred for speed, and there she delivered. On her second voyage, in the fall of 1907, the *Lusitania* broke every record for the fastest hour, day, and passage in transatlantic history. For the first time in ten years, the Germans were thrashed and a British vessel reigned as the swiftest ship on the Atlantic. She flew to New York in four days and twenty hours, then home in four days and twenty-three hours—the first crossings in under five days. "The lion," said a passenger, "is again master of the ocean highway."

*T*he *Lusitania* departed New York on May 1, 1915, for her 202nd trip across the ocean. In a perilous time, she was still carrying 1,257 passengers, 702 officers and crew, and three stowaways: nearly 2,000 people in all. Eight months earlier, the tightening rivalry between Britain and Germany—so vigorously fought over steamships and record passages—had assumed a more grotesque form than anyone could have imagined. The boggling battlefield casualties were already piling up toward a stalemate. At sea, the British blockade of Germany was met by the German declaration of submarine war against all enemy vessels, even merchant liners. A submarine could not warn a merchant vessel of its intent to attack, or wait for an evacuation, or pick up survivors. The old rules for such encounters were off. Instead it was assault by stealth, with no distinction between civilian and military targets—a particularly twentieth-century kind of total war.

To save coal in wartime, the *Lusitania* was running the boilers under just three of her four smokestacks, cutting her speed from twenty-five to twenty-one knots. She was still supposed to be fast enough to outrun any U-boat. As the ship approached the submarine zone that Germany had thrown around Great Britain, a thrill of dread and anticipation rippled through the passengers. What would they encounter? Was it dangerous, or just exciting? Some slept in their clothes, ready for anything. On the bridge, the ship's officers read the Admiralty's warnings, received by wireless, of submarine activity off the coast of Ireland; but the reports inspired no particular concern or evasive action. The *Lusitania* turned parallel to the Irish coast, ten miles off the Old Head of Kinsale, precisely into the sight of a waiting submarine.

The torpedo hit with a heavy, muffled sound, followed rapidly by a larger second explosion that could never be explained. It was a sparkling, sunny day, the sea and sky in gleaming tones of blue, the surface flat and smooth. From somewhere deep in the ship, a column of water shot up through a funnel, followed by showers of debris. With a general awareness of the submarine danger already so widespread through the *Lusitania,* almost everyone knew at once what had happened. The ship quickly listed forward and to starboard. Passengers and crew surged up on deck. The desperate chaos of a dying liner once again played itself out: hopeless attempts to launch lifeboats from the tilting deck, struggles over life jackets, many small dramas of heroism and selfishness, the hasty decision under pressure to stay on the sinking ship or to try the uninviting sea. Whatever was happening down below, it meant a virtually immediate loss of steam and electric power. The modern *Lusitania,* so run by electricity, could not readily close her watertight doors. The lovely elevators stalled between decks, trapping people inside; they screamed and pounded, but nobody could help. The elegant black-and-gold metalwork turned into condemning prison bars.

The bow started to go under. The screws and rudder rose out of the water. Parents threw children from the sharply angled deck down to lifeboats below. The *Lusitania* went into a slow, dignified slide downward at about forty-five degrees to the surface. When the bow hit bottom, the hull pivoted slightly and disappeared—eighteen minutes after taking the torpedo. The cacophony of screams and cries was suddenly stilled. Something exploded down below, like an underwater volcano, sending up a churning, foaming mass of bodies and wreckage. The scene subsided. The water was very cold, perhaps fifty-two degrees; it soon quieted the survivors who were floating about and clinging to random shards of wreckage. The final toll, including the three unidentified stowaways, counted 1,201 deaths.

It is a sad, familiar story, the second most notorious disaster in transatlantic history. The sinking of the *Lusitania* was actually freighted with more significance than the *Titanic* tragedy, which was only an especially horrific accident. The swift Cunarder was sunk on purpose, with evil intent, as an act of war. The shock and recriminations that followed caused political earthquakes in Britain and the United States. It also marked the end of seventy-five years of Cunard luck. For the first time, the line had finally suffered passenger deaths by shipwreck on the Atlantic—and on a vastly more terrible scale than the nineteenth-century catastrophes endured by Junius Smith, Edward Collins, William Inman, Adolph Godeffroy, and Thomas Ismay. When luck ran out, the penalty was doubled and tripled.

*T*he *Mauretania* was the greatest steamship ever built. On the roll of North Atlantic steamers, no other ship so combined technical innovation, speed, beauty, profitability, and longevity. Another great liner might have surpassed her by one or two of these criteria, but none was so uniformly distinguished in all five ways. In addition, the imponderable caprices of ocean travel favored the *Mauretania* during her long career. She was considered a lucky, happy ship, encountering no major disasters, amazingly consistent and predictable in her passages. She almost restored the tradition of Cunard luck that her doomed sister had shattered.

The *Mauretania* was also Andrew Laing's revenge. The man who designed and built the engines of the Fairfield greyhounds for Cunard and the North German Lloyd, the most brilliant engine specialist of his time, could not tractably work with human beings. Even the fine qualities perceived by his friends—the "keen, eager and virile personality," his "vigour and independence of thought," and especially "the single-minded devotion to his life's work"—might, when hardened or irritated by intense shipyard pressures, make him an impossible colleague. After the death of his mentor, William Pearce, in 1888, Laing fought with a series of Fairfield bosses. "He had but little use for personal success," said an admirer. "His joy lay in doing well the work." Clashing with a new ship-yard manager, he was removed from the Fairfield board, then fired as manager of the engine and boiler departments, and finally—in October 1895—fired altogether for, he was told, "obstinacy" and "the great friction you are causing."

Laing switched to the flourishing shipbuilders of the River Tyne, on the northeast coast of England. Hired as general manager of the Wallsend Slipway and Engineering Company, he quickly attracted business from his old clients at the Cunard Line. In a half-dozen years he crafted the engines for six Cunarders built by four different northeast shipyards. His *Ivernia* of 1899, constructed by Swan & Hunter, was—at 600 feet, 13,900 tons, and 10,500 horsepower—the largest ship yet built on the Tyne. Laing tripled his shop's production, extended and improved the machinery, bought two powerful electric traveling cranes and a large horizontal boring machine, and nearly doubled the building area under roof. As his own boss, answering to nobody, Laing found his place. When only four shipbuilders were invited to bid for the special prestige of producing the Cunard world-beaters, Swan and Laing wound up with one of the contracts, and Fairfield—also a bidder—did not. Laing must have especially enjoyed the news.

In a sudden rush, toward the end of the century, northeast shipbuilding had matured into parity with the best shipyards of Britain. As recently as 1881, James Rodger Thomson of Clydebank had complacently declared, "They cannot build first-class passenger steamers on the Tyne." Shipbuilding firms such as the Palmer brothers of Jarrow-on-Tyne were known mainly for cargo steamers, slow and broad of beam, economical to run but not technically ambitious. Swan & Hunter led the Tyneside builders into fast passenger liners. George B. Hunter, the son and grandson of local shipbuilders, also apprenticed under William Pearce during his tenure at Robert Napier's yard on the Clyde. ("Two very interesting and fruitful years," Hunter recalled.) Northeast ship men started a productive tradition of exchanging ideas and talent with Scottish shipbuilders; Laing's arrival was part of this process.

In 1890 George Hunter hired a new naval architect, Edwin De Rusett, and announced his intention to build large passenger and cargo ships of the highest class. De Rusett, forty-eight years old, had been working as naval architect to the Peninsular and Oriental Line, the large British steamship company that ran routes to India and Asia. That De Rusett would shift to Swan & Hunter itself reflected the rising status of northeast shipbuilding. An early advocate of twin screws for ocean steamers, De Rusett shared Edward Harland's insistence that ship designers focus on how the oceangoing experience felt to passengers. "I have travelled a great deal in all sorts of liners for the purpose of finding out what passengers really need," he said. "I think where we naval architects are at a disadvantage is that we do not as a body travel enough. . . . We need to see absolutely what is going on afloat, to get amongst the passengers, speak to them and learn their ideas. . . . When designing put yourself in the place of those who travel and of those who work the ship."

By 1902, with De Rusett bringing his passenger-level perspective to issues of naval architecture, Swan & Hunter had become one of the major shipbuilders in Great Britain. The thirty-six-acre yard at Wallsend, employing up to 2,500 men, included two covered building berths 500 feet long. Its newest Cunarder, the *Carpathia,* would later become famous as the ship that raced to rescue the survivors of the *Titanic*. In 1903 Swan merged with the nearby shipbuilding firm of Wigham Richardson, doubling its facilities. Combined with the affiliated engine building talents of Andrew Laing, the Tyneside shipyard was amply qualified to win the contract that became the *Mauretania*.

For this most significant job, Hunter and De Rusett ran an elaborate series of model tests in the Northumberland Dock, on the north side of the Tyne. They built an unusually large model of the hull, 47.5 feet long, on a scale of 1:16 instead of the more typical 1:48. The model was propelled by four battery-driven electric motors, adjustable for speeds between 150 and 950 revolutions per minute, and four screws. To test its skin resistance, the hull was also towed from a dynamometer on shore. The tests went on for two years, through five hundred speed trials and multiple variations in the size, proportions, and placements of the four screws. No oceangoing vessel had ever used four propellers; the technical mysteries seemed endless. When all the data were sifted, they led to raising the propeller speed of the new ship from 145 to 175 rpms, placing the two forward screws seventy-nine feet ahead of the rear ones to minimize water disturbances, and increasing the distance of the screws outward from the hull. To reduce the effects of wind resistance on the superstructure—a new concern given the ship's expected speed—the deckhouses at the bow and stern were stepped back in layers, bottom to top, and the sides were rounded off: a kind of streamlining, seldom as yet addressed for ocean liners.

An architect named Harold Peto was, against all probabilities, engaged to work on the *Mauretania*'s interiors. Known mainly for building and decorating patrician English country houses, he had recently quit his firm and essentially retired. He had never done a ship. When Lord Inverclyde went to see him at his home in Yorkshire, Peto insisted that, if hired, he would not be supervised by the entire Cunard board. ("He said that everyone would be different and each would think that his taste was the best.") As a professional architect, Peto would only accept his usual commission arrangement instead of the flat fee that Cunard offered to him. "The sum is apt to look large," he instructed the ship men, "before one has got to work and the endless detail work unrolls itself, nor is it possible for the client (or architect for that matter either) to tell beforehand, all it will involve. The commission basis is on the whole I consider <u>fairer</u>, being universally recognized and having stood the test of time." Inverclyde worried about spending even more money on the new liners, but he did admire Peto's work; "as they are special ships," he decided, "I think they are worthy of being put into a good Architect's hands." Peto agreed to design and supervise the decoration of all the first-class public rooms and fifty of the best staterooms, and to appraise Swan's plans for the second-class accommodations—all for a commission of 5 percent, to yield at least £3,500 but not more than £4,000. "In conceding practically all the points raised by you," the Cunard secretary wrote Peto, "my Directors have taken into account your assurance that you will keep the cost of the work down as far as possible." (The architect later found more of his "endless detail work" and got additional money.)

Peto gave the *Mauretania* the look of an English country house, solid and sober, full of natural wood grains and shades of brown. She seemed less light and bright than the *Lusitania;* to modern tastes, not as handsome. Along with the Italian Renaissance styles that had dominated steamship interiors since the 1880s, Peto added eighteenth-century references to Louis XVI and the Georgian furniture and decorations of the Adam brothers of Scotland. The first-class smoking room, aft on the boat deck, featured generous polished-wood wall panels in subtle diamond patterns, surrounded by inlaid borders. The wagon-headed dome and ornamented plaster frieze, all in restful vellum, were bracketed at the ends by harbor paintings of "Old Liverpool" and "Old New York." The lounge, on the boat deck between the third and fourth funnels, was perhaps the prettiest room on the ship. Decorated by the Melliers of London, it had dark, burnished mahogany wall panels within gold moldings, and long ceiling beams held up by fluted columns of mahogany capped and based in gold leaf. Chandeliers of gilt bronze and crystal hung over the carpet of blush-pink roses on a pearly gray background. In the center of the ceiling, a circular dome was defined by tracery patterns of dark ironwork. The crisp contrasts of white, gold, and dark brown in such a large room, eighty feet long and fifty-three feet wide, left a bold impression, sharp and powerful but not overwhelming. In the library, on the boat deck just aft of the second funnel, the wall panels at first glance looked like marble ingrained with silver but actually were of sycamore stained a delicate gray. The outside walls of the lounge and library were bowed outward to allow bay windows—an unusual effect on a ship.

At the front of the deckhouse on the promenade deck, a closed gangway with windows allowed passengers an unobstructed view forward even in the worst weather. (It recalled the Social Hall on William Pearce's *Arizona* of 1879.) From there, looking past the cargo machinery toward the bow, travelers could thrill to the bucking, exploding sight of the *Mauretania* driving ahead at full speed into a North Atlantic headwind and storm. The promenades along the sides were enclosed, as on the *Kaiser Wilhelm der Grosse*, back to the third funnel, allowing walkers to take their constitutionals undrenched. The open promenade, one level up on the boat deck, was in theory protected by its height above the ocean surface. The ship was so long that she needed a second bridge, at the stern, to guide docking maneuvers. Yet size did not spoil her external looks. Seen from almost any angle, especially from a three-quarter vantage off the bow or stern, the *Mauretania* looked graceful and well proportioned, with none of the clunky blockiness of the British and American liners of the 1890s. The four raked funnels loomed high overhead, but within a pleasing scale to the rest of the ship. Along with the stepped-back deckhouses at the ends, and the narrow wedge of the bow, they imparted a sense of speed and flow even when the liner lay idling at anchor. She was gorgeous.

The seagoing town was split into three vertical neighborhoods for the passengers, with side streets for the different clusters of workers. Third class (the improved name for steerage) lived on the shelter, upper, main, and lower decks, between the foremast and first funnel. The dining saloon, roomy enough to seat 330 people at a time, was paneled in polished ash with teak moldings. The long tables had white tablecloths and revolving chairs; the large windows were paned and arched at the tops. Steerage diners were not accustomed to such touches. The men's smoking room and women's lounge, one deck up, were similarly furnished. Most of the steerage cabins slept four people; others had berths for two, six, or eight. The water closets featured door-action valves, so the toilet flushed itself whenever the stall door was opened—a helpful device for any passengers still adjusting to indoor plumbing. In all, the ship had 278 rooms for up to 1,138 third-class travelers.

First class stretched a long way, from the first funnel to the veranda café just in front of the main mast, far beyond the fourth funnel. The grand stairway, placed between the second and third stacks, was the busy heart of the ship, connecting five decks to the public rooms and cabins fore and aft. To save weight, the iron and bronze of the two elevators were replaced with the difficult new metal aluminum. ("The metal involved more trouble in the working than either iron or bronze," *Engineering* noted, "owing to its strange greasy nature.") Moving downward from the top, the library, lounge, smoking room, and café were on the boat deck. The promenade deck had nothing but staterooms, except for the crest of the dining saloon's dome. On the shelter deck were more staterooms, the nursery, lavatories, the upper gallery of the dining saloon, the ship's hospital, and quarters for the doctors and engineers. On the upper deck, more staterooms and bathrooms, the dining saloon (at the relatively serene center of the ship, to encourage placid digestions and avoid embarrassing scenes), and the galleys for first and second class. The main deck finished up with the final staterooms and bathrooms, oddly near quarters for the firemen and trimmers. The total of 253 first-class cabins and suites could sleep 563 passengers.

Second class had the stern, from the main mast back to the docking bridge. "In catering for this class of passenger," said the *Shipbuilder* magazine, "the Cunard Company may fairly claim to lead the way among the great steamship lines." The lounge reposed in its own house on the boat deck. Beneath it, the main staircase bisected the second-class accommodations. On the promenade deck, the drawing room lay forward of the stairs, the smoking room aft. The shelter deck had cabins (all sleeping two or four people) and bathrooms. The dining room was in front of the stairs on the upper deck. Quite fancy in its Georgian oak panels, it included an intricately carved cornice and parquet floor. A large octagonal opening in the ceiling broke through to the shelter deck and rose to nineteen feet. The massive sideboard and revolving chairs were also of oak, with green carpets and velvet upholstery on the chairs. The main deck had more cabins, above the mailroom and quarters for the stewards and stewardesses on the lower deck. The 133 second-class rooms offered berths for 464 people, bringing the ship's possible total to 2,165 passengers.

The American writer Theodore Dreiser took the *Mauretania* to Liverpool in the fall of 1911. When she reached the Mersey, he joined other passengers in the steam tender that would take them to the dock. Dreiser was a determined cynic, the author of brutally realistic fiction, and not given to any sentimental effusions. He watched the great black liner as the tender pulled away. "It was immense from alongside, a vast skyscraper of a ship," he wrote. "At a hundred feet it seemed not so large, but exceedingly more graceful; at a thousand feet all its exquisite lines were perfect, its bulk not so great, but the pathos of its departing beauty wonderful; at two thousand feet it was still beautiful and large against the granite ring of the harbor. . . . The stacks, in their way gorgeous, took on beautiful proportions. I thought, as we veered in near the pier and the ship turned within her length or thereabouts and steamed out, I had never seen a more beautiful sight."

A bit swifter than her sister—thanks to the small improvements of Edwin De Rusett and Andrew Laing—the *Mauretania* quickly set new transatlantic records in both directions, averaging over twenty-five knots all the way across. She was so fast that competing steamship lines did not even try to surpass her. "There is a limit to all development," said Gustav H. Schwab, the North German Lloyd's New York agent, in 1909. "The limit for steamships has almost been reached." During the remaining years before the war, the White Star and Hamburg-American lines built bigger but slower ships. On her maiden voyage, the *Titanic* did not charge through that iceberg field with any hope of establishing a new record; she only wished to prove herself not *too* much slower than the *Mauretania.* "In good weather and in bad," wrote one of her court historians, "she had a grace of action which made most other ships look like labouring tugs."

In the history of transatlantic steam, even the most advanced ships had held the speed records for only a few years. The *Scotia,* the last Cunard paddle wheeler, had established the longest previous tenure at the top, retaining the westbound standard of eight days and three hours for almost nine years until bumped by one of the first set of White Star ships. The *Mauretania* went undefeated for twenty-two years, until 1929, when the new Lloyd liner *Bremen* finally beat her. The Cunarder was retired five years later, after twenty-seven years and some 2.1 million miles on the North Atlantic. Crossing the ocean for the last time, she still managed to average 24.4 knots. Her fittings and fixtures were auctioned off in the spring of 1935. A few months later she was condemned to ship breakers in Scotland. For her final passage, steaming up from Southampton, she was ordered to dawdle at 15 knots. On her own, so it seemed, she kept speeding up to 18 knots.

Some of her remains wound up in a pub in Bristol, the port city in southwestern England that launched the first real transatlantic steamer, the *Great Western,* in 1838. Over the years, this public house has gone through various managers and names but is still generally known to locals as the Mauretania Pub. It's within an old, imposing building at the foot of Park Street, not far from the docks. A few years ago, the mahogany woodwork was inexplicably painted purple, but new and wiser management had the original finish restored. A visitor in the spring of 2000 saw many traces of the great steamship. Over a door, a large embossed *M.* Over another door, the rampant lion and globe of the Cunard logo. In a barroom by the main entrance, mahogany panels and columns—once again gleaming forth, rich and elegant—and bronze-and-crystal chandeliers, all saved from the exquisite first-class lounge. In the main dance hall, behind a video screen at one end, a fireplace from one of the public rooms, framed by brightly painted vertical tiles. In the center of the dance hall's ceiling, one of the ship's domes, now wired for sound and lights.

At nine o'clock on a Saturday night, the Mauretania Pub comes to life with strobe lights and the heavy thumps of dance music. It feels like an odd time warp, the pulsing disco lights reflecting in the old seagoing mahogany: Saturday night fever meets the ghost of Sam Cunard. The menus and bar coasters feature a reminding picture of the ship. The current manager, Roger Durbin, says he has five hundred people there on a typical weekend night. They are mostly young, full of sass and spirit. But they don't trash the place, Durbin explains. They know where it came from.

Notes

MANUSCRIPT MATERIALS

Adams Family Papers, Massachusetts Historical Society, Boston

Admiralty Papers, Public Record Office, Kew

Minute Book of American Steamship Company, Pennsylvania Railroad Company Collection, Urban Archives, Temple University

Frances Elizabeth Appleton Papers, Longfellow National Historic Site, Cambridge, Massachusetts

Francis T. Baring Papers, Hampshire Record Office, Winchester, Hampshire

Joel Barlow Papers, Houghton Library, Harvard University

Diary of Lorenza Stevens Berbineau, Francis Cabot Lowell II Papers, Massachusetts Historical Society, Boston

Phyllis Blakeley Collection, Public Archives of Nova Scotia, Halifax

Bliss Family Papers, Public Archives of Nova Scotia, Halifax

Diary of Sarah Hickling Bradford, Gamaliel Bradford Papers, Houghton Library, Harvard University

Francis B. C. Bradlee Collection, Phillips Library, Peabody Essex Museum, Salem, Massachusetts

Brown Brothers Papers, New-York Historical Society, New York

John Brown Papers, Upper Clyde Shipbuilders, Archives and Business Records Centre, University of Glasgow

Isambard Kingdom Brunel Collection, Special Collections, University of Bristol

Isambard Kingdom Brunel Papers, Public Record Office, Kew

Diary of Elizabeth Rogers Mason Cabot, Massachusetts Historical Society, Boston

Journal of Henry G. Carey, Massachusetts Historical Society, Boston

Grover Cleveland Papers, Manuscript Division, Library of Congress

William Cramp and Sons Papers, Special Collections, Temple University

Minute Book of William Cramp and Sons Ship and Engine Building Company, Library of Independence Seaport Museum, Philadelphia

Cunard Archives, Special Collections, University of Liverpool

Samuel Cunard Letters to Viscount Canning, National Archives of Canada, Ottawa

Cunard Papers, Halifax (forty-four microfilmed documents), Public Archives of Nova Scotia, Halifax

Letters of Mary Leiter Curzon to George Curzon, Oriental and India Office, British Library, London

Clinton E. Dawkins Letterbooks, Manuscripts, Guildhall Library, London

Daniel Dow Papers, Library of Merseyside Maritime Museum, Liverpool

John Elder Papers, Archives and Business Records Centre, University of Glasgow

John Ericsson Papers, American Swedish Historical Museum, Philadelphia

Fairfield Company Papers, Upper Clyde Shipbuilders, City Archives, Mitchell Library, Glasgow

Joshua Field Papers, Science Museum Library, London

Robert Bennet Forbes Papers, Massachusetts Historical Society, Boston

Jane Cunard Francklyn Papers, Public Archives of Nova Scotia, Halifax

Guion Line Papers, Liverpool Record Office, Brown Library, Liverpool

Benjamin Harrison Papers, Manuscript Division, Library of Congress

Diary of George S. Hillard, Massachusetts Historical Society, Boston

Hollingworth Collection, Archives, University of Bath Library

Diary of Frederic Hudson, Special Collections, Free Public Library, Concord, Massachusetts

Inman Company Papers, Liverpool Record Office, Brown Library, Liverpool

Diary of Thomas H. Ismay, National Maritime Museum, Greenwich

Laird Family Letters, Library of Merseyside Maritime Museum, Liverpool

Journals of Amory A. Lawrence, Massachusetts Historical Society, Boston

Diary of Amos A. Lawrence, Massachusetts Historical Society, Boston

William S. Lindsay Papers, National Maritime Museum, Greenwich

Charles Mac Iver Papers, Special Collections, University of Liverpool

William McKinley Papers, Manuscript Division, Library of Congress

Diary of James Howard Means, Massachusetts Historical Society, Boston

J. P. Morgan Papers, Pierpont Morgan Library, New York

J. P. Morgan, Jr., Papers, Pierpont Morgan Library, New York

Napier Collection, Archives and Business Records Centre, University of Glasgow

Napier Papers, Museum of Transport, Glasgow

Charles A. Parsons Papers, Science Museum Library, London

Penn Central Railroad Collection, Pennsylvania State Archives, Harrisburg

James K. Polk Papers, Manuscript Division, Library of Congress

Diary of Alfred S. Roe, Rare Books and Manuscripts, New York Public Library

Theodore Roosevelt Papers, Manuscript Division, Library of Congress

Engineering Notebooks of John Scott Russell, Science Museum Library, London

Letterbook of John Osborne Sargent, New-York Historical Society, New York

Diary of Henry H. Seaver, Massachusetts Historical Society, Boston

J. and G. Thomson Papers, Upper Clyde Shipbuilders, Archives and Business
Records Centre, University of Glasgow
Tod & Macgregor Papers, Archives and Business Records Centre, University of
Glasgow
Martin Van Buren Papers, Manuscript Division, Library of Congress
Journal of Pelham S. Warren, Massachusetts Historical Society, Boston
Diary of Andrew C. Wheelwright, Massachusetts Historical Society,
Boston
Ralph B. Whittier Papers, Public Archives of Nova Scotia, Halifax
Journal of J. J. Garth Wilkinson, Houghton Library, Harvard University
Journal of Samuel W. Woodhouse, Historical Society of Pennsylvania,
Philadelphia
Peter Wright & Sons Collection, Historical Collections, Baker Library, Harvard
Business School
George Renny Young Papers, Public Archives of Nova Scotia, Halifax
William Young Papers, Public Archives of Nova Scotia, Halifax

ABBREVIATIONS OF SOURCES

ADM Admiralty Papers, Public Record Office, Kew
BBR Isambard Kingdom Brunel Collection, Special Collections, University
of Bristol
BON N. R. P. Bonsor, *North Atlantic Seaway* (1955–79)
CAL Cunard Archives, Special Collections, University of Liverpool
CFI *Reports from Committees: Colonization from Ireland,* Parliamentary Papers
(1847), vol. 6
CMC *Report from the Select Committee on Mail Contracts,* Parliamentary Papers
(1868–69), vol. 6
CMI Charles Mac Iver Papers, Special Collections, University of Liverpool
CPS *Report from the Select Committee on Contract Packet Service,* Parliamentary
Papers (1849), vol. 12
DAB *Dictionary of American Biography*
DCB *Dictionary of Canadian Biography*
DNB *Dictionary of National Biography*
ENG *Engineering* (London)
FLH *Fifth Report of the Select Committee on the Roads from London to Holyhead;
and . . . Steamboats,* Parliamentary Papers (1822), vol. 417
GAA Robert Gardiner and Basil Greenhill, *The Advent of Steam: The Merchant
Steamship Before 1900* (1993)
HBM *Report from the Select Committee on Halifax and Boston Mails,* Parliamentary
Papers (1846), vol. 12
IES *Transactions of the Institution of Engineers and Shipbuilders in Scotland*
ILN *Illustrated London News*
INA *Transactions of the Institution of Naval Architects* (London)
JFI *Journal of the Franklin Institute* (Philadelphia)

LIH William S. Lindsay, *History of Merchant Shipping and Ancient Commerce* (1876)

LTI *London Times*

MEL *Marine Engineer* (London)

MEM *Marine Engineer and Motorship Builder* (London)

MEN *Marine Engineer and Naval Architect* (London)

MNA *Correspondence Relating to the Conveyance of Mails (North America),* Parliamentary Papers (1859), vol. 22

NCA *National Cyclopedia of American Biography*

NCG Napier Collection, Archives and Business Records Centre, University of Glasgow

NEI *Transactions of the North-East Coast Institution of Engineers and Shipbuilders* (Newcastle)

NGN *Nautical Gazette* (New York)

NML *Nautical Magazine* (London)

NPT Napier Papers, Museum of Transport, Glasgow

NSC *Novascotian* (Halifax)

NYT *New York Times*

PBC Phyllis Blakeley Collection, Public Archives of Nova Scotia, Halifax

PCR Penn Central Railroad Collection, Pennsylvania State Archives, Harrisburg

PRL *Papers Relating to Loss of the Steam Ship "Atlantic,"* Parliamentary Papers (1873), vol. 60

PTC *First Report from the Select Committee on Packet and Telegraphic Contracts,* Parliamentary Papers (1860), vol. 14

RCU Royal Commission on Unseaworthy Ships, *Preliminary Report of the Commissioners, Minutes of the Evidence, and Appendix,* Parliamentary Papers (1873), vol. 36

RSC *Reports of the Immigration Commission: Steerage Conditions,* Senate Document 753, 61 Cong. 3 Sess. (1911)

RSR *Reports of the Immigration Commission: Statistical Review of Immigration, 1820–1910,* Senate Document 756, 61 Cong. 3 Sess. (1911)

SNA *Transactions of the Society of Naval Architects and Marine Engineers* (New York)

SOT *The Standard Oil Trust,* Hearings, 50 Cong. 1 Sess. (1888)

TCS *Report of an Investigation into the Treatment and Condition of Steerage-Passengers,* Senate Document 23, 43 Cong. 1 Sess. (1874)

The key statistics of transatlantic steamships are in some dispute; in many cases, errors have crept into the secondary literature which have then been passed along from one book to the next. In general, I relied on primary sources that were contemporary with the ship under discussion. For the first fifteen Cunarders, from the *Britannia* of 1840 to the *Persia* of 1855, I drew from an authoritative chart compiled by the Robert Napier firm and published in *Machinery of the British and North*

American Royal Mail Steam Ship "Arabia" and of the West India Royal Mail Steam Ship "La Plata" (ca. 1855); I consulted a copy of this esoteric volume at the Rare Books Room of the National Library of Scotland in Edinburgh. I also used and compared the following tables of the major Atlantic steamers: William S. Lindsay, *History of Merchant Shipping and Ancient Commerce* (1876), opposite 4: 238; *Transactions of the Institution of Naval Architects* (London, 1887), p. 164; F. E. Chadwick et al., *Ocean Steamships* (1891), p. 78; *Engineering* (London), August 2, 1907; *Transactions of the Institution of Engineers and Shipbuilders in Scotland* (1908), opposite p. 46; David Budlong Tyler, *Steam Conquers the Atlantic* (1939), appendix 6; and Robert Gardiner and Basil Greenhill, *The Advent of Steam: The Merchant Steamship Before 1900* (1993), pp. 20, 27, and 105. The historical section of the Cunard Line's Web site, www.cunard.com, was helpful, but it must be checked against other sources.

For the succession of transatlantic speed records, I relied mainly on Hans Deketele's Web site on Blue Riband champions, www.blueriband.com. This site properly ranks the record-breaking passages by the average speed of the voyage instead of by the elapsed time from point to point; those points varied with the era and steamship line. I do not refer to "the Blue Riband" in my text because that phrase was not used during the period, 1838 to 1910, that I am treating; and I employ feminine pronouns for ships because that was then the common usage.

PROLOGUE: THE NORTH ATLANTIC OCEAN AND THE *BRITANNIA*

ix   rogue wave: Laura Ricard in *Yankee,* July 1997.

x   winter gales: Sir James Bisset with P. R. Stephenson, *Tramps and Ladies: My Early Years in Steamers* (1959), pp. 226–27.

x   icebergs: A. C. Spectorsky, *The Book of the Sea* (1954), pp. 230–32; William H. MacLeish, *The Gulf Stream: Encounters with the Blue God* (1989), p. 195; Dale E. Ingmanson and William J. Wallace, *Oceanography: An Introduction* (fourth edition, 1989), p. 395.

x   Gulf Stream: M. Susan Lozier in *Science,* July 18, 1997; Ingmanson and Wallace, *Oceanography,* pp. 154–55; Nelson Hogg in *Oceanus,* summer 1992; Ellen J. Prager with Sylvia A. Earle, *The Oceans* (2000), pp. 66, 97–99; John Purdy, *Memoir, Descriptive and Explanatory, to Accompany the New Chart of the Atlantic Ocean* (third edition, 1817), pp. 70–73; Nathaniel Bowditch, *The New American Practical Navigator* (eleventh edition, 1839), p. 124.

xi   spinning icebergs: Stanley Rogers, *The Atlantic* (1930), p. 25.

xi   bombs: MacLeish, *Gulf Stream,* pp. 190–91.

xi   fogs: Alastair Cooper, *The Times Atlas of the Oceans* (1983), p. 53.

xi   "These horrid fogs": Purdy, *Memoir,* p. 179.

xi   geometric forms: NML, October 1866.

xii   bird resembling sail, and mirages: ibid.

xii   "The uncertainty": Purdy, *Memoir,* p. 179.

xii   Sable Island: Frederick A. Talbot, *Steamship Conquest of the World* (1912), pp. 299–300.

xii   *Britannia*'s expected arrival date: *Boston Atlas,* July 7, 1840.

xii   *Britannia*'s arrival: *Boston Advertiser,* July 18, 1840; *Boston Post,* July 20, 1840.

xiii "No event": Ezra S. Gannett, *The Arrival of the Britannia: A Sermon* (1840), p. 6.

xiii description of *Britannia:* "Specifications of Steamer *Britannia*," by Robert Napier, in NPT; *New York Herald,* July 7, 1840; *Boston Advertiser,* August 6, 1840; deck plans of *Britannia* in *The Cunard Line and the World's Fair, Chicago, 1893* (1893), p. 12; model of *Britannia* on current display at Merseyside Maritime Museum, Liverpool.

xiv   "All these rooms": *Boston Advertiser,* August 6, 1840.

xiv   "the consummation": ibid.

xiv   "She is truly": *Boston Evening Journal,* July 20, 1840.

xv   "Mr. Cunard": *Boston Courier,* July 24, 1840.

xv   Cunard Festival: *Boston Evening Journal,* July 22, 1840; *Boston Post,* July 18, 22, 27, 1840; *Boston Atlas,* July 22, 1840; *Boston Evening Transcript,* July 22, 1840.

xvii Cunard description: F. Lawrence Babcock, *Spanning the Atlantic* (1931), pp. 12–13; *The Provincial: or Halifax Monthly Magazine,* January 1853.

xvii dinner for engineers: *Boston Post,* July 28, 1840.

xvii departure of *Britannia: Boston Advertiser,* August 1, 3, 10, 1840.

xviii "The Atlantic to America": PTC, p. 266.

xix   "I originated": ibid., p. 255.

xviii "A beautiful": CPS, p. 133.

1: THE SAILING PACKETS

3   packet lines: Robert Greenhalgh Albion, *Square-Riggers on Schedule: The New York Sailing Packets to England, France, and the Cotton Ports* (1938).

3   Jeremiah Thompson: Conrad P. Wright, "The Origin and Early Years of the Trans-Atlantic Packet Lines of New York, 1817–1835" (Ph. D. dissertation, Harvard University, 1931), pp. 21, 24, 40–43, 66; DAB, 9: 461.

4   "In order to furnish": Frank Staff, *The Transatlantic Mail* (1956), p. 60.

4   "These ships": ibid.

4   *Pacific* voyage and turnaround: Wright, "Packet Lines," pp. 116–17.

5   Thompson's gifts: [Matthew Maury] in *Southern Literary Messenger,* January 1839.

5   "Such steadiness": *Niles' Weekly Register,* February 24, 1821.

5   *Emerald* record: Basil Lubbock, *The Western Ocean Packets* (1925), p. 26; *Boston Recorder,* March 13, 1824.

5   packets as largest and finest ships: John G. B. Hutchins, "The Rise and Fall of the Building of Wooden Ships in America, 1607–1914" (Ph. D. dissertation, Harvard University, 1937), p. 255.

5   "We have never": Wright, "Packet Lines," p. 465.

5   "In the comfort": *Liverpool Albion,* February 13, 1826.

6 packet design: Maury in *Messenger,* January 1839; Carl C. Cutler, *Greyhounds of the Sea* (1960 edition), pp. 94–95.

6 Collins and Dramatic Line: Albion, *Square-Riggers,* pp. 43–45, 123–25; Edward W. Sloan in *Log of Mystic Seaport,* spring 1988; Sloan in *Gulf Coast Historical Review,* spring 1990.

6 average trips, first ten years: *Niles' Weekly Register,* November 29, 1828.

6 average trips by 1839: *New York Herald,* February 29, 1840.

7 "A sea voyage": Journal of William Young, April 21, May 6, 1839, William Young Papers.

7 "ship sense": Lubbock, *Packets,* p. 2; Patrick O'Brian, *Master and Commander* (Norton edition, 1990), p. 69.

7 captain's pay: G. W. Sheldon in *Harper's Magazine,* January 1884; Maury in *Messenger,* January 1839.

8 "clever mulattoes": James Fenimore Cooper, *Homeward Bound; or, The Chase* (Mohawk edition, 1896), p. 66.

8 sailors' jobs: Ralph Waldo Emerson, *The Journals and Miscellaneous Notebooks,* eds Alfred R. Ferguson et al. (1964), 4: 103.

8 "We are pensioners": ibid., 4: 105.

8 *President* voyage, 1831: James Stuart, *Three Years in North America* (1833), 2: 543–44.

8 Fanny Appleton: Journal of Frances Elizabeth Appleton, November 16–December 6, 1835, Frances Elizabeth Appleton Papers.

9 typical cabin passengers: Albion, *Square-Riggers,* p. 230.

9 "We endeavored": John Maxtone-Graham, *The Only Way to Cross* (1972), p. 18.

10 "These are": Emerson, *Journals,* 4: 238–39.

10 Kemble watching men: Frances Anne Butler [Kemble], *Journal* (1835), 1: 23.

10 menagerie: Arthur H. Clark, *The Clipper Ship Era* (1910), p. 39.

10 meals on *Europe:* Tyrone Power, *Impressions of America* (1836), 1: 19–22.

10 steerage: Albion, *Square-Riggers,* pp. 79, 248; Robert Greenhalgh Albion, *The Rise of New York Port [1815–1860]* (1939), p. 340; Robert C. Leslie, *Old Sea Wings, Ways, and Words, in the Days of Oak and Hemp* (1890), pp. 226–29.

10 odors: Basil Greenhill and Ann Gifford, *Travelling by Sea in the Nineteenth Century: Interior Design in Victorian Passenger Ships* (1972), p. 14.

11 "This snuggery": Power, *Impressions,* 1: 22.

11 sunsets and nights: Diary of Benjamin Millward, March 1854, in *Annals of Iowa,* fall 1977; Frances Appleton Journal, November 17, 1835.

11 Anna Eliot Ticknor: from her diary, June 1835, in *Atlantic Monthly,* July 1927.

11 "About as big": Frederick Marryat, *A Diary in America, with Remarks on Its Institutions* (1839), p. 13.

11 berth: Clement Cleveland Sawtell, *Captain Nash DeCost and the Liverpool Packets* (1955), p. 43.

11 "Oh for a bed!": Butler, *Journal,* 1: 18.

11 noises: John Grantham, *Iron Ship-Building: With Practical Illustrations* (1858), p. 91; Greenhill and Gifford, *Travelling,* pp. 13–14; Millward Diary, March 19, 1854.

12 Harriet Martineau: *Harriet Martineau's Autobiography*, ed. Maria Weston Chapman (1877), 1: 331–32.

12 Martineau on *United States*: Harriet Martineau, *Retrospect of Western Travel* (1838) 1: 13–35.

15 packet schedules, late 1830s: Maury in *Messenger*, January 1839.

15 *Albion* and *Crisis*: ibid.; Albion, *Square-Riggers*, pp. 203, 213.

15 London newspaper, 1834: Albion, *Square-Riggers*, p. 98.

15 committee of Parliament: NML, October 1836.

16 "For strength": Maury in *Messenger*, January 1839

2: STEAM ON WATER

17 "It is impossible": *Quarterly Review*, March 1830.

18 Newcomen engine: R. A. Buchanan, *The Power of the Machine: The Impact of Technology from 1700 to the Present* (1992), pp. 48–50.

18 "It takes": Henry Petroski, *Remaking the World: Adventures in Engineering* (1997), p. 119.

18 James Watt: James Patrick Muirhead, *The Life of James Watt* (Appleton edition, 1859), pp. 33–52; Margaret C. Jacob, *Scientific Culture and the Making of the Industrial West* (1997), pp. 116–30.

19 eureka on Glasgow green: Eric Robinson and A. E. Musson, *James Watt and the Steam Revolution* (1969), p. 42.

19 "I can think": Muirhead, *Watt*, p. 78.

19 "Jamie is": ibid., p. 81.

19 "a midwife": Robinson and Musson, *Watt*, pp. 62–63; and see ibid., pp. 9–15.

19 "The people": ibid., p. 88.

19 Watt resisting improvements: Louis C. Hunter, *A History of Industrial Power in the United States, 1780–1830*, volume 2, *Steam Power* (1985), pp. 161–65.

19 "I do not think": Robinson and Musson, *Watt*, p. 96.

20 Early devices: H. Philip Spratt, *The Birth of the Steamboat* (1958), pp. 40, 45–46, 54.

20 "My natural turn": Symington memoir, 1829, quoted by Robert Cochrane in *Cassier's Magazine*, October 1907.

20 William Symington and Patrick Miller: ibid.; *Scots Magazine*, November 1788; James Nasmyth, *An Autobiography* (1883), pp. 27–31; Spratt, *Birth*, pp. 48–50; and see Patrick Miller in *Edinburgh New Philosophical Journal*, July 1827.

21 Symington in 1790s: R. H. Campbell, *Carron Company* (1961), pp. 214–15.

21 Forth and Clyde Canal: *Scots Magazine*, September 1816.

21 Symington and Dundas: JFI, January 1833; NML, July 1834; and see W. S. Harvey and G. Downs-Rose, *William Symington: Inventor and Engine Builder* (1980), pp. 117–41. I should note that I harbor a family link to this matter: my wife, Alexandra Dundas Todd, is a great-great-great granddaughter of Thomas, Lord Dundas of Kerse; but I don't believe that this distant tie influenced my reading of the historical evidence here.

21   "The nice and effectual": *Glasgow Advertiser,* June 29, 1801.

21   *Charlotte Dundas:* NML, August 1833; Harvey and Downs-Rose, *Symington,* pp. 127–34.

22   "This experiment": Symington memoir in *Cassier's,* October 1907.

22   "This so affected": ibid.

22   "amidst a very large": *Glasgow Herald and Advertiser,* January 7, 1803.

22   "the very appropriate": *Glasgow Herald and Advertiser,* April 4, 1803.

22   letter from proprietor: *Glasgow Herald and Advertiser,* April 8, 1803.

23   *Charlotte Dundas* later: Henry W. Dickinson, *Robert Fulton: Engineer and Artist* (1913), p. 181.

23   Robert Fulton: Cynthia Owen Philip, *Robert Fulton: A Biography* (1985).

24   summer 1802: ibid., pp. 129–34.

24   Symington re Fulton: Symington memoir in *Cassier's,* October 1907.

24   "*silent and steady*": Joel Barlow to Robert Fulton, July 28, 1802, Joel Barlow Papers.

25   "immediately": Philip, *Fulton,* pp. 139–40.

25   West dinner party: Joseph Farington, *The Farington Diary* (1922–23), 2: 33–34.

25   Fulton's lies and forgeries: Philip, *Fulton,* pp. 174–77, 263–65, 343–44; Alice Crary Sutcliffe, *Robert Fulton and the "Clermont"* (1909), p. 137 and opposite p. 140; Dickinson, *Fulton,* p. 268.

26   Weir affidavit: Bennet Woodcroft, *A Sketch of the Origin and Progress of Steam Navigation* (1848), pp. 65–67.

26   "I want your": Robert Fulton to Joel Barlow, June 28, 1811, Barlow Papers.

26   Fulton's first plans: Dickinson, *Fulton,* opposite p. 152.

27   *Clermont* machinery: Richard Shelton Kirby et al., *Engineering in History* (1956), p. 253.

27   "restless volatile": Edward Morris, *The Life of Henry Bell* (1844), p. 158.

28   "I was not": ibid., p.17.

28   Henry Bell at Carron Works: NML, August 1833.

28   Clyde undredged: W. J. Millar in IES (1881), pp. 49–50.

28   *Comet:* Spratt, *Birth,* p. 87; MEN, January 1909 and August 1912.

28   local boys: MEM, October 1926.

28   road coaches stopped: *Scots Magazine,* June 1813.

29   John Thomson: Millar in IES (1881), pp. 51–53; MEN, August 1912.

29   "The merchant": letter from Henry Bell in *Scots Magazine,* February 1817.

29   *Glasgow:* FLH, p. 224.

29   *Glasgow* passage to London: J. C. Delametherie and Isaac Weld in *Scots Magazine,* March 1816.

31   Forty-eight steamers by 1822: FLH, pp. 224–27; and see George Dodd, *An Historical and Explanatory Dissertation on Steam-Engines and Steam-Packets* (1818), pp. xvii-xviii, 140.

31   shipbuilding on the Clyde: Michael S. Moss and John R. Hume, *Workshop of the British Empire: Engineering and Shipbuilding in the West of Scotland* (1977), p. 36; James Cleland, *The Former and Present State of Glasgow* (1840), pp. 43–44.

31 David Napier: James R. Napier in IES (1865–66), pp. 92–105.

31 "Although then": *David Napier: Engineer: 1790–1869: An Autobiographical Sketch with Notes,* ed. David Bell (1912), pp. 16–17.

32 "Not having been": ibid., p. 17.

32 Napier and hull design: John Scott Russell, *On the Nature, Properties, and Applications of Steam, and on Steam Navigation* (1841), pp. 246–47.

32 *Rob Roy* to Greenock: *Edinburgh Magazine,* July 1818.

32 "I was the first": FLH, p. 233.

33 Napier's lack of restraint: James Napier, *Life of Robert Napier* (1904), pp. 23–24, 246.

33 Robert Napier: *Memoirs and Portraits of One Hundred Glasgow Men Who Have Died During the Last Thirty Years* (1886), 2: 241–44; Napier, *Robert Napier.*

33 born with hammer in hand: Napier, *Robert Napier,* p. 3.

34 "He was a man": *One Hundred Glasgow Men,* 1: 117.

35 "read Nature's laws": Robert Mansel in IES (1880), p. 5.

35 "A very simple": memoir of Eleanor Bristow Laird, Laird Family Letters.

3: SHIPS AS ENTERPRISE: SAMUEL CUNARD OF HALIFAX

39 "Those who have": Samuel Cunard to Charles Mac Iver, November 22, 1859, Charles Mac Iver Papers.

39 Thones Kunders: Henry C. Conrad, *Thones Kunders and His Children* (1891).

40 "an hospitable": ibid., p. 9; and see Cunard genealogical notes by William Otis Sawtelle, ca. 1929, in file PR 3.1/1a, CAL.

40 oral tradition: Kay Grant, *Samuel Cunard: Pioneer of the Atlantic Steamship* (1967), p. 12.

40 Quakers and Revolution: Anne M. Ousterhout, *A State Divided: Opposition in Pennsylvania to the American Revolution* (1987), pp. 4–5, 104, 125, 282–85.

41 Robert Cunard: Lorenzo Sabine, *Biographical Sketches of Loyalists in the American Revolution* (1864), 2: 346.

41 "Chagrined as your": Marion Gilroy, *Loyalists and Land Settlement in Nova Scotia* (1937), p. 148.

41 "All our golden": Neil MacKinnon, *This Unfriendly Soil: The Loyalist Experience in Nova Scotia, 1788–1791* (1986), p. 69.

42 Margaret Murphy: Grant, *Cunard,* p. 13; typed notes, file 3011–24, PBC.

42 Pennsylvanians in Shelburne: Wilbur H. Siebert, *The Loyalists of Pennsylvania* (1920), p. 100.

42 Eight hundred pounds in cash: *Halifax Mail,* February 31, 1931, in file 3010–31, PBC.

42 stories of Margaret's drinking: Margaret MacDonald to John Pakington, April 10, 1852, with notes by Blakeley, file 3012–5, PBC; W. S. MacNutt, *New Brunswick: A History: 1784–1867* (1963), p. 216; author's interview with Ross Graves, September 4, 2000.

42 Sam working early: Grant, *Cunard,* pp. 22–23.

43   "I have never": CFI, p. 70.

43   "'Tis true": Samuel Cunard to Jane Cunard Francklyn, December 23, 1863, Jane Cunard Francklyn Papers.

43   Cunard in Boston: Grant, *Cunard*, p. 25.

44   "As all around": John Young quoted by Phyllis Blakeley, "Halifax in the Year 1814" (lecture, August 6, 1985), file 3014–4, PBC.

44   Susan Duffus: Grant, *Cunard*, pp. 29–32.

44   Margaret in Pleasant Valley: *Truro Weekly News,* November 21, 1957, and Ross Graves to Phyllis Blakeley, May 6, 1974, file 3011–11, PBC; Ralph B. Whittier to George McLaren, January 18, 1967, and other materials in file 23, Ralph B. Whittier Papers.

45   Abraham's retirement: note in file 3010–1, PBC; and his will in file 3020–23, ibid.

45   "If you think": Sam Cunard to Thomas McCulloch, September 7, 1815, file 3011–20, PBC.

45   "The masters": ibid.

45   Halifax in 1820s: John McGregor, *British America* (1833), 2: 324–29; Terrence M. Punch, *Nova Scotia Vital Statistics from Newspapers, 1769–1812* (1981), 2: 55–63; Blakeley, "Halifax in the Year 1814."

46   old settlers and Loyalists: MacKinnon, *Unfriendly Soil,* pp. 99–105, 161–62; McGregor, *British America,* 2: 304.

46   "Nova Scotia approaches": McGregor, *British America,* 2: 317, 331; and see James S. Buckingham, *Canada, Nova Scotia, New Brunswick* (1843), p. 341.

46   Brunswick Street: James H. Raddall, *Halifax: Warden of the North* (1965), pp. 42–43.

46   Province Building: McGregor, *British America,* 2: 327–28.

46   local government: V. L. O. Chittick, *Thomas Chandler Haliburton: A Study in Provincial Toryism* (1924), pp. 74–76.

47   "I have always": Samuel Cunard to Robert Napier, April 5, 1839, NPT.

47   change of name: *Nova-Scotia Royal Gazette,* May 5, 1824.

47   office and warehouse: note on back of photograph of the building, Maritime Museum of the Atlantic, Halifax.

48   "Our pretensions": Samuel Cunard to East India Company, February 11, 1825, file 3011–20, PBC.

48   first tea ship: *Acadian Recorder,* November 17, 1924, file 3010–31, PBC.

48   "I also had intended": NSC, May 3, 1826.

48   "I have always": Samuel Cunard to George R. Young, August 7, 1838, file 3017–12, PBC.

49   "His conduct": William Blowers Bliss to Henry Linden, May 15, 1826, Bliss Family Papers.

49   "Those only": NSC, February 7, 1828.

49   "I have been": William Cunard to Jane Cunard Francklyn [April 1865], Francklyn Papers.

50   steps, spring 1825: NSC, April 6, May 4, 1825.

50 "It does seem": NSC, June 22, 1826.

50 "We are entirely": Samuel Cunard to Ross & Primrose, October 28, 1829, file H 1, CAL.

51 steam ferry: notes in file 3014-2, PBC.

51 "thus depriving": NSC, March 18, 1830.

51 *Royal William:* GAA, p. 15; Archibald Campbell in *Transactions of the Literary and Historical Society of Quebec* (1891), pp. 30–62; materials in file 3018-12, PBC; *Montreal Gazette,* May 3, 1831; Robert Ker in *Canadian Magazine,* May 1907; *Mariner's Mirror,* January 1948.

51 William Simmons as shipbuilder: James Cleland, *Enumeration of the Inhabitants of the City of Glasgow* (1832), p. 159.

51 "As I had": James Goudie to Archibald Campbell, February 17, 1891, in *Transactions* (1891), p. 37.

52 maiden voyage: *Montreal Gazette,* August 27, 1831.

52 "Her beautiful": *Acadian Recorder,* September 3, 1831, file 3018-19, PBC.

52 sending her to England: NSC, October 12, 1831.

52 "While at this port": NSC, December 12, 1831.

52 reduced rates: NSC, April 19, 1832.

53 stockholders accused Cunard: NSC, March 21, 1833.

53 "She was neglected": Samuel Cunard to Rupert D. George, May 7, 1833, file 3011-20, PBC.

53 sale of *Royal William:* Ker in *Canadian Magazine,* May 1907.

53 departure of *Royal William:* ibid.

53 "We were very": John McDougall to William King, November 16, 1833, file 3018-12, PBC.

54 "We sincerely hope": NSC, November 11, 1830.

54 "I think he": Lewis Bliss to Henry Bliss, February 5, 1831, file 3010-20, PBC.

54 "He is the most": Lewis Bliss to Henry Bliss, January 28, 1831, ibid.

55 "As it is": William Blowers Bliss to Henry Bliss, November 27, 1832, file 3010-2, PBC.

55 "If it should": Samuel Cunard to Charles Mac Iver, January 1, 1858, CMI.

55 "that he did not": Edward Cunard to Jane Cunard Francklyn, April 25, 1865, Francklyn Papers.

4: SHIPS AS ENGINEERING: ISAMBARD KINGDOM BRUNEL

56 "Indeed, all things": *Mechanics' Magazine,* May 6, 1837.

57 early history of engineering: Hans Straub, *A History of Civil Engineering* (MIT edition, 1964), pp. 111–20, 146–53, 172; Richard Shelton Kirby et al., *Engineering in History* (1956), pp. 327–28.

57 first British engineers: Samuel Smiles, *Lives of the Engineers* (1874), 2: 183–217, 261–83, and 3: 101–110; Donald Cardwell, *The Norton History of Technology* (1994), pp. 225–26.

57 Sapperton Tunnel: *The Works of Isambard Kingdom Brunel: An Engineering Appreciation,* ed. Alfred Pugsley (1976), p. 38.

57   "Civil Engineering is": W. H. G. Armytage, *A Social History of Engineering* (1976 edition), p. 123.

58   Henry Maudslay: Samuel Smiles, *Industrial Biography: Iron-Workers and Tool-Workers* (1863), pp. 198–207; Philip Banbury, *Shipbuilders of the Thames and Medway* (1971), pp. 198–99.

58   "the beautiful machine": James Nasmyth, *An Autobiography* (1883), p. 123.

58   railroad as extension of mines, canals, and mail coaches: Cardwell, *Norton History,* p. 234.

59   Industrial Revolution before railroad in Britain: Derek Beales, *From Castlereagh to Gladstone, 1815–1885* (1969), pp. 108–9.

59   George Stephenson: L. T. C. Rolt, *The Railway Revolution: George and Robert Stephenson* (1962).

59   Fanny Kemble and the locomotive: Frances Ann Kemble, *Records of a Girlhood* (Henry Holt edition, 1879), pp. 278–83. (Rolt, *Railway Revolution,* pp. 191–92 directed me to this source.)

60   "Verily is ours": ILN, July 16, 1842.

60   dissenting engineers: L. T. C. Rolt, *Victorian Engineering* (1970), p. 25; Nasmyth, *Autobiography,* pp. 163–65.

60   "If we would": William Fairbairn, *Useful Information for Engineers* (1860), p. 205.

60   worn-out engineers: [F. R. Conder], *Personal Recollections of English Engineers* (1868), pp. 61–62, 173–74.

60   Buchanan suggestion: R. A. Buchanan in *The Great Engineers: The Art of British Engineers, 1837–1987,* ed. Derek Walker (1987), pp. 84–88.

61   "I believe": *Diaries of Sir Daniel Gooch,* ed. Theodore Martin (1892), p. 37.

61   "Drawing is": Nasmyth, *Autobiography,* pp. vii, 57.

61   Isambard Kingdom Brunel: the best biography is R. A. Buchanan, *Brunel: The Life and Times of Isambard Kingdom Brunel* (2002); it became available after I had already finished writing the Brunel sections of this book.

61   "the benumbing effect": *Works of Brunel,* ed. Pugsley, p. 22.

61   "No one can": John Pudney, *Brunel and His World* (1974), p. 53.

62   "Brunel's billiard table": Derrick Beckett, *Brunel's Britain* (1980), p. 38.

62   Box Tunnel: E. T. MacDermot, *History of the Great Western Railway* (1927), 1: 128–31.

62   birthday legend: Beckett, *Brunel's Britain,* p. 64.

62   queen's first railway trip: ILN, June 18, 1842.

62   "Free from": Celia Brunel Noble, *The Brunels: Father and Son* (1938), p. 148.

62   Albert flew: A. M. Acworth in *Cornhill Magazine,* September 1897.

62   "Even to shake": Noble, *Brunels,* p. 210.

62   five feet four: James Dugan, *The Great Iron Ship* (1953), p. 20.

62   jokes and pranks: Isambard Brunel, *The Life of Isambard Kingdom Brunel, Civil Engineer* (1870), pp. 75, 97.

62   "It is at once": Conder, *Recollections,* p. 271.

63   "Mr. Brunel passed": LTI, May 2, 1843; and ibid., May 6, 9, 15, 16, 22, 1843.

63   "Stephenson is": Noble, *Brunels,* p. 164.

63   "Admit him": Conder, *Recollections,* pp. 404–5.

63   "usefulness": I. K. Brunel to John Scott Russell, November 30, 1849, BBR.

64   "He was the very": *Quarterly Review,* July 1862.

64   "The most useful": *Works of Brunel,* ed. Pugsley, p. 147.

64   "But what cannot": Brunel, *Brunel,* p. 260.

64   Brunel testifying: *Fraser's Magazine,* March 1871.

64   doubting talent for writing: I. K. Brunel to Thomas Coates, February 26, 1841, BBR.

64   folk hero: Henry Petroski, *Remaking the World: Adventures in Engineering* (1997), p. 131. (In 2002 Brunel placed second—ahead of Shakespeare and Darwin, among others—in a BBC poll to determine the greatest Briton of all time.)

65   Marc Brunel: *Mechanics' Magazine,* December 22, 1849; Paul Clements, *Marc Isambard Brunel* (1970).

65   "Yes, I am": Richard Beamish, *Memoir of the Life of Sir Marc Isambard Brunel* (1862), p. 322.

66   "your firm": I. K. Brunel to Maudslay & Field, March 24, 1853, BBR.

66   "One sadly loses": I. K. Brunel to William Froude, April 7, 1847, BBR.

67   "Take them": I. K. Brunel to P. J. Palmer, December 2, 1848, BBR.

67   Thames Tunnel: *Works of Brunel,* ed. Pugsley, pp. 30–31; Kemble, *Records,* pp. 120–21.

67   "No one has": Brunel, *Brunel,* p. 22.

67   Marc was criticized: *Mechanics' Magazine,* June 9, October 13, 1827.

68   "My self-conceit and love": L. T. C. Rolt, *Isambard Kingdom Brunel: A Biography* (1959), p. 89; Cynthia Gladwyn in *Proceedings of the Institution of Civil Engineers,* September 1971.

68   praised his coolness: *Mechanics' Magazine,* January 19, 1828.

68   "Tunnel is now": Rolt, *Brunel,* pp. 40–41.

68   "The time is not": ibid., pp. 44–45.

69   "The most eventful": private diary, December 26, 1835, BBR.

69   Mary Horsley: Rolt, *Brunel,* pp. 91–98.

69   first four months: Brunel correspondence, July 29–November 19, 1836, I. K. Brunel Papers, Public Record Office, Kew.

69   "It is an understood": *Works of Brunel,* ed. Pugsley, p. 41.

70   "The Box Tunnel": *The Birth of the Great Western Railway,* ed. Jack Simmons (1971), p: 89.

70   "With all his": ibid., p. 42.

70   Great Experimental Railway: *Quarterly Review,* July 1862.

70   blunt conversation: *Great Western,* ed. Simmons, pp. 52, 61, 64.

70   "Why not make": Brunel, *Brunel,* p. 233.

71   "Great improvements": report, January 1836, printed in NML, January 1838.

71   "known as a man": Christopher Claxton, *Logs of the First Voyage . . . by the Great Western* (1838), p. ii.

71   "Mr. Patterson drew": Claxton letter in LTI, September 22, 1859.

72   Maudslay, Sons and Field: Banbury, *Shipbuilders,* p. 198.

72 "No vessel ever": FLH, p. 154.

72 "I have not": James Napier, *Life of Robert Napier* (1904), p. 104.

72 "The distance is limited": JFI, July 1836.

72 side-lever engine: Banbury, *Shipbuilders,* pp. 199–200; John Scott Russell, *On the Nature, Properties, and Applications of Steam, and on Steam Navigation* (1841), p. 268; Andrew Murray, *The Theory and Practice of Ship-Building* (1861), p. 112.

72 Napier changed: J. W. C. Haldane, *Steamships and Their Machinery* (1893), pp. 212–13.

73 *Savannah:* Frank O. Braynard, *S. S. Savannah: The Elegant Steam Ship* (1963).

73 at least five other steamships: H. Philip Spratt, *Transatlantic Paddle Steamers* (second edition, 1967), pp. 19–25.

73 London group, 1824: LTI, June 28, 1824; NSC, October 12, 1825.

73 "almost certain": *Boston Independent Chronicle and Patriot,* September 28, 1825; *Niles' Weekly Register,* October 22, 1825.

74 Junius Smith's plans: *New York Herald,* October 19, 1838.

74 "I shall not": E. LeRoy Pond, *Junius Smith: A Biography of the Father of the Atlantic Liner* (1927), p. 34.

74 "The patience": ibid., p. 52.

74 Macgregor Laird: DNB 11: 407–8.

74 *London Times* announced: LTI, November 10, 1835; "Prospectus of the British and American Steam Navigation Company," NPT.

74 "Job's patience": Pond, *Smith,* p. 63.

74 Brunel and three tenders: Brunel, *Brunel,* pp. 235–36.

75 Field's cams and boilers: John Rennie in NML, July 1847; GAA, pp. 163–65.

75 Brunel checking on progress: I. K. Brunel to Christopher Claxton, April 4, December 3 and 18, 1837, BBR.

75 "There are but few": I. K. Brunel to Christopher Claxton, November 21, 1838, BBR.

75 Brunel wanted to make ship larger: "Draft History," BBR.

75 "took the liberty": NML, September 1837.

75 Smith's engine problems: Junius Smith, 1844, in *Magazine of American History,* November 1882; Napier, *Robert Napier,* pp. 114–16.

75 "The steamer is going": Pond, *Smith,* p. 88.

76 Field and cycloidal wheel: John Rennie in NML, July 1847.

76 "Her size": *Mechanics' Magazine,* September 23, 1837.

76 *Great Western: Mechanics' Magazine,* April 7, 1838; Denis Griffiths, *Brunel's 'Great Western'* (1985), pp. 27–28, 63; MEM, April 1928; Philip Hone, *The Diary of Philip Hone,* ed. Allan Nevins (1927), 1: 319; model at Merseyside Maritime Museum, Liverpool.

77 "floating palace": NSC, June 27, 1839.

77 fire: LTI, April 2 and 3, 1838; Brunel, *Brunel,* pp. 242–43.

78 "I hope the Vessel": I. K. Brunel to Christopher Claxton, April 3, 1838, BBR.

78 "the father of Atlantic steam": Pond, *Smith,* p. 182.

78 voyage of *Sirius:* T. Sheppard in *Mariner's Mirror,* January 1937; MEL, June 1909; LTI, April 11, 1838.

78   Brunel provided chart: I. K. Brunel to G. Aircy, May 28, 1855, BBR.

78   voyage of *Great Western:* Grahame E. Farr in *Mariner's Mirror,* April 1938; *The Logs of the First Voyage . . . by the Great Western* (1838).

79   passengers noticed differences: ibid.; [Thomas Chandler Haliburton], *The Letter-Bag of the Great Western; or, Life in a Steamer* (1840).

79   "How this glorious": ibid., p. 28.

79   203 tons of coal left: NML, June 1838.

80   loss on first voyage: Griffiths, *'Great Western,'* p. 43.

80   131 passengers to New York: LTI, September 10, 1838.

80   "Here we have": *Mechanics' Magazine,* April 20, 1839; next five quotations from ibid., April 20, May 11, August 10, June 15, August 10, 1839.

81   "They unfortunately let": *David Napier: Engineer: 1790–1869: An Autobiographical Sketch with Notes,* ed. David Bell (1912), p. 39.

82   Laird diary: typescript, July 12–26, 1839, Laird Family Letters.

83   steamship race: LTI, August 15, 1839; NSC, August 14, 1839.

83   "The *British Queen* was": HBM, p. 24.

83   1839 season: *New York Herald,* February 28, 1840.

83   "Is it not reasonable": *Mechanics' Magazine,* May 16, 1840.

5: THE CUNARD LINE

84   "Altho I am": Samuel Cunard to Viscount Canning, March 11, 1853, National Archives of Canada.

84   "The plan was": HBM, p. 29.

85   Claxton wrote Wood: Christopher Claxton to Charles Wood, December 13, 1838, BBR.

85   "Many of the government": NML, October 1837.

86   Cunard and Prince Edward Island: files 3017-1, 3017-2, and 3017-10, PBC.

86   "People fear him": DCB, 8: 956.

86   "You seem afraid": Samuel Cunard to George Renny Young, August 7, 1838, file 3017-12, PBC.

87   Cunard broke arm: Brenton Halliburton to Enos Collins, March 5, 1831, file 3012-5, PBC.

87   packets redesigned: Robert Greenhalgh Albion, *Square-Riggers on Schedule: The New York Sailing Packets to England, France, and the Cotton Ports* (1938), pp. 17–18.

87   "Almost every year": PTC, p. 269.

87   "I lost five": ibid.

88   "I have always": David Budlong Tyler, *Steam Conquers the Atlantic* (1939), p. 80.

88   Cunard's plan: CPS, p. 133.

88   "I submitted": ibid.

88   "I hereby offer": Samuel Cunard to Charles Wood, February 11, 1839, ADM 1/4497.

89 word was passing: Thomas H. Brooking to George Renny Young, February 28, 1839, George Renny Young Papers.

89 Melvill recommended: James Napier, *Life of Robert Napier* (1904), p. 124.

89 "I shall want": Samuel Cunard to William Kidston & Sons, February 28, 1839, file H 1, CAL.

89 Vulcan Foundry: NML, September 14, 1839; *Penny Magazine,* September 1843.

90 "He appears": Robert Napier to Reid Irving & Co., March 12, 1839, NPT.

90 "From the frank": Napier, *Robert Napier,* p. 134.

90 "You have no idea": Samuel Cunard to Robert Napier, March 21, 1839, NPT.

91 "I am sorry": Napier, *Robert Napier,* p. 137.

91 "I was quite prepared": ibid., p. 136.

91 £14,000 to complete: Thomas H. Brooking to Samuel Cunard, March 26, 1839, file 720–2, Young Papers; signed agreement, March 26, 1839, in file 3017–10, PBC.

91 "The truth is": Robert Napier to Samuel Cunard, March 28, 1839, NPT.

92 "I have no wish": Robert Napier to Robert Rodger, April 1, 1839, NPT.

92 George Burns and the Mac Ivers: Edwin Hodder, *Sir George Burns, Bart.: His Time and Friends* (1890), pp. 160–63.

92 "the vessels would": Robert Napier to Samuel Cunard, April 2, 1839, NPT.

92 "I have several": Samuel Cunard to Robert Napier, April 4, 1839, NPT.

93 "I dare say": HBM, p. 23.

93 shares in final agreement: Francis E. Hyde, *Cunard and the North Atlantic, 1840–1973* (1975), p. 13.

93 "I had the whole": HBM, p. 28.

93 "Remarkable for the great": John Scott Russell in INA (1861), p. 145.

94 "two strong bilge-pieces": "Specification of Steamer '*Britannia*'" and signed contract with amendments, NPT.

94 coal compartments: Elias H. Derby, *Two Months Abroad* (1844), p. 4.

94 averages, first two years: NML, August 1842.

94 Dickens account: Charles Dickens, *American Notes* (Fawcett edition, 1961), pp. 15–37.

97 "Sneers": ILN, October 22, 1842.

98 Appleton account: Frances Appleton to Emmeline Austin, May 2, 10, 15, 1841, and to Nathan Appleton, May 2, 14, 15, 1841, Frances Elizabeth Appleton Papers.

99 *President:* NML, January 1840; LTI, December 9, 1839, and July 13, 1840; "Glances at Atlantic Steam Navigation, 1838–1841," pp. 41–43, Joshua Field Papers; *Mechanics' Magazine,* April 11, 18, 1840.

100 "so that the springing": Jacob B. Moore in *Merchant's Magazine,* October 1840.

100 "What on earth": I. K. Brunel to Thomas Guppy, November 23, 1840, BBR.

101 fractured engine frame: James R. Napier in IES (1865–66), p. 101.

101 oral tradition: note by Eleanor Bristow Laird, n. d., Laird Family Letters.

101 *London Times* noted: LTI, April 7, 1841; and see NGN, February 8, 1873.

101 "I do not altogether": E. LeRoy Pond, *Junius Smith: A Biography of the Father of the Atlantic Liner* (1927), p. 217.

102 "At the time": J. C. Arnell, *Steam and the North Atlantic Mails: The Impact of the Cunard Line and Subsequent Steamship Companies on the Carriage of Transatlantic Mails* (1986), p. 76.

102 Admiralty agreed: ADM 12/373: 76, March 5–23, 1840.

102 Cunard's guess, 1839: Samuel Cunard to Robert Napier, April 11, 1839, NPT.

102 £15,355 loss: ADM 12/387: 76, May 13, 1841.

102 "The day must": David Mac Iver to Samuel Cunard, April 16, 1841, Cunard Papers, Halifax.

103 "with pleasure": ADM 12/387: 76, June 7, 1841.

103 new contract: ibid., August 25, 1841.

103 "It would have been": Glasgow paper quoted in LTI, September 28, 1841.

103 borrowed £15,000: document 3, Cunard Papers, Halifax.

103 "Sam and I": Journal of William Young, June 29, 1839, William Young Papers; and see ibid., May 30, June 16, 1839.

103 "They have been": Samuel Cunard to Charles Walton, September 17, 1844, file 3010–27, PBC.

104 loan of £45,000: Bank of Nova Scotia minutes, April 5–11, 1842, file 3010–12, PBC.

104 Cunards on Prince Edward Island: Ian Ross Robertson, *The Prince Edward Island Land Commission of 1860* (1988), pp. 54–56, 75, 87–88, 105–6, 141, 209, 213.

104 "We left there": Margaret Macdonell, *The Emigrant Experience: Songs of Highland Emigrants in North America* (1982), p. 123.

104 debts and assets, 1842: file 3010–31, PBC.

104 hiding from server and flight from Liverpool: account by Duncan Gibb, file 3010–28, PBC.

104 loss of £26,400: ADM 12/411: 21, March 10, October 25, 1843.

105 "with regularity": Arnell, *Steam*, p. 88.

105 loan to partners paid: George Burns to Samuel Cunard, March 14, 18, April 16, 1846, Cunard Papers, Halifax.

105 Cunard Line paying dividends: HBM, p. 28.

105 "There is great": Diary of John Quincy Adams, September 2, 1840, and Adams to Louisa Adams, September 2, 1840, Adams Papers.

105 loss of *Columbia:* NSC, July 10, 1843; ILN, July 29, 1843.

106 "Up to the last": A. T. Squarey to David Mac Iver [II], February 6, 1874, CMI.

106 Mac Ivers: ENG, January 8, 1886; *Clan Iver,* ed. H. Mac Iver (1912), file 3015–3, PBC.

107 "When you go home": RCU 2: 331.

107 "It will be obvious": "Remarks on the Sailing Department," file PR 3.1/23a, CAL.

107 "A cheerful acquiescence": regulations of 1840, *Official Guide and Album of the Cunard Steamship Company* (1878), p. 44.

108 "wanton and extravagant": Charles Mac Iver to Officers of the Mess, January 20, 1847, file D 138/2/3, CAL.

108   "Mac Iver's letters": *The Liverpool Exchange Portrait Gallery* (1884), p. 105, file 3015–3, PBC.

108   orders in 1848: "Notes for Captain Long," file D 42/PR3/1/35, CAL.

109   Mac Iver's inspection: [W. H. Rideing] in *Appleton's Journal,* October 1876.

109   "The highest court": RCU 2: 332.

109   Appleton visited Cunards: Frances Appleton to Emmeline Austin, May 10, 1841, Appleton Papers.

110   "I have been": CFI, p. 65.

110   "What a slow": Frederic S. Cozzens, *Acadia: or, A Month with the Blue Noses* (1859), p. 20.

110   "the lassitude": Isabella Lucy Bird, *The Englishwoman in America* (1856), p. 17.

110   country estate: Kay Grant, *Samuel Cunard: Pioneer of the Atlantic Steamship* (1967), pp. 149–50.

110   "shy, silent": Frances Ann Kemble, *Records of a Girlhood* (Henry Holt edition, 1879), p. 176.

110   Great Western's profits: Denis Griffiths, *Brunel's 'Great Western,'* (1985), p. 150.

110   "We are quite": HBM, p. 21.

6: THE COLLINS LINE

112   "so far": Samuel Cunard to Charles Mac Iver, May 1, 1847, CMI.

113   *Hibernia: Practical Mechanic and Engineer's Magazine* (Glasgow), May 1843; ILN, September 10, 1842, and October 7, 1843.

113   *Cambria:* LTI, December 10, 1844.

113   Robert Steele (the son): Mark Howard in *Northern Mariner/Marin du Nord,* July 1992.

114   *Eclipse:* FLH, p. 200.

114   *United Kingdom: Mechanics' Magazine,* August 12, 1826.

114   "in his silent": Robert Mansel in IES (1880), p. 5.

114   "I came to England": HBM, p. 28.

114   "I saw that": ibid., p. 25.

114   "I pressed": CPS, p. 134.

114   "and I could not": HBM, p. 25.

114   "I would have": CPS, p. 134.

115   Napier's advice: Robert Napier to George Burns, May 2, 1846, enclosed with Thomas Casswell to Robert Napier, April 30, 1861, NCG.

115   final cost of £90,000: Samuel Cunard to Charles Wood, December 20, 1849, file 3011–20, PBC.

115   "Although chastely": NSC, April 10, 1848; and see J. C. Arnell, *Steam and the North Atlantic Mails: The Impact of the Cunard Line and Subsequent Steamship Companies on the Carriage of the Transatlantic Mails* (1986), pp. 186–87.

115   machinery cost £50,000: HBM, p. 24.

116   Edward Knight Collins: Ralph Whitney in *American Heritage,* February 1957; Edward W. Sloan in *Log of Mystic Seaport,* spring 1988; and see Sloan's articles

on Collins in *Gulf Coast Historical Review,* spring 1990; *American Neptune,* summer 1991; *International Journal of Maritime History,* June 1992; *Log of Mystic Seaport,* winter 1992; *Northern Mariner/Marin du Nord,* January 1995; and *Research in Maritime History,* 1998.

117 "The *Herald* is": [Isaac Clark Pray], *Memoirs of James Gordon Bennett and His Times* (1855), p. 216.

117 "Put in three more": Frederic Hudson, *Journalism in the United States, from 1690 to 1872* (1873), pp. 457–58.

117 Collins submitted plan: E. K. Collins to Martin Van Buren, February 18, 22, 1841, Martin Van Buren Papers.

117 "Can it be": E. K. Collins to John M. Niles, February 22, 1841, Van Buren Papers; and see Collins to Charles H. Haswell, April 6, 1842, box 22, John Ericsson Papers.

118 Mac Iver sent: David Mac Iver to Samuel Cunard, April 19, 1841, Cunard Papers, Halifax.

118 Edward Mills: *Merchants' Magazine,* July 1846.

118 *Washington* and *Hermann:* Cedric Ridgely-Nevitt, *American Steamships on the Atlantic* (1981), pp. 128–31.

118 Polk administration: E. K. Collins to James K. Polk, April 17, 1847, James K. Polk Papers.

118 William H. Brown: Robert Greenhalgh Albion, *The Rise of New York Port [1815–1860]* (1939), p. 292; Ridgely-Nevitt, *Steamships,* pp. 98, 122; Codman Hislop, *Eliphalet Nott* (1971), p. 352; New York shipbuilders listed in *Hunt's Merchants' Magazine,* February 1849.

119 *Southerner:* Ridgely-Nevitt, *Steamships,* pp. 98–101.

119 "like the side": letter from J. E. G., *Hunt's Merchants' Magazine,* May 1849.

119 Faron on *Niagara:* Charles B. Stuart, *The Naval and Mail Steamers of the United States* (second edition, 1853), p. 123.

120 "well known": *New York Herald,* February 2, 1849.

120 Collins liners: Stuart, *Steamers,* pp. 199–207; JFI, May 1850 and June 1851.

120 passenger rooms: *The Diary of Philip Hone,* ed. Allan Nevins (1927), 1: 884; *New York Herald,* April 24 and July 1, 1850; text and four illustrations in ILN, May 25, 1850; *Chambers's Edinburgh Journal,* June 29, 1850.

121 costs of liners: Senate Report 267, by Thomas J. Rusk, 32 Cong. 1 Sess; *Congressional Globe,* 32 Cong. 1 Sess., p. 1163.

121 Brown Brothers: John Crosby Brown, *A Hundred Years of Merchant Banking* (1909); John A. Kouwenhoven, *Partners in Banking: An Historical Portrait of a Great Private Bank: Brown Brothers Harriman & Co., 1818–1868* (1968).

121 Brown's investment in Cunard Line: Francis E. Hyde, *Cunard and the North Atlantic, 1840–1973* (1975), p. 13.

121 "certainly a strange": Sloan in *International Journal of Maritime History,* June 1992.

122 "On no question": Joseph Shipley to Stewart Brown, September 1, 1844, Brown Brothers Papers.

122   James Brown: Brown, *Hundred Years,* pp. 230–36.

122   secret cartel: Hyde, *Cunard,* pp. 40–41; Sloan in *International Journal of Maritime History,* June 1992.

122   "such arrangements": Sloan in *Research in Maritime History* (1998).

123   "A better sea-cook": William H. G. Kingston, *Western Wanderings; or, A Pleasure Tour in the Canadas* (1856), 1: 4.

123   "Rich enough": Benjamin Silliman, *A Visit to Europe in 1851* (1853), 1: 14.

124   *Pacific* run under ten days: LTI, May 24, 1851.

124   averages in 1851: *Hunt's Merchants' Magazine,* August 1852.

124   "The United States have": John S. C. Abbott in *Harper's,* June 1852.

124   "They are": Lauchlan B. Mackinnon in *Harper's,* July 1853.

125   strains on wooden hulls: Frank C. Bowen, *A Century of Atlantic Travel, 1830–1930* (1930), p. 57; John H. Morrison, *History of American Steam Navigation* (1903), pp. 412–13.

125   "not yet in the exact": *New York Herald,* July 1, 1850.

125   "In power of engines": NYT, January 17, 1852.

125   "The proprietors": Samuel Cunard to Francis T. Baring, February 8, 1851, Francis T. Baring Papers.

125   average loss of $17,000: David Budlong Tyler, *Steam Conquers the Atlantic* (1939), p. 390.

126   John O. Sargent: NCA (1907), p. 432.

126   "What I now write": E. K. Collins to John O. Sargent, January 12, 1852, box 22, Ericsson Papers.

126   "Had not the ships": ibid.

126   "Keep quiet": John O. Sargent to E. K. Collins, February 4, 1852, box 22, Ericsson Papers.

126   "Others may manage": E. K. Collins to John O. Sargent [February 1852], ibid.

127   "I was apprehensive": John O. Sargent to E. K. Collins, February 12, 1852, ibid.

127   "I shall endeavor": E. K. Collins to John O. Sargent, February 13, 1852, ibid.

127   "These persons": *Congressional Globe,* 32 Cong. 1 Sess., p. 658.

127   "Many were the salutes": *New York Herald,* March 3, 1852.

127   "Every body was": *Washington Daily National Intelligencer,* March 5, 1852.

127   "In this": *Congressional Globe,* 32 Cong. 1 Sess., p. 1714.

127   "corrupting influences": ibid., appendix, p. 787.

127   "a bare inspection": ibid., p. 1304.

128   "I hope Collins": William E. Bowen to Brown Brothers, July 25, 1852, Brown Brothers Papers.

128   three years earlier: Silliman, *Visit,* 2: 461.

128   *Arctic* sinking: Alexander Crosby Brown, *Women and Children Last: The Loss of the Steamship* Arctic (1961); NYT, October 12–18, 1854.

129   "He was a sensitive": Nathaniel Hawthorne, *The English Notebooks,* ed. Randall Stewart (1941), p. 92.

130   engine room men with knives: Charles C. Nott to Mrs. R. S. Minturn, April 19, 1912, and to Mrs. John Crosby Brown, May 27, 1912, Brown Brothers Papers.

131   deaths: I computed these totals from the lists in Brown, *Women and Children,* pp. 213–18; Brown gives a different final toll on p. 166. My total of 109 women and children includes 7 members of the crew.

131   "Oh, what a manly": *New York Express* quoted in LTI, November 14, 1854.

132   responses of James and Eliza Brown: Brown, *Women and Children,* p. 159.

132   insurance of $540,000: Ridgely-Nevitt, *Steamships,* p. 161.

132   eight Cunard ships to Crimea: *Official Guide and Album of the Cunard Steamship Company* (1878), p. 5; ADM 12/607: 76, January 1, June 27, November 17, 1855.

132   transatlantic mail revenues: Pliny Miles, *The Advantages of Ocean Steam Navigation* (1857), p. 48.

133   *Arabia*: *New York Herald,* January 3, 12, 1853.

133   "goes bowling down": *Daily News* quoted in NSC, November 4, 1850.

133   cartel renewed: Hyde, *Cunard,* pp. 41–44.

133   "We do not attempt": Samuel Cunard to Francis T. Baring, February 8, 1851, Baring Papers.

134   "The *Atlantic* is fitted": ILN, July 26, 1851.

134   Cunard got permission: ADM 12/543: 76, March 27 and April 8, 1851; ADM 12/523: 25a, August 12, 1850; and ADM 12/527: 76, September 7, 1850.

134   "Nothing seems to": Hawthorne, *Notebooks,* pp. 11 and 21.

134   "It is our duty": LTI, September 18, 1852.

135   departure of *Pacific:* LTI, March 20, 1856.

135   Eldridge and Lyle to *Pacific:* NSC, November 12, 1855, and March 31, 1856.

135   "The handles": NYT, March 15, 1856.

135   Judkins said: NYT, March 21, 1856.

135   "who was quite": George Templeton Strong, *The Diary of George Templeton Strong,* ed. Allan Nevins and Milton Halsey Thomas (1952), 2: 260.

136   Collins quietly applied: NYT, April 2, 1858.

136   "No good reason": LTI, March 22, 1856.

136   "A little longer": NYT, April 14, 1856.

136   divers found in 1991: Edward W. Sloan in *Bermuda Journal of Archaeology and Maritime History* (1993), pp. 84–92.

137   "I think Collins": *Congressional Globe,* 34 Cong. 3 Sess., p. 1106.

137   "The wise course": F. A. Hamilton to Joseph Shipley, December 8, 1857, Brown Brothers Papers.

138   "An odor": *Harper's Weekly,* March 10, 1860.

138   "The whole capital": PTC, p. 267.

138   "I knew that": ibid., p. 253.

138   "America is weeping": LTI, May 18, 1859.

7: DISTINGUISHED FAILURES

141   "In the marine boiler": Robert Duncan in NGN, February 23, 1887.

141   wrought iron: Allen Andrews, *Wonders of Victorian Engineering* (1978), p. 50.

141 "that all-civilizing": Samuel Smiles, *Industrial Biography: Iron-Workers and Tool-Workers* (1863), p. 331.

142 "It answers": LTI, May 18, 1865.

142 *Aaron Manby:* NML, July 1842; H. W. Brady in *Syren and Shipping Illustrated,* January 6, 1954 (reprinted by Steamship Historical Society, 1954); GAA, p. 22.

142 "To the present": William Pole, *The Life of Sir William Fairbairn, Bart.* (1877), p. 62; and see INA (1875), pp. 263–65.

143 "It is from": INA (1860), p. 81.

143 John Laird: INA (1875), pp. 266–67; ILN, July 27, 1861; Kenneth Warren, *Steel, Ships and Men: Cammell Laird, 1824–1993* (1998), pp. 27–30.

143 "being a less draught": *Liverpool Mercury* quoted in LTI, October 17, 1829.

143 series of iron steamships: NML, November 1833, May, November 1837.

143 *Garry Owen:* John Grantham, *Iron Ship-Building: With Practical Illustrations* (1858), pp. 12–13.

143 Charles Wye Williams and watertight compartments: LTI, September 20, 1837; Charles Lungley in INA (1861), p. 246.

144 "It is no difficult": NML, November 1837.

144 "There is no": E. LeRoy Pond, *Junius Smith: A Biography of the Father of the Atlantic Liner* (1927), pp. 160–61.

144 early problems with iron: GAA, pp. 22–24; Fred Walker, *Song of the Clyde: A History of Clyde Shipbuilding* (1984), pp. 41–42.

144 limitations of paddles: GAA, pp. 16–17.

145 two dozen tinkerers: John Bourne, *A Treatise on the Screw Propeller* (1852), pp. 8–21.

145 Francis P. Smith: ILN, April 26, 1856, and June 5, 1858; Bourne, *Treatise,* pp. 84–87; DNB 18: 442–44.

146 *Archimedes:* NML, June 1839, November 1840.

146 explosion on *Archimedes:* LTI, June 6, 1839.

146 Henry Wimshurst: Henry Wimshurst in NML, May 1876; NML, September 1877.

146 "Everywhere the vessel": Bourne, *Treatise,* p. 86.

147 plans for second steamer: Isambard Brunel, *The Life of Isambard Kingdom Brunel, Civil Engineer* (1870), pp. 246–47.

147 Brunel asked Grantham: I. K. Brunel to John Grantham, November 17, 1838, BBR.

147 "If these points": I. K. Brunel to Christopher Claxton, November 21, 1838, BBR.

148 "It is most important": Annual Report of Great Western Steam Ship Company, March 26, 1840, BBR.

148 "I have long since": I. K. Brunel to H. Booth, July 4, 1842, BBR.

148 "The result": I. K. Brunel to C. A. Caldwell, July 4, 1840, BBR.

148 long report: I. K. Brunel to Great Western directors, October 1, 1840, BBR.

149 "If all goes well": Celia Brunel Noble, *The Brunels: Father and Son* (1938), pp. 171–72.

150   "Remember when I urged": I. K. Brunel to Maudslay & Field, May 21, 1841, BBR.

150   "These are moving": I. K. Brunel to Joshua Field, September 15, 1843, BBR.

151   tests on eight screws: Thomas Richard Guppy in *Proceedings of Institution of Civil Engineers* (1845), pp. 151–85.

151   "We must stick": I. K. Brunel to Thomas Guppy, [August 1843?], BBR.

151   four-bladed screw was announced: ILN, July 15, 1843.

151   "sick of the continued": *Mechanics' Magazine,* November 19, 1842.

152   "It must be granted": LTI, September 15, 1842.

152   article of sixteen pages: *Mechanics' Magazine,* November 10, 1842.

152   Albert at launch: ILN, July 22, 29, 1843.

152   description of *Great Britain:* NML, July 1842; JFI, November 1843; ILN, February 1, 15, 1845; Christopher Claxton, *History and Description of the Steam-Ship Great Britain* (1845), pp. 5–18.

152   "the impressions of fragility": *Chambers's Edinburgh Journal,* August 2, 1845.

153   "There is little": ILN, February 15, 1845.

153   maiden voyage: ILN, August 30, 1845; LTI, August 30, 1845.

153   "a mere tremulous": NML, December 1845.

153   excessive rolling: ILN, September 20, 1845.

153   second round-trip: *Mechanics' Magazine,* November 1, 22, 1845.

153   "intolerable" rolling: LTI, October 30, 1845.

154   extensive renovations: *Mechanics' Magazine,* March 7, May 30, 1846; Denis Griffiths et al., *Brunel's Ships* (1999), p. 83.

154   "Decidedly more in accordance": *Manchester Guardian* quoted in LTI, April 30, 1846.

154   "a leviathanism": *Mechanics' Magazine,* March 7, 1846.

154   grounding: LTI, October 24, 26, November 2, 3, 4, 1846; E. C. B. Corlett in *Mariner's Mirror,* May 1975. Re Hosken's claim of a faulty chart, see J. P. Younghusband to McMurdo, Rathbone, & Martin, October 24, 1846, BBR.

154   "I have returned": Brunel, *Brunel,* pp. 264–65; and see I. K. Brunel to Christopher Claxton, December 11, 1846, BBR.

155   costs: Graham Farr in *Mariner's Mirror,* January 1950; Annual Report of Great Western Steam Ship Company, March 2, 1848.

155   sold for £25,000: LTI, March 14, 1849.

155   Brunel had to ask Patterson: I. K. Brunel to William Patterson, February 8, 1850, BBR.

155   John Scott Russell: INA (1882), pp. 258–60; George S. Emmerson, *John Scott Russell: A Great Victorian Engineer and Naval Architect* (1977).

156   "Steam was made": [J. S. Russell] in *Foreign Quarterly Review,* October 1832.

156   steam road carriage: notebook 1: 130–39, 142–45, 148–55, Engineering Notebooks of John Scott Russell; *Mechanics' Magazine,* February 21, 28, 1835.

157   "The little released": J. S. Russell in INA (1860), p. 197.

157   "By these improvements": J. S. Russell in JFI, November 1848; and see NML, November 1836.

157 "Naval architecture is": J. S. Russell, *On the Nature, Properties, and Applications of Steam, and on Steam Navigation* (1841), pp. 293, 259.

158 "If we could": J. S. Russell in INA (1860), p. 172.

158 "I am afraid": J. S. Russell to James R. Napier, January 31, 1860, NCG.

158 Russell and Royal Society: G. P. Mabon in *Journal of the Royal Society of Arts,* February 1967.

158 Brunel recruited: I. K. Brunel to J. S. Russell, April 13, 1848, BBR; and see Brunel to Russell, November 11, 1849, BBR.

158 Russell persuaded Albert: Mabon in *Journal,* February 1967.

158 Russell alienated: G. P. Mabon in *Journal of the Royal Society of Arts,* March 1967.

159 Russell was denied: John R. Davis, *The Great Exhibition* (1999), p. 17.

159 late in 1851: Brunel memo re origin of *Great Eastern,* February 25, 1854, BBR.

159 problem of coal supply: I. K. Brunel to Thomas Basley, July 1, 1852, BBR.

160 "Every body seems": I. K. Brunel to Richard Potter, June 23, 1852, BBR.

160 "With respect to": I. K. Brunel to Eastern Steam Navigation Company, July 21, 1852, BBR.

160 "The wisest and safest": I. K. Brunel to J. S. Russell, July 13, 1852, BBR.

160 "How I wish": I. K. Brunel to J. S. Russell, December 1, 1852, BBR.

161 "Everybody understands": I. K. Brunel to C. Manby, November 22, 1853, BBR.

161 "Of course I": I. K. Brunel to Samuel Beale, February 6, 1855, BBR.

161 "I have slaved": I. K. Brunel to J. S. Russell, November 23, 1853, BBR.

161 estimate of £492,750: I. K. Brunel to Eastern Steam Navigation directors, December 21, 1853, BBR .

162 Brunel refused to accept: John Yates to I. K. Brunel, August 7, 1854, and Brunel to Yates, August 16, 1854, BBR.

162 "I am by no means": I. K. Brunel to John Yates, November 16, 1854, BBR.

162 "you are the father": L. T. C. Rolt, *Isambard Kingdom Brunel: A Biography* (1959), p. 246.

162 Cunard had toyed: HBM, p. 25.

162 Admiralty would not allow: ADM 12/523: 25a, June 19, 1850.

163 Robert Mansel: MEL, January 1, 1906.

163 Napier's diagonal bracings: INA (1880), p. 265.

163 "The effect is that": ILN, February 9, 1856.

163 ran into iceberg: *North American Review,* October 1864.

163 176 tons a day: MNA, p. 42.

163 Mac Iver aghast: Brown, Shipley to Brown Brothers (New York), June 2, 1859, Brown Brothers Papers.

164 Brunel discovered: I. K. Brunel to Directors, January 15, 1856, Hollingworth Collection.

164 "Look the facts": I. K. Brunel to J. S. Russell, November 11, 1855, BBR.

164 "Do you disregard": I. K. Brunel to J. S. Russell, January 12, 1856, BBR.

164 late change: in November 1856, Brunel was still planning to launch on ways of pine or oak: memo for Directors, November 14, 1856, BBR. The first order for

iron rails for the launch was made in July 1857, with delivery a month later: Thomas E. Marsh to I. K. Brunel, July 17, August 6, 1857, BBR.

164   "I am anxious": I. K. Brunel to H. T. Hope, September 19, 1857, BBR.

165   "It has just been": J. S. Russell to James R. Napier, October 30, 1857, NCG.

165   Brunel warned: LTI, October 24, 1857.

165   launch: ILN, November 7, 1857; *Mechanics' Magazine,* December 19, 1857.

165   "We were ruined": J. S. Russell to Mr. Ayrton, December 12, 1874, file 8/4, William S. Lindsay Papers; and see J. S. Russell, *The Modern System of Naval Architecture* (1865), 1: 633.

165   "I wish you would": I. K. Brunel to Blake, August 30, 1859, BBR.

166   ship wagged excessively: ILN, June 16, 1860.

166   rolled too much: NML, October 1862; William B. Forwood, *Recollections of a Busy Life* (1910), pp. 67–68; *The Works of Isambard Kingdom Brunel: An Engineering Appreciation,* ed. Alfred Pugsley (1976), p. 155.

166   "a warning": William John in INA (1887), p. 148.

8: EMIGRATION AND THE INMAN LINE

168   potato fungus: Cecil Woodham-Smith, *The Great Hunger: Ireland, 1845–1849* (1962), pp. 94–101; William E. Fry and Stephen B. Goodwin in *BioScience,* June 1997.

169   emigrants to United States: Kerby A. Miller, *Emigrants and Exiles: Ireland and the Irish Exodus to North America* (1985), p. 291.

169   most from central areas: ibid., p. 293.

169   steamboats to Liverpool: Terry Coleman, *Going to America* (1972), p. 57.

169   "To Liverpool": *Irish Emigrant Ballads and Songs,* ed. Robert L. Wright (1975), p. 115.

169   emigrant ships: *Chambers's Edinburgh Journal,* April 13, 1844; Woodham-Smith, *Hunger,* pp. 225–26; Miller, *Emigrants,* p. 292.

170   "Before we were": *Ballads,* ed. Wright, p. 104.

170   *Washington: Chambers's Edinburgh Journal,* July 12, 1851.

171   fifty ships lost in six years: Edward Laxton, *The Famine Ships: The Irish Exodus to America, 1846–51* (1996), p. 90.

171   annual totals: Arnold Schrier, *Ireland and the American Emigration, 1850–1900* (1958), p. 157.

171   "passage money": Miller, *Emigrants,* p. 293.

171   nonenforcement of regulations: *Littell's Living Age,* September 14, 1850; Miller, *Emigrants,* pp. 353–54.

172   "You could not": CFI, p. 86.

172   David Tod: History of Tod & Macgregor, typescript (1991), file 239/1/11, Tod & Macgregor Papers; *David Napier: Engineer: 1790–1869: An Autobiographical Sketch with Notes,* ed. David Bell (1912), pp. 120–21; IES (1858–59), pp. 661–3.

173   John Macgregor: IES (1881), p. 70.

173   steeple engine: J. W. C. Haldane, *Steamships and Their Machinery* (1893), p. 236; J. S. Russell, *The Modern System of Naval Architecture* (1865), 1: 508–9.

173   their own workshop: History of Tod & Macgregor.

174   "extraordinary speed": LTI, July 17, 1841.

174   Thomas Assheton Smith: DNB, 18: 542–43; John E. Wilmot, *Reminiscences of the Late Thomas Assheton Smith* (1862).

174   "the most princely": *Glasgow Mail* quoted in NSC, May 26, 1851.

174   own version of Russell's theories: Wilmot, *Smith,* pp. 86–96.

175   "What struck me": ibid., p. 84. Napier's *Fire King,* built for Smith in 1840, was called "the fastest steamer afloat" (LTI, May 26, 1840).

175   chains and sides of *Great Britain: Mechanics' Magazine,* January 10, 1842.

175   *Fire Queen*'s steeple engine: John Bourne, *A Treatise on the Screw Propeller* (1852), p. 203 and plate 6.

176   *Vesta:* dimensions from vessel list in Tod & Macgregor Papers.

176   "I hope": Colin M. Castle, *A Legacy of Fame: Shipping and Shipbuilding on the Clyde* (1990), pp. 42–43.

176   *City of Glasgow:* LTI, January 29, April 19, 1850.

176   Tod modified: Bourne, *Treatise,* p. 205 and plate 8; Denis Griffiths, *Power of the Great Liners: A History of Atlantic Marine Engineering* (1990), pp. 24–25.

177   Archibald Gilchrist: *Steamship,* February 1900.

177   steam up within two weeks: *New York Herald,* May 4, 1850.

177   "The internal fittings": ibid.

177   "We look upon": LTI, April 19, 1850.

177   "It is to be hoped": *Liverpool Chronicle* quoted in NML, May 1850.

177   maiden trip: *New York Herald,* May 4, 1850.

178   "No steerage": *Philadelphia Evening Bulletin,* December 18, 1850.

178   "I prefer": William Inman to William S. Lindsay, December 7, 1874, file 8/1, William S. Lindsay Papers; and see *Liverpool Mercury,* July 4, 1881; DNB, 10: 457; *Dictionary of Business Biography,* 3: 427–33; PTC, pp. 212–28; CMC, pp. 125–35.

178   Inman family: Charles Chapman, *The Ocean Waves: Travels by Land and Sea* (1875), pp. 57–58.

179   "very fond": Inman to Lindsay, December 7, 1874.

179   Richardsons: Jane Marion Richardson, *Six Generations of Friends in Ireland (1655 to 1890)* (1894), pp. 218, 223–24.

179   "closely watched": Inman to Lindsay, December 7, 1874.

179   Inman investors: list of shareholders, August 1882, Inman Company Papers.

180   John Grubb Richardson: *A Biographical Dictionary of Irish Quakers,* ed. Richard S. Harrison (1997), pp. 87–88.

180   "strongly impressed": Charlotte Fell Smith, *James Nicholson Richardson of Bessbrook* (1925), p. 13.

180   "enabling us": ibid., p. 15.

180   Quakers and potato blight: Christine Kinealy, *This Great Calamity: The Irish Famine, 1845–52* (1994), pp. 158–61.

180   "A limited number": *Philadelphia Evening Bulletin,* April 9, 1852.

181   "I had great fear": Inman to Lindsay, December 7, 1874.

181 "Our company was": PTC, p. 228.

181 "I then learned": Inman to Lindsay, December 7, 1874.

182 480 on board: ILN, May 13, 1854.

182 breaking in new captain: *Philadelphia Evening Bulletin,* April 3, 1854.

182 "Some anxiety": Richardson Brothers letter in LTI, April 22, 1854, noting situation in Philadelphia as of April 8.

183 "probable loss": Coleman, *Going to America,* p. 241.

183 "The most painful": ILN, May 13, 1854.

183 Inman concluded: PTC, p. 221.

184 *City of Philadelphia: Philadelphia Evening Bulletin,* September 26, 30, and October 2, 1854.

185 dissolution of partnership: printed notice, November 2, 1854, Inman Company Papers.

185 French offered better terms: CMC, p. 134; *The Inman Line: Official Tourist Guide to Paris* [1878], pp. 5–6.

185 "We made": PTC, p. 220.

185 "We had the Cunard": CMC, p. 129.

185 averages in 1852: B. F. Isherwood in JFI, March 1853.

186 "They answer": CPS, p. 135.

186 *City of Baltimore* beat *Africa:* H. Littledale to I. K. Brunel, March 3, 1857, BBR.

186 Cunard-Inman arrangement: PTC, pp. 219–20.

186 "The Cunard Company quieted": ibid., p. 220.

186 "There was an agreement": ibid., p. 273.

186 "I have always": ibid., p. 255.

187 "I do not like": ibid., p. 261.

187 "We should not": ibid., p. 263.

187 "The mail steamers": LTI, December 30, 1856.

188 "When we first": RCU 2: 337.

188 "I really thought": Inman to Lindsay, December 7, 1874.

188 Burns and Mac Iver objected: George Burns to Samuel Cunard, April 17, 1841, Cunard Papers, Halifax.

189 "most heartily willing": George Burns to Charles Mac Iver, September 3, 1856, CMI.

189 "So long as": Charles Mac Iver to Samuel Cunard, September 4, 1856, ibid.

189 "I have the most": Samuel Cunard to Charles Mac Iver, January 1, 1858, ibid.

189 "I regret to say": Samuel Cunard to Charles Mac Iver, April 24, 1858, ibid.

189 "We should not forget": Samuel Cunard to Charles Mac Iver, June 5, 1858, ibid.

190 "We should assist": Samuel Cunard to Charles Mac Iver, November 9, 1858, ibid.

190 John Napier: IES (1900), pp. 361–63.

190 *Scotia:* specifications for her engine and boilers, NPT; LTI, November 8, 1861.

190 "In model she is.": *Liverpool Albion,* July 1, 1858.

190 Mansel acknowledged: IES (1879), p. 153.

190   *China:* ILN, April 12, 1862; MEM, March 1928; GAA, p. 167.

190   *Cuba:* ILN, December 10, 1864.

190   Cunard moved into town: Kay Grant, *Samuel Cunard: Pioneer of the Atlantic Steamship* (1967), p. 165.

191   "I feel quite": Samuel Cunard to Jane Cunard Francklyn, April 3, 1865, Jane Cunard Francklyn Papers.

191   "He has been": Edward Cunard to Jane Cunard Francklyn, April 25, 26, 1865, ibid.

191   "The Cunard Line": J. S. Russell, *The Modern System of Naval Architecture* (1865), 1: 644.

191   *City of Paris:* LTI, February 26, 1866; MEM, September 1926.

191   eight days, four hours: LTI, December 2, 1867.

192   "Smooth as": Henry W. Bellows, *The Old World in Its New Face* (1870), 1: 14, 18.

192   "the dernier resort": Henry Morford, *Over-Sea; or, England, France and Scotland as Seen by a Live American* (1867), p. 49; next three quotations from ibid., pp. 31, 34, 352.

193   "good and abundant": Samuel Amos in *Harper's,* October 1865.

193   complaints re steerage food: NYT, July 8, 1869.

193   "I do not make": NYT, April 21, 1868.

193   George Thomson: *Memoirs and Portraits of One Hundred Glasgow Men Who Have Died During the Last Thirty Years* (1886), pp. 321–22; James Napier, *Life of Robert Napier* (1904), pp. 86–88; Anthony Slaven, *The Development of the West of Scotland: 1750–1960* (1975), p. 129.

194   Burns and Russell: IES (1886), pp. 219, 221.

194   earlier Thomson ships: ILN, September 22, 1855, and October 24, 1857; James and George Thomson to J. E. Robinaw, July 24, 1860, J. and G. Thomson Papers.

194   brother Robert and Cunard: Fred M. Walker, *Song of the Clyde: A History of Clyde Shipbuilding* (1984), p. 146.

194   "We must confess": James and George Thomson to British & North American Royal Mail Steam Packet Company, October 14, 1859, Thomson Papers.

194   "The best proof": James and George Thomson to J. E. Robinaw, October 3, 1860, Thomson Papers.

194   *Russia:* LTI, May 31, 1867; ILN, September 7, 1867.

195   "Well, I'll tell": *New York Herald,* February 20, 1869.

195   race: ILN, March 6, 1869; *Harper's Weekly,* April 3, 1869.

195   "The excitement": NML, March 1869.

9: LIFE ON A STEAMER

196   "I was in": Henry Wadsworth Longfellow, *The Letters of Henry Wadsworth Longfellow,* ed. Andrew Hilen (1966), 2: 496.

197   "certain sense": Jacob Abbott in *Harper's,* July 1870.

197   machinery felt smoother: Diary of James Howard Means, April 3, 1856.

197 "I turn over": ibid. Means included a rough sketch of the room.

198 "and who combined": Isabella Lucy Bird, *The Englishwoman in America* (1856), p. 10.

198 saloon: Abbott in *Harper's*, July 1870.

198 "The five meals": Diary of Amos Lawrence, May 28, 1867.

199 "Steam is": Samuel Laing, *Observations on the Social and Political State of the European People* (1850), p. 1.

199 "ministers of": J. Jay Smith, *A Summer's Jaunt Across the Water* (1846), 2: 222.

199 Bird noticed: Bird, *Englishwoman*, pp. 8–9.

200 "for purposes": George Wilkes, *Europe in a Hurry* (1852), p. 3.

200 Douglass incident: Eliot Warburton, *Hochelaga; or, England in the New World* (1846), 2: 359–61; LTI, April 13, 1847; ILN, April 17, 1847.

200 "No one can": LTI, April 14, 1847.

200 Sunday policy: NSC, December 9, 1841, and October 14, 1850; Caroline Kirkland, *Holidays Abroad; or Europe from the West* (1849), 2: 330; *New York Independent*, September 25, 1850; Lawrence Diary, May 1, 1870; William C. Beecher and Samuel Scoville, *A Biography of Rev. Henry Ward Beecher* (1888), pp. 350–51.

200 "I am under": Samuel Cunard to Charles Mac Iver, May 1, 1847, CMI.

201 Presbyterian preachers: Richard Henry Dana, Jr., *The Journal of Richard Henry Dana, Jr.,* ed. Robert F. Lucid (1968), 2: 697; Arthur Hugh Clough, *The Correspondence of Arthur Hugh Clough,* ed. Frederick L. Mulhauser (1957), 2: 326; Journal of Amory A. Lawrence, July 14, 1867.

201 "a motley crew": Diary of George S. Hilliard, July 19, 1847.

201 "I told him": Lawrence Diary, November 24, 1839.

202 toasted the ladies: ibid., November 29, 1839.

202 "Oh, Stewardess": Cornelia Adair, *My Diary* (1965), p. 5.

202 "A certain regimen": Lydia H. Sigourney, *Pleasant Memories of Pleasant Lands* (third edition, 1856), p. 370.

202 "I sat down": Means Diary, April 4, 1856.

203 "very cold": Diary of Elizabeth Rogers Mason Cabot, October 11, 1847.

203 "Awful life": ibid., October 16, 1847.

203 "So long as": Horace Greeley, *Glances at Europe* (1852), p. 10.

203 "There was": Adeline Trafton, *An American Girl Abroad* (1872), p. 16.

204 "And I have loved": *Childe Harold's Pilgrimage*, canto 4, stanza 184.

204 "Let me assure you": Harriet Beecher Stowe, *Memories of Foreign Lands* (1854), pp. 1–6.

204 "Our most poetic": John S. C. Abbott in *Harper's*, June 1852.

205 "The sunshine": William H. G. Kingston, *Western Wanderings or, A Pleasure Tour in the Canadas* (1856), 1: 12.

205 "As to": Henry W. Bellows, *The Old World in Its New Face* (1870), 1: 18.

205 "The infinite": Dana, *Journal*, 2: 693.

206 "The absence": Bellows, *Old World*, 1: 17.

206 "The dampness": James Russell Lowell, *Fireside Travels* (1904), pp. 122–23.

207   "love lane": William Tod Helmuth, *A Steamer Book! A Picturesque Account of a City on the Sea* (1880), pp. 56–57.

207   "If the gentleman": Abbott in *Harper's,* July 1870.

207   "Cock-Tail Club": Diary of Andrew C. Wheelwright, May 26, 1847.

208   provisioned in 1848: LTI, March 22, 1848.

208   early Cunard cuisine: Robert Bennet Forbes to Thomas G. Cary, March 4, 1841, Robert Bennet Forbes Papers.

208   "wretched": Journal of Pelham S. Warren, April 8, 1848.

208   "The elegance": William Chambers, *Things as They Are in America* (New York edition, 1854), p. 17.

209   "I can see": Stowe, *Memories,* p. 10.

209   Bird overheard: Bird, *Englishwoman,* p. 457.

210   "the unwashed": Henry Morford, *Over-Sea; or, England, France and Scotland as Seen by a Live American* (1867), p. 38.

210   summer twilight: Dana, *Journal,* 2: 708–10.

210   "The sky seemed": Charles Lyell, *A Second Visit to the United States* (1849), 2: 274.

211   "A night of marvellous": Journal of J. J. Garth Wilkinson, September 19, 1869.

211   "Give me some": Journal of Henry G. Carey, September 7, 1866.

211   "With rosy": Wilkinson Journal, August 27, 1869; next quotation, ibid.

211   "A notion appears": letter in NML, August 1873.

212   old custom: Carey Journal, April 21, 1866.

212   tricks for children: Diary of Lorenza Stevens Berbineau, July 17, 1851.

212   deck daily routine: Dana, *Journal,* 2: 696

213   "From this scene": Abbott in *Harper's,* June 1852.

213   superstitions: *Harper's,* January 1858; *Old Ocean's Ferry,* ed. John Colgate Hoyt (1900), pp. 130–32; J. D. Jerrold Kelley in *Century,* July 1894.

213   "always a head": Trafton, *American Girl,* p. 21.

213   "No refinement": Joshua Field to I. K. Brunel, July 22, 1852, BBR.

214   two daily tons: Wilkinson Journal, September 25, 1869.

214   rum ration: Cunard maintained the daily rum allowance the longest of any major line, replacing it in 1882 with coffee; Board Minutes, September 5, 1882, CAL; NYT, December 19, 1882.

214   "It is incredible": Wilkinson Journal, August 30, 1869.

214   Walter McCrossan: Board Minutes, December 14, 1884, CAL.

214   "You will please": *New York Herald,* February 17, 1851.

215   Irwin killed: John O. Choules, *Young Americans Abroad* (1851), p. 22.

215   "Stoke freely": N. P. Burgh, *Modern Marine Engineering* (1867), p. 375.

215   priming: NML, July 1871.

216   fixed steam pipe: Notebook re engineers, November 1842, CAL.

216   "The atmosphere": NML, February 1875.

217   captain and navigation: see sketch of captain's cabin and instruments, Abbott in *Harper's,* July 1870.

217   "In fact": ibid.

217   "a degree": Lauchlan B. Mackinnon in *Harper's,* July 1853.

218   "I tried": Charles Algernon Dougherty in *Harper's*, August 1886.

218   "I'm afraid": Erwin Hodder, *Sir George Burns, Bart.: His Time and Friends* (1890), p. 481.

218   "an inferior animal": Adair, *Diary*, pp. 124–25.

218   dinner for Lott: ILN, February 27, 1864.

218   Theodore Cook: LTI, March 19, 28, 1889.

218   Charles H. E. Judkins: *Cunard Magazine*, November 1919.

218   "Few clergymen": Dana, *Journal*, 2: 816.

218   "a dove-like": NGN, March 29, 1873.

219   Burns observed: John Burns in *Good Words*, April 1887.

219   Appleton and Judkins: Frances Appleton to Emmeline Austin, May 2, 10, 11, 14, 1841, and to Nathan Appleton, May 2, 4, 1841, Frances Elizabeth Appleton Papers.

219   "One long disgust": Ralph Waldo Emerson, *The Journals and Miscellaneous Notebooks*, ed. Alfred R. Ferguson (1964), 10: 332.

220   "There is nothing": Lowell, *Travels*, p. 122.

220   "It's awfully": *The Letters and Private Papers of William Makepeace Thackeray*, ed. Gordon N. Ray (1946), 1: 108.

220   "But I could": Newman Hall, *From Liverpool to St. Louis* (1870), p. 4.

221   "See": Denis Griffiths, *Brunel's 'Great Western'* (1985), p. 96.

221   Mackinnon on *Baltic:* Lauchlan B. Mackinnon in *Harper's*, July 1853.

221   "Fine weather": Michael D. Buckley, *Diary of a Tour in America* (1889). p. 6.

221   "like a thousand": Carey Journal, September 5, 1866.

222   "Rocked in": ibid., April 17, 1866.

222   "It was then": Sigourney, *Memories*, p. 374.

222   Maury maintained: Matthew F. Maury in NML, January 1855.

223   *Britannia*, 1846: Lyell, *Visit*, 2: 272–74.

223   *Caledonia*, 1847: Wheelwright Diary, May 21, 1847.

223   "The captain looked": *Littell's Living Age*, June 29, 1854.

223   "Captain Luce": LTI, October 30, 1854.

223   "the national wail": *Harper's*, December 1854.

224   "Whether they like": Dana, *Journal*, 2: 699.

224   steam whistles: John Maxtone-Graham, *The Only Way to Cross* (1972), p. 179.

224   "No one who": Abbott in *Harper's*, June 1852.

224   "Very flattering": Warburton, *Hochelega*, 2: 358.

225   "We are all": Thackeray, *Letters*, 1: 892.

10: THE WHITE STAR LINE

229   "Comfort at sea": Samuel Smiles, *Men of Invention and Industry* (1885), p. 312.

230   "You cannot have": INA (1879), p. 173.

230   "The machinery": INA (1894), p. 35.

230   "We have": INA (1879), p. 149.

231   Edward Harland: ENG, January 3, 1896; Smiles, *Men*, pp. 284–318.

231 "I was slow": Smiles, *Men,* p. 286.

231 "surprisingly mechanical": ibid., p. 286.

232 "I was now": ibid., p. 288.

232 Britannia Bridge: ILN, January 13, June 30, 1849; Smiles, *Men,* p. 292; Nathan Rosenberg and Walter G. Vincenti, *The Britannia Bridge: The Generation and Diffusion of Technological Knowledge* (1978).

233 "The sons": INA (1879), p. 150.

233 "Some were there": Smiles, *Men,* p. 292.

233 Harland and Schwabe: Michael Moss and John R. Hume, *Shipbuilders to the World: 125 Years of Harland and Wolff, Belfast, 1861–1986* (1986), p. 14.

234 "I found the": Smiles, *Men,* p. 294.

234 "They were": ibid., p. 295.

234 Belfast shipyards: Sidney Pollard and Paul Robertson, *The British Shipbuilding Industry, 1870–1914* (1979), p. 66.

235 Harland and Liverpool: Moss and Hume, *Shipbuilders,* p. 18.

235 "Nonetheless": Smiles, *Men,* p. 301.

236 "In this way": ibid., p. 302.

236 "All long": John Grantham in INA (1866), p. 35.

236 *Persian:* Smiles, *Men,* pp. 304–5.

237 "After a great": ibid., p. 307.

237 "My idea": INA (1868), p. 24.

237 "An unassuming": Robert MacIntyre in *Cassier's Magazine,* April 1897.

237 Harland and wardrobe: Herbert Jefferson, *Viscount Pirrie of Belfast* (1947), p. 50.

237 Harland and shipyard: ibid., p. 50; James Douglas in *Strand Magazine,* November 1919.

238 "He has been": ENG, January 3, 1896.

238 "by a sort": LTI, November 25, 1899.

238 Thomas Henry Ismay: Roy Anderson, *White Star* (1964), pp. 40–41; *Dictionary of Business Biography,* 3: 455–61.

238 Ismay diary: Diary of Thomas H. Ismay, January–October 1856; quotations from January 11, 13, 17, July 17, 1856.

239 price £1,000: Anderson, *White Star,* p. 39.

239 Harland's plans: *Oceanic* plans, September 3, 1869, file H3, CAL.

240 *Oceanic:* NYT, March 4, 30, 1871; *Appleton's,* September 30, 1871.

241 cabin placement: John P. Eaton and Charles A. Haas, *Falling Star: Misadventures of White Star Line Ships* (Norton edition, 1990), p. 34.

241 Murray on Collins ships: *New York Herald,* March 30, 1871.

241 "The management": NML, April 1876.

241 next five ships: *White Star Line of Ocean Steamers* (1872), pp. 6, 17.

241 "Her lines": NYT, March 30, 1871.

241 "great Yankee improvement": *White Star Line,* p. 14.

241 "The visitor's ideas": ibid. p. 15.

242 "A masterpiece": *New York Herald,* March 30, 1871.

242   "For the first": Smiles, *Men,* p. 310.

242   "So far": NYT, September 26, 1871.

243   Ismay sent protest: Eaton and Haas, *Falling Star,* p. 17.

243   *Adriatic* record controversy: LTI, May 27, 28, June 6, 8, 1872.

243   averages, 1872: PRL, p. 164. (This chart transposes westward and eastward voyages.)

243   Digby Murray: IES (1906), p. 475; LTI, January 8, 1906.

244   "I always": PRL, p. 296.

244   "bluff, hearty": *Liverpool Daily Courier,* June 13, 1871.

244   *Republic:* MEM, July 1931.

244   formal report: PRL, pp. 153–54.

245   "Everything recommended": PRL, pp. 89, 291, 296.

245   Bradley S. Osbon: Albert Bigelow Paine, *A Sailor of Fortune: Personal Memoirs of B. S. Osbon* (1906), pp. 301–2; and see NYT, July 30, 1890.

245   "We do claim": NGN, January 25, 1873.

246   number on board *Atlantic*: PRL, pp. 10–11.

246   usually consumed: ibid., pp. 82, 134.

246   Williams fired for drinking: Eaton and Haas, *Falling Star,* pp. 17–18.

247   "We had": PRL, p. 488.

247   "In plain": ibid., p. 273. On the habit of underestimating, see ibid., p. 107.

247   "We must": ibid., p. 172.

247   all agreed: ibid., pp. 33–34.

248   *Atlantic* disaster: NYT, April 2–8, 1873; *New York Tribune,* April 2–8, 1873.

249   "What are we": PRL, p. 23.

250   585 dead: N. R. P. Bonsor, *North Atlantic Seaway* (1955), I: 255.

250   "Our prophecies": NGN, April 5, 1873.

251   "This vessel": LTI, April 10, 1873.

251   Harland inspection: PRL, pp. 269–78.

251   "She was": ibid., p. 75.

252   Cunarders extended: J. and G. Thomson to George Burns, August 22, September 14, 1871, J. and G. Thomson Papers.

252   *City of Berlin:* NGN, March 31, 1875; NYT, April 6, 1875.

252   stories: Henry Liggins in INA (1877), p. 65.

253   Froude studies: INA (1876), pp. 181–86.

253   "It is to": ENG, March 1, 1878.

253   "Other builders": MEL, October 1885.

11: COMPETITION AND INVENTION

255   Pennsylvania Railroad: *The Pennsylvania Railroad: Its Place in History, 1846–1996,* ed. Dan Cupper (1996), pp. 4–6; Alfred D. Chandler, *The Visible Hand: The Managerial Revolution in American Business* (1977), pp. 151–55.

255   Standard Oil secretly conspired: Ron Chernow, *Titan: The Life of John D. Rockefeller, Sr.* (1998), pp. 135–38, 201–3.

256   Randall scheme: *Prospectus of the Philadelphia and European Steamship Company* (1859); Board Minutes, March 21, 1860, PCR.

256   "If the traffic": Board Minutes, April 1, 1863, PCR.

257   20 to 35 percent cheaper: John G. B. Hutchins, *The American Maritime Industries and Public Policy, 1789–1914* (1941), p. 452.

257   "All New York": Augustus C. Buell, *The Memoirs of Charles H. Cramp* (1906), p. 93.

257   "The screw-propeller": Charles H. Cramp in SNA (1909), p. 154.

257   Cramp shipyard: NGN, July 12, 1879; DAB, 2: 499–500; Gail E. Farr and Brett F. Bostwick, *Shipbuilding at Cramp & Sons* (1991), pp. 9–12; David B. Tyler, *The American Clyde: A History of Iron and Steel Shipbuilding on the Delaware from 1840 to World War I* (1958), pp. 23–28; Thomas R. Heinrich, *Ships for the Seven Seas: Philadelphia Shipbuilding in the Age of Industrial Capitalism* (1997), pp. 51–52.

257   "Our yard": Charles H. Cramp in *Proceedings of the Numismatic and Antiquarian Society of Philadelphia* (1904–6), pp. 175–88.

258   Clement A. Griscom: *Harper's Weekly*, April 4, 1891; NCA, 4: 186–87; DAB, 4: 6–7.

258   Griscom and Betsy Ross: *Munsey's*, December 1902.

258   "Very well": *Philadelphia Inquirer*, November 11, 1912.

259   "Nominal Friend": Philip S. Benjamin, *The Philadelphia Quakers in the Industrial Age, 1865–1920* (1976), pp. 57–58.

259   "The dominant personality": Lawrence Perry in *World's Work*, December 1902.

259   "what the devil": Lloyd C. Griscom, *Diplomatically Speaking* (1940), p. 9.

259   Wright submitted: Board Minutes, March 18, 1870, PCR.

259   Griscom and oil: Edward Needles Wright in *Quaker History*, autumn 1967.

260   mainly railroad investors: Heinrich, *Ships*, p. 57.

260   family story: Griscom, *Diplomatically*, p. 10.

260   Antwerp and oil exports: oil reports, volumes 43 and 54, Peter Wright & Sons Collection.

260   "We expatriated": Clement Griscom to J. S. Clarkson, September 5, 1891, Benjamin Harrison Papers.

260   fall of 1871: Clement Griscom to Washington Butcher, September 30, 1871, James A. Wright to J. E. Thomson, September 26, 1871, Board Files 48, PCR; Board Minutes, September 27, November 8, 1871, PCR.

260   *Vaderland:* Vernon E. W. Finch, *The Red Star Line and International Mercantile Marine Company* (1988), pp. 23–24.

260   maiden voyage: NGN, February 22, 1873.

261   ASC financing: Board Minutes, November 23, 1870, PCR.

261   lowest offer: William Henry Flayhart III, *The American Line (1871–1902)* (2000), p. 20.

261   "Mr. Wilson": Charles H. Cramp in SNA (1910), pp. 54–55.

261   "No problem": Minute Book of the William Cramp and Sons Ship and Engine Building Company, April 30, 1894; and see SNA (1894), pp. 316–17.

261 Cramp yard: NGN, February 22, 1873.

262 "Nothing could": ibid.

262 *Pennsylvania:* NGN, July 5, 1873; drawings of *Indiana* in *Harper's Weekly,* April 10, 1880.

262 "superior to that": NGN, October 28, 1874.

262 ASC asked to withdraw: Minute Book of American Steamship Company, November 28, 1874; Board Minutes, November 25, 1874, PCR.

263 Allan Line: NGN, December 21, 1882; George Henry Preble, *A Chronological History of the Origin and Development of Steam Navigation* (1883), pp. 390–95.

263 Anchor Line: Preble, *History,* pp. 342–44.

264 Allan Line losses: Henry Fry, *The History of North Atlantic Steam Navigation* (1896), pp. 144–46.

264 Anchor Line losses: NGN, September 18, 1880.

264 National Line: NML, December 1875; Frank C. Bowen, *The Sea: Its History and Romance* (1924), 4: 193; Preble, *History,* pp. 350–54.

264 "They have": *The Education of Henry Adams* (1918), p. 156.

265 "Really a marvel": Henry Morford, *Over-Sea; or, England, France and Scotland as Seen by a Live American* (1867), p. 49.

265 profit of £120,000: LTI, July 29, 1871.

265 Stephen Guion: *Liverpool Daily Post,* December 21, 1885; NGN, December 24, 1885.

265 Brown Brothers offer: Brown, Shipley to Brown Brothers, New York, June 17, 1859, Brown Brothers Papers.

265 "What we are": *Cause of the Reduction of American Tonnage and the Decline of Navigation Interests* (1870), p. 34.

266 "Practically speaking": NML, January 1876.

266 Palmer and Jordan shares: NYT, August 2, 1873.

266 service to three categories: Jacob Abbott in *Harper's,* July 1870.

266 Jordan: Philip Banbury, *Shipbuilders of the Thames and Medway* (1971), p. 160.

266 Liverpool schedule: *Our Ocean Highways: A Condensed Universal Hand Gazeteer,* ed. J. Maurice Dempsey and William Hughes (1871), pp. clxii-clxv.

266 1873 traffic: *Westminster Review,* April 1, 1874.

267 William John Macquorn Rankine: David F. Channell in *Technology and Culture,* January 1982; J. Stephen Jeans, *Western Worthies* (1872), pp. 104–8; *Memoirs and Portraits of One Hundred Glasgow Men Who Have Died During the Last Thirty Years* (1886), 2: 269–72; INA (1873), pp. 235–38.

268 "His outflow": William Froude to James R. Napier, May 23, 1874, NCG.

268 Two songs: *Blackwood's Magazine,* December 1862.

269 INA founding: INA (1860), pp. xv-xx.

269 "France is snatching": INA (1863), p. 71.

269 "I regret": INA (1867), p. 227.

269 new members, 1867: ibid., p. 297.

269 "Nothing can": INA (1871), p. 269.

269 "Whenever Mr. Scott Russell": INA (1876), p. 54.

269   "Your duty": INA (1893), p. 235.

270   "It was very": INA (1870), pp. 82–83.

270   "The feeling": ibid., p. 87.

270   "We have": ibid., p. 90.

270   William Froude: INA (1879), pp. 264–69; DNB, 7: 731–32.

270   "When I became": William Froude to James R. Napier, February 22, 1875, NCG.

271   "Our doubts": David K. Brown in *Engineers and Engineering: Papers of the Rolt Fellows,* ed. R. Angus Buchanan (1996), p. 182.

271   Froude assisted Brunel: William Froude to I. K. Brunel, January 4, May 8, 12, 15, 28, August 9, 16, 21, 1857, BBR.

271   ship models: INA (1908), pp. 234–35; *Engineers,* ed. Buchanan, p. 201.

271   "All this": William Froude to James R. Napier, September 21, 1872, NCG.

271   "simply to <u>know</u>": William Froude to James R. Napier, October 16, 1872, NCG.

272   "I can": INA (1874), pp. 153, 61.

272   compound engines: GAA, pp. 98, 169–70; R. A. Buchanan, *The Power of the Machine: The Impact of Technology from 1700 to the Present* (1992), p. 144.

273   John Elder: W. J. Macquorn Rankine, *A Memoir of John Elder: Engineer and Builder* (1871); DNB, 6: 589–90.

273   "He was handsome": *One Hundred Glasgow Men,* 1: 121.

274   first transatlantic compound engines: NML, January, March 1876.

274   Cramp tour, 1871: Buell, *Cramp,* p. 114.

274   Alexander Carnegie Kirk: IES (1893), pp. 320–22; INA (1893), pp. 238–39; ENG, October 14, 1892.

274   "keeping an eye": Mark Twain, *Mark Twain's Letters,* ed. Edgar Marquess Branch et al., (1988), 5: 224.

275   "It is rather": *New York Tribune,* January 27, 1873.

275   "Livy darling": Twain, *Letters,* ed. Branch, 5: 473.

276   "The Cunard Company": *English Mechanic and World of Science,* November 12, 1875.

276   "Every year": *New York Tribune,* November 19, 1875.

276   "I have the": RCU 2: 329.

276   "There is": Charles Mac Iver to David Mac Iver, August 17, 1874, Charles Mac Iver Papers.

277   "The next thing": E. H. Hoblyn in *Cunard Magazine,* July 1919.

277   "The growing wants": *Dictionary of Scottish Business Biography* (1986), 2: 267.

277   "A splendid stroke": John G. Hughes to Archibald Bryce, March 11, 1880, Guion Line Papers.

12: SHIPS AS BUILDINGS: TWO CYCLES TO CUNARD

279   *Britannic:* NYT, March 14, 1874; NGN, July 15, 1874; MEL, September 1, 1903; MEM, July 1929 and August 1930.

279   *Germanic:* NGN, May 26, 1875; NML, May 1875; NYT, May 29, 1875.

279 saloon and public rooms: drawings and descriptions in *The White Star Line of Mail Steamers: Official Guide* [1877], pp. 6–10.

282 indicated horsepower: *Shipbuilding, Theoretical and Practical,* ed. W. J. Macquorn Rankine (1866), pp. 28, 277.

282 "Taking them": John R. Ravenhill in INA (1877), p. 288.

282 "Of the eleven": J. P. Morgan to wife, April 20, 1880, J. P. Morgan Papers; Jean Strouse, *Morgan: American Financier* (1999), pp. 192–93.

282 best times, summer 1891: MEL, October 1891.

282 over 2 million miles: *Quarterly Review,* January 1900.

282 William Pearce: David Pollock, *Modern Shipbuilding and the Men Engaged in It* (1884), after p. 30; ENG, December 21, 1888; MEL, January 1, 1889; *Dictionary of Scottish Business Biography* (1986) 1: 229–30.

283 "We have always": MEL, May 1, 1882.

283 "In spite of": ibid.

283 Pearce recklessly predicted: LTI, September 22, 1884.

284 "the compensating": LTI, September 22, 1883.

284 *Montana* and *Dakota* boilers: NML, November 1873; MEM, May 1927; GAA, p. 171.

284 "Find me X": Thomas Dykes in *Fortnightly Review,* April 1886.

284 £140,000: Memorandum of contracts, *Arizona,* in Ledger UGD 39 1/1, John Elder Papers. The ship's actual cost was £130,171.

284 £200,000 each: Ismay, Imrie to William S. Lindsay, December 3, 1874, file 8/1, William S. Lindsay Papers.

285 "similar to": John G. Hughes to Barrows, July 14, 1877, Guion Line Papers.

285 "has brought": John G. Hughes to J. Price, May 31, 1877, Guion Papers.

285 *Arizona* engine: IES (1881), plate 60; MEM, March 1929; Denis Griffiths, *Power of the Great Liners: A History of Atlantic Marine Engineering* (1990), pp. 42–43.

285 *Arizona: Arizona* specifications, May 22, 1878, Guion Papers; NGN, April 26, September 27, 1879; MEL, July 1, 1879; text and drawings in *Harper's Weekly,* July 5, 1879; ENG, July 9–September 3, 1880.

286 "This arrangement": NGN, April 26, 1879.

286 minor problems: John G. Hughes to A. D. Bryce, June 20, July 2, 1879, Guion Papers. (Bryce later changed his name to Bryce-Douglas; on his career, see NGN, October 11, 1883; ENG, April 10, 1891; MEL, May 1, 1891.)

286 "not very alarming": John G. Hughes to A. D. Bryce, October 7, 1879, Guion Papers.

286 easy in her motions: NGN, December 25, 1880.

286 ran into iceberg: NYT, November 10, 12, 1879; *New York Herald,* November 10, 1879; ILN, November 29, 1879.

287 damage to *Arizona*: John G. Hughes to Elder Company, November 20, 24, 26, 29, 1879, Guion Line Papers.

287 "In case": John G. Hughes to A. D. Bryce, July 8, 1879, Guion Papers; next four quotations from ibid., March 11, 15, June 7, 1880.

288 *Alaska:* NYT, October 24, 1880, November 10, 1881; NGN, April 22, 1882.

288   "greyhound": Dykes in *Fortnightly Review,* April 1886.

288   electric ship lighting: Andrew Jamieson in IES (1882), pp. 73–83; ENG, April 11, 18, 1884.

289   "Let it be": John G. Hughes to John S. Raworth, January 13, 1882, Guion Papers.

289   Cunard office lights: Board Minutes, May 2, 1888, CAL.

289   "the best known": *Marine Engineering,* August 1899.

289   "So great is": NYT, August 12, 1883.

290   "The speed will": John Burns letter in LTI, November 5, 1879.

290   "Science is": NML, April 1881.

291   slow building of other Cunarders: Board Minutes, September 7, 13, 27, December 13, 1881, CAL.

291   "It seemed": J. S. Hubbard to Cunard secretary, July 30, 1883, Chairman's Letter Books, CAL.

291   increase order to two ships: *Harper's Weekly,* July 14, 1888.

291   no Guion ship for eighteen months: John Burns to David Jardine (telegram), July 6, 1883, Chairman's Letter Books, CAL.

291   "Directors give": A. P. Moorhouse to John Burns (telegram), July 6, 1883, Chairman's Letter Books, CAL.

291   £616,000: A. P. Moorhouse to Elder Company, July 19, 1883, Chairman's Letter Books, CAL.

291   richest pact: ENG, July 20, 1883.

291   scramble to raise money: Board Minutes, November 13, December 18, 1883, February 19, June 12, 1884, CAL.

291   "probably the best-known": ENG, December 21, 1888.

292   "Now the verdure": William H. Rideing in *Scribner's,* April 1889.

292   "It is doubtful": Pollock, *Shipbuilding,* p. 155; and see ENG, November 15, 1867.

292   steel and shipbuilding: INA (1865), p. 169; (1868), pp. 22–23; (1875), pp. 140–41; *Shipbuilding,* ed. Rankine, pp. 174–75; LTI, January 20, 1880; P. F. Maccallum in IES (1882), pp. 31–51; J. H. Biles in MEL, October 1885.

293   Andrew Laing: INA (1931), pp. 344–47; Reginald W. Skelton in MEM, December 1932.

294   "an astonishing": *Glasgow Bailie* quoted in *Seaboard,* June 5, 1890.

294   *Umbria:* LTI, October 6, 1884; MEL, November 1, 1884; NGN, November 13, 20, 1884; ENG, June 19, 1885.

294   *Etruria:* LTI, September 22, 1884; ILN, April 11, 1885; NYT, May 5, 1885; [Charles Beadle], *A Trip to the United States in 1887* [1887], pp. 2–8; MEM, May 1928.

294   "The lamp is": MEL, December 1, 1884.

295   Speed's trip: recalled by J. G. Speed in *Harper's Weekly,* August 29, 1891.

295   "When is this": William Parker in INA (1888), p. 41.

295   "We must march": William John in INA (1887), p. 148.

296   ASC losses: William Henry Flayhart III, *The American Line (1871–1902)* (2000), p. 75; report by H. H. Houston, January 26, 1883, Board Files 49, PCR; and see NYT, December 4, 1883.

296   National Transit: Harold F. Williamson and Arnold R. Daum, *The American Petroleum Industry: The Age of Illumination, 1859–1899* (1959), pp. 452–58; Assistant Secretary to Clement Griscom, November 20, 1882, Secretary's Letter Books, PCR; Griscom testimony, SOT, pp. 395–96; memorandum of agreement, May 25, 1881, signed by C. A. Griscom and John D. Rockefeller, Special Collections, Van Pelt Library, University of Pennsylvania.

296   Inman available: report by George E. Holt, May 1875, Inman Company Papers; *Liverpool Daily Courier,* May 1, 1882; LTI, July 30, 1883.

297   "The Inman Line": NYT, May 28, 1885.

297   "Our Company": Clement Griscom to J. S. Clarkson, September 5, 1891, Benjamin Harrison Papers.

297   "hostility": Board Minutes, May 27, 1885, PCR.

297   £600,000: James A. Wright to George B. Roberts, January 21, 1887, Board Files 49, PCR.

297   Benjamin Brewster: Ralph W. Hidy and Muriel E. Hidy, *History of Standard Oil Company of New Jersey: Pioneering in Big Business, 1882–1911* (1955), pp. 28, 77–78; NCA, 22: 374–75.

297   "no interest": *Liverpool Daily Post,* October 19, 1886.

298   contract with Lairds: NYT, March 24, 1887.

298   need to expand yard: LTI, May 6, 1887.

298   James Rodger Thomson: ENG, March 20, 1903.

298   John Gibb Dunlop: MEN, October 1913.

298   John Harvard Biles: INA (1934), pp. 456–58; DNB (1931–1940), pp. 78–79.

299   Biles re twin screws: INA (1887), pp. 172–73.

299   "Naval architects": INA (1903), pp. 28–29.

299   "We have": James A. Wright to George B. Roberts, April 1887, Board Files 49, PCR.

299   "It is believed": Board Minutes, April 13, 1887, PCR.

299   *City of New York* and *City of Paris:* MEL, April 1, 1888; NYT, January 8, July 24, 1888, April 14, 1889; LTI, July 25, 1888, December 21, 1887; ENG, June 22, August 3, 1888; INA (1894), pp. 36–37; (1896), pp. 191–92; MEM, August 1924.

301   "All of the": Richard Harding Davis in *Harper's Weekly,* August 20, 1892.

302   "From a commercial": MEL, October 1885.

302   William James Pirrie and Walter H. Wilson: Michael Moss and John R. Hume, *Shipbuilders to the World: 125 Years of Harland and Wolff, Belfast, 1861–1986* (1986), pp. 41–48; *Dictionary of Business Biography,* 4: 702–4.

302   *Teutonic* and *Majestic:* NYT, January 20, 1889, April 11, 1890; LTI, July 31, 1889; ENG, January 25, 1889; MEL, February 1, 1889; T. Fitz-patrick, *A Transatlantic Holiday* (1891), pp. 14–15.

302   "Certainly the most": *Seaboard,* August 15, 1889.

303   "A somewhat": ENG, January 25, 1889.

304   Brown sent plans: Minutes of Executive Committee, June 6, 1888, April 24, 1889, CAL.

304   "So far": A. P. Moorhouse to John Burns, August 27, 1888, Chairman's Letter Books, CAL.

305   Brown soon reported: Minutes of Executive Committee, September 18, 1889, CAL.

305   "have instructions": John Burns to Vernon Brown, March 27, 1889, Secretary's Letter Books, CAL.

305   "They do not": A. P. Moorhouse to John Burns, October 2, 1891, Chairman's Letter Books, CAL.

305   attention to rivals: A. P. Moorhouse to John Burns, December 18, 23, 1891, Chairman's Letter Books, CAL; Board Minutes, June 29, 1892, CAL; Minutes of Executive Committee, June 8, 1892, CAL.

306   "will no doubt": Harland letter in LTI, September 13, 1892.

306   Laing's bonus: Robert Barnwell to Andrew Laing, April 14, 1891, Secretary's Letter Books, Fairfield Company Papers.

306   Laing's new engine: ENG, April 21, 1893.

306   "avoid friction": Board Minutes, November 23, 1892, Fairfield Company Papers.

306   *Campania* and *Lucania:* ENG, April 21, 1893; MEL, October 1, 1893; LTI, April 17, 1893; NYT, November 20, 1892, May 13, October 8, 1893; NML, November 1892.

308   list of fastest passages: MEL, October 1894.

308   "We continually": *Architectural Review,* October 1909.

308   Dawpool: Andrew Saint, *Richard Norman Shaw* (1976), pp. 261–65.

309   "Here then": ENG, June 22, 1888.

13: SHIPS AS TOWNS: OFFICERS, CREW, STEERAGE

310   ship as town: William H. Rideing in *Appleton's,* December 1878.

311   "There were": Samuel G. S. McNeil, *In Great Waters: Memoirs of a Master Mariner* (1932), p. 24.

312   "I always liked": ibid., p. 101.

312   "pitched and lurched": ibid., p. 104.

312   "There is no": ibid., p. 120.

313   William H. P. Hains: Alfred T. Story in *Strand Magazine,* June 1897; Charles Algernon Dougherty in *Harper's,* August 1886; T. N. Charles in *Sea Breezes* (Liverpool), March 1992.

313   "sort of Bunthorne": NGN, September 10, 1887.

313   Hains and Nilsson: Story in *Strand,* June 1897.

313   Samuel Brooks: NML, July 1904.

313   "urbane, genial": Rideing in *Appleton's,* December 1878.

314   "Whom all": ibid.

314   "He is a favorite": Dougherty in *Harper's,* August 1886.

314   Brooks and Langtry: Lillie Langtry, *The Days I Knew* (1925), p. 178.

314   "It depends": Dougherty in *Harper's,* August 1886; next three quotations from ibid.

314   William McMickan: NYT, September 11, 1881.

315   "He likes": Dougherty in *Harper's,* August 1886.

315   Horatio McKay: NML, February 1896, June 1908; *Cunard Magazine,* December 1919.

315   "The much dreaded": McKay letter in NML, May 1881.

315   Daniel Dow: Daniel Dow Papers, NML, October 1912, December 1913; *New York World,* March 13, 1914.

315   passed by *Aurania: Liverpool Journal of Commerce,* April 16, 1904.

315   "He is a genial": *Liverpool Journal of Commerce,* June 1, 1907.

315   "You can't imagine": *Liverpool Sunday Chronicle,* July 25, 1915.

315   "I like": Alfred T. Story in *Strand Magazine,* August 1897.

316   top Cunard salaries: Board Minutes, July 7, 1880, October 14, 1890, CAL.

316   "You remain away": *Liverpool Sunday Chronicle,* July 25, 1915.

316   open bridge: James Bisset with P. R. Stephenson, *Tramps and Ladies: My Early Years in Steamers* (1959), pp. 208–9; McNeil, *Waters,* p. 104.

316   White Star ships: Bertram Hayes, *Hull Down: Reminiscences of Wind-Jammers, Troops and Travellers* (1925), p. 19.

316   "My God": Story in *Strand,* August 1897.

317   "Having ploughed": Charles Terry Delaney in *Atlantic,* May 1910.

317   Grace's death: NGN, October 27, 1886.

317   crew of *Campania:* ENG, April 21, 1893.

318   "Men in steamers": Bisset, *Tramps,* p. 28.

318   seamen's loss of function: William H. Rideing in *Cosmopolitan,* April 1892.

318   "We are all": Story in *Strand,* June 1897.

319   daily schedule: John Burns in *Good Words,* April 1887.

319   provisions on *Etruria:* ibid.

319   "the type": Violet Jessop, *Titanic Survivor* (1997), p. 89.

319   "Is the ship": Langtry, *Days,* p. 178.

320   "Pampered and spoiled": Delaney in *Atlantic,* May 1910.

320   "No place": Jessop, *Survivor,* p. 117.

320   "misconduct" and "impropriety": Minutes of Executive Committee, September 2, 1886, and January 18, 1899, CAL.

320   "There's some": Rideing in *Cosmopolitan,* April 1892.

321   coaling: S. Howard Smith in *Cassier's,* August 1897; NML, April 1886 and September 1897; McNeil, *Waters,* pp. 216–17.

321   descending to engine room: Rideing in *Appleton's,* December 1878.

321   engine room: Rideing in *Cosmopolitan,* April 1892; J. D. Jerrold Kelly, *The Ship's Company and Other Sea People* (1897), pp. 22–23; NYT, October 14, 1900; W. C. Rogers in *World's Work,* March 1931; A. D. Blue in *Sea Breezes* (Liverpool), May 1988.

322   *Majestic* in 1890: NYT, April 12, 1890.

322   "They were": Bisset, *Tramps,* pp. 19, 218.

322   brawl on *Montana: New York Tribune,* January 23, 1877.

322   "The average engine": NGN, June 24, 1882.

322   *Etruria,* 1889: Minutes of Executive Committee, November 27, 1889, CAL.

322   *St. Louis,* 1898: LTI, January 2, 3, 1899.

323   *Lucania* fireman: Minutes of Executive Committee, January 24, 1900, CAL.

323   White Star firemen, 1905: NYT, October 12, 1905.

323   *Ultonia* firemen: Charles T. Spedding, *Reminiscences of Transatlantic Travellers* (1926), pp. 30–31.

323   "Drink and women": Herbert Hartley and W. B. Courtney in *Collier's,* August 29, 1931.

323   "I can account": NYT, April 14, 1890.

323   Cunard incentives: note, November 6, 1889, Secretary's Letter Book; Minutes of Executive Committee, June 27, 1900; and Board Minutes, October 9, 1901, CAL.

323   salaries of engineers and officers: NML, February 1887.

323   "Captains should": letter in MEL, October 1, 1885.

324   "The older": letter in MEL, February 1, 1890.

324   "Leaving port": Robert S. Riley in MEL, August 1904.

324   "A fact which": NML, May 1909.

325   "His word should": E. O. Murphy in *Steamship,* November 1892.

326   "Putting aside": *Catholic World,* February and March, 1873.

327   "Much of the cruelty": TCS, p. 12.

327   "a number of seats": ibid., p. 111.

328   "The treatment of": ibid., p. 145.

328   "Another fact": ibid., pp. 146–47.

328   "sin, full of": *Pall Mall Gazette* quoted in NGN, May 14, 1881.

328   "The ships are": Stephen Gwynn, *Charlotte Grace O'Brien* (1909), p. 88.

328   "The single women": ibid., pp. 65–6.

329   Thomas I. Wharton: Thomas Wharton, *"Bobbo" and Other Stories* (1897), pp. i–xxvii.

329   "It seems": Thomas Wharton in *Lippincott's,* February 1885.

330   statistics of demographic shift: RSR, p. 10.

330   Cunard official, 1888: *New York Tribune,* July 27, 1888.

331   "The naturally dirty": Minutes of Executive Committee, September 12, 1894, CAL.

331   German lines and White Star refused: ibid., August 29, September 12, 1894.

331   "Never travel": Wharton in *Lippincott's,* February 1885.

331   number on board *Utopia:* LTI, March 20, 1891.

331   160 life belts: LTI, March 27, 1891.

331   *Utopia* sinking: NYT, March 19, 21, April 12, 13, 1891; LTI, March 18–21, 25–27, 1891.

332   "The Italians were": NYT, March 19, 1891.

332   562 dead: LTI, March 18, 1891.

332   law waived: NYT, April 13, 1891.

333   *Celtic: New York Tribune,* September 27, 28, 1904, and April 18, 1906.

333   *Vaderland: New York Tribune,* January 27, 1905.

333   "Sweeping is": RSC, p. 7.

333   "The indifferent": ibid., p. 8.

333   "a congestion": ibid., p. 10.

334   "The stewardess": ibid., p. 27.

334   "The atmosphere": ibid., p. 22.

334   "The average steerage": Edward A. Steiner, *On the Trail of the Immigrant* (1906), p. 41.

14: ANGLO-AMERICANS

336   Americans seeking past: Allison Lockwood, *Passionate Pilgrims: The American Traveler in Great Britain, 1800–1914* (1981), p. 288.

337   "It was the": Henry Adams, *The Letters of Henry Adams,* ed. J. C Levenson et al. (1982–88), 4: 377.

337   decided not to send: ibid., 3: 229.

337   "He is queer": *The Letters and Friendships of Sir Cecil Spring Rice,* ed. Stephen Gwynn (1929), 1: 81.

337   Adams on *Siberia:* Adams, *Letters,* 2: 142–46.

338   "Nothing short": ibid., 2: 416.

338   "The voyage was": ibid., 4: 3; and see *The Education of Henry Adams* (1918), pp. 318–19.

338   "until I saw": Adams, *Letters,* 4: 5.

339   "The ocean": Inman Line, *Official Hand-Book of Information* (1888), p. 24.

339   "There are people": Ella W. Thompson, *Beaten Paths; or, A Woman's Vacation* (1874), pp. 11–12.

339   "Ocean travel": David R. Locke, *Nasby in Exile; or, Six Months of Travel* (1882), pp. 19, 34.

339   "A torpid": Julian Ralph in *Harper's Weekly,* December 14, 1889.

339   "Its very silence": *Cornhill Magazine,* June 1894.

340   "Nobody is": Martin Morris in *Nineteenth Century,* September 1896.

340   "The novice aboard": *McClure's Magazine,* March 1895.

340   man on *City of Paris:* Peter J. Hamilton, *Rambles in Historic Lands* (1893), p. 285.

341   *Germanic* and heavy gale: NYT, June 3, 1885.

341   "We had one": [Charles Beadle,] *A Trip to the United States in 1887* [1887], pp. 8–11.

342   rpms and piston speeds: W. H. White, *A Manual of Naval Architecture* (fifth edition, 1900), pp. 552–53.

342   twin screws and vibration: ENG, December 19, 1890; GAA, pp. 117, 121.

342   *Campania* trials: William B. Forwood, *Recollections of a Busy Life* (1910), pp. 179–80.

342   Bain reported: James Bain to Cunard secretary, May 2, 1893, Chairman's Letter Books, CAL; and see *Seaboard,* August 17, 1893.

343   "It is better": George W. Smalley in *New York Tribune,* March 9, 1890.

343   "They insist": letter in LTI, October 29, 1889.

343   "a rival steamer": William H. Rideing in *Cosmopolitan,* April 1892.

344   *Teutonic* maiden and race: NYT, August 16, 1889.

344   race later that season: NYT, October 10, 1889.

345   "Great interest": NYT, August 14, 1890; next quotation from ibid.

345   two weeks later: NYT, August 27, 1890.

345   "There is nothing": ENG, September 5, 1890.

346   gamblers in smoking room: Julian Ralph in *Harper's Weekly,* December 14, 1889; W. B. Lord in NML, August and November 1901.

346   "He was uneasy": Locke, *Nasby,* p. 30.

346   Hen Rice and Robert Solomon: NYT, November 27, 28, 1883.

346   "of an entirely": Minutes of Executive Committee, December 18, 1883, CAL.

347   "the noisy scenes": letter in LTI, February 1, 1887.

347   "This Smoking Room": Minutes of Executive Committee, March 22, 1887, CAL.

347   casualty on *Westernland: New York Tribune,* August 19, 1887.

347   "The Atlantic steamers": C. W. Kennedy in *North American Review,* June 1890; and rebuttal by H. Parsell, ibid., July 1890.

348   "The Ocean Service": *Official Guide and Album of the Cunard Steamship Company* (1876), p. 78.

348   "from 1840": *London Society,* July 1880.

348   rogue wave, 1905: MEL, November 1905; LTI, October 16, 1905.

348   death of Cogswell: NSC, May 13, 1867.

348   six steerage deaths: Minutes of Executive Committee, March 13, April 25, 1888, December 18, 1889, April 8, 1891, November 28, 1894, November 20, 1902, CAL.

348   *Oregon* sinking: NYT, March 15, 26, 1886; NML, June 1886; Minutes of Executive Committee, March 16–October 12, 1886, CAL.

349   *Umbria* breakdown: *Harper's Weekly,* January 14, 1893; ENG, January 20, March 17, 1893; *Spectator,* January 7, 1893; Minutes of Executive Committee, March 29, 1893, CAL.

350   "In the Cunard": NGN, July 5, 1873.

350   Cunard drills and order: *North American Review,* October 1864; LTI, November 17, 1875; W. Fraser Rae, *Columbia and Canada* (1877), pp. 17–18; John Burns in *Good Words,* April 1887.

350   "I found prevailing": LIH, 4: 241–42.

350   "Never was discipline": Henry Fry, *The History of North Atlantic Steam Navigation* (1896), pp. 105–6.

350   Cunard food and service: Benjamin Robbins Curtis, *Dottings Round the Circle* (1876), pp. 327–29; Minutes of Executive Committee, March 27, 1901, October 31, November 7, 1907, CAL.

350   "Very sorry": Morley Roberts in *Murray's Magazine,* January 1891.

350   "'Cunard luck'": NML, July 1891; and see MEL, August 1902.

350   "Your flag": L. L. Rees, *We Four: Where We Went and What We Saw in Europe* (1880), p. 21.

351   "London never": Adams, *Letters,* 2: 363.

351   "We ought not": ibid., 2: 527.

351   "The impetus": NML, September 1875.

352   "Each successive": James B. Pond, *A Summer in England with Henry Ward Beecher* (1887), p. 67.

352   Langham Hotel: John W. Forney, *Letters from Europe* (1867), p. 79; Morford's *Short-Trip Guide to Europe* (1870), pp. 337–38; Curtis Guild, *Abroad Again* (1877), pp. 47–48.

352   "Gradually the Langham": Forney, *Letters,* p. 81.

352   "how so small": Thompson, *Paths,* p. 269.

352   "feeing": Lockwood, *Pilgrims,* p. 320.

352   "I am so": Mary H. Krout, *A Looker On in London* (1899), p. 43.

352   children in streets: Charles Carroll Fulton, *Europe Viewed Through American Spectacles* (1874), p. 279.

353   refreshing qualities to extremes: Martin Morris, *Transatlantic Traits* (1897), pp. 111–12.

353   "There is not": Adams, *Letters,* 2: 346.

353   "If they": Julian Hawthorne, *Shapes That Pass: Memories of Old Days* (1928), 146.

353   "Nothing in London": Mary Elizabeth Blake, *A Summer Holiday in Europe* (1890), p. 197.

353   "I thought": Mary L. Ninde, *We Two Alone in Europe* (1886), p. 13.

353   "In London": Adelaide L. Harrington, *The Afterglow of European Travel* (1882), p. 291.

354   "the pet": Martin Duberman, *James Russell Lowell* (1966), p. 334.

354   "Just now": Bret Harte, *Selected Letters of Bret Harte,* ed. Gary Scharnhorst (1997), p. 267.

354   meeting on *Etruria:* Edmund Morris, *The Rise of Theodore Roosevelt* (1979), p. 358.

354   "An intelligent": Adams, *Letters,* 3: 62.

354   George W. Smalley: Joseph J. Mathews, *George W. Smalley: Forty Years a Foreign Correspondent* (1973).

354   "well-bred men": George W. Smalley, *London Letters and Some Others* (1891), 2: 111.

354   Smalley's appearance: Hawthorne, *Shapes,* p. 128.

354   Langtry noticed: Langtry, *Days,* p. 63.

355   "A singular": Henry James, *Henry James Letters,* ed. Leon Edel (1975), 2: 42.

355   "He is more": Theodore Roosevelt, *The Selected Letters of Theodore Roosevelt,* ed. H. W. Brands (2001), p. 116.

355   cheaper American grain: W. W. Nevin, *Vignettes of Travel* (1881), p. 83.

355   society had to admit: Price Collier, *England and the English from an American Point of View* (1909), p. 368; David Cannadine, *The Decline and Fall of the British Aristocracy* (1990), pp. 199–202.

355   Prince of Wales and new rich: ibid., p. 346.

355   "He's more": Bret Harte, *The Letters of Bret Harte,* ed. Geoffrey Bret Harte (1926), p. 178.

355   "He began": Smalley, *London,* 2: 71.

356   "They affect": Adams, *Letters,* 4: 194.

356   "where all": *Spring Rice,* ed. Gwynn, 1: 61.

356   Newport copied: Emily Hope Westfield in *Cosmopolitan,* August 1902.

356   "which they talk about": *Spring Rice,* ed. Gwynn, 1: 116.

356   J. P. Morgan: Jean Strouse, *Morgan: American Financier* (1999), p. 330.

356   Morgan and Vanderbilt: Bertram Hayes, *Hull Down: Reminiscences of Wind-Jammers, Troops and Travellers* (1925), p. 163.

356   "They are so": M. E. W. Sherwood in *North American Review,* June 1890.

356   "The American girl": Smalley, *London,* 2: 70–71.

357   Vanderbilt and duke: Consuelo Vanderbilt Balsan, *The Glitter and the Gold* (1952), pp. 45–51.

357   seventy heiresses: Charles S. Campbell, Jr., *Anglo-American Understanding, 1898–1903* (1957), p. 9.

357   "For the most": Westfield in *Cosmopolitan,* August 1902.

357   "No one could": Abigail Adams Homans, *Education by Uncles* (1966), p. 47.

357   "My breakfast-girls": Adams, *Letters,* 3: 199.

357   Endicott family: DAB, 3: 158–59; *Harper's Weekly,* November 24, 1888.

357   Mary Endicott: Diana Whitehill Laing, *Mistress of Herself* (1965).

357   "very English": *Spring Rice,* ed. Gwynn, 1: 88.

358   "Chamberlain amused": Adams, *Letters,* 3: 106.

358   "Our Mary": ibid., 3: 569.

358   Mary Victoria Leiter: Nigel Nicholson, *Mary Curzon* (1977), pp. 27–31.

358   "My young women": Adams, *Letters,* 3: 239.

358   "We have had": Mary Leiter to George Curzon, November 5, 1890, Mary Leiter Curzon Papers.

359   "Another of my": Adams, *Letters,* 4: 271.

359   "Suits me": ibid., 3: 570–71.

359   "England always presents": ibid., 4: 605.

360   "The most astonishing": Campbell, *Anglo-American,* p. 49.

360   guests at banquet sang: A. V. Dicey in *Atlantic,* October 1898.

360   American pop songs in London: E. V. Lucas, *A Wanderer in London* (1906), p. 243.

15: GERMANS

361   *Hohenzollern:* ILN, August 19, 1893; Lamar Cecil, *Wilhelm II: Prince and Emperor, 1859–1900* (1989), p. 292.

361   Naval review: ILN, August 10, 1889.

362   *Teutonic* guns: LTI, July 31, 1889.

362   according to *Times* account: LTI, August 5, 1889.

362   "We must have": Bertram Hayes, *Hull Down: Reminiscences of Wind-Jammers, Troops and Travellers* (1925), pp. 67–68.

363   Hanseatic League: Philippe Dollinger, *The German Hansa,* trans. and ed. D. S. Ault and S. H. Steinberg (1970), pp. 141–42, 365–69.

363   "this perennial": Michael Stürmer, *The German Empire, 1870–1918* (2000), p. 45.

363   Hamburg and Prussia compared: Lamar Cecil, *Albert Ballin: Business and Politics in Imperial Germany* (1967), pp. 5–6.

363   Hamburg and Bremen refused to join Zollverein: Wilson King, *Chronicles of Three Free Cities: Hamburg, Bremen, Lubeck* (1914), p. 302.

364   "a regular connection": Otto J. Seiler, *Bridge Across the Atlantic: The Story of Hapag-Lloyd's North American Liner Services* (second edition, 1991), p. 8.

364   "Since a similar": ibid., p. 13.

364   "if we do not": ibid.

364   *Borussia* maiden: *New York Herald,* June 18, 1856.

364   *Hammonia* maiden: ibid., July 19, 1856.

365   *Austria* disaster: NYT, September 27, 28, 30, 1858; LTI, October 11, 15, 1858.

366   471 dead: *New York Herald,* September 28, 1858.

366   "We leave it": letter in LTI, October 15, 1858.

367   Bremen trade with United States: *Seventy Years: North German Lloyd Bremen, 1857–1927* [1928], pp. 5–8.

367   Bremen's medieval traces: King, *Chronicles,* p. 148.

367   Hermann H. Meier: ibid., p. 149; Edwin Drechsel, *Norddeutscher Lloyd Bremen, 1857–1970* (1994), 1: 2–3.

367   "We cannot": *Seventy Years,* p. 30.

367   first four ships: Drechsel, *Lloyd,* 1: 13–16.

368   rivalry between German lines: Frank Broeze in *International Journal of Maritime History,* June 1991.

368   early problems: Drechsel, *Lloyd,* 1: 21; Seiler, *Bridge,* pp. 15–16.

368   "foreign bridgehead": Otto Pflanze, *Bismarck and the Development of Germany* (1990), 2: 529.

368   "the land of writers": Gordon A. Craig, *Germany 1866–1945* (1978), p. 57.

368   "what it means": ibid.

369   "Then they": William Leonard Gage, *A Leisurely Journey* (1886), p. 91.

369   rise of German industry: Craig, *Germany,* pp. 186–87, 248–49; W. O. Henderson, *The Rise of German Industrial Power, 1834–1914* (1975), pp. 140–41; Sturmer, *Empire,* pp. xiv, 87.

370   first Lloyd *Schnelldampfer:* Cecil, *Ballin,* p. 22.

370   *Elbe:* NYT, July 7, 8, September 11, 1881.

370   switch of mail to *Elbe:* Minutes of Executive Committee, September 27, 1881, March 6, 1884, CAL.

370   *Aller:* NGN, April 24, May 12, 19, 1886; MEL, May 1, 1886; Diary of Henry H. Seaver, October 3, 1907.

370   "The main saloon": NGN, May 19, 1886.

370   *Lahn:* NGN, December 24, 1887; MEM, February 1928; Drechsel, *Lloyd,* 1: 120, 310.

371   Johannes Georg Poppe: John Malcolm Brinnin and Kenneth Gaulin, *Grand Luxe: The Transatlantic Style* (1988), p. 7.

371   "As the Germans": NYT, March 14, 1885.

371   "Today, the day": Drechsel, *Lloyd,* 1: 95.

371   German shipbuilding: Henderson, *Rise,* pp. 198–201.

372   Lloyd's first German-built ships: *The History of the Norddeutscher Lloyd* [ca. 1903], pp. 17–22; NGN, September 29, 1886.

372   "I am the only": Bernhard Huldermann, *Albert Ballin,* trans. W. J. Eggers (1922), p. 236.

372   Ballin's favorite newspaper: Cecil, *Ballin,* p. 137.

373   Ballin's personality: ibid., pp. 27–30; Huldermann, *Ballin,* pp. 301–13.

373   "One needed": *Memoirs of Count Bernstorff,* trans. Eric Sutton (1936), pp. 101–2.

373   urging of Bismarck and Wilhelm: Huldermann, *Ballin,* p. 196.

374   fastest maiden: NYT, May 20, 1889.

374   30,000 admirers: NYT, May 23, 1889.

374   "Her lines": NYT, May 20, 1889.

374   *Furst Bismarck:* NYT, May 17, 1891; ENG, April 1, 1892.

374   "Everything about": Cy Warman in *Idler,* March 1895; and see John Henry Barrows, *A World-Pilgrimage* (1897), pp. 12–16.

375   "I should not": Mark Twain, *Collected Tales, Sketches, Speeches, and Essays* (1992), 2: 81–84.

375   fastest *Elbe* passage: NYT, April 12, 1891.

376   *Elbe* sinking: LTI, January 31, February 1–6, 1895; NYT, January 31, February 1, 11, June 18, 1895; *Harper's Weekly,* February 9, 1895; NML, July 1895.

377   "A more shocking": *Seaboard,* February 7, 1895.

377   mate found in default: NYT, June 18, 1895.

377   German tonnage overtook: Henderson, *Rise,* p. 202.

377   "The jealousy": *New York Tribune,* February 24, 1885.

377   passengers to New York, 1880s: NML, April 1892.

378   Lloyd largest steamship company in world: NML, March 1895.

378   "The British people": INA (1897), p. xxix.

378   indirect benefits: Winthrop L. Marvin in *Review of Reviews,* March 1903.

378   Ballin pointed out: letter in LTI, August 31, 1901.

378   rudders from Krupp: ENG, April 21, 1893.

379   Vulcan shipyard: ENG, July 7, September 8, 1899.

379   Robert Zimmermann: MEN, February 1912.

380   "One hears": NGN, May 9, 1901.

380   "With infinite": Ray Stannard Baker in *McClure's,* September 1900.

380   *Wilhelm:* ENG, October 1, 1897, March 4, 11, 25, April 8, 1898; MEL, October 1, 1897, January 1, 1898; Gustav H. Schwab in *Cassier's Magazine,* January 1898; *New York Tribune,* September 26, 28, October 3, 1897; MEM, February 1924.

381   distinct regions: *Norddeutscher Lloyd Bremen: History and Organisation* [ca. 1910]. pp. 62–64.

382   "A mechanical dish": ENG, March 11, 1898.

383   *Deutschland:* ENG, March 23, November 23, December 28, 1900; MEL, December 1, 1900.

384   "Instead of": German journal quoted in *Marine Engineering,* June 1901.

384   "A source of": NGN, June 23, 1904.

385   "The swiftest Cunarders": LTI, November 6, 1900.

16: THE TWO FINEST CUNARDERS

387   Clement Griscom: *Philadelphia Inquirer,* November 11, 1912; Lawrence Perry in *World's Work,* December 1902; Dexter Marshall in *Cosmopolitan,* May 1903.

387   "Our household": Lloyd C. Griscom, *Diplomatically Speaking* (1940), p. 9.

388   "We would": Clement Griscom to J. S. Clarkson, September 5, 1891, Benjamin Harrison Papers.

388   Congress allowed: LTI, May 10, 1892.

388   Harrison raised flag: NYT, February 23, 1893.

388   "whom her friends": NYT, March 8, 1893.

389   railroad had sunk $3 million: H. W. Schotter, *The Growth and Development of the Pennsylvania Railroad Company* (1927), p. 231.

389   $6-million mortgage: William Henry Flayhart III, *The American Line (1871–1902)* (2000), pp. 169–70.

389   half these bonds bought by Standard Oil: Minutes of Executive Committee, March 14, 1894, CAL.

389   Griscom and Brewster on Cramp board: Minute Book, July 9, September 27, 1890, William Cramp and Sons Ship and Engine Building Company.

389   "The generation": Charles Cramp in SNA (1894), p. 45.

389   United States rebuilding navy: Charles Cramp in *North American Review,* January 1892.

389   "In no other": Charles Cramp in *North American Review,* April 1894.

389   *St. Louis* and *St. Paul:* ENG, November 16, 1894, June 14, 21, 1895; MEL, December 1, 1894, July 1, 1895; *Seaboard,* April 11, June 6, 1895; NYT, May 30, 1895; Harper's Weekly, October 26, 1895.

389   Biles consulting: MEL, January 1893.

389   Lewis Nixon: *Cassier's,* September 1897.

390   "These ships": Charles Cramp in SNA (1893), p. 17.

390   "They have": *Seaboard,* November 8, 1894.

390   Bitter's work: *Harper's Weekly,* June 15, 1895.

390   "an unusually large": ENG, November 16, 1894.

390   "Instead of breaking": NYT, June 23, 1895.

391   steam pipe burst: NYT, December 19, 1895.

391   *St. Paul* aground: NYT, January 26, 1896.

391   *St. Louis* and *St. Paul* set new records: *Philadelphia Evening Bulletin,* September 9, 1897.

391   "Heave over": NYT, July 10, 1897.

391   *St. Louis* as favorite of Adams: Edward Chalfant, *A Biography of Henry Adams* (2001), 3: 254.

391   Bob Hope emigrated: Peter Morton Coan, *Ellis Island Interviews: In Their Own Words* (1997), pp. 78–79.

391 Hanna fortune: Herbert Croly, *Marcus Alonzo Hanna: His Life and Work* (1912), pp. 56–61.

392 "It's just": Thomas Beer, *Hanna* (1929), p. 218.

392 "My presentation": Mark Hanna to William McKinley, November 13, 1899, William McKinley Papers.

392 Hanna stayed with Griscoms: Margaret Leech, *In the Days of McKinley* (1959), p. 536.

392 "I know": Clement Griscom to William McKinley, June 26, 1900, McKinley Papers.

392 J. P. Morgan: Jean Strouse, *Morgan: American Financier* (1999).

392 "A man": Bernhard Huldermann, *Albert Ballin,* trans. W. J. Eggers (1922), p. 47.

392 "He must": Ray Stannard Baker in *McClure's,* October 1901.

393 "I took": J. P. Morgan to R. J. Cortis, February 3, 1881, J. P. Morgan Papers.

393 mortgage bond issue: Thomas R. Navin and Marian V. Sears in *Business History Review,* December 1954.

393 Griscom and Morgan conferred at Aix-les-Bains: Griscom, *Diplomatically,* pp. 127–28.

393 by fall 1900: Navin and Sears in *Business History Review,* December 1954.

393 William James Pirrie and Bruce Ismay: ibid.

394 Morgan explained: memorandum by S. M. Prevost, July 11, 1901, file 27/17, Presidential Correspondence of A. J. Cassatt, PCR.

394 "We had had": Samuel Rea to A. J. Cassatt, August 16, 1901, file 27/18, ibid.

394 "Why, Mr. Cassatt": memorandum by S. M. Prevost, July 10, 1901, file 27/17, ibid.

394 "I will tell": ibid.

394 "It is not": A. J. Cassatt to Clement Griscom, [July 1901], ibid.

394 "because he thought": S. M. Prevost to A. J. Cassatt, July 12, 1901, ibid.

394 shift to Morgan: Navin and Sears in *Business History Review,* December 1954.

395 "Our object": Winthrop L. Marvin in *Review of Reviews,* December 1902.

395 Ismay's sabotage: Clinton Dawkins to J. P. Morgan, Jr., November 22, 1901, to Charles Steele, February 22, April 12, June 21, 25, July 1, 1902, to W. J. Pirrie, April 21, 1902, Clinton E. Dawkins Letterbooks; J. P. Morgan, Jr., to George Perkins, April 19, 1902, J. P. Morgan, Jr., Papers.

395 "torn by two": Clinton Dawkins to Charles Steele, October 29, 1902, Clinton E. Dawkins Letterbooks.

395 "He is a calamity": Clinton Dawkins to Charles Steele, November 18, 1902, ibid.

395 "My arms are": Clinton Dawkins to Charles Steele, October 16, 1903, ibid.

396 "It appeared": William B. Forwood, *Recollections of a Busy Life* (1910), p. 178.

396 huge subsidy to Cunard Line: Board Minutes, September 30, 1902, CAL.

396 Charles Parsons: Rollo Appleyard, *Charles Parsons: His Life and Work* (1933).

396 father of Parsons: DNB, 15: 425–26.

397 turbine and electricity: C. A. Parsons in NEI (1888), pp. 127–34.

398 "curiously vertical": Appleyard, *Parsons,* p. 281.

398 "He has not": J. H. Biles, *The Steam Turbine as Applied to Marine Purposes* (1907), p. 2.

398   "avoid as far": Charles Parsons to Lawrence Parsons, January 30, 1894, Charles A. Parsons Papers.

398   "we are following": Charles Parsons to Lawrence Parsons, February 2, 1894, Parsons Papers.

398   *Turbinia:* C. A. Parsons in NML, November 1897.

398   "an almost complete": Charles Parsons to Lawrence Parsons, October 3, 1894, Parsons Papers.

398   heated water and arc lamp: Charles Parsons in INA (1897), p. 234.

399   engines of *Turbinia:* ibid., p. 285.

399   "like a very": Charles Parsons in IES (1901), p. 219.

399   "If you believe": Leonard Outhwaite, *The Atlantic: A History of an Ocean* (1957), p. 308.

399   "At the cost": LTI, June 28, 1897; and see ENG, July 2, 1897, and MEL, August 1897.

399   "like a whale": *Marine Engineering* (New York), August 1897.

399   "Here, there": ibid.

400   *Cobra* and *Viper:* Charles Parsons to Lawrence Parsons, November 21, 1899, June 22, 1900, Parsons Papers.

400   "It would be difficult": ENG, November 10, 1899.

400   *Viper* grounded: LTI, August 5, 1901.

400   *Cobra* sank: LTI, September 20, October 5, 11, 15, 1901.

400   Bain's report: Minutes of Executive Committee, August 14, 28, 1901, CAL.

400   Parsons at board meeting: Board Minutes, December 11, 1901, January 9, February 6, 20, 1902, CAL.

400   "that the Board": Board Minutes, October 23, 1902, CAL.

401   Cunard asked for tenders: ibid.

401   initial specs: Leonard Peskett in INA (1914), pp. 175–76.

401   George A. Burns: ENG, October 13, 1905; MEL, November 1, 1905; NML, November 1905.

401   "a man of": Forwood, *Recollections,* p. 182.

401   hull tests: ENG, August 2, 1907.

402   *Queen:* Charles Parsons in INA (1903), p. 290.

402   Bain's report: Board Minutes, July 16, 29, August 20, September 17, 1903, CAL.

402   turbine commission: ENG, April 1, 1904, August 2, 1907.

402   "Every point": INA (1907), p. 110.

402   report was leaked and published: *Glasgow Herald,* March 26, 1904; Board Minutes, February 18, 25, March 24, April 7, 1904, CAL.

403   four smokestacks: Board Minutes, July 2, 1904, CAL.

403   Brown bought Thomson: Anthony Slaven in *Research in Maritime History* (1993), p. 175.

403   Brown firm: MEL, January 1, 1905.

403   *Lusitania* team: ENG, August 2, 1907.

403   William J. Luke: INA (1934), pp. 463–65; W. J. Luke in INA (1907), pp. 65–77.

403   remaking yard: Board Minutes, May 18, September 16, 1904, May 2, August 16, 1905, May 21, 1906, John Brown Papers.

403   "My inclination": Inverclyde to A. D. Mearns, June 3, 1905, file s7, CAL.

403   James Miller: Inverclyde to A. D. Mearns, April 14, 1904, and James Miller to Mearns, January 30, 1905, ibid.; *Architectural Review,* June 1900.

403   Miller was urged: signed agreement, August 23, 1905, and James Miller to Inverclyde, September 13, 1905, file S7, CAL.

403   *Lusitania:* ENG, July 19, August 2, 1907; NML, December 1907; MEN, September 1, 1907.

404   "It may be": Cunard Line, *Royal Mail Express Steamer "Lusitania"* (1907), p. 6.

404   *Lusitania's* vibrations: Minutes of Executive Committee, July 10, 1907, CAL.

405   "the result": ibid., July 17, 1907; and see ibid., August 28, 1907, March 19, May 14, July 9, 1908, February 11, 1909.

405   "The lion": NYT, October 11, 1907.

405   last voyage of *Lusitania:* I drew from the fullest recent account, Diana Preston's *Lusitania: An Epic Tragedy* (2002).

407   Laing's fine qualities: Reginald W. Skelton in MEM, December 1932.

407   "He had": ibid.

407   "obstinacy": R. Barnwell to Donald Currie, August 22, 1895, Secretary's Letterbook, Fairfield Company Papers.

407   "the great friction": R. Barnwell to Andrew Laing, October 12, 1895, ibid.

407   engines for six Cunarders: ENG, November 8, 1907.

407   *Ivernia:* MEL, October 1899; *Steamship,* October 1899.

408   Laing and turbines: Andrew Laing in MEM, June 1929.

408   Northeast shipbuilding: George B. Hunter and Edwin W. De Rusett in IES (1909), pp. 323–46.

408   "They cannot build": NGN, August 6, 1881.

408   "Two very interesting": IES (1909), p. 345.

408   Hunter hired De Rusett: *Steamship,* July 1890.

408   Edwin De Rusett: LTI, May 25, 1921.

408   De Rusett as advocate of twin screws: E. W. De Rusett in INA (1876), pp. 288–97.

408   "I have travelled": NEI (1902), pp. 263–64.

408   Swan shipyard: ENG, August 1902.

409   Wigham Richardson: ENG, April 24, 1908; NEI (1908), pp. 368–71.

409   model tests: E. W. De Rusett in *Shipbuilder,* November 1907; and see De Rusett in NGN, July 25, 1907; in INA (1907), pp. 308–10; and in *Cassier's,* November 1908.

409   "He said": Inverclyde to A. D. Mearns, May 25, 1905, file s7, CAL.

409   "The sum": Harold Peto to A. D. Mearns, June 27, 1905, ibid.

409   "as they are": Inverclyde to A. D. Mearns, June 29, 1905, ibid.

410   "In conceding": A. D. Mearns to Harold Peto, September 13, 1905, ibid.

410   *Mauretania:* ENG, November 8, 1907; MEN, January 1, 1908.

411   seagoing town: cutaway illustration of *Mauretania* in ILN, March 26, 1927.

411   "The metal": ENG, November 8, 1907.

412   "In catering": *Shipbuilder,* November 1907.

412   "It was immense": Theodore Dreiser in *Century,* August 1913.

413   "There is a limit": *North German Lloyd Bulletin,* September 1909.

413   "In good weather": Humfrey Jordan, *Mauretania: Landfalls and Departures of Twenty-Five Years* (1936), p. 144.

413   a visitor saw: on my research trip to Bristol, in April 2000, to use the Brunel Collection.

Acknowledgments

$\mathcal{A}$s a freelance historian, I am especially dependent on the kindness of librarians. I have once again drawn on the uncommonly
rich library resources of the Boston area. Most of my research was done at
the Boston Public Library in Copley Square. I am particularly grateful to
Scott Cornwall, the now-retired head of stack services in the Research
Library, and all the people who have brought me so many books for so
many years; Henry Scannell and others in Microtext; Sinclair Hitchings
of the Print Department; and Roberta Zonghi and the staff in Rare
Books. At the Massachusetts Historical Society, Peter Drummey again
allowed me the use of the comprehensive MHS catalog that he carries
around in his head. I have also spent months at the various nearby
libraries of Tufts University, the Massachusetts Institute of Technology
(especially in the ancient nineteenth-century engineering journals of the
Retrospective Collection), and Harvard University. At Harvard, I am
most beholden to John Collins and many others in the Government
Documents and Microforms section of Lamont Library.

My manuscript research depended on three archives in particular.
Special thanks go to Garry Shutlak and Lois Yorke of the Public Archives
of Nova Scotia in Halifax; Michael Richardson of Special Collections at
the University of Bristol; and Maureen Watry of Special Collections at
the University of Liverpool. Elsewhere in Britain, I was notably helped by
Alexandra Robertson of the Museum of Transport in Glasgow, Karen
Howard of the Merseyside Maritime Museum in Liverpool, Lizzie
Richmond of the University of Bath Library, and Mandy Taylor of the
Science Museum Library in London. Among the American archives I
used, I owe particular thanks to Margaretha Talerman of the American

Swedish Historical Museum in Philadelphia, Anita B. Israel of the Longfellow National Historic Site in Cambridge, Massachusetts, Sharon Nelson of the Pennsylvania State Archives in Harrisburg, Christopher T. Baer of the Hagley Museum and Library in Wilmington, Megan Fraser of the Independence Seaport Museum in Philadelphia, and Kurt Hasselbalch of the Hart Nautical Collections at MIT.

Priscilla Griscom sent me helpful clippings and information about her distant kinsman Clement Griscom. David T. Moore of the Germantown Historical Society, Edward W. Chichirichi of the Historical Society of Delaware, and Jayne Larion helped me sort through the Cunard family genealogy. My thanks also to Rick Archbold, Ann Thomas, and Edward W. Sloan for various research tips.

Of friends and relations, Scott-Martin Kosofsky and Lanier Smythe did their best to penetrate my ignorance about the history of architecture and interior design. Alan Andres drew on his astonishing private collection of books and media to give me a videotape of a British TV program about Isambard Brunel. Deirdre Dundas Newton clarified the Dundas family line for me. My wife, Alexandra Dundas Todd, appreciates the peculiar circumstances of living with a writer, and she again sustained me through a book project with her love and understanding. My editors—Hugh Van Dusen in New York and Michael Fishwick in London—have, by their bracing blends of criticism and enthusiasm, rebutted the familiar notion that book editors today no longer edit. Robin Straus, my agent for more than two decades, brought this book and me to HarperCollins, and she gave me encouragement and shrewd advice at every step along the way.

This book's dedication reflects my gratitude to the late Phyllis Blakeley of Halifax, Nova Scotia. During her long and distinguished career as an archivist and historian at the Public Archives of Nova Scotia, she worked at intervals on the biography of Samuel Cunard that she planned to write in retirement. An interim report, her 10,000-word entry on Cunard that was published in the *Dictionary of Canadian Biography* in 1976, is the fullest sketch of his career in print. Dr. Blakeley became ill and died of cancer within a year of retiring, and never got to write her book. A meticulous archivist to the end, she assembled her Cunard materials and left them to the Halifax archive. These thick files were invaluable to me for the third and fifth chapters of this book. I am glad that so much of Phyllis Blakeley's skillful research will now, seventeen years after her death, finally be published.

Index